Biology of Rove Beetles (Staphylinidae)

Oliver Betz • Ulrich Irmler •
Jan Klimaszewski
Editors

Biology of Rove Beetles (Staphylinidae)

Life History, Evolution, Ecology and Distribution

Springer

Editors
Oliver Betz
Institut für Evolution und Ökologie,
Abteilung für Evolutionsbiologie der
Invertebraten
Universität Tübingen
Tübingen, Germany

Ulrich Irmler
Institut für Ökosystemforschung,
Abteilung für Angewandte Ökologie
Universität Kiel
Kiel, Germany

Jan Klimaszewski
Natural Resources Canada,
Canadian Forest Service,
Laurentian Forestry Centre
Québec, QC, Canada

ISBN 978-3-030-09943-5 ISBN 978-3-319-70257-5 (eBook)
https://doi.org/10.1007/978-3-319-70257-5

Printed on acid-free paper

This Springer imprint is published by the registered company Springer International Publishing AG part of Springer Nature.
The registered company address is: Gewerbestrasse 11, 6330 Cham, Switzerland

Contents

Introduction to the Biology of Rove Beetles 1

Ulrich Irmler, Jan Klimaszewski, and Oliver Betz

Abstract

A short overview of the evolution of rove beetles is given to introduce to the reader the topics contained in the book. Additional subjects are distribution, behavior, ecology, bioindication, ecomorphology, and development. The morphological diversity of rove beetles is presented based on the diversity of their ecological niches. Representative rove beetles are briefly described in order to illustrate the wide range of their niches with regard to food resources, habitats, and geographic zones.

The megadiverse rove beetles (Coleoptera: Staphylinidae) constitute one of the largest beetle families having a tremendous heterogeneity and comprising more than 63,657 species known worldwide. Their evolutionary history has lasted more than 200 million years and dates back to the Triassic. Their small body size (most species measure 2–8 mm in length) allows them to enter minute spaces, and they are typical elements of soil biota, where they live in the litter and in deeper soil layers and prey upon a variety of soil animals. However, other habitat preferences and feeding types are widely known among rove beetles making them a highly interesting study group for comparative research.

In terms of their classification, monophyletic Staphylinidae belong to the Staphylinoidea within the staphyliniform beetles, which is the largest infraorder of mostly non-phytophagous beetles. The closest relatives of the Staphylinidae are the Ptiliidae + Hydraenidae (Ptiliid group) and the Leiodidae + Agyrtidae (Leiodid group). Currently, the Staphylinidae comprise 32 subfamilies and 167 tribes.

Because of their worldwide distribution, their ecological significance, and their behavioral and ecomorphological diversity, staphylinids are nowadays becoming an increasingly investigated insect group in the fields of evolution and ecology. As shown in this book, they are also widely used as meaningful bioindicators of environmental conditions in applied sciences such as forest research and conservation.

U. Irmler (✉)
Department of Applied Ecology, Institute for Ecosystem Research, University of Kiel, Kiel, Germany
e-mail: uirmler@ecology.uni-kiel.de

J. Klimaszewski
Natural Resources Canada, Canadian Forest Service, Laurentian Forestry Centre, Québec, QC, Canada
e-mail: jan.klimaszewski@canada.ca

O. Betz
Evolutionsbiologie der Invertebraten, Institut für Evolution und Ökologie, Universität Tübingen, Tübingen, Germany
e-mail: oliver.betz@uni-tuebingen.de

O. Betz et al. (eds.), *Biology of Rove Beetles (Staphylinidae)*,
https://doi.org/10.1007/978-3-319-70257-5_1

Staphylinids occur in nearly all ecosystems of the world, from the Arctic to the humid tropical zones, and from marine habitats to high alpine ecosystems in everlasting snow areas. Although their morphology seems to be highly uniform with short elytra, their body is extremely flexible and predisposes them to inhabit an enormous variety of microhabitats. Certainly, no other group of Coleoptera has been so successful as the Staphylinidae in living in such an enormous number of diverse habitats. A large number of staphylinid species are found on the soil surface; these species represent the most common type of Staphylinidae with their elongated shape and flexible body, which allows them to inhabit all kinds of crevices in the soil (Fig. 1.1b).

However, some species show a high affinity for deep soil layers. They have lost their eyes and have reduced legs and antennae (Fig. 1.1a), whereas others that live in natural holes or in burrows and nests of mammals have developed longer legs and antennae (Fig. 1.1l). Species living in vegetation have broad adhesive tarsi, which enable them to climb on stems and leaves (Fig. 1.1c). They often have a broad body, and their elytra are longer than those of the species mentioned so far (Fig. 1.1d). The species living under the bark of deadwood are dorsoventrally flattened, permitting them to exploit the narrow habitat between bark and log (Fig. 1.1h). Here, the stage of decay provides fixed bark of different densities, a feature that is responsible for the adaptation of the species to these habitats.

A specific attribute of Staphylinidae is their affinity to certain microhabitats. In contrast to most other coleopteran families, Staphylinidae live together with a large variety of mammals, birds, and social insects. They are either adapted to living with a single host species or with small species groups. One group of Amblyopinina lives ectodermally on Neotropical mammals (Fig. 1.1k). In particular, ants, other social Hymenoptera, and termites are favored host groups of social insects for the Staphylinidae (Fig. 1.1i, j). The wide range of ecological resources offered by these host species is exploited by the Staphylinidae. For some host species, the staphylinid guest species have beneficial effects, e.g., when they feed on their waste products, but for others, they are harmful, e.g., when they feed on their eggs, larvae, adults, or stored food.

Several groups of Staphylinidae are highly specialized to specific food such as fungi (Fig. 1.1g), or they hunt specific prey such as Collembola (Fig. 1.1f) or oribatid mites (preyed upon by the large group of Scydmaeninae). For other subfamilies, their relationships with plant species, e.g., for species of the genus *Eusphalerum* that occur on flowers, are mostly unknown.

The high diversity of the Staphylinidae targets them for research in order to gain answers to two of the main ecological questions, namely, the value and the function of biodiversity in ecosystems. Unfortunately, little work has been undertaken in this field because of the difficulties in the identification of rove beetles and the insufficient number of scientists working with them. However, the diversity of Staphylinidae may be representative of the whole biodiversity of a single ecosystem, because these beetles occupy numerous niches related not only to the soil and the vegetation but also to the diversity of other arthropods and mammals.

Another open ecological question concerns the indicator value of rove beetles for specific environmental factors such as soil quality, soil structure, and pollution. Although some investigators have found only weak relationships between the soil quality and the occurrence of staphylinid species, others have revealed extremely close associations with soil types.

This book covers three main topic areas, i.e., (1) phylogeny, systematics, and zoogeography; (2) ecology, conservation, and biotic interactions; and (3) biology, reproduction, and development.

Phylogeny, systematics, and zoogeography: these issues at the level of genus or tribe have always been among the major activities of staphylinidologists. A review of the higher phylogeny of Staphylinidae based on molecular data (Vladimir Gusarov) is followed by another on the fossil history of Staphylinoidea (Stylianos Chatzimanolis). Other chapters deal with the zoogeography and biodiversity of Neotropical

Fig. 1.1 Staphylinid species representing various types of ecological groups: *Geomitopsis* (**a**), subterranean; *Philonthus* (**b**), soil surface; *Eusphalerum* (**c**), vegetation; *Tachyporus* (**d**), vegetation; *Phytosus* (**e**), tunneling; *Stenus* (**f**), optically oriented predator; *Gyrophaena* (**g**), fungus dweller; *Anomognathus* (**h**), subcortical, living under bark; *Lomechusa* (**i**), associated with ants; *Pseudomimeciton* (**j**), living with army ants; *Edrabius* (**k**), ectodermal on mammals; *Rheochara* (**l**), living in holes within mull (i.e., a type of humus)

Staphylinidae (Ulrich Irmler and Angelico Asenjo), and adventive species in Canada (Jan Klimaszewski and Adam Brunke). Finally, a general overview on the current stage of knowledge of saw-lipped rove beetles (Euaesthetinae), including biogeography is provided (Dave Clarke).

Ecology, conservation, and biotic interactions: after a discussion of the general effect of environmental parameters important for explaining the distribution patterns of Staphylinidae (Ulrich Irmler and Erhard Lipkow), more specific contributions concerning Nearctic rove beetles associated with the nests of vertebrates (Adam Brunke and J. Buffam) or on the use of staphylinids as bioindicators in forestry and agroecosystems (Jan Klimaszewski, Adam Brunke, Tim Work, Lisa Venier) are presented. This part of the book is concluded by a contribution providing an updated list of the parasites of Staphylinidae (J. Howard Frank).

Biology, reproduction, and development: this section first provides a review of our current knowledge on the biology of the megadiverse Steninae (Oliver Betz, Lars Koerner, Konrad Dettner), followed by a contribution concerning the biology of mite-feeding ant-like stone beetles (Scydmaeninae) (Paweł Jałoszyński). The next chapter provides new information about the evolution of genitalia in various staphylinids (Shun-Ichiro Naomi). The final contribution deals with the biology and diversity of immature ant-like stone beetles (Scydmaeninae) (Paweł Jałoszyński).

The chapters in this book are written by authors actively engaged in studying staphylinids, and each chapter offers a synthesis of the current knowledge on a variety of topics and suggests future directions for research. The case studies show the potential of staphylinids in evolutionary research and demonstrate their fascinating biology in topics such as predator-prey interactions, chemical communication, and reproduction. Moreover, the methodological sections in the single contributions demonstrate the way that researchers have designed their studies and attained their results.

Both experienced scientists and beginners will thus find the diversity of subjects covered by this book intriguing and inspiring for continuing or initiating their own research on these fascinating beetles.

Part I

Phylogeny, Systematics and Zoogeography

Phylogeny of the Family Staphylinidae Based on Molecular Data: A Review

2

Vladimir I. Gusarov

Abstract

Molecular phylogeny of staphylinid beetles as published in 36 papers is reviewed. For every paper, the markers used, the size of the analyzed dataset, and the taxonomic focus are listed. Availability of sequence and genomic data for the entire family and all subfamilies is summarized. The current knowledge of staphylinid phylogeny is presented in a supertree. Phylogenetic relationships between the Staphylinidae and other staphylinoid families, among and within staphylinid subfamilies, are discussed.

2.1 Introduction

The history of molecular phylogenetic studies of the family Staphylinidae goes back some 20 years. During that time, DNA sequences were used to address staphylinid phylogeny at different levels: from relationships to other members of Staphylinoidea and from subfamily-level phylogeny of the entire family (e.g., McKenna et al. 2015a) to phylogenies of genera or species groups (e.g., Song and Ahn 2017) and phylogeographic studies within species (e.g., Chatzimanolis and Caterino 2007). As of June 2017 virtually all published studies used one or few molecular markers obtained through PCR and Sanger sequencing (one study (Timmermans et al. 2016) used the Roche/454 platform). These datasets proved helpful in resolving some phylogenetic questions and failed in others. While sequencing additional markers one by one may improve the resolution in some clades, there is no doubt that within a few years phylogenetic studies based on genome-wide sampling of hundreds of genes will appear and provide answers to many questions where the use of a few markers proved insufficient. Meanwhile, it is interesting and appropriate to assess how far we have advanced so far in our understanding of the staphylinid phylogeny. Among the published analyses of staphylinid phylogeny, there are those relying only on molecular data and those combining molecular and morphological datasets. The focus of this chapter is on how well the phylogeny of staphylinids can be resolved based on molecular data alone.

2.2 Markers

The genes used in staphylinid phylogenetic studies are listed in Table 2.2. The first works relied mostly on mitochondrial genes and nuclear RNA genes (Ballard et al. 1998; Maus et al. 2001; Chatzimanolis and Caterino 2007; Hunt et al. 2007; Leschen et al. 2008; Grebennikov and

V. I. Gusarov (✉)
Natural History Museum, University of Oslo, Oslo, Norway
e-mail: vladimir.gusarov@nhm.uio.no

O. Betz et al. (eds.), *Biology of Rove Beetles (Staphylinidae)*,
https://doi.org/10.1007/978-3-319-70257-5_2

Newton 2009; Thomas 2009; Ahn et al. 2010). Recently, with new primers available (Wild and Maddison 2008), several nuclear single-copy protein-coding genes are becoming commonly used, particularly wingless (*wg*), topoisomerase I (*TP*), arginine kinase (*AK*), and carbamoyl phosphate synthetase domain of CAD (*CAD*). A major advantage of using protein-coding genes is that their sequences are much easier to align unambiguously compared to RNA gene sequences. For this reason, some recent studies avoided RNA-coding genes altogether (Song and Ahn 2014; Schomann and Solodovnikov 2017).

As of June 8, 2017, the total of 15,447 staphylinid DNA sequences are available in GenBank (excluding environmental and genomic sequences). BOLD database includes 29,099 staphylinid specimens with DNA sequences, among them 26,651 specimens (2122 species) with DNA barcodes (BOLD Systems 2017). The total of 139 complete and partial mitochondrial genomes are available in GenBank. Genome or transcriptome data are available for 14 staphylinid species (in GenBank). All this information is a formidable resource for phylogenetic research on the Staphylinidae, but the taxon coverage is very uneven, as described in the next section.

2.3 Taxon Coverage

Currently, 32 extant subfamilies are recognized within the family Staphylinidae (Bouchard et al. 2011). The size of the subfamilies varies greatly, ranging from a single (extant) species (e.g., in Solieriinae) to more than a dozen thousand species (Aleocharinae). Table 2.1 lists the number of genera and species and the number of sequences available in GenBank for every subfamily. Only Protopselaphinae lack any data. To compare coverage among the subfamilies, the number of sequences per genus and per species was calculated for every subfamily (Table 2.1). Interestingly, judging by this ratio, most of the small subfamilies are represented better than the larger subfamilies, i.e., the diversity of the largest subfamilies, such as Aleocharinae, Pselaphinae, and Staphylininae, has been sampled poorly.

In the larger subfamilies the majority of sequences are *CO1* sequences, mostly as a result of DNA barcoding efforts. If only non-*CO1* sequences in GenBank are counted (Table 2.1), the underrepresentation of some large subfamilies is even more striking (Fig. 2.1). The average based on the entire family is 0.062 sequences per species. Among the larger subfamilies, Paederinae, Scydmaeninae, Osoriinae, and Leptotyphlinae are particularly poorly represented.

2.4 Phylogeny

In this section, results of published phylogenetic studies are discussed. Among the three most widely used approaches, maximum parsimony, maximum likelihood, and Bayesian, the last two consistently outperform the first. This becomes evident when comparing the trees based on the same datasets but obtained by different analyses (e.g., Elven et al. 2010; Osswald et al. 2013; Brunke et al. 2016). For this reason, the discussion below is based on results of the maximum likelihood and Bayesian analyses. Trees resulting from these two kinds of analyses are normally presented with support values listed for every node/clade, posterior probability in Bayesian analyses, and bootstrap values in the maximum likelihood analyses. The higher the support, the more confident one can be in the result. As demonstrated in simulation studies where the true phylogeny is known, higher support values indicate that the resulting tree is closer to the true phylogeny (Erixon et al. 2003). Comparison among the trees obtained by different analyses of the same or overlapping datasets (real datasets for which the true phylogeny is unknown) indicates that the clades recovered with higher statistical support are more consistent across analyses than those that have low support. For this reason, in this review a conservative

Table 2.1 The number of genera and species (based on Thayer 2005 and O'Keefe 2005) and the number of sequences (with and without *CO1* sequences; total counts and ratios per genus and species) available in GenBank for every staphylinid subfamily

	Number of		All nucleotide sequences in GenBank			Non-*CO1* nucleotide sequences in GenBank		
			Number of sequences			Number of sequences		
Subfamily	genera	species	total	per genus	per species	total	per genus	per species
Aleocharinae	1151	12851	4879	4.24	0.38	1490	1.29	0.12
Apateticinae	2	25	6	3	0.24	3	1.5	0.12
Dasycerinae	1	17	7	7	0.41	6	6	0.35
Empelinae	1	1	2	2	2	2	2	2
Euaesthetinae	26	762	45	1.73	0.059	22	0.85	0.029
Glypholomatinae	2	8	13	6.5	1.63	12	6	1.5
Habrocerinae	2	22	26	13	1.18	10	5	0.45
Leptotyphlinae	44	521	1	0.023	0.0019	1	0.023	0.0019
Megalopsidiinae	1	164	3	3	0.018	3	3	0.018
Micropeplinae	6	80	191	31.83	2.39	7	1.17	0.088
Microsilphinae	1	4	2	2	0.5	2	2	0.5
Neophoninae	1	1	5	5	5	4	4	4
Olisthaerinae	1	2	8	8	4	5	5	2.5
Omaliinae	117	1396	793	6.78	0.57	47	0.40	0.034
Osoriinae	103	2083	53	0.51	0.025	39	0.38	0.019
Oxyporinae	1	95	49	49	0.52	19	19	0.2
Oxytelinae	47	1975	579	12.32	0.29	73	1.55	0.037
Paederinae	221	6101	490	2.22	0.080	87	0.39	0.014
Phloeocharinae	7	52	21	3	0.40	16	2.29	0.31
Piestinae	7	104	11	1.57	0.11	8	1.14	0.077
Proteininae	11	190	99	9	0.52	33	3	0.17
Protopselaphinae	1	8	0	0	0	0	0	0
Pselaphinae	1200	9110	494	0.41	0.054	234	0.20	0.026
Pseudopsinae	4	54	8	2	0.15	6	1.5	0.11
Scaphidiinae	45	1451	630	14	0.43	110	2.44	0.076
Scydmaeninae	82	4600	64	0.78	0.014	22	0.27	0.0048
Solieriinae	1	1	2	2	2	2	2	2
Staphylininae	320	6876	1848	5.78	0.27	563	1.76	0.082
Steninae	2	2246	561	280.5	0.25	88	44	0.039
Tachyporinae	39	1518	2427	62.23	1.60	228	5.85	0.15
Tachyporinae*[a]		1518				56		0.037
Trichophyinae	1	16	16	16	1	5	5	0.31
Trigonurinae	1	10	1	1	0.1	1	1	0.1
STAPHYLINIDAE	3449	52344	15447	4.48	0.29	3233	0.94	0.062

[a]The number and ratio of tachyporine sequences without *Vatesus* sequences are listed in a separate row labeled Tachyporinae* (Tachyporinae with an asterisk)

threshold is selected: the posterior probability $PP \geq 0.99$ and bootstrap support $BS \geq 75$. The groups that have support values below the threshold are not considered as supported and not discussed. These thresholds are not repeated every time, and for the purpose of the discussion below, "supported clade" means "the support value is not below the threshold," while "no support for a group" means "the support value is below the accepted threshold."

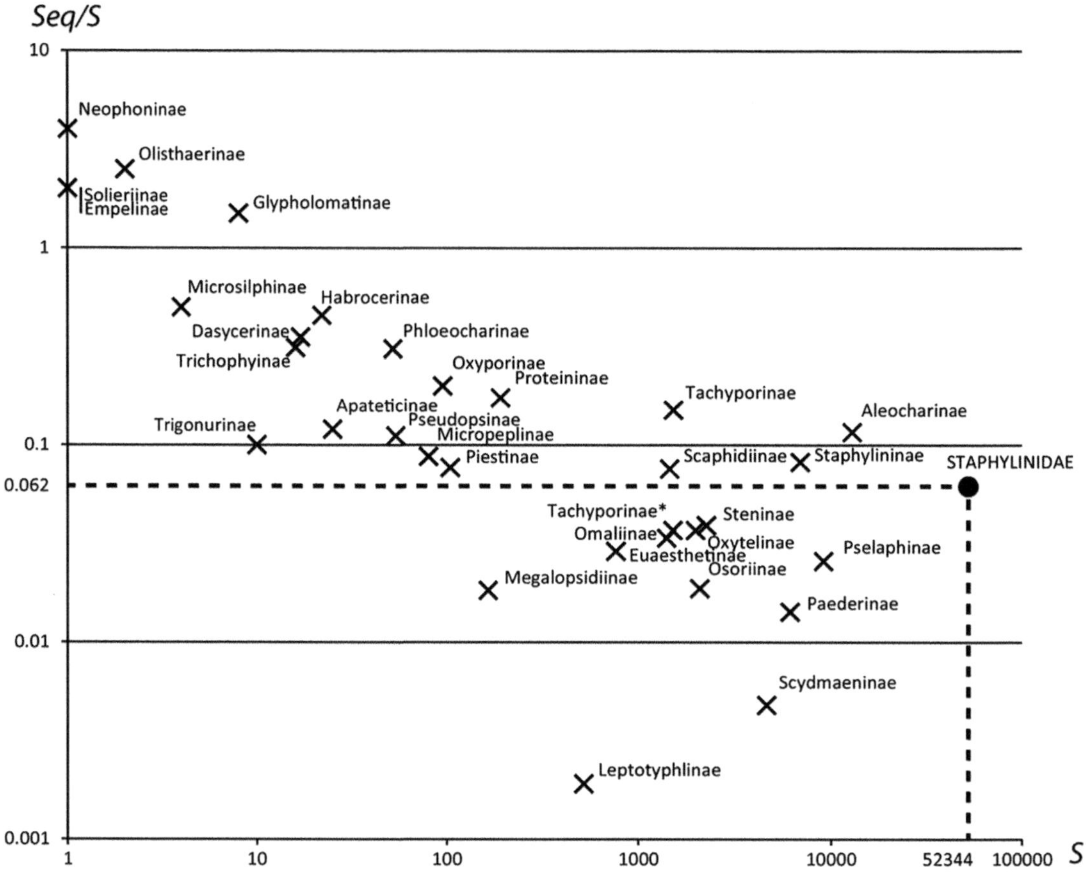

Fig. 2.1 Number of sequences per species (*Seq/S*) as a function of the number of species in a taxon (*S*). The plot shows all the subfamilies (crosses) except Protopselaphinae and the family as a whole (filled circle). Tachyporinae* denotes the point with all *Vatesus* sequences excluded

2.4.1 What Is a Sister Group of the Staphylinidae?

Several family-level studies addressed the phylogenetic position of the Staphylinidae within Coleoptera (Caterino et al. 2005; Hunt et al. 2007; McKenna et al. 2015a, b; Timmermans et al. 2016).

Caterino et al. (2005) in their Bayesian analyses based on *18S* sequences recovered Silphidae and the staphylinid subfamily Phloeocharinae as a well-supported clade, (Phloeocharinae, (*Oxelytrum*, *Necrophila*)), both nodes with $PP = 1$. However, the position of the (Phloeocharinae, Silphidae) clade within Staphylinoidea and family-level relationships in general remained unresolved.

Hunt et al. (2007) in their Bayesian analysis based on three molecular markers (*CO1*, *16S*, and *18S*) did not resolve the relationships between the Staphylinidae and other members of Staphylinoidea.

Grebennikov and Newton (2009) analyzed the relationships among 21 subfamilies of Staphylinidae and 5 other families of Staphylinoidea using *18S* sequences. Their published Bayesian tree, based on ClustalX alignment with all positions included, did not resolve the relationships between Staphylinidae and other families except suggesting a clade that included all members of Ptiliidae and Aleocharinae ($PP = 1$) and a broader clade that included (Ptiliidae + Aleocharinae) and 11 other subfamilies of Staphylinidae ($PP = 0.99$).

According to the brief description of methods in that paper, the Bayesian analysis included only a single run. As a result, the posterior probabilities listed as support values for the clades in the tree are likely to be an overestimation.

McKenna et al. (2015a) analyzed the phylogeny of Staphyliniformia, based on 282 species and 2 nuclear genes, *28S* and *CAD*. In their Bayesian analysis they found Silphidae nested within Staphylinidae in a well-supported clade ($PP = 1$). The relationships of the monophyletic (Staphylinidae + Silphidae) to the three other clades within the monophyletic Staphylinoidea ($PP = 1$) remained unresolved. These three clades are ((Leiodidae without *Colon*), Agyrtidae) ($PP = 1$), (Ptiliidae + Hydraenidae) ($PP = 0.99$), and *Colon*. In maximum likelihood analysis of the same dataset, there was no support for monophyletic (Staphylinidae + Silphidae), and the relationships among Staphylinidae and other families of Staphylinoidea remained unresolved.

McKenna et al. (2015b) analyzed the relationships among 367 species representing 172 of 183 extant families of beetles using 8 nuclear genes, including 6 single-copy protein-coding genes. However, their analyses did not identify a sister group of the Staphylinidae.

Timmermans et al. (2016) used complete or partial mitochondrial genomes of 245 taxa to analyze the phylogeny of Coleoptera. The focus of the study was on suborder and superfamily-level relationships, and the results of most analyses were not described below superfamily level. Some analyses supported the clade consisting of Staphylinidae and Silphidae, but did not resolve the relationships of this clade to other families of Staphyliniformia.

To conclude, based on McKenna et al. (2015a) and Timmermans et al. (2016), (Staphylinidae + Silphidae) was found to be a monophyletic group in which the Silphidae may be nested within the Staphylinidae. The sister group of the (Staphylinidae + Silphidae) clade is unknown.

2.4.2 Subfamily-Level Phylogeny

A study addressing phylogenetic relationships among the staphylinid subfamilies ideally should include multiple representative taxa from all the subfamilies (unless a subfamily is small and includes only one or few very similar species) and the closely related families as outgroup. Few studies meet this requirement. Most studies addressing the family-level relationships within Coleoptera (Caterino et al. 2005; Hunt et al. 2007; McKenna et al. 2015b; Timmermans et al. 2016) include too few staphylinid subfamilies to say much about subfamily-level relationships.

Ballard et al. (1998) analyzed relationships among 14 subfamilies of the Staphylinidae using a morphological and molecular datasets. Their molecular dataset included two mitochondrial markers, *12S* and *cyt B*, and did not succeed in resolving the relationships among the subfamilies when analyzed alone. These two genes may evolve too fast to be of much use at the subfamily level, and more recent studies did not use them.

Caterino et al. (2005) recovered monophyletic Aleocharinae (represented by three tribes) and Scydmaeninae (represented by two tribes), but none of the other subfamilies represented by more than one taxon were monophyletic. In addition to the (Phloeocharinae, Silphidae) ($PP = 1$) clade mentioned above, they found two more subfamily-level clades: (Oxyporinae, Steninae) ($PP = 1$) and (unidentified Tachyporinae, unidentified Paederinae) ($PP = 0.99$). Another member of Paederinae, *Lathrobium*, did not group with the latter clade, casting doubts on its credibility.

Hunt et al. (2007) included in their analyses representatives of 19 staphylinid subfamilies, each represented by a single species. Their Bayesian analysis did not resolve the relationships among the subfamilies except recovering two clades: (Oxyporinae, Steninae) and (Aleocharinae, Proteininae).

In their Bayesian analysis (see previous section), Grebennikov and Newton (2009) recovered monophyletic Euaesthetinae (3 genera),

Steninae (2 genera, 4 species), Scydmaeninae (5 genera), Tachyporinae (2 species, 1 identified to subfamily only), Micropeplinae (1 genus, 2 species), and Paederinae (2 genera, 3 species, one of which was identified to subfamily only). Relationships among the subfamilies remained largely unresolved except a clade that included all members of Ptiliidae and Aleocharinae (*PP* = 1) and a larger clade that included (Ptiliidae + Aleocharinae) and subfamilies Apateticinae, Glypholomatinae, Habrocerinae, Micropeplinae, Omaliinae, Oxytelinae, Paederinae, Piestinae, Proteininae, Scaphidiinae, and Staphylininae. As pointed out in the previous section, support values in the published tree (Fig. 12 in Grebennikov and Newton 2009) are likely overestimated.

Zhang and Zhou (2013) analyzed the relationships within the family Staphylinidae using three markers (*CO1*, *28S*, and *wg*) in an analysis that included 15 subfamilies and family Silphidae. Out of 11 subfamilies represented by more than 1 species, only 4 were recovered as monophyletic: Steninae (2 genera), Paederinae (4 genera, 5 species), Oxyporinae (2 species of *Oxyporus*), and Proteininae (2 species of *Megarthrus*). As far as relationships among subfamilies are concerned, not much has been achieved: either there was no support or members of different subfamilies grouped together making the subfamilies non-monophyletic. Unfortunately, only the topology without branch length is presented in the published trees making it more difficult to assess the causes of problematic groupings. At least in part the unfeasible results seem to be caused by errors. For example, *Ochthephilus emarginatus* (Oxytelinae) was found nested within the clade that includes all three members of Aleocharinae, making both subfamilies non-monophyletic. BLAST results indicate that while the *CO1* and *wg* sequences of *O. emarginatus* are similar to those of some other oxytelines, its *28S* sequence is most similar to some species of *Aleochara*. A comparison of aligned *28S* sequences of *Ochthephilus* and *Aleochara* demonstrates that the *28S* sequence of *O. emarginatus* (JX878725) is extremely different from that of its congener *O. planus* (KJ845013) and much more similar to the sequences of some species of *Aleochara*. Apparently, JX878725 is not the true *28S* sequence of *O. emarginatus*, and this error misplaced the species in the tree.

McKenna et al. (2015a) included in their study 31 out of 32 subfamilies of Staphylinidae and used 2 nuclear genes, *28S* and *CAD*. Their Bayesian analysis found Silphidae nested within Staphylinidae in a clade (*PP* = 1) that included Leptotyphlinae, Oxyporinae, Steninae, Euaesthetinae, Solieriinae, Scydmaeninae, Aleocharinae, Neophoninae, Dasycerinae, Pselaphinae, and parts of Tachyporinae (Tachyporini + Deropini), Osoriinae (Thoracophorini + Eleusinini + Leptochirini), and Phloeocharinae (*Charhyphus*). Within that clade, Silphidae grouped together with (Tachyporini + Deropini). Support for that grouping was *PP* = 0.98 which is very close to passing the support threshold accepted in this review.

Out of 18 subfamilies represented in the analyses of McKenna et al. (2015a) by more than one species, only 12 were recovered as monophyletic: Aleocharinae, Dasycerinae, Euaesthetinae, Habrocerinae, Oxytelinae, Paederinae, Proteininae, Pselaphinae, Scaphidiinae, Scydmaeninae, Staphylininae, and Steninae. Monophyly of two subfamilies (Omaliinae and Glypholomatinae) was neither rejected nor supported (Fig. 2.2). Four subfamilies (Osoriinae, Phloeocharinae, Piestinae, and Tachyporinae) were found to be non-monophyletic (Fig. 2.2).

The focus of McKenna et al. (2015b) was on the entire order Coleoptera, and the analysis included only nine species of Staphylinidae representing seven subfamilies. Scydmaeninae represented by three species were recovered as monophyletic, but the relationships among the included subfamilies remained unresolved.

The mitochondrial genome dataset of Timmermans et al. (2016) included only six subfamilies of Staphylinidae and was not designed to test the monophyly of subfamilies or relationships among them.

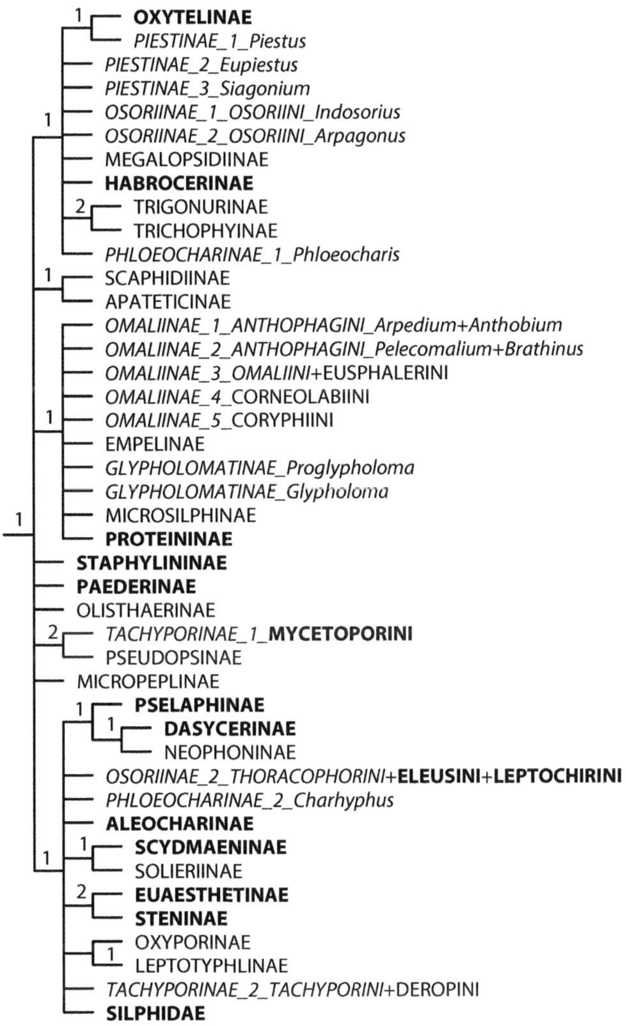

Fig. 2.2 Subfamily-level phylogeny of Staphylinidae. The tree summarizes the phylogeny of the (Staphylinidae + Silphidae) clade presented in Fig. 3 of McKenna et al. (2015a) by lumping together all members of every subfamily (if monophyly of subfamily is supported) or by lumping together all subfamily members that form monophyletic group. Subfamily and tribe names are set in bold font if they were represented in the analyses by more than one species and found to be monophyletic. Subfamily and tribe names set in italic font denote the taxa that were not recovered as monophyletic. Support value of 1 at the nodes is assigned to the clades supported only in Bayesian analysis (Fig. 3 in McKenna et al. 2015a). Support value of 2 at the nodes is assigned to the clades supported in both Bayesian and maximum likelihood analysis (Figs. 3 and 4 in McKenna et al. 2015a).

In conclusion, all studies but one (McKenna et al. 2015a) did not succeed in resolving the relationships among staphylinid subfamilies. McKenna et al. (2015a) found several monophyletic groups within the (Staphylinidae + Silphidae) clade, established sister group relationships for several subfamilies, confirmed the monophyly for some subfamilies, and rejected for a few others (Fig. 2.2). The four traditionally recognized groups of subfamilies (Omaliine, Oxyteline, Tachyporine, and Staphylinine groups: Thayer 2005) were not confirmed. However, some of the clades of McKenna et al. (2015a) overlap with the traditional groups. One of these clades included most of the subfamilies of the Omaliine group (Omaliinae, Empelinae, Glypholomatinae, Microsilphinae, and Proteininae) and could become a redefined Omaliine group. The Micropeplinae were found as a member of the basal polytomy and could potentially join the redefined Omaliine group. Three subfamilies (Pselaphinae, Dasycerinae, and Neophoninae) were recovered within another major clade and have to be excluded from the Omaliine group. Another major clade included subfamilies Oxytelinae, Piestinae, Osoriinae (without Thoracophorini + Eleusinini + Leptochirini), and Trigonurinae—all four are members of the

Oxyteline group (Thayer 2005)—and subfamilies Trichophyinae, Habrocerinae, Phloeocharinae (*Phloeocharis*, but not *Charhyphus*), and Megalopsidiinae. These eight subfamilies could become a redefined Oxyteline group. A clade consisting of Scaphidiinae and Apateticinae was found as one of the lineages in the basal polychotomy and could potentially join the Oxyteline group. The results of McKenna et al. (2015a) do not support any equivalent of Tachyporine and Staphylinine groups, because Staphylininae were a member of the basal polytomy, while most of the remaining subfamilies formed a clade that included members of both traditional Staphylinine and Tachyporine groups. The fact that even eight nuclear genes (McKenna et al. 2015b) did not resolve the relationships among the included staphylinid subfamilies suggests that much more data may be needed to fully resolve high-level relationships within the Staphylinidae.

2.4.3 Staphylinid Supertrees

Considering that a number of studies have addressed the same question of subfamily-level relationships in the Staphylinidae (Ballard et al. 1998; Caterino et al. 2005; Hunt et al. 2007; Grebennikov and Newton 2009; Zhang and Zhou 2013; McKenna et al. 2015a, b; Timmermans et al. 2016), it is tempting to analyze the data obtained in these studies together, to increase taxon coverage in the tree. Unfortunately, the above studies used different markers (Table 2.2), and combining all sequence data in a single supermatrix would result in so much missing data that the analyses would become pointless. An alternative approach could be the construction of a supertree based on the source trees produced by individual studies. Many different methods have been developed (Bininda-Emonds 2004; Akanni et al. 2015), the most popular being MRP (matrix representation with parsimony analysis) (Baum 1992; Ragan 1992) and its modifications. Simulations suggest that the capacity of MRP to find the true tree depends on the number of source trees and the overlap among them (Bininda-Emonds and Sanderson 2001; Ross and Rodrigo 2004). Further, there are important general issues concerning source trees used in supertree construction (Bininda-Emonds et al. 2004), including the non-independence of datasets and the validity of source trees. In a situation where source trees from relatively few (8) different studies considerably differ in the number of taxa included (from 6 staphylinid subfamilies to 31, from 9 staphylinid species to 129), in the number of markers used (from 1 to 8; from 765 to 12271 bp), and in resolution and credibility of results (see discussion above), a better strategy is to select the best tree as the basis for supertree construction.

For the purposes of supertree construction, the Bayesian analysis tree from McKenna et al. (2015a) was selected as the basis since it is covering more superfamilies and species and providing better resolution at subfamily level than any other available subfamily-level tree (Ballard et al. 1998; Caterino et al. 2005; Hunt et al. 2007; Grebennikov and Newton 2009; Zhang and Zhou 2013; McKenna et al. 2015b; Timmermans et al. 2016). First, all nodes with support below the threshold accepted in this review were collapsed. Next, molecular phylogenetic trees obtained in the studies addressing relationships within more restricted groups of the Staphylinidae (subfamilies, tribes, and groups of genera) were used to replace the corresponding clades in the tree of McKenna et al. (2015a). The resulting supertree is presented in Fig. 2.3. Phylogenies of four large clades, each collapsed to a single terminal branch in Fig. 2.3, are shown separately: subfamily Staphylininae, Fig. 2.4; subfamily Paederinae, Fig. 2.5; subfamily Aleocharinae, Fig. 2.6 (with two alternative hypotheses compared in Fig. 2.7), and pselaphine supertribe Clavegiritae, Fig. 2.8. Some congeneric species (e.g., many species of *Hyperomma* (Paederinae)) were pruned unless they were important for interpretation of relationships among genera or species groups.

The resulting supertree (Figs. 2.3–2.8) provides a summary of the current knowledge about staphylinid phylogeny and makes it

Table 2.2 Focal taxa and molecular datasets analyzed in published studies on staphylinid phylogeny

	Number of staphylinid		Focal taxon			Mitochondrial genes							Nuclear genes										
Paper	Species	Subfamilies	Name	Rank	Dataset size, bp	Number	*16S*	*12S*	*CO1*	*CO2*	*cyt B*	mt genome	Number	*18S*	*28S*	*H3*	*wg*	*TP*	*AK*	*CAD*	*EF1-α*	*AS*	*PEPCK*
Ballard et al. (1998)	19	14	Staphylinidae	Family	765	2		+			+												
Maus et al. (2001)	59	1	*Aleochara*	Genus	2022	2			+	+													
Caterino et al. (2005)	33	20	Staphyliniformia	Series	2191								1	+									
Chatzimanolis and Caterino (2007)	3	1	*Sepedophilus castaneus*	Species	826	1			+														
Hunt et al. (2007)	53	19	Coleoptera	Order	3610 or 3784	2	+		+				1	+									
Ikeda et al. (2008)	3	1	Silphidae	Family	2592	1	+						3		+		+						+
Leschen et al. (2008)	1	1	*Brachynopus scutellaris*	Species	820	1			+														
Grebennikov and Newton (2009)	58	21	Staphylinidae	Family	ca. 2100								1	+									
Thomas (2009)	21	1	Aleocharinae	Subfamily	ca. 770	2	+	+															
Ahn et al. (2010)	21	1	Liparocephalini	Tribe	ca. 3510	3		+	+	+			1	+									
Chatzimanolis et al. (2010)	51	1	Staphylinini	Tribe	3164	1			+				3		+		+	+					
Elven et al. (2010)	81	1	Athetini	Tribe	3865	3	+		+	+			1	+									
Fresneda et al. (2011)	7	1	Leptodirini	Tribe	ca. 3000	3	+		+		+		2	+	+								
Elven et al. (2012)	113	1	Athetini, Geostibini, and Lomechusini	Tribes	3786	3	+		+	+			1	+									
Jeon et al. (2012)	37	1	*Cafius*	Genus	ca. 3954	4	+		+	+			1	+	+								
Koerner et al. (2013)	44	1	Steninae	Subfamily	807	1			+														
Osswald et al. (2013)	110	1	Aleocharinae	Subfamily	4599	3	+		+	+			3	+	+			+					
Song and Ahn (2013)	64	1	*Aleochara*	Genus	ca. 2021	2			+	+													
Zhang and Zhou (2013)	102	15	Staphylinidae	Family	2413	1			+				2		+		+						
Chatzimanolis (2014)	34	1	Xanthopygina	Subtribe	5548	1			+				5		+		+	+	+	+			
Parker and Grimaldi (2014)	51	1	Clavigeritae	Supertribe	5944	2	+		+				3		+		+						

(continued)

Table 2.2 (continued)

Paper	Number of staphylinid		Focal taxon		Dataset size, bp	Mitochondrial genes							Nuclear genes										
	Species	Subfamilies	Name	Rank		Number	*16S*	*12S*	*CO1*	*CO2*	*cyt B*	mt genome	Number	*18S*	*28S*	*H3*	*wg*	*TP*	*AK*	*CAD*	*EF1-α*	*AS*	*PEPCK*
Song and Ahn (2014)	6	1	*Aleochara fucicola* group	Species group	2500	2			+	+			3				+			+	+		
Caterino et al. (2015)	2	1	*Thinopinus pictus* and *Hadrotes crassus*	Species	826	1			+														
Grebennikov and Smetana (2015)	ca. 30	1	Micropeplinae	Subfamily	658	1			+														
Lang et al. (2015)	22	1	Steninae	Subfamily	ca. 2100	2	+		+				1			+							
McKenna et al. (2015a)	129	31	Staphyliniformia	Series	3430								2		+					+			
McKenna et al. (2015b)	9	7	Coleoptera	Order	8377								8	+	+		+		+	+	+	+	+
von Beeren et al. (2016a)	4	1	*Vatesus*	Genus	1603	1			+				2				+			+			
von Beeren et al. (2016b)	5	1	*Tetradonia*	Genus	1595	1			+				2				+			+			
Brunke et al. (2016)	90	1	Staphylinini	Tribe	4370	1			+				5		+		+	+	+	+			
Parker (2016)	50	1	Pselaphinae	Subfamily	949								1		+								
Serri et al. (2016)	23	1	*Stenus*	Genus	658	1			+														
Timmermans et al. (2016)	10	6	Coleoptera	Order	12271	15						+											
Chani-Pose et al. (2017)	133	1	Philonthina	Subtribe	4458	1			+				5		+		+	+	+	+			
Marujama and Parker (2017)	181	1	Aleocharinae	Subfamily	4312	2	+		+				3	+	+			+					
Schomann and Solodovnikov (2017)	31	1	Paederinae	Subfamily	ca. 3380	1			+				4				+	+	+	+			

"Dataset size" indicates the number of nucleotides in the alignment used in the analyses (if the alignment details were not provided, the average total length of all the used markers is listed). Gene abbreviations are as follows: 16S rRNA (*16S*), 12S rRNA (*12S*), cytochrome oxidase subunit 1 (*CO1*), cytochrome oxidase subunit 2 (*CO2*), cytochrome *b* (*cyt B*), 18S rRNA (*18S*), 28S rRNA (*28S*), histone *H3* (*H3*), wingless (*wg*), topoisomerase I (*TP*), arginine kinase (*AK*), and carbamoyl phosphate synthetase domain of *CAD* or rudimentary (*CAD*), elongation factor 1-alpha (*EF1-α*), alpha-spectrin (*AS*), phosphoenolpyruvate carboxykinase (*PEPCK*). Small tRNA gene sequences and short fragments of flanking genes (e.g., NADH dehydrogenase subunit 1) are not included. "Number of species" indicates the number of *staphylinid* species included in the analyses

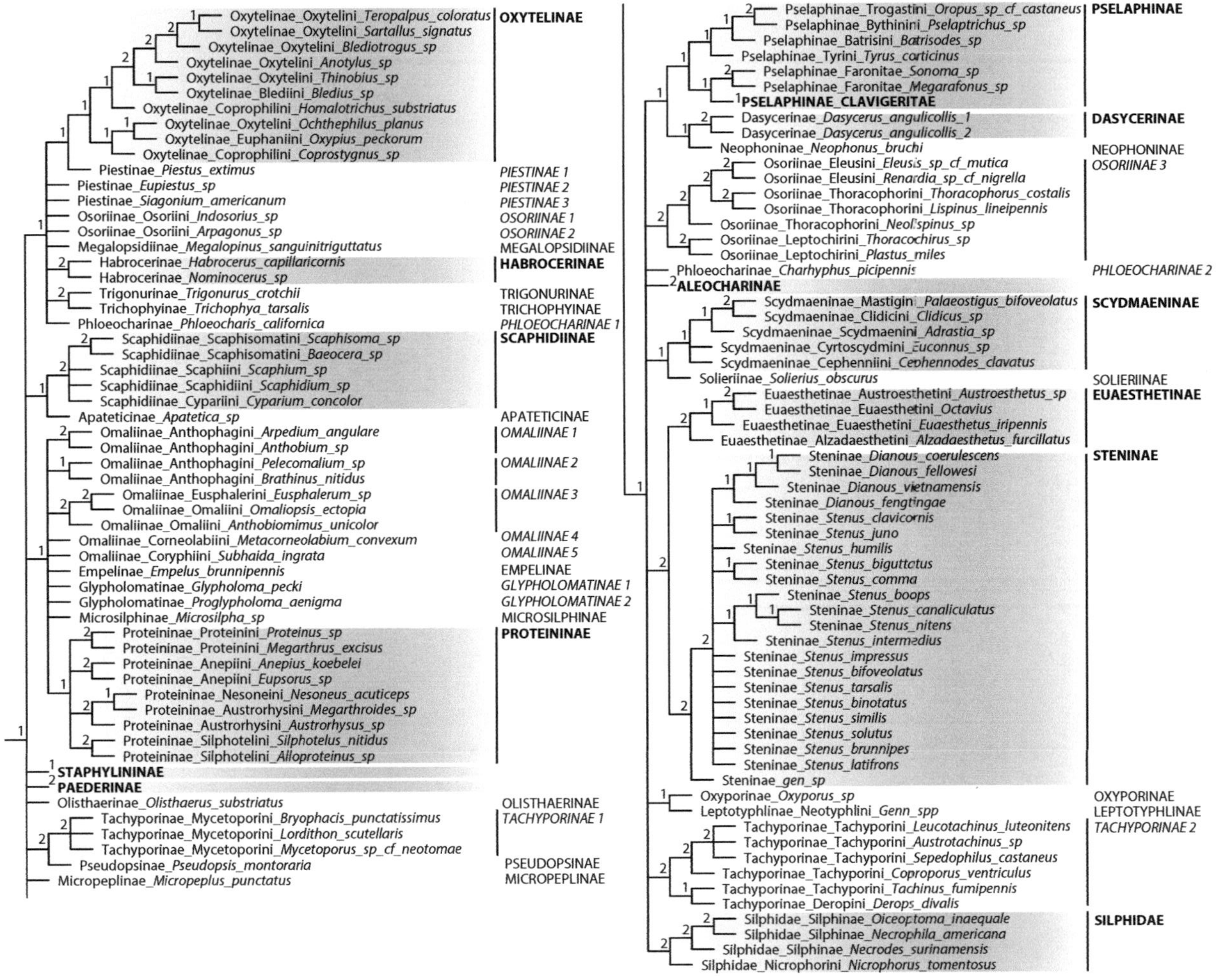

Fig. 2.3 Supertree summarizing the phylogeny of Staphylinidae. The tree was obtained from the (Staphylinidae + Silphidae) clade of the Bayesian analysis tree in McKenna et al. (2015a) by replacing the (*Stenus* + *Dianous*) clade with the equivalent clade from Fig. 3 in Lang et al. (2015). All unsupported nodes were collapsed. Four clades (subfamilies Staphylininae, Paederinae and Aleocharinae, and pselaphine supertribe Clavigeritae) are collapsed to a single terminal. Their detailed phylogeny is presented in Figs. 2.4–2.8. The names of monophyletic taxa are set in bold font, and the corresponding clades are shaded. The names of the subfamilies represented by more than one species and not recovered as monophyletic are set in italic font. Support values at the nodes indicate the number of model-based analyses of molecular datasets (Bayesian or maximum likelihood) in one or more studies supporting the relevant nodes

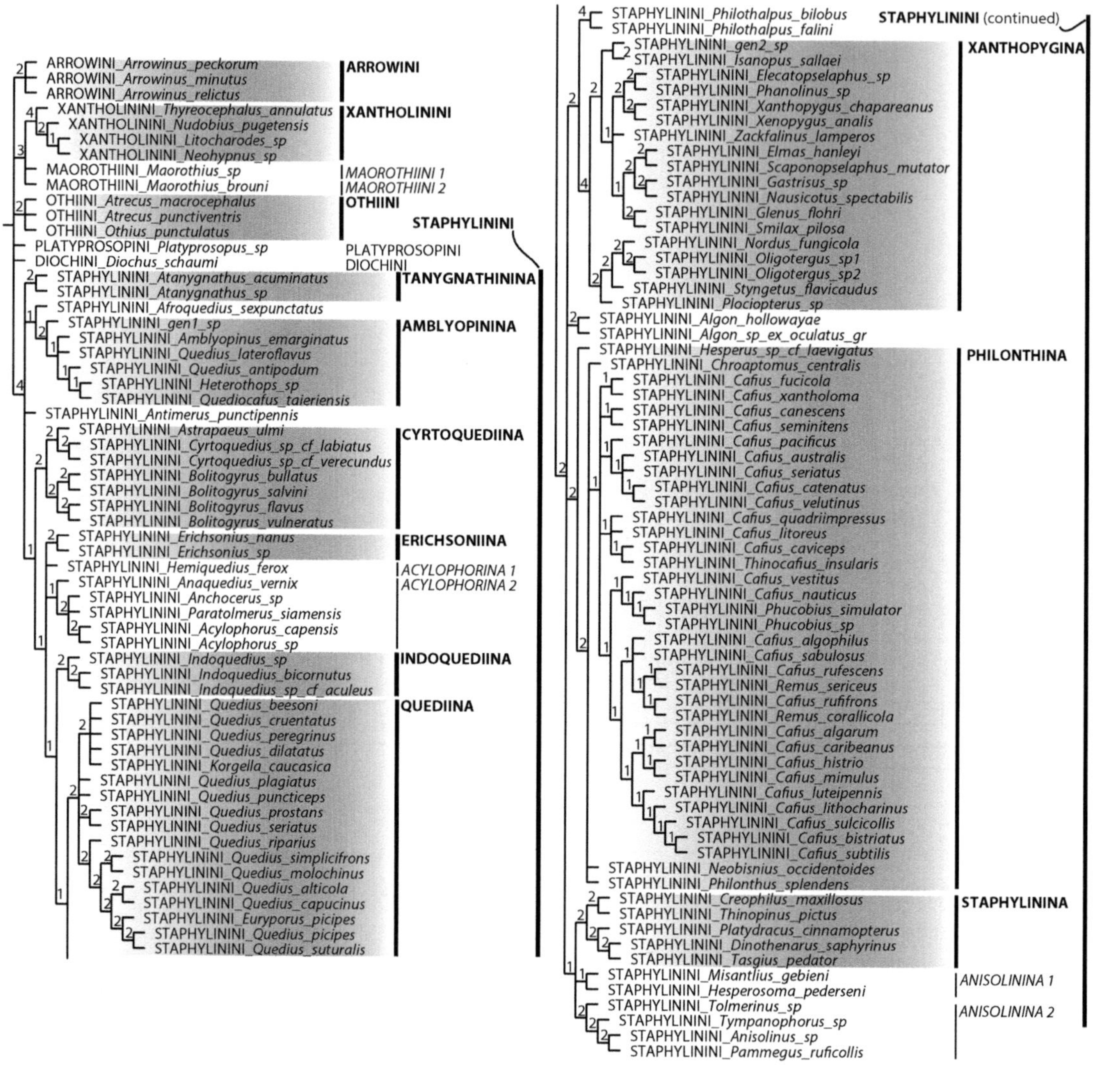

Fig. 2.4 Fragment of the staphylinid supertree for the subfamily Staphylininae. The tree was obtained from the Bayesian analysis tree in McKenna et al. (2015a) by replacing the Staphylinini and Xantholinini clades with the equivalent clades from Fig. 1 in Brunke et al. (2016). In three genera (*Arrowinus*, *Maorothius*, and *Atrecus*), the species from both studies were combined. The Xanthopygina clade was then replaced with the equivalent clade from Fig. 2 in Chatzimanolis (2014). Within Philonthina, the *Cafius seminitens* terminal branch was replaced by clade Z in Fig. 3 of Jeon et al. (2012). All unsupported nodes were collapsed. The names of monophyletic (sub)tribes are set in bold font, and the corresponding clades are shaded. The names of the (sub)tribes represented by more than one species and not recovered as monophyletic are set in italic font. Support values at the nodes indicate the number of model-based analyses of molecular datasets (Bayesian or maximum likelihood) in one or more studies supporting the relevant nodes

possible to compare resolution in different taxa. While some taxa appear as well-resolved clades (e.g., subfamily Oxytelinae, staphylinine subtribe Xanthopygina), others are represented as polytomies (e.g., (*Stenus* + *Dianous*)) or parts of polytomies (e.g., subfamily Omaliinae). Most taxa fall in between and combine polytomies with well-resolved nodes. Phylogenetic relationships within subfamilies are briefly discussed in the next section.

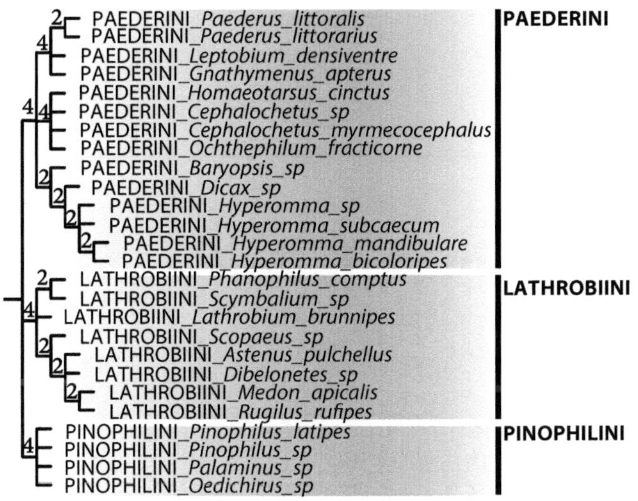

Fig. 2.5 Fragment of the staphylinid supertree for subfamily Paederinae. The tree was obtained from the Bayesian analysis tree in McKenna et al. (2015a) by replacing the Lathrobiini clade with the equivalent clade from Fig. 1 in Schomann and Solodovnikov (2017) and by replacing the *Hyperomma* terminal branch with the Dicaxina clade (with some species of *Hyperomma* pruned). In three taxa (Pinophilini, Paederina, Cryptobiina), the species from both studies were combined. All unsupported nodes were collapsed. Support values at the nodes indicate the number of model-based analyses of molecular datasets (Bayesian or maximum likelihood) in two studies supporting the relevant nodes

2.4.4 Phylogeny Within Subfamilies

Subfamily Protopselaphinae has not been included in any molecular phylogenetic study.

Many of the staphylinid subfamilies (Apateticinae, Dasycerinae, Empelinae, Glypholomatinae, Habrocerinae, Microsilphinae, Neophoninae, Olisthaerinae, Solieriinae, Trichophyinae, and Trigonurinae) are very small and include only one or two extant genus/genera with very few species. For these subfamilies, interesting phylogenetic questions mostly go beyond the limits of individual subfamily.

Some subfamilies have been represented in phylogenetic studies by one or very few species, and their phylogeny remains largely unknown (Leptotyphlinae, Megalopsidiinae, Micropeplinae, Oxyporinae, and Pseudopsinae).

Several subfamilies have been shown to be non-monophyletic (Osoriinae, Phloeocharinae, Piestinae, Tachyporinae). It is safe to conclude that the current Tachyporinae include two unrelated clades (Fig. 2.3): (1) Tachyporinae proper that should include the tribes Deropini, Megarthropsini, Tachyporini, and Vatesini and (2) tribe Mycetoporini that should be elevated to the rank of subfamily. Subfamilies Osoriinae, Phloeocharinae, and Piestinae require more extensive taxon sampling and better resolution.

Monophyly of Omaliinae is neither confirmed nor rejected, and phylogenetic relationships within this large and diverse subfamily remain largely unknown.

Phylogeny of the monophyletic subfamily Oxytelinae was analyzed only within a broad context of Staphylinidae (McKenna et al. 2015a). Relationships among the ten included species representing ten different genera were well resolved (Fig. 2.3) suggesting that even few markers could be sufficient for answering many questions about phylogeny of this subfamily. Both tribes represented by more than one species (Oxytelini and Coprophilini) were found to be non-monophyletic.

Phylogeny of the monophyletic Scaphidiinae has not been analyzed in much detail (only five

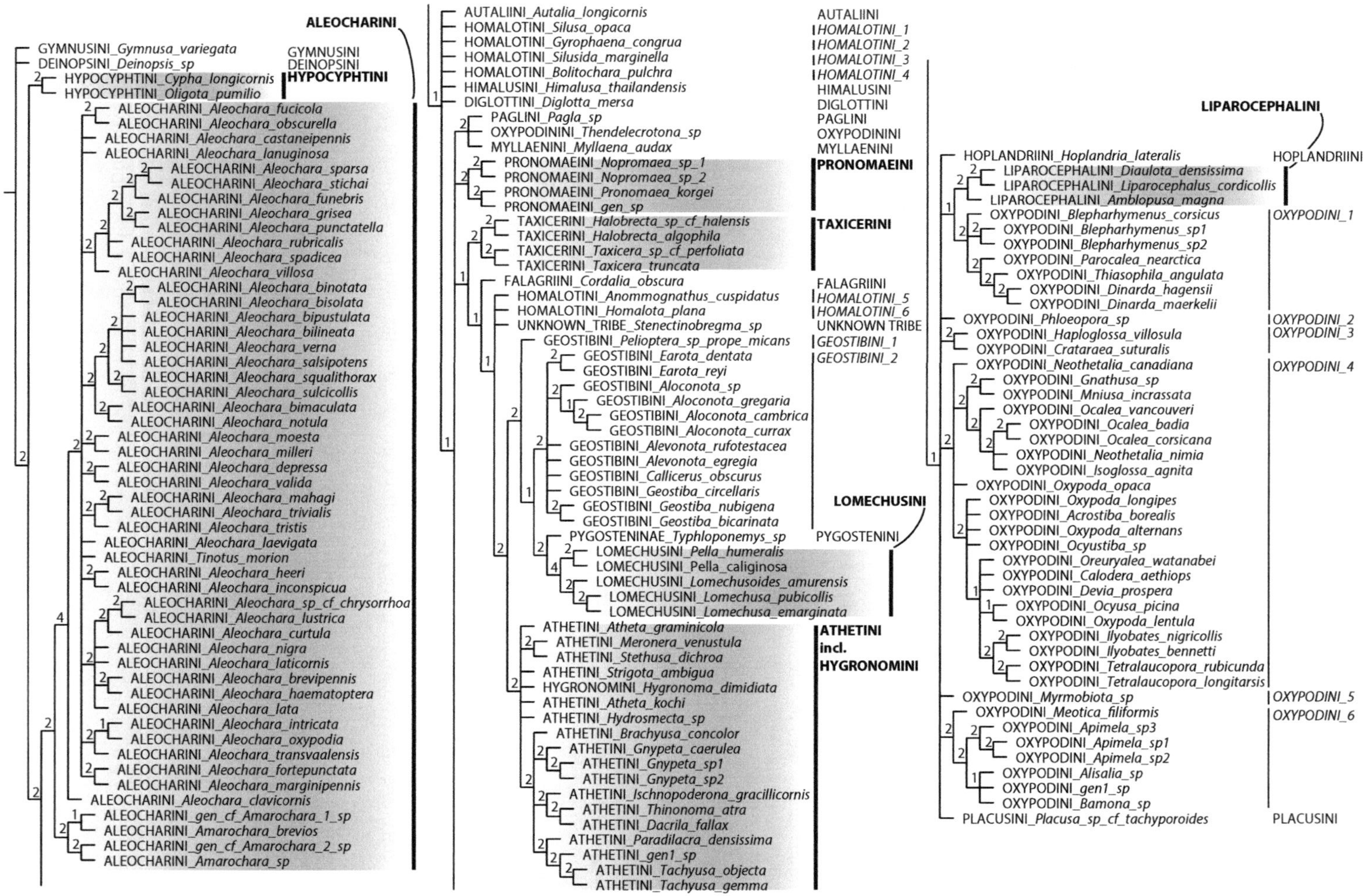

Fig. 2.6 Fragment of the staphylinid supertree for subfamily Aleocharinae. The tree was obtained by replacing all members of the Aleocharinae clade in the Bayesian analysis tree in McKenna et al. (2015a) with the aleocharine tree from Fig. 2 in Osswald et al. (2013). The Geostibini and (*Pella* + *Lomechusa*) clades of Osswald et al. (2013) were then replaced with the equivalent clades from Fig. 2 in Elven et al. (2012). The *Aleochara* clade of Osswald et al. (2013) was replaced with the equivalent clade from Fig. 2 in Song and Ahn (2013) with some species pruned. All unsupported nodes were collapsed. The names of monophyletic tribes are set in bold font and the corresponding clades are shaded. The names of the tribes represented by more than one species and not recovered as monophyletic are set in italic font. Support values at the nodes indicate the number of model-based analyses of molecular datasets (Bayesian or maximum likelihood) in one or more studies supporting the relevant nodes

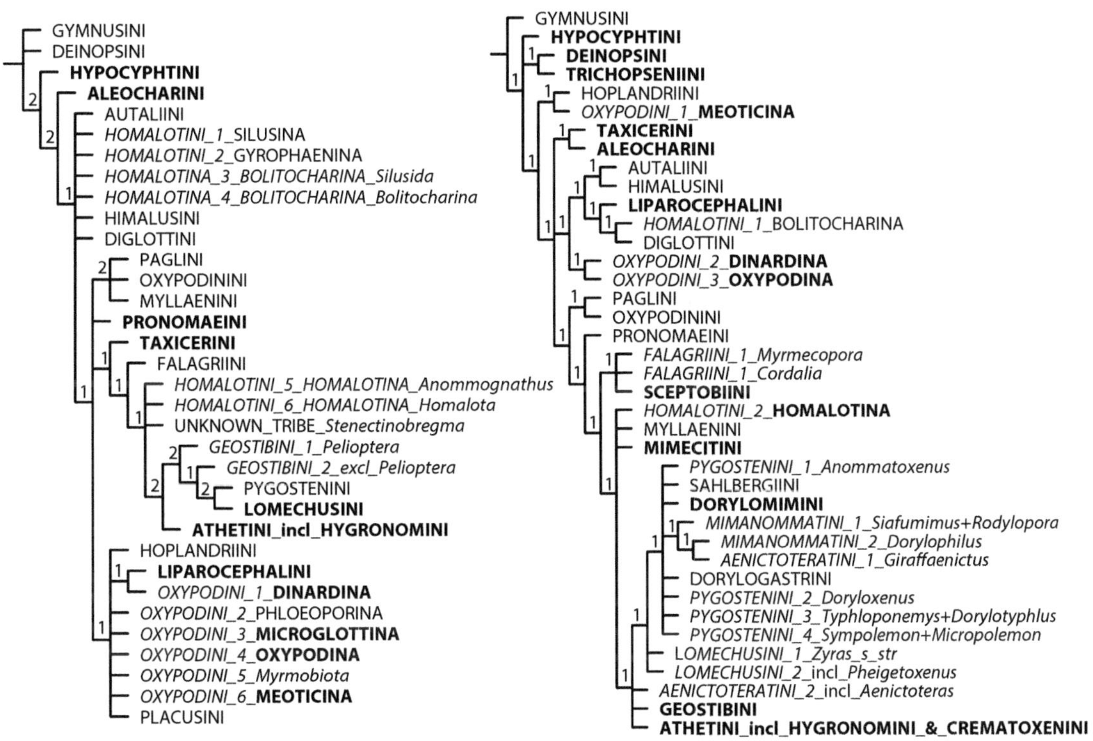

Fig. 2.7 Comparison of the tribe-level phylogenies of subfamily Aleocharinae based on (left) Fig. 2.6 in this chapter and (right) Fig. S2 in Maruyama and Parker (2017). The trees are obtained by collapsing monophyletic groups to tribe or the highest possible level below tribe, if a tribe is not supported as monophyletic. The names of monophyletic (sub)tribes are set in bold font. The names of the (sub)tribes represented by more than one species and not recovered as monophyletic are set in italic font. Support values calculated as in Figs. 2.3–2.6

species included). Relationships among the four tribes have not been resolved (Fig. 2.3).

The monophyletic Proteininae were represented by all five tribes, but relationships among them remained unresolved (Fig. 2.3).

The monophyletic Scydmaeninae were represented by five out of ten tribes (one species each). More extensive taxon sampling is needed to address relationships within the subfamily.

The monophyletic Euaesthetinae were represented by three tribes (four species) out of six. A better taxon coverage is needed to investigate the relationships within the subfamily.

Within the monophyletic Steninae, the genera (*Dianous* + *Stenus*) were recovered as monophyletic group (McKenna et al. 2015a), but the relationships between the monophyletic *Dianous* and different groups of species within *Stenus* are unknown. Two alternatives are compatible with the results of published phylogenetic analyses (Koerner et al. 2013; Lang et al. 2015) (given the support threshold accepted in this chapter: see Sect. 2.4 above): (1) *Dianous* and *Stenus* are sister clades; (2) *Dianous* is nested within *Stenus*.

Phylogeny of the megadiverse monophyletic subfamily Staphylininae (6000+ species) was recently analyzed in a series of studies addressing relationships among all the tribes of the subfamily or particular clades (tribes, subtribes, or genera) within it (Chatzimanolis et al. 2010; Jeon et al. 2012; Chatzimanolis 2014; Brunke et al. 2016; Chani-Pose et al. 2017). Many monophyletic clades within Staphylininae have been recognized (Fig. 2.4), but there are still taxa unassigned to subtribes, and some unresolved relationships remain (both in the backbone and within well-supported

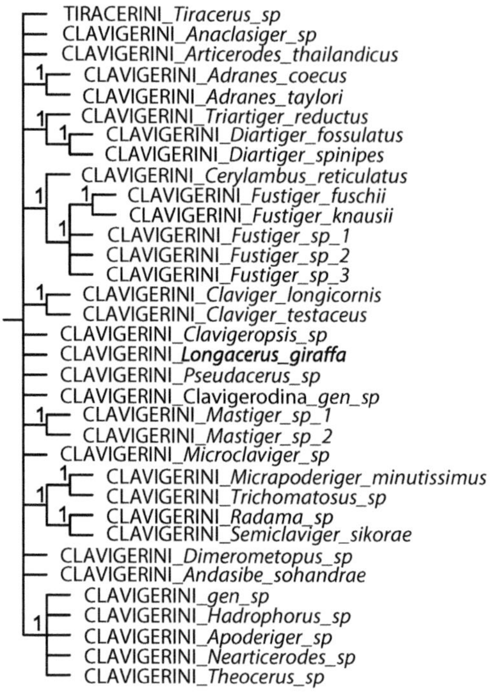

Fig. 2.8 Fragment of the staphylinid supertree for pselaphine supertribe Clavigeritae. The tree was obtained from Fig. S2 in Parker and Grimaldi (2014) by collapsing all unsupported nodes

clades). The results of Chani-Pose et al. (2017) are not included in Fig. 2.4, because the only presented analysis of the molecular dataset alone (Fig. 4, B in Chani-Pose et al. 2017) does not show support values, making it impossible to apply the threshold accepted in this review.

Phylogeny of the monophyletic subfamily Paederinae is summarized in Fig. 2.5 based on McKenna et al. (2015a) and Schomann and Solodovnikov (2017). Relationships among the three included tribes, Paederini, Lathrobiini, and Pinophilini, are unknown. Within Paederini, the relationships among (Paederina + Dolicaonina), Cryptobiina, and Dicaxina are also unknown. The diversity of Lathrobiini has not been sampled sufficiently to test the monophyly of the subtribes within and the relationships among them. No representative of the tribe Cylindroxystini has been included in molecular analyses.

The phylogeny of the largest staphylinid subfamily Aleocharinae (monophyletic; 12,000+ species) is summarized in Fig. 2.6 based on Elven et al. (2012), Osswald et al. (2013), and Song and Ahn (2013). In Fig. 2.7 (left), the same tree is presented in a different way: the monophyletic groups of Fig. 2.6 have been collapsed to tribe or the highest possible level below tribe, if a tribe is not supported as monophyletic. It is clear that much work is needed to improve the resolution of the tree, extend coverage to all subtribes, and increase the taxon sampling in the larger clades, particularly by adding non-Holarctic taxa. Recent analyses by Maruyama and Parker (2017) were based on the widest ever selection of aleocharine taxa. The aleocharine sequences available in GenBank were complemented with newly obtained sequences for many myrmecophile taxa, bringing the total number of staphylinid species in the analyses to 181. The tribe-level results of the analyses in Maruyama and Parker (2017) are presented in Fig. 2.7 (right), in comparison with the tribe-level tree based on Osswald et al. (2013). Although the two trees are based on largely overlapping dataset, they differ in their tree topology, in the relationships of the most basal nodes, and in the position and monophyly of particular tribes. Detailed discussion of the causes of the difference between the two trees is outside the scope of this review and would require an extensive sensitivity analyses. Here, it is sufficient to mention that the two studies (Osswald et al. 2013; Maruyama and Parker 2017) used different methods to align the ribosomal gene sequences (based on the secondary structure vs. using MAFFT), treated loop regions of the RNA genes differently (deleting most of the ambiguously aligned loop regions vs. keeping the entire sequences), used different sets of outgroup taxa (given the uncertainty about the sister group of Aleocharinae, five different subfamilies including four genera of Tachyporinae vs. just one subfamily, Tachyporinae, with three genera), had different number of independent runs in the Bayesian analyses (four vs. two), etc.

Taxon sampling in the megadiverse monophyletic subfamily Pselaphinae (9000+ species) has been rather limited. Only seven species have been included in the analyses of McKenna et al. (2015a). Some of these species appeared in the phylogenetic tree as very long terminal branches (Fig. 5 in McKenna et al. 2015a) likely affecting the tree topology within the clade. Parker and Grimaldi (2014) analyzed in detail the phylogeny of the tribe Clavigeritae (Fig. 2.8). The relationships among genera remained largely unresolved. Thirty six species of Pselaphinae have been included in the molecular phylogenetic analysis of Parker (2016). This analysis was based on a single marker (*28S*) and resulted in a topology (Fig. S1, A in Parker (2016)) different from the one in the study of McKenna et al. (2015a).

2.5 Concluding Remarks

As a result of 20 years effort, many questions about phylogeny of the family Staphylinidae have been answered, but even more questions remain to be addressed. Together with Silphidae, Staphylinidae form a monophyletic group, but the exact relationships between the two families and their relationships to the other staphylinoid families are unknown. Monophyly was confirmed for a few staphylinid subfamilies, rejected for others, but for some still remains untested. Although a few subfamily-level clades within Staphylinidae have been firmly established, the relationships among staphylinid subfamilies remain largely unknown. Phylogenetic analyses of individual subfamilies or their subclades have covered just a small fraction of their diversity.

Molecular phylogenies have been used to revise staphylinid classification (Elven et al. 2012; Osswald et al. 2013; Brunke et al. 2016; Schomann and Solodovnikov 2017), interpret evolution of morphological characters (Koerner et al. 2013; Marujama and Parker 2017) and ecological adaptations (Ahn et al. 2010; Maruyama and Parker 2017), date clade origin (Zhang and Zhou 2013; Parker 2016; Maruyama and Parker 2017), discriminate among closely related species (von Beeren et al. 2016a, b), and address many other questions. To ensure that the answers to all these interesting questions are reliable, a well-resolved and highly supported phylogeny of the family Staphylinidae is needed.

References

Ahn K-J, Jeon M-J, Branham MA (2010) Phylogeny, biogeography and the stepwise evolutionary colonization of intertidal habitat in the Liparocephalini based on morphological and molecular characters (Coleoptera: Staphylinidae: Aleocharinae). Cladistics 26 (4):344–358. https://doi.org/10.1111/j.1096-0031.2009.00290.x

Akanni WA, Wilkinson M, Creevey CJ et al (2015) Implementing and testing Bayesian and maximum-likelihood supertree methods in phylogenetics. R Soc Open Sci 2(140436):1–9. https://doi.org/10.1098/rsos.140436

Ballard JWO, Thayer MK, Newton AF Jr et al (1998) Data sets, partitions, and characters: philosophies and procedures for analyzing multiple data sets. Syst Biol 47 (3):367–396. https://doi.org/10.1080/106351598260770

Baum BR (1992) Combining trees as a way of combining data sets for phylogenetic inference, and the desirability of combining gene trees. Taxon 41(1):3–10. https://doi.org/10.2307/1222480

Bininda-Emonds ORP (ed) (2004) Phylogenetic supertrees. Combining information to reveal the tree of life, Computational biology series, vol 4. Kluwer Academic, Dordrecht

Bininda-Emonds ORP, Sanderson MJ (2001) Assessment of the accuracy of matrix representation with parsimony analysis supertree construction. Syst Biol 50 (4):565–579. https://doi.org/10.1080/10635150120358

Bininda-Emonds ORP, Jones KE, Price SA et al (2004) Garbage in, garbage out. Data issues in supertree construction. In: Bininda-Emonds ORP (ed) Phylogenetic supertrees. Combining information to reveal the tree of life, Computational biology series, vol 4. Kluwer Academic, Dordrecht, pp 267–280

BOLD Systems (2017) The barcode of life data systems. http://www.boldsystems.org. Accessed 7 June 2017.

Bouchard P, Bousquet Y, Davies AE et al (2011) Family-group names in Coleoptera (Insecta). ZooKeys 88:1–972. https://doi.org/10.3897/zookeys.88.807

Brunke AJ, Chatzimanolis S, Schillhammer H et al (2016) Early evolution of the hyperdiverse rove beetle tribe Staphylinini (Coleoptera: Staphylinidae: Staphylininae) and a revision of its higher classification. Cladistics 32(4):427–451. https://doi.org/10.1111/cla.12139

Caterino MS, Hunt T, Vogler AP (2005) On the constitution and phylogeny of Staphyliniformia (Insecta: Coleoptera). Mol Phylogenet Evol 34(3):655–672. https://doi.org/10.1016/j.ympev.2004.11.012

Caterino MS, Chatzimanolis S, Richmond MP (2015) On the origins of the insect fauna of California's Channel islands: a comparative phylogeographic study of island beetles. Monogr West North Am Nat 7 (1):276–296. https://doi.org/10.3398/042.007.0121

Chani-Posse MR, Brunke AJ, Chatzimanolis S et al (2017) Phylogeny of the hyper-diverse rove beetle subtribe Philonthina with implications for classification of the tribe Staphylinini (Coleoptera: Staphylinidae). Cladistics. https://doi.org/10.1111/cla.12188

Chatzimanolis S (2014) Phylogeny of xanthopygine rove beetles (Coleoptera) based on six molecular loci. Syst Entomol 39(1):141–149. https://doi.org/10.1111/syen.12040

Chatzimanolis S, Caterino MS (2007) Toward a better understanding of the "transverse range break": lineage diversification in southern California. Evolution 61 (9):2127–2141. https://doi.org/10.1111/j.1558-5646.2007.00186.x

Chatzimanolis S, Cohen IM, Schomann A et al (2010) Molecular phylogeny of the mega-diverse rove beetle tribe Staphylinini (Insecta, Coleoptera, Staphylinidae). Zool Scr 39(5):436–449. https://doi.org/10.1111/j.1463-6409.2010.00438.x

Elven H, Bachmann L, Gusarov VI (2010) Phylogeny of the tribe Athetini (Coleoptera: Staphylinidae) inferred from mitochondrial and nuclear sequence data. Mol Phylogenet Evol 57(1):84–100. https://doi.org/10.1016/j.ympev.2010.05.023

Elven H, Bachmann L, Gusarov VI (2012) Molecular phylogeny of the Athetini–Lomechusini–Ecitocharini clade of aleocharine rove beetles (Insecta). Zool Scr 41(6):617–636. https://doi.org/10.1111/j.1463-6409.2012.00553.x

Erixon P, Svennblad B, Britton T et al (2003) Reliability of Bayesian posterior probabilities and bootstrap frequencies in phylogenetics. Syst Biol 52(5):665–673. https://doi.org/10.1080/10635150390235485

Fresneda J, Grebennikov VV, Ribera I (2011) The phylogenetic and geographic limits of Leptodirini (Insecta: Coleoptera: Leiodidae: Cholevinae), with a description of *Sciaphyes shestakovi* sp.n. from the Russian Far East. Arthropod Syst Phyl 69(2):99–123

Grebennikov VV, Newton AF (2009) Good-bye Scydmaenidae, or why the ant-like stone beetles should become megadiverse Staphylinidae sensu latissimo (Coleoptera). Eur J Entomol 106 (2):275–301. 10.14411/eje.2009.035

Grebennikov VV, Smetana A (2015) DNA barcoding and regional diversity of understudied Micropeplinae (Coleoptera: Staphylinidae) in southwest China: phylogenetic implications and a new *Micropeplus* from Mount Emei. Zootaxa 3919(3):583–599. 10.11646/zootaxa.3919.3.8

Hunt T, Bergsten J, Levkanicova Z et al (2007) A comprehensive phylogeny of beetles reveals the evolutionary origins of a superradiation. Science 318 (5858):1913–1916. https://doi.org/10.1126/science.1146954

Ikeda H, Kagaya T, Kubota K et al (2008) Evolutionary relationships among food habit, loss of flight, and reproductive traits: life-history evolution in the Silphinae (Coleoptera: Silphidae). Evolution 62 (8):2065–2079. https://doi.org/10.1111/j.1558-5646.2008.00432.x

Jeon M-J, Song J-H, Ahn K-J (2012) Molecular phylogeny of the marine littoral genus *Cafius* (Coleoptera: Staphylinidae: Staphylininae) and implications for classification. Zool Scr 41(2):150–159. https://doi.org/10.1111/j.1463-6409.2011.00510.x

Koerner L, Laumann M, Betz O et al (2013) Loss of the sticky harpoon – COI sequences indicate paraphyly of *Stenus* with respect to *Dianous* (Staphylinidae, Steninae). Zool Anz 252(3):337–347. https://doi.org/10.1016/j.jcz.2012.09.002

Lang C, Koerner L, Betz O et al (2015) Phylogenetic relationships and chemical evolution of the genera *Stenus* and *Dianous* (Coleoptera: Staphylinidae). Chemoecology 25(1):11–24. https://doi.org/10.1007/s00049-014-0171-4

Leschen RAB, Buckley TR, Harman HM et al (2008) Determining the origin and age of the Westland beech (*Nothofagus*) gap, New Zealand, using fungus beetle genetics. Mol Ecol 17(5):1256–1276. https://doi.org/10.1111/j.1365-294X.2007.03630.x

Maruyama M, Parker J (2017) Deep-time convergence in rove beetle symbionts of army ants. Curr Biol 27 (6):920–926. https://doi.org/10.1016/j.cub.2017.02.030

Maus C, Peschke K, Dobler S (2001) Phylogeny of the genus *Aleochara* inferred from mitochondrial cytochrome oxidase sequences (Coleoptera: Staphylinidae). Mol Phylogenet Evol 18(2):202–216. https://doi.org/10.1006/mpev.2000.0874

McKenna DD, Farrell BD, Caterino MS et al (2015a) Phylogeny and evolution of Staphyliniformia and Scarabaeiformia: forest litter as a stepping stone for diversification of nonphytophagous beetles. Syst Entomol 40(1):35–60. https://doi.org/10.1111/syen.12093

McKenna DD, Wild AL, Kanda K et al (2015b) The beetle tree of life reveals that Coleoptera survived end-Permian mass extinction to diversify during the Cretaceous terrestrial revolution. Syst Entomol 40 (4):835–880. https://doi.org/10.1111/syen.12132

O'Keefe ST (2005) 11.5. Scydmaenidae Leach, 1815. In: Beutel RG, Leschen RAB (eds) Coleoptera, beetles. Vol 1: Morphology and systematics (Archostemata, Adephaga, Myxophaga, Polyphaga partim). Handbook of zoology. Vol IV Arthropoda: insecta. Part 38. Walter de Gruyter, Berlin, pp 280–288

Osswald J, Bachmann L, Gusarov VI (2013) Molecular phylogeny of the beetle tribe Oxypodini (Coleoptera: Staphylinidae: Aleocharinae). Syst Entomol 38 (3):507–522. https://doi.org/10.1111/syen.12011

Parker J (2016) Emergence of a superradiation: pselaphine rove beetles in mid-Cretaceous amber from Myanmar and their evolutionary implications. Syst Entomol 41(3):541–566. https://doi.org/10.1111/syen.12173

Parker J, Grimaldi DA (2014) Specialized myrmecophily at the ecological dawn of modern ants. Curr Biol 24 (20):1–7. https://doi.org/10.1016/j.cub.2014.08.068

Ragan MA (1992) Phylogenetic inference based on matrix representation of trees. Mol Phylogenet Evol 1(1):53–58. https://doi.org/10.1016/1055-7903(92)90035-F

Ross AH, Rodrigo AG (2004) An assessment of matrix representation with compatibility in supertree construction. In: Bininda-Emonds ORP (ed) Phylogenetic supertrees. Combining information to reveal the tree of life, Computational biology series, vol 4. Kluwer Academic, Dordrecht, pp 35–63

Schomann AM, Solodovnikov A (2017) Phylogenetic placement of the austral rove beetle genus *Hyperomma* triggers changes in classification of Paederinae (Coleoptera: Staphylinidae). Zool Scr 46 (3):336–347. https://doi.org/10.1111/zsc.12209

Serri S, Frisch J, von Rintelen T (2016) Genetic variability of two ecomorphological forms of *Stenus* Latreille, 1797 in Iran, with notes on the infrageneric classification of the genus (Coleoptera, Staphylinidae, Steninae). ZooKeys 626:67–86. https://doi.org/10.3897/zookeys.626.8155

Song J-H, Ahn K-J (2013) Molecular phylogeny reveals multiple origins of seashore colonisation in the genus *Aleochara* Gravenhorst (Coleoptera: Staphylinidae: Aleocharinae). Invertebr Syst 27(2):239–244. https://doi.org/10.1071/IS12040

Song J-H, Ahn K-J (2014) Species delimitation in the *Aleochara fucicola* species complex (Coleoptera: Staphylinidae: Aleocharinae) and its phylogenetic relationships. Zool Scr 43(6):629–640. https://doi.org/10.1111/zsc.12077

Song J-H, Ahn K-J (2017) Species trees, temporal divergence and historical biogeography of coastal rove beetles (Coleoptera: Staphylinidae) reveal their early Miocene origin and show that most divergence events occurred in the early Pliocene along the Pacific coasts. Cladistics. https://doi.org/10.1111/cla.12206

Thayer MK (2005) 11.7. Staphylinidae Latreille, 1802. In: Beutel RG, Leschen RAB (eds) Coleoptera, beetles. Vol 1: Morphology and systematics (Archostemata, Adephaga, Myxophaga, Polyphaga partim). Handbook of zoology. Vol IV Arthropoda: insecta. Part 38. Walter de Gruyter, Berlin, pp 296–344

Thomas JC (2009) A preliminary molecular investigation of Aleocharine phylogeny (Coleoptera: Staphylinidae). Ann Entomol Soc Am 102 (2):189–195. https://doi.org/10.1603/008.102.0201

Timmermans MJTN, Barton C, Haran J et al (2016) Family-level sampling of mitochondrial genomes in Coleoptera: compositional heterogeneity and phylogenetics. Genome Biol Evol 8(1):161–175. https://doi.org/10.1093/gbe/evv241

von Beeren C, Maruyama M, Kronauer DJC (2016a) Cryptic diversity, high host specificity and reproductive synchronization in army ant-associated *Vatesus* beetles. Mol Ecol 25(4):990–1005. https://doi.org/10.1111/mec.13500

von Beeren C, Maruyama M, Kronauer DJC (2016b) Community sampling and integrative taxonomy reveal new species and host specificity in the army ant-associated beetle genus *Tetradonia* (Coleoptera, Staphylinidae, Aleocharinae). PLoS One 11(11): e0165056. https://doi.org/10.1371/journal.pone.0165056

Wild AL, Maddison DR (2008) Evaluating nuclear protein-coding genes for phylogenetic utility in beetles. Mol Phylogenet Evol 48(3):877–891. https://doi.org/10.1016/j.ympev.2008.05.023

Zhang X, Zhou H-Z (2013) How old are the rove beetles (Insecta: Coleoptera: Staphylinidae) and their lineages? Seeking an answer with DNA. Zool Sci 30 (6):490–501. https://doi.org/10.2108/zsj.30.490

A Review of the Fossil History of Staphylinoidea

3

Stylianos Chatzimanolis

Abstract
The superfamily Staphylinoidea includes the families Hydraenidae, Ptiliidae, Agyrtidae, Leiodidae, Silphidae, and Staphylinidae and currently includes more than 400 described fossils. The geological history of the superfamily ranges from the Triassic to Cenozoic, and the lineage is well represented in multiple significant Lagerstätte. During the last several years, many new important fossils have been described, and for most Staphylinidae subfamilies, their geological history extends into the Cretaceous. I summarize the most important fossil discoveries for families and subfamilies from compressions/impressions and amber inclusions. I discuss the diversity of Staphylinoidea though time, and I provide a summary phylogeny of Staphylinoidea with fossil constraints, updated from previously published works. Additionally, I discuss the apparent bradytely in many lineages and possible future steps to expand our knowledge of fossil Staphylinoidea.

S. Chatzimanolis (✉)
Department of Biology, Geology and Environmental Science, University of Tennessee at Chattanooga, Chattanooga, TN, USA
e-mail: stylianos-chatzimanolis@utc.edu

O. Betz et al. (eds.), *Biology of Rove Beetles (Staphylinidae)*,
https://doi.org/10.1007/978-3-319-70257-5_3

3.1 Introduction

The superfamily Staphylinoidea is composed of the families Agyrtidae, Leiodidae, Hydraenidae, Ptiliidae, Silphidae, and Staphylinidae. Currently there are almost 70,000 species described in these families, including more than 400 fossils (see Table 3.1). The status and relationships between the families were recently reviewed by McKenna et al. (2015) and Newton (2016). Besides the families listed above, the family Jacobsoniidae and the extinct family Ptismidae have been discussed as potentially belonging in Staphylinoidea (Yamamoto et al. 2017; Kirejtshuk et al. 2016), and these two subfamilies are discussed briefly below in Sect. 3.3.

In the past, four informal groups of subfamilies (omaliine, tachyporine, oxyteline, and staphylinine groups; Lawrence and Newton 1982; Thayer 2016) were recognized in Staphylinidae. However, the usefulness of these informal groups may have come to an end, since none of these are monophyletic (McKenna et al. 2015). While the phylogeny of McKenna et al. (2015) is not ideal (it was based only on two molecular markers: 28S rDNA and the nuclear protein-coding gene CAD), it is the only nearly complete molecular phylogeny of Staphylinoidea available in terms of subfamily coverage. It is also much more resolved and/or inclusive than previous phylogenies of Staphylinidae/ Staphylinoidea (e.g., Newton and Thayer 1995;

Table 3.1 Number of extinct and extant species of Staphylinoidea

Families/superfamily	Number of extinct species	Number of extant species
Hydraenidae	8	1962
Ptiliidae	6	776
Agyrtidae	9	72
Leiodidae	10	4167
Silphidae	21	187
Staphylinidae	374	62,480
Staphylinoidea	428	69,644

Data provided by Alfred Newton (Newton unpublished database) and are valid as of October 17, 2016

Hansen 1997; Ballard et al. 1998; Caterino et al. 2005; Grebennikov and Newton 2009, 2012) and certainly more realistic than previously published "phylogenetic schema" (Thayer 2005; Newton 2011; Chatzimanolis et al. 2012; Thayer 2016). Given that these informal groups of subfamilies are not monophyletic, I have decided not to use them in this review and by doing so perhaps drawing the ire of my fellow rove beetle systematists. I have used the phylogeny produced by McKenna et al. (2015) as the basis for Fig. 3.4 (that maps the earliest fossils for all major staphylinoid lineages) and as a guide on how to list the various families.

Below I provide a quick summary of the early paleontological works in Staphylinoidea, and then I review the most important fossils for each of the families/subfamilies. If there has been a recent review of the paleontological history for a particular group (e.g., Scydmaeninae by Jałoszyński 2016), then I reference the review and spent less time on that group. Following that section, I briefly discuss the diversity of fossils through time and stasis in the fossil record and provide brief remarks on future directions.

3.2 History of Staphylinoid Paleontology

The first fossil rove beetle was described by Gravenhorst in 1806 (Herman 2001). For the next 150 years or so, the record of new staphylinoid fossil descriptions was scant with just a few fossils being described. Early on, the most fossils, by far, were described by Samuel H. Scudder, perhaps the most prominent and influential American insect paleontologist of the nineteenth century. These fossils were from the Early Eocene Green River Formation in Wyoming/Utah/Colorado and the late Eocene-Oligocene Florissant Formation in Colorado (Scudder 1876, 1878, 1890, 1900). And while Scudder was instrumental in bringing all these fossils to light, many of his descriptions and identifications are typical of the era: short, incomplete, and inaccurate. Another paleontologist working in Florissant around the same time was H. F. Wickham who also described several taxa (Wickham 1913a, b). Figures 3.1 and 3.2c illustrate five of the species described by Scudder and Wickham from Florissant and Green River.

Other important early fossil contributions (individual fossils discussed below in Sect. 3.3) were made by Giebel (1856) and Oustalet (1874) who described fossils from Aix-en-Provence, France, and Heer (1847, 1862) who described Miocene insects from Oeningen, Germany, and the Firkanten Formation of Svalbard (Heer 1870). von Heyden and von Heyden (1866) described several taxa from the Rott Formation of Germany, and Schaufuss (1890) described many species from Baltic amber.

The last part of the twentieth century saw many important fossil discoveries from Asia. These included taxa described by Tikhomirova from the Jurassic of Karatau, Kazakhstan (Tikhomirova 1968), by Ponomarenko from the former Soviet Union (Ponomarenko 1977, 1980, 1985), by Ryvkin from the Russian Federation (1985, 1988, 1990), and Zhang from China (Zhang 1988, 1989).

One important characteristic of all the previously listed scientists is that they were (with few

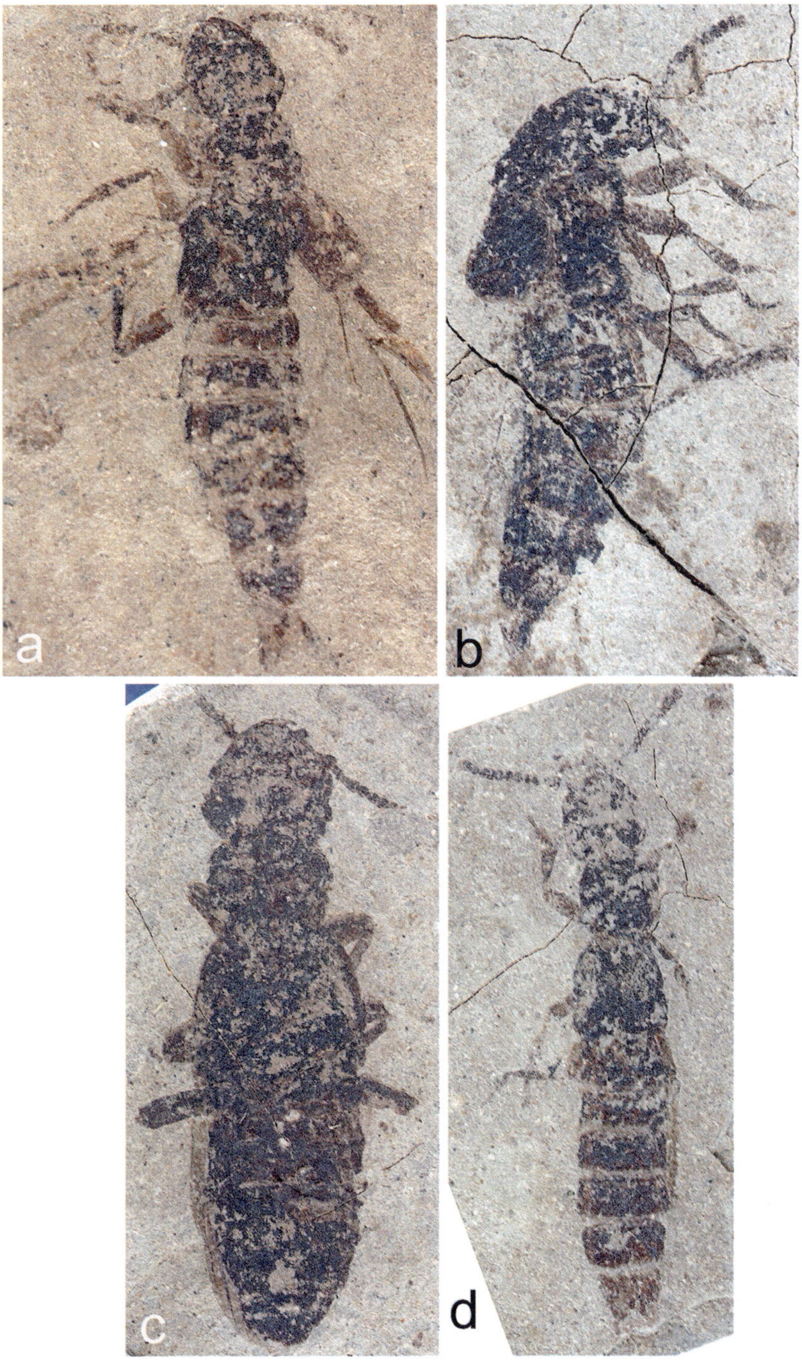

Fig. 3.1 Compression fossils of Staphylininae from Florissant, Colorado. (**a**) *Philonthus marcidulus* Scudder, USNM 1529, (**b**) *Staphylinus lesleyi* Scudder, USNM 1519; Herman (1986) suggested that this species probably belongs in *Bledius* Leach; (**c**) *Staphylinus vetulus* Scudder, USNM1537, (**d**) *Staphylinus vulcanus* Wickham, USNM 59636. While all these specimens were labeled as "types," they are syntypes because Scudder and Wickham designated multiple specimens in the type series without specifying one as holotype (see text for more details)

exceptions) paleontologists who did not work with Recent taxa. The past generations of rove beetle systematists spent very little time describing new fossil taxa or critically reviewing taxa described by the early paleontologists. There are many reasons why this happened: the false dichotomy between paleontologists and neontologists was much more prevalent in the

Fig. 3.2 Compression fossils of Paederinae from Green River. (**a**) Cryptobiina, USNM 58386; (**b**) Pinophilini, USNM 53181; (**c**) *Lathrobium absessum* Scudder, USNM 18593b; (**d**) *Palaminus*? USNM53181

past, and one had to be a paleontologist to look at a fossil; rove beetle taxonomy (even today) is a tangled mess, and it is hard to assign fossil taxa to higher-level groups; there are many extant species still awaiting descriptions, and most systematists prefer those than fossils; many modern techniques such as confocal microscopy or synchrotron microtomography that allow visualization of fossils were not available; and the absence of modern systematic methodology allows the incorporation of fossil taxa into phylogenetic analyses.

Another reason that might have discouraged the previous generation of rove beetle systematists to work with fossils is the frequent mistakes in identifying and assigning fossils to higher groups by the early paleontologists. A particular problem with fossils is that it is easy to fall into the trap of assigning similarities between fossils or between fossils and Recent taxa based on the reduction or loss of structures. This is especially true because without a proper phylogenetic analysis of all the subfamilies (such as the one produced by McKenna et al. 2015), it is hard to figure out which structures are apomorphies or plesiomorphies. Many paleontologists in the past had tried to fit fossils into Recent genera (Engel pers. comm.), rather than describing new fossil genera. For example, many fossils described by paleontologists such as Heer, Scudder, or Wickham in the nineteenth or twentieth century were placed incorrectly in Recent taxa and that created several problems: first, the geological age of the Recent taxa was incorrectly expanded by several million years (but see discussion below on stasis). Second, the generic limits were altered producing in some cases paraphyletic groupings; and, third, as mentioned above, these fossil taxa became unappealing to neontologists. Some of these early paleontologists even placed taxa in the wrong orders, describing, for example, earwigs as rove beetles (e.g., see Chatzimanolis and Engel 2010, for an attempt to correct some of these mistakes).

In recent years, there has been an important increase in the rate and quality of new fossil rove beetle descriptions. Herman (2001) in the introduction of his Catalogue of Staphylinidae mentioned that there had been 173 fossil taxa described between 1806 and 1999. Seventeen years later, the number of described fossil taxa has more than doubled with 374 described species as of October 17, 2016 (see Table 3.1). Many of these new descriptions incorporate high-quality images done with confocal microscopy or synchrotron X-ray microtomography, and we slowly start to see the incorporation of rigorous phylogenetic techniques in fossil studies (e.g., Solodovnikov et al. 2013).

3.3 Paleontological Record of Staphylinoidea

The purpose of this section is not to provide a complete catalogue of fossil staphylinoids; rather I attempt to synthesize some of the previous key works. Readers interested in more complete listings are encouraged to consult Herman (2001), Fossilworks (http://fossilworks.org), Mitchell (2013), and the forthcoming online catalogue of staphylinoids by Newton (unpublished). The age range (from Fossilworks) is given for the first time when a particular Lagerstätte is mentioned. However, the discussion below does not include Quaternary fossils or subfossils.

Besides the families listed below, two additional families might belong in this superfamily, the families Jacobsoniidae and the extant Ptismidae. Yamamoto et al. (2017) described a fossil Jacobsoniidae from Burmese amber (99.7–94.3 Ma) as the oldest member of this family. Ptismidae was described from Lebanese amber (130–124.45 Ma; Kirejtshuk et al. 2016) in Staphyliniformia and possibly in Staphylinoidea, although the fossil is rather difficult to interpret without a proper phylogenetic analysis.

3.3.1 Hydraenidae

Four species have been described in the genus *Ochtebiites* Ponomarenko (Ponomarenko 1977, 1980, 1985), and species in this genus are known from the Jurassic of Kazakhstan (Karatau Formation, 164.7–155.7 Ma), the Jurassic of the Russian Federation (189.6–155.7 Ma), and the Cretaceous of Mongolia (125.45–112.6 Ma). Whether or not these taxa truly belong in this family is not clear since the descriptions and drawings are rather vague. Additional species have been described in the extant genus *Ochthebius* Leach from the Rott Formation of Germany (28.4–23.03 Ma; von Heyden and von Heyden 1866) and the Firkanten Formation of Svalbard (66–58.7 Ma; Heer 1870).

3.3.2 Ptiliidae

While there are many undescribed ptiliids in amber from all major amber Lagerstätten (Shockley and Greenwalt 2013; see also for review of all described taxa), just six specimens have been described to species. Four of the described species are from Baltic (37.2–33.9 Ma) or Rovno (54.8–33.7 Ma) amber, while one species was described from the Rott Formation of Germany and another from the Kishenehn Formation of Montana (46.2–43.5 Ma).

3.3.3 Agyrtidae

A surprising number of fossil Agyrtidae (nine) have been described given that there are only 72 extant species known (see Table 3.1). The oldest fossils belong in the genus *Mesecanus* Newton (replacement name for *Mesagyrtes* Ponomarenko) from the Jurassic of the Russian Federation and the Jurassic of China (Hanshan Formation 171.6–164.7 Ma). Other fossils are known from the Turga Formation of the Russian Federation (125.45–122.46 Ma), the Chijinqiao Formation of China (125.45–112.6 Ma), the Eocene of Germany (Geiseltal, 48.6–40.4 Ma), and Baltic/Bitterfeld amber. A review of the fossil Agyrtidae was provided by Newton (1997) who questioned the placement of all fossil taxa currently in the family except *Mesecanus*.

3.3.4 Leiodidae

The oldest taxon in Leiodidae, *Mesagyrtoides fulvus* Perkovsky, is known from the Jurassic of Mongolia (Shar Teg: 152.1–145 Ma), but its placement in the family was recently disputed by Perreau (2012). Wickham (1913a) described a couple of species from the Florissant Formation in Colorado (37.2–33.9 Ma), while the majority of other taxa have been described from Baltic/Rovno (Perreau 2012) and Dominican amber (20.43–13.65 Ma; Perkovsky 2000).

3.3.5 Silphidae

Until recently, fossil Silphidae were only known from the Cenozoic. Several taxa were known from the Florissant Formation, and some of these taxa were described in the extant genus *Silpha* Linnaeus by Scudder and Wickham. Other Cenozoic taxa included species from the Geiseltal Formation of Germany, the Oligocene of France (Caylux, 28.4–23.03 Ma), the Miocene of Croatia (Radoboj, 12.7–11.6 Ma), and the Miocene of Germany (Oeningen, 12.7–11.6 Ma).

Recently, several Mesozoic Silphidae were reported by Cai et al. (2014a) from three different Lagerstätten: the Jurassic Daohugou Formation of China (164.7–155.7 Ma), the Cretaceous Yixian Formation of China (125.45–122.46 Ma), and the Burmese amber. While these taxa have not been described to species yet, many are well-preserved and undoubtedly will be eventually described.

3.3.6 Staphylinidae

Given the number of taxa in Staphylinidae, I will first discuss below the taxa that have been described as Staphylinidae *incertae sedis* and then provide details about the fossil history of each subfamily. Previous attempts to synthesize parts of the geological history of the family were made by Herman (2001), Cai and Huang (2010), Chatzimanolis and Engel (2011, 2013), Chatzimanolis et al. (2012), and Peris et al. (2014a).

The oldest described staphylinoid beetle is *Leehermania prorova* Chatzimanolis et al. from the Triassic Cow Branch Formation (221.5–205.5 Ma; Chatzimanolis et al. 2012). While some disagreement exists regarding the placement of *Leehermania* (Grebennikov and Newton 2012), the fossil has been used to calibrate several recent molecular phylogenies of Coleoptera as the oldest rove beetle (e.g., Misof et al. 2014; Toussaint et al. 2017). Of course, Chatzimanolis et al. 2012) could have been wrong about the placement of the fossil in

Staphylinidae, but without additional specimens showing its ventral view, the arguments in favor of the placement of the fossil in Hydroscaphidae as presented by Grebennikov and Newton (2012) are weak and rely on plesiomorphies (e.g., small size, short elytra, lack of abdominal paratergites, short antennae and legs).

Jurassic Staphylinidae *incertae sedis* are known from Karatau, Kazakhstan, and China. The genera *Tunicopterus* Tikhomirova and *Sulcelytrinus* Tikhomirova (Tikhomirova 1968) were recently reviewed by Solodovnikov et al. (2013) in a context of a phylogenetic analysis, and it was determined that they were not preserved well enough to assign them to a subfamily. Cai and Huang (2010) reviewed the Jurassic Staphylinidae from China and indicated that there are many more well-preserved genera awaiting formal description. Many Cretaceous compression fossils have been described as Staphylinidae *incertae sedis*, and several of those will be discussed below along with the subfamilies that are closely related to.

The various subfamilies are presented below in the order they appear on Fig. 3.4 which is based on the sister group relationships in the latest staphylinoid phylogeny by McKenna et al. (2015). Fossils formally described to species (non-Quaternary) are not known from the following subfamilies: Empelinae, Habrocerinae, Leptotyphlinae, Microsilphinae, Neophoninae, Protopselaphinae, Pseudopsinae, and Trichophyinae.

Tachyporinae Yamamoto (2016a) recently reviewed the Mesozoic Tachyporinae. The oldest Tachyporinae is known from the Jurassic of Jiulongshan, China (164.7–155.7 Ma). It is a specimen illustrated in Hong (1983) as "*Protostaphylinus mirus*" Lin, but according to Cai and Huang (2010) and Yamamoto (2016a), it is a different taxon than the one described by Lin (1976). Other Jurassic tachyporines are known from Karatau, Kazakhstan (Tikhomirova 1968), and the Talbragar beds in Australia (155.7–150.8 Ma; Cai et al. 2013a). This last fossil, *Protachinus minor* Cai et al., is significant because it provides a good early calibration point for the tribe Tachyporini. There are multiple known Cretaceous compression fossils (e.g., Yue et al. 2009; Cai et al. 2011), but only a couple in Cretaceous amber: *Mesotachyporus puer* Gusarov from New Jersey amber (94.3–89.3 Ma; Gusarov 2000) and *Procileoporus burmiticus* Yamamoto from Burmese amber (Yamamoto 2016a). Several taxa are known from the Cenozoic, including compression fossils from the Green River Formation (50.3–46.2 Ma; Chatzimanolis pers. obs.), Florissant (Scudder 1900), and amber fossils from Baltic and Dominican amber (e.g., Pasnik and Kubisz 2002; Yamamoto and Takahashi 2016).

Oxyporinae The oldest Oxyporinae are known from the Yixian Formation where three taxa have been described: *Cretoxyporus extraneus* Cai and Huang, *Protoxyporus grandis* Cai and Huang (Cai and Huang 2014a), and *Oxyporus yixianus* Solodovnkov and Yue (Yue et al. 2011). Additional fossils are known from the Cenozoic, but only *Oxyporus impressus* Piton (from Menat, France, 58.7–55.8 Ma; Piton 1940) and *Oxyporus vulcanus* von Heyden and von Heyden (from the Rott Formation; von Heyden and von Heyden 1866) definitely belong in *Oxyporus* according to Yue et al. (2011). The remaining species (*Oxyporus blumenbachi* Gravenhorst, *Oxyporus seuberti* Heer, and *Oxyporus stiriacus* Scudder) should be reexamined and probably removed from Oxyporinae (Yue et al. 2011).

Steninae Clarke and Chatzimanolis (2009) reviewed the geological history of Steninae. Two species of *Stenus* are known from the Mesozoic, *Stenus inexpectatus* Schlüter from French amber (99.7–94.3 Ma; Schlüter 1978) and *Stenus imputribilis* Ryvkin from Obeshchayushchiy, Russian Federation (84.9–70.6 Ma; Ryvkin 1988). Several other species of fossil *Stenus* are known from the Cenozoic, with many described by Puthz (2010). Cai et al. (2014b) described the only non-*Stenus* Steninae from the Eocene of France (Alès-Monteils, 37.2–33.9 Ma; *Eocenostenus fossilis* Cai et al.).

Euaesthetinae As above, Clarke and Chatzimanolis (2009) reviewed the geological history of Euaesthetinae. While several taxa have been attributed to this family (see Puthz 2008 and Fig. 3.3b on this chapter), only two taxa have been formally described: *Nordenskioldia pentatarsus* Lefebvre et al. (from Lebanese amber; Lefebvre et al. 2005) and *Octavius electrospinosus* Clarke and Chatzimanolis (from Burmese amber; Clarke and Chatzimanolis 2009).

Solieriinae The fossil Solieriinae all belong in the genus *Prosolierius* Thayer et al. Three species are known from Burmese amber (Thayer et al. 2012) and one species from Spanish amber (112.6–109 Ma; Peris et al. 2014a), while another undescribed species is known from Lebanese amber (Thayer et al. 2012).

Scydmaeninae The geological history of Scydmaeninae was recently reviewed by Jałoszyński (2016) and Jałoszyński and

Fig. 3.3 Compression fossils of Staphylinidae from Crato, Brazil. (**a**) The oldest known specimen of Cryptobiina (Paederinae), AMNH SA46271; (**b**) a possible Paederinae, AMNH SA46246

Perkovsky (2016). A large number of fossils have been described from most major amber deposits with many more taxa remaining undescribed. The oldest taxa are known from Spanish amber, and all supertribes are known from the Cretaceous (Cai and Huang 2016; Jałoszyński and Peris 2016; Jałoszyński et al. 2017).

Aleocharinae More than 20 species of fossil Aleocharinae have been described (Yamamoto et al. 2016). The oldest taxa are known from Burmese amber from where two species were recently described: *Cretodeinopsis aenigmatica* Cai and Huang (Cai and Huang 2015a) and *Mesosymbion compactus* Yamamoto et al. (Yamamoto et al. 2016). That latter fossil is of particular importance because it had specialized morphological adaptations to coexist with social insects. The remaining taxa are known from Cenozoic amber and compression deposits. While many more taxa exist in Burmese, Baltic, and Dominican ambers (Chatzimanolis pers. obs.), the taxonomic chaos of recent taxa prohibits the rapid description of new fossil species. Unfortunately, for the same reason, all fossil taxa described in Recent genera should be reevaluated, particularly the taxa described by Scudder (1876, 1890) and Wickham (1913b) from Eocene compression fossils.

Phloeocharinae The only fossil species known from this subfamily is *Phloeocharis agerata* Chatzimanolis et al. described from New Jersey amber (Chatzimanolis et al. 2013).

Osoriinae Cai and Huang (2015b) described a fossil species, *Mesallotrochus longiantennatus* Cai and Huang, from Burmese amber. Other species are known from the Eocene of India (Cambay amber, 55.8–48.6 Ma; Ortega-Blanco et al. 2013), Dominican amber (Irmler 2003), and Shanwang, China (15.97–11.608 Ma; Zhang 1989).

Dasycerinae Yamamoto (2016b) described the only known fossil, *Protodasycerus aenigmaticus* Yamamoto, from Burmese amber. According to Yamamoto (2016b), the fossil seems to combine many features of Dasycerinae and Neophoninae.

Pselaphinae There are multiple species of Pselaphinae known from the Cenozoic and particularly from Baltic/Bitterfeld/Rovno amber and Dominican amber, either described or simply mentioned in the literature. Most of the species in Baltic amber were described by Schaufuss (1890), and Poinar and Poinar (1999) mentioned several species of Pselaphinae in Dominican amber. More recently, Peris et al. (2014a) described two species of Pselaphinae from Spanish amber, Parker and Grimaldi (2014) a species from Cambay amber, and Parker (2016) other species from Burmese amber.

Micropeplinae The oldest fossil of the subfamily was described by Cai and Huang (2014b) from Burmese amber. Three other species have been described in *Micropelpus* Latreille, two from Lava Camp Mine of Alaska (7.246–2.588; Mathews 1970) and another one from Willershausen clay pit in Germany (3.6–2.588; Gersdorf 1976).

Olisthaerinae Both fossils known of these peculiar beetles are from the Jurassic: Ryvkin (1985) described *Anicula inferna* Ryvkin from Novospasskoye of the Russian Federation (161.2–155.7 Ma) and Cai et al. (2015) described *Protolisthaerus jurassicus* Cai et al. from the Daohugou Formation of China. Cai et al. (2015) expressed some concern that *Anicula* might not belong in this subfamily but did not officially remove it. Another genus, *Megolisthaerus* Solodovnikov and Yue (Yue et al. 2010b), was described as closely related to Olisthaerinae, but Cai and Huang (2013b) transferred it to Staphylininae.

Paederinae The oldest described Paederinae are known from the Yixian Formation in the genus *Mesostaphylinus* Zhang (Solodovnikov et al. 2013). Other undescribed Mesozoic Paederinae are known from Spanish amber (Peris et al.

2014a) and from the Crato Formation of Brazil (122.46–112.6 Ma; Fig. 3.3a depicts the oldest record, still not formally described, for the subtribe Cryptobiina, A. Schomann pers. comm.). Multiple species have been described from Florissant, Colorado (e.g., Scudder 1890, 1900) and Baltic amber (e.g., Pasnik and Kubisz 2002), and most of those have been described (incorrectly) in the genus *Lathrobium* Gravenhorst. Other notable Cenozoic taxa are known from the Aix-en-Provence, France (11.6–7.2 Ma; Oustalet 1874), the Rott Formation of Germany (von Heyden and von Heyden 1866), and Shanwang, China (Zhang 1989).

Many new taxa are awaiting description from Burmese amber, Green River (see Fig. 3.2), and Kishenehn Formations in North America (Chatzimanolis pers. obs.). However, much work is needed on the classification of Paederinae before these taxa can be reliably described to an existing genus or subtribe (see also discussion below). The genus *Apticax* Schomann and Solodovnikov was described from the Crato Formation and was initially placed as closely related to Paederinae + Staphylininae (Schomann and Solodovnikov 2012). A later phylogenetic analysis (Solodovnikov et al. 2013) indicated that the genus could not be placed with confidence in any subfamily.

Staphylininae Solodovnikov et al. (2013) described many new Mesozoic genera (*Paleothius* Solodovnikov and Yue, *Cretoprosopus* Solodovnikov and Yue, *Thayeralinus* Solodovnikov and Yue, *Paleowinus* Solodovnikov and Yue, and *Durothorax* Solodovnikov and Yue) from the Yixian Formation, China. Other Cretaceous taxa include the genus *Cretoquedius* Ryvkin (Ryvkin 1988), originally described from Obeshchayushchiy, Russian Federation, but with additional taxa described from the Yixian Formation, China (Solodovnikov et al. 2013). The genus *Laostaphylinus* Zhang from the Laiyang Formation, China (125.45–112.6 Ma; Zhang, 1988), does not belong in the Staphylininae + Paederinae lineage according to Solodovnikov et al. (2013). A species described in *Quedius* Stephens by Cai and Huang (2013c) from the Yixian Formation likely belongs in *Cretoquedius* and could be a synonym of one of the taxa described by Solodovnikov et al. (2013). But even if that is not the case, the taxonomic concept of *Quedius* has changed dramatically over the last few years (Solodovnikov 2006; Brunke et al. 2016), and assigning that fossil to *Quedius* is problematic to say the least.

Many fossils have been described in this subfamily, at least in part because early paleontologists tried to describe fossils into Recent common genera. As such, many taxa from Florissant Colorado have been described in *Leptacinus* Erichson, *Philonthus* Curtis, *Quedius*, and *Xantholinus* Dejean (e.g., Scudder 1890, 1900; see Fig. 3.1). A common problem with these taxa is that multiple specimens were included in the type series, and sometimes these specimens are not conspecific. Needless to say, the generic assignment for all these fossils needs to be reevaluated. Other notable Cenozoic taxa include a species of *Tympanophorus* Nordmann from the Kishenehn Formation (Brunke et al. 2017); a species of *Platydracus* from Florissant (Cai et al. 2014d), Colorado; a species of *Algon* from the Miocene of China (Zhang 1989); and a species of *Leistotrophus* from the Green River Formation (Scudder 1876). Brunke et al. (2017) indicated that the *Leistotrophus* fossil is misidentified and does not belong in Staphylinina but most likely in either Acylophorina, Cyrtoquediina, Indoquediina, or Quediina. Chatzimanolis and Engel (2011, 2013) reviewed the taxa described in Baltic and Dominican amber.

Proteininae Vetuproteinus cretaceus Cai et al. (Cai et al. 2016) was recently described as the first definitive fossil representative of this subfamily from Burmese amber. Other fossils known include an undescribed *Proteinus* Latreille from Baltic amber (Cai et al. 2016).

Omaliinae Many Mesozoic fossils are known from this subfamily. Ryvkin (1985) described

Morsum abdominale Ryvkin from the Middle Jurassic of Kubekovo, Russian Federation (171.6–164.7 Ma). Other Jurassic records are known from Daohugou China (Cai and Huang 2013d), from the Karatau Formation of Kazakhstan (Tikhomirova 1968), and from Novospasskoe, Russia (161.2–155.7 Ma; Ryvkin 1985). Cretaceous Omaliinae are known from Daya (150.8–145.5 Ma) and Turga (125.45–122.46 Ma), Russian Federation, and were described by Ryvkin (1990). Recently, Peris et al. (2014b) described an Omaliini from French Cretaceous amber. Cenozoic records for the subfamily include many taxa, including species described from Baltic amber (Schaufuss 1890, Zanetti et al. 2016) and from Florissant, Colorado (e.g., Scudder 1900, Wickham 1913b).

Glypholomatinae Only two fossil species are known, and both belong in the Jurassic genus *Juroglypholoma* Cai et al. *Juroglypholoma antiquum* Cai et al. was described from the Daohugou Formation of China (Cai et al. 2012) and *Juroglypholoma talbragarense* Cai et al. from the Talbragar beds in Australia (Cai et al. 2013a).

Scaphidiinae The oldest Scaphidiinae are known from the Jurassic Formation of Solnhofen in Germany (150.8–145.5 Ma) and were described in the extinct genus *Scaphidiopsis* Handlirsch (Weyenbergh 1869). Other species were described from Oeningen, Germany, by Heer (1847, 1862).

Apateticinae While there are no formal fossil species described in Apateticinae, Newton (1997) suggested that *Miosilpha necrophiloides* Wickham (originally attributed to either Silphidae or Agyrtidae) from Florissant resembles the extant *Apatetica* Westwood and should be placed in that family. *Mesoapatetica aenigmatica* Cai et al. from Daohugou China was described as closely related to Apateticinae and Trigonurinae and more specifically as "a basal member of the Apateticinae and Trigonurinae clade" (Cai et al. 2014c). Unfortunately, the latest staphylinoid phylogeny of McKenna et al. (2015) does not support the hypothesis of a sister group relationship between these two subfamilies, and the fossil is interpreted here (see Fig. 3.4) as a stem group of the clade containing Apateticinae, Trigonurinae, and other subfamilies. Grebennikov and Newton (2012) and Cai and Huang (2013a) discussed the presence of Apateticinae in the Daohugou Formation but it is unclear if they were referring to *Mesoapatetica* or other fossils.

Trigonurinae The oldest known fossil is *Abolescus glabratus* Tikhomirova (Tikhomirova 1968) from the Jurassic Karatau Formation of Kazakhstan that was moved to this subfamily from Piestinae by Grebennikov and Newton (2012). Other fossils include *Triguna lata* Ryvkin (Ryvkin 1990) from Unda, Russian Federation (150.8–145.5 Ma), and *Kovalevia onokhoica* Ryvkin from Onokhoy, Russian Federation (125.45–112.6 Ma).

Megalopsidiinae The only fossil known is *Megalopinus extinctus* Yamamoto and Solodovnikov (Yamamoto and Solodovnikov 2016) from Burmese amber.

Piestinae Yue et al. (2016) recently described two species in the genus *Paleosiagonium* Yue et al. from the Yixian Formation of China. A fossil described much earlier, *Siagonium crassum* (Giebel), from Aix-en-Provence, France (Giebel 1856) has received little attention since the original description and might belong here.

Oxytelinae The fossil history of Oxytelinae was recently reviewed by Lü et al. (2017). Thirty-two Oxytelinae fossils have been described, ranging from 157.3 Ma to 5.3 Ma (not including the Quaternary taxa). All but two of these taxa, *Dolichoxenus newtoni* Engel and Chatzimanolis (Engel and Chatzimanolis 2009) and *Prajna tianmiaoae* Lü et al. (Lü et al. 2017), are compression fossils with the majority described from Florissant, Green River, or Chaomidian, China. While some of the compression fossil Oxytelinae

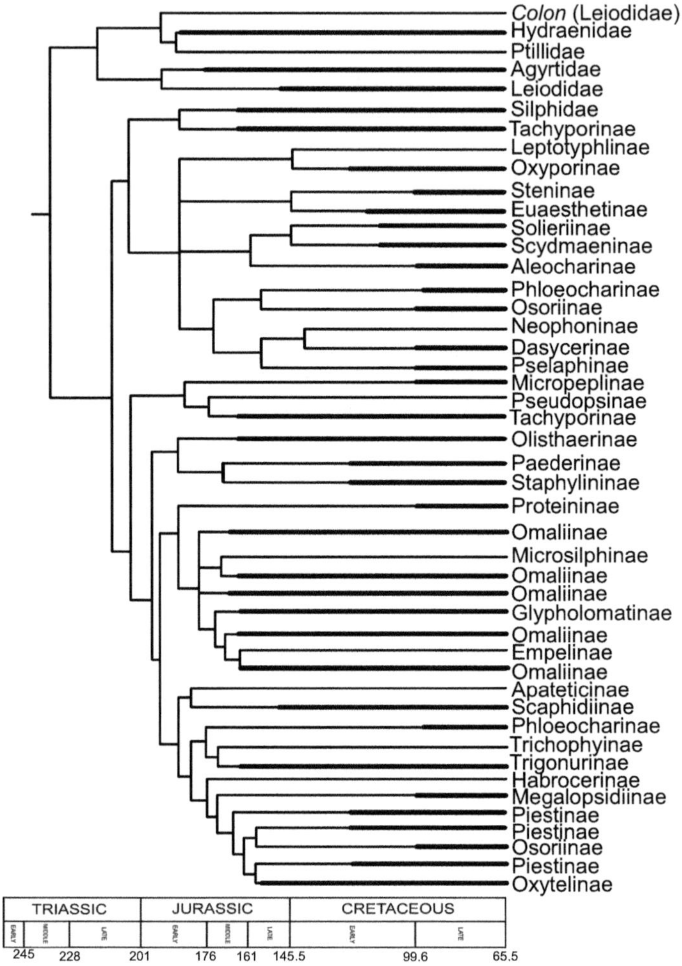

Fig. 3.4 Summary phylogeny of Staphylinoidea based on the Bayesian analysis of McKenna et al. (2015). The presence of Mesozoic fossils (based on records mentioned in the text) is indicated by thicker black bars. In the few instances where subfamilies were poly- or paraphyletic, all clades were treated as having the same geological history. Fossil taxa that were described as incertae sedis in Staphylinidae (such as *Leehermania*) are not included in this figure. Protopselaphinae was not included in the analysis of McKenna et al. (2015), but no fossils are known from the subfamily. See text (Sect. 3.4) for more details

are well-preserved (*Sinoxytelus:* Yue et al. 2010a; Cai et al. 2013b), others were poorly illustrated and described (especially taxa described in the nineteenth to twentieth century) and cannot really be used effectively in any comparative studies without a thorough review of the materials. Additionally, the status of all taxa described by Scudder (e.g., 1878, 1890, 1900) in Recent genera needs to be reevaluated.

†*Protactinae* This is the only extinct subfamily in Staphylinidae, containing two species from Oeningen, Germany (Heer 1847), in the genus *Protactus* Heer. The specimens are rather incomplete, and according to Yue et al. (2010b), they probably belong to a Recent subfamily.

3.4 Diversity Through Time

The origin of Staphylinidae by the Late Triassic is supported both by fossils (*Leehermania*, Chatzimanolis et al. 2012; other undescribed taxa mentioned in Fraser et al. 1996) and divergence estimates from molecular analyses (e.g., Toussaint et al. 2017). Given the phylogeny presented in McKenna et al. (2015), the oldest fossils known for the various subfamilies (presented in Sect. 3.3 and mapped on Fig. 3.4) and the various *incertae sedis* fossils in Staphylinidae, it is likely that almost all subfamilies currently recognized were present by Middle to Late Jurassic. While the timing of the Jurassic cladogenetic events presented in

Fig. 3.4 is hypothetical, it is likely that they all occurred during that period given the *incertae sedis* fossils for Staphylinidae known from that period. One potential problem with Middle Jurassic taxa is that it is hard to tell if these fossils belong in an existing subfamily or if they belong in a stem groups (e.g., see Cai et al. 2014c on the placement *Mesoapatetica aenigmatica* from Daohugou China).

As mentioned above, of all the existing subfamilies of rove beetles, we do not have fossils (formally described, non-Quaternary) for these: Empelinae, Habrocerinae, Leptotyphlinae, Microsilphinae, Neophoninae, Protopselaphinae, Pseudopsinae, and Trichophyinae. It will be probably hard to find Leptotyphlinae fossils (the extant members of this subfamily are so minute that it is difficult to see them with the naked eye), but given the long evolutionary history of Staphylinidae, it is probably just a matter of time before fossils for all the other subfamilies have been discovered. The current number of described fossils for the subfamilies Paederinae, Staphylininae, and Tachyporinae is higher than in most other subfamilies, but this is probably due to taxonomic bias of the work done by early paleontologists.

The majority of fossils have been described from Cenozoic formations, but there are some interesting surprises in the Cretaceous. For example, the subfamilies Olisthaerinae and Trigonurinae are only known from Jurassic and Jurassic/Cretaceous, respectively, with no known fossils from the Cenozoic. Other subfamilies with a Cretaceous-only record (no fossil Cenozoic taxa) include Apateticinae, Dasycerinae, Euaesthetinae, Glypholomatinae, Megalopsidiinae, Phloeocharinae, and Solieriinae. A lot of Cenozoic fossils still remain undescribed, and it is likely that we will discover fossil taxa for all these subfamilies. For example, there are yet to be described fossil Euaesthetinae and Megalopsidiinae from the Green River Formation (Chatzimanolis pers. obs.).

Recent fossil discoveries have altered our views on the biogeography and evolution of certain lineages. It was hypothesized that at least some subfamilies in Staphylinidae and certainly several in the "omaliine" group could have a Gondwanan distribution (e.g., Newton 1985). However, the discovery of several fossils (Jurassic Glypholomatinae from China: Cai et al. 2012; Solieriinae from Burmese, Lebanese and Spanish amber: Thayer et al. 2012; Peris et al. 2014a; and Proteininae from Burmese amber: Cai et al. 2016) have busted the Gondwanan origin hypothesis of several extant austral-endemic groups, at least at the subfamily level. It appears that rove beetles have the opposite problem of plants: recently, Ruhfel et al. (2016) concluded that the Malpighiales plants are too young to have a Gondwanan vicariant distribution, while it seems that rove beetles are too old for such distribution since all subfamilies were likely present before the break up of Pangea.

3.5 Stasis in the Fossil Record

In recent years many taxa from Cenomanian/Albian ambers have been described in Recent genera, and the authors of these papers have attributed this phenomenon to bradytely, i.e., slow rates of morphological change. Clarke and Chatzimanolis (2009) described a species of the Recent genus *Octavius* Fauvel from Burmese amber and were one of the first papers to discuss bradytely in Staphylinoidea. Other examples of bradytely from Cretaceous include species of *Stenus* Latreille from French amber (Schlüter 1978) and from Obeshchayushchiy, Russian Federation (Ryvkin 1988); undescribed species of *Nicrophorus* Fabricius from Burmese amber (Cai et al. 2014a); a species of *Phloeocharis* Mannerheim from New Jersey amber (Chatzimanolis et al. 2013); a species of *Eutheia* from Taimyr amber (Jałoszyński and Perkovsky 2016), and a species of *Megalopinus* Eichelbaum from Burmese amber (Yamamoto and Solodovnikov 2016). Other taxa that perhaps exhibit bradytely are several species of *Oxyporus* Fabricius (but several with uncertain placement, see section above) and the Jurassic *Protolisthaerus* (Cai et al. 2015).

In addition to the taxa mentioned above, there are other more recent fossils from the Cenozoic that have been described in Recent genera, mainly from Green River and Florissant Formations and Baltic amber by earlier paleontologists. However, given that most of these taxa need to be reexamined to verify their taxonomic position, it is hard to identify concrete examples of bradytely. Another problem is that the taxonomic concept of genera frequently changes, and taxa that were described in one Recent genus may actually belong in another. An example was recently mentioned by Yamamato et al. (2016), where a species described in *Homalota* Mannerheim from Green River by Scudder (Scudder 1890) is a member of what we will identify today as *Atheta* Thomson.

As a hypothesis for the explanation of bradytely in staphylinoid genera, Clarke and Chatzimanolis (2009) proposed "the continuous presence of mesic habitats over geological time." For better or worse, all subsequent papers dealing with bradytely (see references above) have used this as an explanation for the phenomenon. While the hypothesis of Clarke and Chatzimanolis (2009) is possible, we should consider the percentage of rove beetles living in mesic habitats, the long-term availability of other types of habitats, and the fossilization process. The "mesic habitat" hypothesis has not been statistically tested, and perhaps we should do so before we perpetuate this hypothesis.

We also need to consider the following paradox: how is it possible that we discuss bradytely for a group of animals such as Staphylinidae that has (given the time of origin and the current number of species) some of the highest speciation rates? Clarke and Chatzimanolis (2009) argued that perhaps some lineages of Staphylinidae exhibit bradytely, while others exhibit tachytely (rapid evolutionary change), or differential fossilization, or even lower extinction rates. The latter hypothesis was suggested for taxa such as *Octavius* or *Stenus* that exhibit bradytely and have high speciation rates with ~250 and ~2900 described species (see Chap. 11), respectively (Clarke and Chatzimanolis 2009), but again without any statistical tests. Perhaps, it is due time that we move into the age of statistical testing and try to decipher the tempo of evolution (see Voje 2016) given all these examples of stasis, our recent molecular analyses of Staphylinoidea, and the available diversification rates.

3.6 Future Directions

While we have made significant headway in understanding the staphylinoid fossil biota, there is still a long way to go before we can achieve a complete picture. Below is a list of items that I believe can help us make significant progress.

Review the Work of Early Paleontologists We certainly need people who have access and are able to reexamine all fossils described by Scudder, Tikhomirova, and Wickham, just to name a few. Many of these fossils were published with short descriptions, incomplete illustrations, and, as I have mentioned multiple times already above, in the wrong genera. These reviews should include lectotype designations since multiple specimens were included in the original type series that are not conspecific. Ideally, these taxa should be reexamined with modern imaging techniques such as SEM or microCT to be able to distinguish as many characters as possible.

Incorporate Fossils into Phylogenetic Analyses Whenever possible, we should try to move away from single-species descriptive works. Instead, we should describe new fossil species in the context of revisionary works and place them into a phylogenetic context. Attempting to discuss relationships without a phylogenetic analysis produces just-so-stories, and it should not be acceptable anymore. Solodovnikov et al. (2013), Jałoszyński (2015), and Yamamoto et al. (2016) greatly enhanced their fossil discoveries by providing phylogenetic analyses that included both fossils and Recent taxa.

Explore New or Neglected Lagerstätten Many new important taxa are currently being described from Burmese amber and the Yixian Formation (see references above). However, there are other Lagerstätten that have received little attention in the last few years or have recently become available. These include the Jurassic Formations in China (Cai and Huang 2010) and Cambay amber (Rust et al. 2010; Ortega-Blanco et al. 2013) but also formations in North America such as the Kishenehn Formation of Montana (Greenwalt and Labandeira 2013; Greenwalt et al. 2016). The Eocene formations of Florissant and Green River received a lot of attention by Scudder, but many new fossils have been uncovered recently. Unfortunately, it will be rather difficult to work with these fossils before a thorough review of the existing taxa from these localities is completed. These Eocene N. America formations are particularly important from a conservation/climate change perspective because they can provide data on the fauna of N. America under an elevated temperature scenario. Additionally, while Eocene formations may not be as exciting for discovering key innovations at the subfamily level, they can be really useful in figuring out biogeographic or genus-level group relationships (Brunke et al. 2017).

Bridge the Gap Between Paleontologists and Neontologists If possible, we need rove beetle systematists who work with both Recent and extinct materials, systematists who are comfortable revising Recent genera and at the same time describing fossil taxa. Typically, paleontologists tend to be generalists (i.e., describing fossils in many orders) and do not have the understanding of plasticity and variation of characters (or character polarity) within rove beetles.

Produce Generic Revisions for all Staphylinoid Subfamilies Although I have no illusion that this is a formidable task, and unlikely to be completed anytime soon, we will never be able to correctly classify many fossils unless we sort out the classification of extant taxa. For example, many fossil Cryptobiina Paederinae (see Figs. 3.2 and 3.3a) from Green River and Crato Formations remain undescribed because the higher-level classification of Cryptobiina is in disarray and placing these fossil taxa is impossible. Of course, one can start creating new genera for all these fossils, but this is not a good practice. Similarly, there are many fossil Aleocharinae (Chatzimanolis pers. observ.) in every major Lagerstätte that are hard or impossible to place because of the current issues with the Aleocharinae classification.

Acknowledgments I am grateful to Alfred Newton for providing data from his unpublished database of Staphylinoidea used in Table 3.1, to the editors of this volume for allowing me to participate, and to Adam Brunke for reviewing a previous version of this chapter. I am also grateful to the curatorial staff at American Museum of Natural History, the Museum of Comparative Zoology, Harvard, and Smithsonian Institution for access to their specimens.

References

Ballard JWO, Thayer MK, Newton AF et al (1998) Data sets, partitions, and characters: philosophies and procedures for analyzing multiple data sets. Syst Biol 47:367–396

Brunke AJ, Chatzimanolis S, Schillhammer H et al (2016) Early evolution of the hyperdiverse rove beetle tribe Staphylinini (Coleoptera: Staphylinidae: Staphylininae) and a revision of its higher classification. Cladistics 32(4):427–451

Brunke AJ, Schillhammer H, Chatzimanolis S (2017) The first fossil rove beetle from the middle Eocene Kishenehn Formation (North America) provides evidence for ancient Eocene relicts within the hyperdiverse Staphylinini (Coleoptera: Staphylinidae: Staphylininae). J Syst Paleontol 15(12):1015–1025. https://doi.org/10.1080/14772019.2016.1266402

Cai CY, Huang DY (2010) Current knowledge on Jurassic staphylinids of China (Insecta, Coleoptera). Earth Sci Front 17:151–153

Cai CY, Huang DY (2013a) Rove beetles roving in deep time: Chinese Mesozoic fossils tell a story (Coleoptera: Staphylinidae). In: Reitner J, Qun Y, Yongdong W, Reich M (eds) Paleobiology and geobiology of fossil Lagerstätten through earth history. Universitätsverlag Göttingen, Göttingen, pp 25–26

Cai CY, Huang DY (2013b) *Megolisthaerus*, interpreted as staphylinine rove beetle (Coleoptera: Staphylinidae) based on new early Cretaceous material from China. Cretac Res 40:207–211

Cai CY, Huang DY (2013c) A new species of small-eyed *Quedius* (Coleoptera: Staphylinidae: Staphylininae) from the early Cretaceous of China. Cretac Res 44:54–57

Cai CY, Huang DY (2013d) *Sinanthobium daohugouense*, a tiny new omaliine rove beetle (Coleoptera: Staphylinidae) from the middle Jurassic of China. Can Entomol 145(5):496–500

Cai CY, Huang DY (2014a) Diverse oxyporine rove beetles from the early Cretaceous of China (Coleoptera: Staphylinidae). Syst Entomol 39:500–505

Cai CY, Huang DY (2014b) The oldest micropepline beetle from Cretaceous Burmese amber and its phylogenetic implications (Coleoptera: Staphylinidae). Naturwissenschaften 101(10):813–817

Cai CY, Huang DY (2015a) The oldest aleocharine rove beetle (Coleoptera, Staphylinidae) in Cretaceous Burmese amber and its implications for the early evolution of the basal group of hyper-diverse Aleocharinae. Gondw Res 28:1579–1584

Cai CY, Huang DY (2015b) The oldest osoriine rove beetle from Cretaceous Burmese amber (Coleoptera: Staphylinidae). Cretac Res 52:495–500

Cai CY, Huang DY (2016) *Cretoleptochromus archaicus* gen. et sp. nov., a new genus of ant-like stone beetles in upper Cretaceous Burmese amber (Coleoptera, Staphylinidae, Scydmaeninae). Cretac Res 63:7–13

Cai CY, Huang DY, Solodovnikov A (2011) A new species of *Hesterniasca* (Coleoptera: Staphylinidae: Tachyporinae) from the early Cretaceous of China with discussion of its systematic position. Insect Syst Evol 42:213–220

Cai CY, Huang DY, Thayer MK, Newton AF (2012) Glypholomatine rove beetles (Coleoptera: Staphylinidae): a southern hemisphere recent group recorded from the middle Jurassic of China. J Kansas Entomol Soc 85(3):239–244

Cai CY, Yan EV, Beattie R, Wang B, Huang DY (2013a) First rove beetles from the Jurassic Talbragar fish bed of Australia (Coleoptera, Staphylinidae). J Paleontol 87:650–656

Cai CY, Yan EV, Vasilenko DV (2013b) First record of *Sinoxytelus* (Coleoptera: Staphylinidae) from the Urey locality of Transbaikalia, Russia, with discussion on its systematic position. Cretac Res 41:237–241

Cai CY, Thayer MK, Engel MS et al (2014a) Early origin of parental care in Mesozoic carrionbeetles. Proc Natl Acad Sci USA 111:14170–14174

Cai CY, Clarke DJ, Huang DY et al (2014b) A new genus and species of Steninae from the late Eocene of France (Coleoptera, Staphylinidae). Alcheringa 38:557–562

Cai CY, Huang DY, Newton AF et al (2014c) *Mesapatetica aenigmatica*, a new genus and species of rove beetles (Coleoptera, Staphylinidae) from the middle Jurassic of China. J Kansas Entomol Soc 87 (2):219–224

Cai CY, Newton AF, Huang DY et al (2014d) A new species of *Platydracus* Thomson, 1858 (Coleoptera, Staphylinidae, Staphylininae) from the upper Eocene Florissant beds, Colorado, USA. Palaeoworld 23:321–332

Cai CY, Beattie R, Huang DY (2015) Jurassic olisthaerine rove beetles (Coleoptera: Staphylinidae): 165 million years of morphological and probably behavioral stasis. Gondw Res 28(1):425–431

Cai C, Newton AF, Thayer MK et al (2016) Specialized proteinine rove beetles shed light on insect–fungal associations in the Cretaceous. Proc R Soc B 283:20161439

Caterino MS, Hunt T, Vogler AP (2005) On the constitution and phylogeny of Staphyliniformia (Insecta: Coleoptera). Mol Phylogenet Evol 34:655–672

Chatzimanolis S, Engel MS (2010) *Laasbium* Scudder: a genus of tertiary earwigs, not rove beetles, and the classification of Florissant fossil Dermaptera (Insecta). Ann Zool 60(1):101–108

Chatzimanolis S, Engel MS (2011) A new species of *Diochus* from Baltic amber (Coleoptera, Staphylinidae, Diochini). ZooKeys 138:65–73

Chatzimanolis S, Engel MS (2013) The fauna of Staphylininae in Dominican amber (Coleoptera: Staphylinidae). Ann Carnegie Mus 81:281–294

Chatzimanolis S, Grimaldi DA, Engel MS et al (2012) *Leehermania prorova*, the earliest staphyliniform beetle, from the Late Triassic of Virginia (Coleoptera: Staphylinidae). Amer Mus Novitates 3761:1–28

Chatzimanolis S, Newton AF, Soriano C et al (2013) Remarkable stasis in a phloeocharine rove beetle from the late Cretaceous of New Jersey (Coleoptera: Staphylinidae). J Paleontol 7(2):177–182

Clarke D, Chatzimanolis S (2009) Antiquity and long-term morphological stasis in a group of rove beetles (Coleoptera: Staphylinidae): description of the oldest *Octavius* species, from Cretaceous Burmese amber and review of the 'Euaesthetine subgroup' fossil record. Cretac Res 30(6):1426–1434

Engel MS, Chatzimanolis S (2009) An oxyteline rove beetle in Dominican amber with possible African affinities (Coleoptera: Staphylinidae: Oxytelinae). Ann Carnegie Museum 77(4):425–429

Fraser NC, Grimaldi DA, Olsen PE (1996) A Triassic Lagerstätte from eastern North America. Nature 380:615–619

Gersdorf E (1976) Dritter Beitrag über Käfer (Coleoptera) aus dem Jungtertiär von Willershausen, Bl. Northeim 4226. Geol Jahrb Reihe A 36:103–145

Giebel CG (1856) Die Insecten und Spinnen der Vorwelt mit steter Berücksichtigung der lebenden Insekten und Spinnen. Die Fauna der Vorwelt 2:1–511

Grebennikov VV, Newton AF (2009) Good-bye Scydmaenidae, or why the ant-like stone beetles should become megadiverse Staphylinidae sensu latissimo (Coleoptera). Eur J Entomol 106:275–301

Grebennikov VV, Newton AF (2012) Detecting the basal dichotomies in the monophylum of carrion and rove beetles (Insecta: Coleoptera: Silphidae and Staphylinidae) with emphasis on the oxyteline group of subfamilies. Arthropod Syst Phylogeny 70 (3):133–165

Greenwalt D, Labandeira C (2013) The amazing fossil insects of the Eocene Kishenehn Formation in northwestern Montana. Rocks Miner 88(5):434–441

Greenwalt D, Rose TR, Chatzimanolis S (2016) Preservation of mandibular zinc in a beetle from the Eocene Kishenehn Formation of Montana, USA. Can J Earth Sci 53:614–621

Gusarov VI (2000) *Mesotachyporus puer*, a new genus and species of Cretaceous Tachyporinae (Coleoptera: Staphylinidae) from New Jersey amber. In: Grimaldi D (ed) Studies on fossils in amber, with particular reference to the Cretaceous of New Jersey. Backhuys Publishers, Leiden, pp 255–258

Hansen M (1997) Phylogeny and classification of the staphyliniform beetle families (Coleoptera). Biol Skr 48:1–339

Heer O (1847) Die Insektenfauna der Tertiärgebilde von Oeningen und von Radoboj in Croatien. Verlag von Wilhelm Engelmann, Leipzig

Heer O (1862) Beiträge zur Insektenfauna Oeningens: Coleoptera. Geoadephagen, Hydrocanthariden, Gyriniden, Brachelytren, Clavicornen Lamellicornen und Buprestiden. Natuurkundige Verhandelingen van de Hollandsche Maatschappij der Wetenschappen te Haarlem 16(2):1–90

Heer O (1870) Die Miocene Flora und Fauna Spitzbergens. Kongliga Svenska Vetenskaps-Akademiens Handlingar 8(7):1–98

Herman LH (1986) Revision of *Bledius*. Part IV. Classification of species groups, phylogeny, natural history, and catalogue (Coleoptera, Staphylinidae, Oxytelinae). Bull Am Mus Nat Hist 184:1–368

Herman LH (2001) Catalog of the Staphylinidae (Insecta: Coleoptera). 1758 to the end of the second millennium. Bull Am Mus Nat Hist 265:1–4218

Hong YC (1983) Middle jurassic fossil insects in north China. Geological Publishing House, Beijing

Irmler U (2003) Osoriinae (Coleoptera: Staphylinidae) from Dominican amber. Stuttg Beitr Naturkunde B 342:1–16

Jałoszyński P (2015) A new Eocene genus of ant-like stone beetles sheds new light on the evolution of Mastigini. J Paleontol 89(6):1056–1067

Jałoszyński P (2016) Scydmaeninae leach, 1815. In: Beutel RG, Leschen RAB (eds) Coleoptea, beetles, Vol 1: Morphology and systematics (Archostemata, Adephaga, Myxophaga, Polyphaga partim), 2nd edn. De Gruyter, Berlin, pp 376–385

Jałoszyński P, Peris D (2016) Cretaceous amber inclusions of Spain and Myanmar demonstrate early diversification and wide dispersal of Cephenniitae (Coleoptera: Staphylinidae: Scydmaeninae). Cretac Res 57:190–198

Jałoszyński P, Perkovsky EE (2016) The extant genus *Eutheia* (Coleoptera: Staphylinidae: Scydmaeninae) discovered in upper Cretaceous Taimyr amber. Cretac Res 66:6–10

Jałoszyński P, Brunke AJ, Metscher B et al (2017) *Clidicostigus* gen. nov., the first Mesozoic genus of Mastigini (Coleoptera: Staphylinidae: Scydmaeninae) from Cenomanian Burmese amber. Cretac Res 72:110–116

Kirejtshuk AG, Chetverikov PE, Azar D et al (2016) Ptismidae fam. nov. (Coleoptera, Staphyliniformia) from the lower Cretaceous Lebanese amber. Cretac Res 59:201–213

Lawrence JF, Newton AF (1982) Evolution and classification of beetles. Annu Rev Ecol Syst 13:261–290

Lefebvre F, Vincent B, Azar D et al (2005) The oldest beetle of the Euaesthetinae (Staphylinidae) from early Cretaceous Lebanese amber. Cretac Res 26:207–211

Lin QB (1976) The Jurassic fossil insects from western Liaoning. Acta Palaeontol Sin 15(1):97–116

Lü L, Cai CY, Huang DY (2017) The earliest oxyteline rove beetle in amber and its systematic implications (Coleoptera: Staphylinidae: Oxytelinae). Cretac Res 69:169–177

Matthews JV (1970) Two new species of *Micropeplus* from the Pliocene of western Alaska with remarks on the evolution of Micropeplinae (Coleoptera: Staphylinidae). Can J Zool 48:779–788

Mckenna DD, Farrell BD, Caterino MS et al (2015) Phylogeny and evolution of Staphyliniformia and Scarabaeiformia: forest litter as a stepping stone for diversification of nonphytophagous beetles. Sys Entomol 40(1):35–60

Misof B, Liu S, Meusemann S et al (2014) Phylogenomics resolves the timing and pattern of insect evolution. Science 346:763–767

Mitchell AA (2013) EDNA, the fossil insect database. https://fossilinsectdatabase.co.uk

Newton AF (1985) South temperate Staphylinoidea (Coleoptera): their potential for biogeographic analysis of austral disjunctions. In: Ball GE (ed) Taxonomy, phylogeny, and zoogeography of beetles and ants: a volume dedicated to the memory of Philip Jackson Darlington, Jr. (1904–1983). Junk-Kluwer, Dordrecht, pp 180–220

Newton AF (1997) Review of Agyrtidae (Coleoptera), with a new genus and species from New Zealand. Ann Zool 47(1/2):111–156

Newton AF (2011) Phylogenie und Systematik. In: Assing A, Schülke M (eds) Die Käfer Mitteleuropas. Band 4. Staphylinidae (exklusive Aleocharinae, Pselaphinae und Scydmaeninae), vol 2. Spektrum Akademischer Verlag, Heidelberg, pp 1–4

Newton AF (2016) Staphylinoidea Latreille, 1802. In: Beutel RG, Leschen RAB (eds) Coleoptea, beetles, Morphology and systematics (Archostemata, Adephaga, Myxophaga, Polyphaga partim), vol 1, 2nd edn. De Gruyter, Berlin, pp 315–316

Newton AF, Thayer MK (1995) Protopselaphinae new subfamily for *Protopselaphus* new genus from Malaysia, with a phylogenetic analysis and review of the Omaliine Group of Staphylinidae including Pselaphidae (Coleoptera). In: Pakaluk J, Ślipiński SA (eds) Biology, phylogeny and classification of Coleoptera: papers celebrating the 80th birthday of Roy A. Crowson. Muzeum i Instytut Zoologii PAN, Warsaw, pp 219–320

Ortega-Blanco J, Chatzimanolis S, Singh H et al (2013) The oldest fossil of the subfamily Osoriinae (Coleoptera: Staphylinidae), from Eocene Cambay amber (India). Coleopt Bull 67:304–308

Oustalet E (1874) Recherches sur les insectes fossiles des terrains Tertiaires de la France, deuxième partie, insectes fossiles d'Aix en Provence. Ann Sci Geol 5 (2):1–347

Parker J (2016) Emergence of a superradiation: pselaphine rove beetles in Mid-Cretaceous amber from Myanmar and their evolutionary implications. Sys Entomol 41(3):541–566

Parker J, Grimaldi DA (2014) Specialized myrmecophily at the ecological dawn of modern ants. Curr Biol 24:2428–2434

Pasnik G, Kubisz D (2002) A new genus and new species of Staphylinidae (Coleoptera) from Baltic amber. Eur J Entomol 99:353–361

Peris D, Chatzimanolis S, Delclòs X (2014a) Diversity of rove beetles (Coleoptera: Staphylinidae) in early Cretaceous Spanish amber. Cretac Res 48:85–95

Peris D, Thayer MK, Néraudeau D (2014b) Oldest Omaliini (Coleoptera: Staphylinidae: Omaliinae) discovered in the opaque Cretaceous Amber of Charentes. Ann Entomol Soc Am 107(5):902–910

Perkovsky EE (2000) The first find of leiodid tribe Scotocryptini (Coleoptera: Leiodidae: Leiodinae) in fossil resin (Dominican amber). Paleontol J 34(Suppl 3):S331–S332

Perreau P (2012) Description of a new genus and two new species of Leiodidae (Coleoptera) from Baltic amber using phase contrast synchrotron X-ray microtomography. Zootaxa 3455:81–88

Piton L (1940) Paléontologie du Gisement Éocéne de Menat. Flore et Faune, Puy-de-Dôme, pp 1–303

Poinar G, Poinar R (1999) The amber forest: a reconstruction of a vanished world. Princeton University Press, Princeton

Ponomarenko AG (1977) Adephaga. In: Arnold LV, Zherikhin VV, Nikritin LM, Ponomarenko AG (eds) Mesozoic Coleoptera. Trudy Paleontologicheskogo Instituta Akademiya Nauk, Moscow, pp 17–104

Ponomarenko AG (1980) Novye vidy zhukov mestonakhozhdeniya Manlay. Rannemelovoe Ozero Manlay, Sovmestnaya Sovetsko-Mongol'skaya Paleontologicheskaya Ekspeditsiya Trudy 13:52–56

Ponomarenko AG (1985) Zhestkokrylye iz Yury Sibiri i zapadnoy Mongolii. In: Rasnitsyn AP (ed) Yurskie nasekomye Sibiri i Mongolii. Trudy Paleontologicheskogo Instituta Akademiya Nauk, Moscow, pp 47–87

Puthz V (2008) Über Euaesthetinen aus dem Bernstein (Coleoptera, Staphylinidae). 99. Beitrag zur Kenntnis der Euasethetinen [sic]. Entomol Bl 103(104):59–62

Puthz V (2010) *Stenus* Latreille, 1797 aus dem Baltischen Bernstein nebst Bemerkungen über andere fossile *Stenus*-Arten (Coleoptera, Staphylinidae). Entomol Bl 106:265–287

Ruhfel BR, Bove CP, Philbrick CT et al (2016) Dispersal largely explains the Gondwanan distribution of the ancient tropical clusioid plant clade. Am J Bot 103 (6):1117–1128

Rust J, Singh H, Rana RS et al (2010) Biogeographic and evolutionary implications of a diverse paleobiota in amber from the early Eocene of India. Proc Natl Acad Sci USA 107(43):18360–18365

Ryvkin AB (1985) Beetles of the family Staphylinidae from the Jurassic of Transbaikal [in Russian]. In: Rasnitsyn AP (ed) Jurassic insects of Siberia and Mongolia, vol 211. Trudy Paleontologicheskogo Instituta Akademiya Nauk SSSR, Moscow, pp 88–91

Ryvkin AB (1988) New Cretaceous Staphylinidae (Insecta) from the far East. Paleontol J 22(4):100–104

Ryvkin AB (1990) Family Staphylinidae Latreille, 1802. In: Trudy Paleontologicheskogo. Instituta Akademii Nauk SSSR, vol 239, pp 52–66

Schaufuss LW (1890) System-schema der Pselaphiden, ein Blick in die Vorzeit, in die Gegenwart und in die Zukunft. Tijdschr Entomol 33:101–162

Schlüter T (1978) Zur Systematik und Palökologie harzkonservierter Arthropoda einer Taphozönose aus dem Cenomanium von NW-Frankreich. Berliner Geowiss Abh A 9:1–150

Schomann A, Solodovnikov AY (2012) A new genus of Staphylinidae (Coleoptera) from the lower Cretaceous: the first fossil rove beetles from the Southern Hemisphere. Syst Entomol 37:379–386

Scudder SH (1876) Fossil Coleoptera from the Rocky mountain tertiaries. US Geol Surv Bull 2:77–87

Scudder SH (1878) The fossil insects of the Green River shales. US Geol Surv Bull 4:747–776

Scudder SH (1890) The tertiary insects of North America. Report of the United States geological survey of the territories, 13, pp 1–734

Scudder SH (1900) Adephagous and clavicorn Coleoptera from the tertiary deposits at Florissant, Colorado with descriptions of a few other forms and a systematic list of the non-rhynchophorus tertiary Coleoptera of North America. Monogr US Geol Surv 40:1–148

Shockley FW, Greenwalt D (2013) *Ptenidium kishenehnicum* (Coleoptera: Ptiliidae), a new fossil described from the Kishenehn oil shales, with a checklist of previously known fossil ptiliids. Proc Entomol Soc Wash 115(2):173–181

Solodovnikov AY (2006) Revision and phylogenetic assessment of *Afroquedius* gen. nov. from South Africa: toward new concepts of the genus *Quedius*, subtribe Quediina and reclassification of the tribe Staphylinini (Coleoptera: Staphylinidae: Staphylininae). Ann Entomol Soc Am 99:1065–1084

Solodovnikov AY, Yue Y, Tarasov S et al (2013) Extinct and extant rove beetles meet in the matrix: early Cretaceous fossils shed light on the evolution of a hyperdiverse insect lineage (Coleoptera: Staphylinidae: Staphylininae). Cladistics 29 (4):360–403

Thayer MK (2005) Staphylinidae. In: Beutel RG, Leschen RAB (eds) Handbook of zoology: Coleoptera, Evolution and systematics, Archostemata, Adephaga, Myxophaga, Staphyliniformia, Scarabaeiformia, Elateriformia, vol 1. De Gruyter, Berlin, pp 296–344

Thayer MK (2016) Staphylinidae Latreille, 1802. In: Beutel RG, Leschen RAB (eds) Coleoptea, Beetles, Morphology and systematics (Archostemata, Adephaga, Myxophaga, Polyphaga partim), vol 1, 2nd edn. De Gruyter, Berlin, pp 394–442

Thayer MK, Newton AF, Chatzimanolis S (2012) *Prosolierius*, a new mid-Cretaceous genus of Solieriinae (Coleoptera: Staphylinidae) with three new species from Burmese amber. Cretac Res 34:124–134

Tikhomirova AL (1968) Staphylinid beetles from Jurassic of Karatau (Coleoptera: Staphylinidae). In: Rohdendorf BB (ed) Jurassic insects of Karatau. Akademiya Nauk SSSR, Moscow, pp 139–154

Toussaint EFA, Seidel M, Arriaga-Varela E et al (2017) The peril of dating beetles. Sys Entomol 42(1):1–10

Voje KL (2016) Tempo does not correlate with mode in the fossil record. Evolution 70(12):2678–2689

von Heyden C, von Heyden L (1866) Käfer und Polypen aus der Braunkohle des Siebengebirges. Palaeontographica 15:131–156

Weyenbergh H (1869) Sur les Insectes fossiles du Calcaire Lithographique de la Bavière, qui se trouvent au Musée Teyler. Arch Mus Teyler 2:1–48

Wickham HF (1913a) Fossil Coleoptera from Florissant in the United States national museum. Proc US Nat Mus 45:283–303

Wickham HF (1913b) Fossil Coleoptera from the Wilson Ranch near Florissant, Colorado. Bull Labs Nat His State Univ Iowa 6:3–29

Yamamoto S (2016a) The oldest tachyporine rove beetle in amber (Coleoptera: Staphylinidae): a new genus and species from upper Cretaceous Burmese amber. Cretac Res 65:163–171

Yamamoto S (2016b) The first fossil of dasycerine rove beetle (Coleoptera: Staphylinidae) from upper Cretaceous Burmese amber: phylogenetic implications for the Omaliine group subfamilies. Cretac Res 58:63–68

Yamamoto S, Solodovnikov AY (2016) The first fossil Megalopsidiinae (Coleoptera: Staphylinidae) from upper Cretaceous Burmese amber and its potential for understanding basal relationships of rove beetles. Cretac Res 59:140–146

Yamamoto S, Takahashi Y (2016) *Coproporus electron* sp. nov., the first tachyporine rove beetle in Dominican amber (Coleoptera, Staphylinidae). PalZ 90:629–635

Yamamoto S, Maruyama M, Parker J (2016) Evidence for social parasitism of early insect societies by Cretaceous rove beetles. Nat Comm 7:1–9

Yamamoto S, Takahashi Y, Parker J (2017) Evolutionary stasis in enigmatic jacobsoniid beetles. Gondwana Res 45:275–281

Yue Y, Zhao Y, Ren D (2009) *Glabrimycetoporus amoenus*, a new tachyporine genus and species of Mesozoic Staphylinidae (Coleoptera) from Liaoning, China. Zootaxa 2225:63–68

Yue Y, Zhao YY, Ren D (2010a) Three new mesozoic staphylinids (Coleoptera) from Liaoning, China. Cretac Res 31:61–70

Yue Y, Ren D, Solodovnikov AY (2010b) *Megolisthaerus chinensis* gen. et sp. n. (Coleoptera: Staphylinidae incertae sedis): an enigmatic rove beetle lineage from the early Cretaceous. Insect Sys Evol 41:317–327

Yue Y, Ren D, Solodovnikov AY (2011) The oldest fossil species of the rove beetle subfamily Oxyporinae (Coleoptera: Staphylinidae) from the early Cretaceous (Yixian Formation, China) and its phylogenetic significance. J Sys Palaeontol 9(4):467–471

Yue Y, JJ G, Yang Q, Wang J et al (2016) The first fossil species of subfamily Piestinae (Coleoptera: Staphylinidae) from the lower Cretaceous of China. Cretac Res 63:63–67

Zanetti A, Perreau M, Solodovnikov AY (2016) Two new fossil species of Omaliinae from Baltic amber (Coleoptera: Staphylinidae) and their significance for understanding the Eocene-Oligocene climate. Arthropod Syst Phylogeny 74:53–64

Zhang JF (1988) The late Jurassic fossil Staphylinidae (Coleoptera) of China. Acta Entomol Sin 31:79–84

Zhang JF (1989) Fossil insects from Shanwang, Shandong, China. Shandong Science and Technology Publishing House, Jinan

Biodiversity and Geographic Patterns of Neotropical Staphylinidae

4

Ulrich Irmler and Angelico Asenjo

Abstract

The history of the discovery of the Central and South American Staphylinidae fauna is reported beginning with the start of modern taxonomy in the mid of the eighteenth century up to the present. An overview over the number of genera is given for all Central and South American countries. The subfamily Osoriinae is described in more detail. The similarities of the faunal compositions between countries are analysed, and countries with similar composition are combined to larger regions. Biodiversity and biogeographic peculiarities within the Neotropics and to other continents are described and discussed.

4.1 Introduction

Modern taxonomy and herewith also taxonomic entomology started with the fundamental work of Linné (1758): Systema Naturae per Regna tria Naturae, secundum classes, ordines genera, species, cum characteribus, differentiis, synonymis and locis. Among the 573 Coleoptera described in this book, 19 species of Staphylinidae were included. However, no staphylinid species from the Neotropical region were described except those that were imported to the new continent, e.g. *Creophilus maxillosus* (Linné 1758). The first staphylinid species originating from the Neotropics was described by Olivier (1795), i.e. from the presently recognized genus *Tenodema*. In the first half of the nineteenth century, Gravenhorst (1806), Perty (1830), Laporte (1834–1840) and, in particular, Erichson (1839–1840) added further species. They based their descriptions on species that were accidentally collected or sampled by researchers that mainly looked for plants, mammals or birds, e.g. M. von Wied (von Wied 1825) during his travel to eastern Brazil from Rio de Janeiro to Bahia in 1815–1817 or of J.B. von Spix and C.F.P. von Martius to Brazil from 1817 to 1820, where they travelled from Rio de Janeiro via Goias to the Amazon. Later, R. F. Sahlberg, a Finnish biologist, was especially interested in Coleoptera. During his travel to the New World in 1839–1843, he collected Coleoptera in the region of Rio de Janeiro, Brazil. In 1850 approximately 640 staphylinid species were known from the Neotropical region.

In the second half of the nineteenth century, mainly British researchers travelled through

U. Irmler (✉)
Department of Applied Ecology, Institute for Ecosystem Research, University of Kiel, Kiel, Germany
e-mail: uirmler@ecology.uni-kiel.de

A. Asenjo
Department of Biologia & Zoologia, Instituto de Biociencia, Federal University of Mato Grosso (UFMT), Cuiabá, Mato Grosso, Brazil
e-mail: pukara8@gmail.com

O. Betz et al. (eds.), *Biology of Rove Beetles (Staphylinidae)*,
https://doi.org/10.1007/978-3-319-70257-5_4

Central and South America. One of the most effective collectors was H.W. Bates, who travelled through the Amazon from 1848 to 1859 (Bates 1864). His extensive collection is deposited in the British Museum. Together with the material of other collectors, Sharp (1876) published the fundamental study on the Amazon Staphylinidae with the description of more than 400 species. Another famous and most effective collector was G.C. Champion, who collected especially Coleoptera in Guatemala from 1879 to 1882. The Staphylinidae of his collection were described together with the material of other collectors also by Sharp (1883–1887) in the Biologia Centralis Americana, edited by G.C. Champion. During the second half of the nineteenth century, the Belgium taxonomist A. Fauvel was in close contact with D. Sharp and described many Neotropical species that were collected by M.E. Simon to Venezuela in 1887/1888; by the Suisse scientist E. Bugnion to Colombia, Venezuela and Antillean islands in 1895/1896; and by C. van Volxem who travelled from Brazil to Argentina in 1872. He mainly studied the extensive material from Chile collected by the French scientists A.J.J. Solier and H. de Ronvouloir in the first half of the century. The Russian entomologist V.I. Motschulsky, who was mainly interested in Coleoptera, travelled to many regions of the New World, among these also to the West Indian islands and adjacent regions in 1853, and described 53 new staphylinids (mainly Pselaphinae) from that region. Later, S. Solsky continued Motschulsky's tradition in Russia and described many new Neotropical staphylinid species mainly collected by C. Jelsky and Baron de Nolcken from Colombia and Peru. In the second half of the century, the first South American taxonomists studied Staphylinidae. Among these F. Lynch Arribálzaga (1884) and later C. Bruch published their important works on the Argentinean staphylinids. Other entomologists were interested in specific biological subjects. The Jesuit E. Wasmann and the American scientist W.M. Wheeler made fundamental work in sociobiology of ants and described many Neotropical staphylinids living with ants. At the end of the nineteenth century, the number of described Neotropical staphylinids has been increased tremendously to more than 3490 species.

The first half of the twentieth century was dominated by the taxonomic work of the Austrian lawyer M. Bernhauer. Among the numerous species described by M. Bernhauer, 1024 species originated from Central and South America. Many of the Neotropical species described by M. Bernhauer were collected by C. Bruch, the German/Brazilian entomologist H v. Ihering, the Colombian entomologist H. Fassl, F. Nevermann at Hamburgfarm, Costa Rica, and many other collectors. He also bought staphylinids from insect dealers, e.g. the German A. Bang-Haas. Thus, sometimes detailed information about the collecting localities or the date is not available. In this period, North American scientists began to study the Neotropical fauna most intensively. Beginning with P.J. Darlington, they were mainly interested in zoogeographical aspects and collected systematically in the West Indies. Among these, R. Blackwelder studied the Staphylinidae and collected species at many stations of the different islands from Trinidad in the South and Cuba in the North. He described 141 new species from the West Indian islands with a comprehensive key to identification using also the rich collection of Darlington and others, e.g. British collectors (Blackwelder 1943), and published the first overview on all species of Central and South America (Blackwelder 1944). J.M. Cameron, too, among his large contribution to Staphylinidae from different parts of the world, described new species mainly from the West Indies. A. Bierig, a German artist, immigrated at first to Cuba in 1919 and later to Costa Rica where he became professor at the university. He mainly studied the West Indian and Central American Staphylinidae that he mostly sampled during excursions in these countries sometimes together with F. Nevermann (Puthz 1998). In Cuba, he collected staphylinids together with the Czech entomologists M. Rambousek. He wrote 28 publications on Staphylinidae and described 109 species with excellent graphs. In Germany, H. Wendeler described 152 new species mainly from southern Brazil that he got from Brazilians who emigrated from Germany, e.g. F. Plaumann and S. Ohaus. Many other collectors worked in that time in southern Brazil and sent their material to different museums in the world, e.g. the Czech collector J. Mráz. Much of the material collected by them is still unstudied. By 1950 approximately 6250 species were described from the Neotropics.

While the first half of the twentieth century suffers under the consequences of World Wars I and II, the second half of the century benefits from the increasing mobility by aeroplanes, car traffic and increasing number of roads to regions which were hard to reach in earlier times. The easier travel to Neotropical countries, within the countries and the increasing interest in tropical ecology and biology, had also the effect that several biological stations were established in different countries which were the base for many sampling expeditions. One of the first (1946) was the Smithsonian Tropical Research Institute on Barro Colorado Island, Panama. Many others followed, e.g. in Costa Rica with the Biological Stations La Selva founded by the American botanist L. Holdrige in 1953 or Monteverde Cloud Forest, in Ecuador with several stations, Peru with the Biological Station Panguana founded in 1968 by H.W. Koepcke and his wife M. Koepcke and many stations in other countries founded by other persons. Among taxonomists studying Staphylinidae, O. Scheerpeltz continued the Austrian tradition. He described many new species from the Neotropics that he received from expeditions made by Austrian scientists, e.g. H. Loeffler or G. Topal, or sent to him by scientists studying specific habitats. Several collectors of the first half of the century continued their sampling activity and sent their material to different museums in the world, e.g. F, Plaumann, who collected in Santa Catarina and adjacent regions. New collectors followed, e.g. the Brazilian M. Alvarenga. Later, a large number of German collectors sampled in Peru at the Panguana station or in the Central Amazon region near Manaus in combination with ecological studies, e.g. U. Irmler and J. Adis at the Instituto Nacional de Pesquisas da Amazônia (INPA). S. B. and J. Peck collected throughout the Neotropical region (except Brazil) from late 1960s until present and probably collected more staphylinids than anyone else during their time due to extensive use of Berlese funnels and baited traps. The Canadian entomologist J.M Campbell collected in many Neotropical countries from Brazil to Mexico. R.T. Schuh collected extensively in the region of Belém for the American Museum, New York, A. Newton in Panama and Mexico and together with M. Thayer in Chile and Peru and L. Herman in several countries of South and Central America. At the end of the century, scientists of the Kansas University and Snow Entomological Museum extensively collected Coleoptera in many Neotropical countries, except Brazil, e.g. among others R. Anderson, R.W. Brooks, R. Leschen and J.S. Ashe. In that time a relatively high number of entomologists studied Neotropical Staphylinidae: V. Puthz worked on the subfamilies Steninae, Euaesthetinae and Megalopsidiinae; U. Irmler on Osoriinae; L. Herman on Paederinae and Oxytelinae; A. Newton and M. Thayer generally on generic taxonomy and Omaliinae; J. Klimaszewski, J.S. Ashe and R. Pace on Aleocharinae; J. Asiain, J. Marquez and J.J. Morrone on Xantholininae; and C. Seevers and C.E. Machado-Allison on Amblyopinina. Other entomologists studied Staphylinidae of specific regions, e.g. F. Saiz, who extensively collected and described Staphylinidae of Chile or J. Asiain and J. Marquez who studied the fauna of Mexico. Many more scientists were also involved in the discovery of the Neotropical fauna who could not all be named here. At the end of the twentieth century, approximately 9450 species were described.

Since the beginning of the new millennium, the research on Neotropical staphylinids continued. S. Chatzimanolis studied the subfamily Staphylininae and E. Jiménez-Sanchez and J. Galián Paederinae, J. E. Caron and C. - Ribeiro-Costa Piestinae, and many of the scientists already mentioned continued their work on their specific field of taxonomy. First country-specific keys and checklists of Staphylinidae were provided, e.g. for Mexico (Navarrete-Heredia et al. 2002), and checklists for Colombia (Newton et al. 2005), for Chile and southern Argentina (Thayer and Newton 2005), for Peru (Newton 2015), for Cuba (Peck 2005), for the Lesser Antilles (Peck 2016), and for Brazil (Asenjo et al. 2013) were made. There is also the review of Staphylinidae of Argentina with generic but not species checklist in Chani-Posse and Thayer (2008). Despite of the tremendous success in our knowledge about Neotropical Staphylinidae, there is no comprehensive treatment of staphylinids of any South American country. First efforts to understand the composition of the staphylinid fauna and their geographical pattern of distribution were made by Irmler (2007, 2009a) and Asiain et al. (2015). However, nearly nothing is known about the ecology of the family in South America. The present study will try to give a compilation of the species richness of Staphylinidae in the Neotropics and will provide the geographical pattern of species richness, endemisms, zoogeographic aspects and relationships to other tropical regions of the world.

4.2 Biodiversity Referring to Genera in Subfamilies and Countries

A total of 11,469 species are at present known from the Neotropical region which belong to 1083 genera (Table 4.1). The subfamily Aleocharinae accounts for the highest number of genera, followed by the Pselaphinae. A low

Table 4.1 Diversity of genera of Staphylinidae present in Latin America

	Latin America		West Indies		Central America		South America	
	Tot	Excl	Tot	Excl	Tot	Excl	Tot	Excl
Glypholomatinae	2	1	–	–	–	–	2	1
Microsilphinae	1	0	–	–	–	–	1	0
Omaliinae	16	1	2	0	8	0	9	1
Proteininae	4	2	1	0	2	0	3	2
Micropeplinae	3	0	1	0	3	0	1	0
Neophoninae	1	1	–	–	–	–	1	1
Pselaphinae	259	202	51	16	112	26	184	112
Phloeocharinae	2	0	–	–	2	0	–	–
Tachyporinae	16	4	3	0	11	1	11	1
Habrocerinae	2	1	–	–	–	–	2	1
Aleocharinae	411	256	57	5	145	31	317	170
Scaphidiinae	7	2	–	–	5	0	7	2
Piestinae	3	0	2	0	3	0	2	0
Osoriinae	47	20	16	1	27	2	41	15
Oxytelinae	18	6	12	1	13	0	14	2
Oxyporinae	1	0	–	–	1	0	1	0
Megalopsidiinae	1	0	1	0	1	0	1	0
Steninae	1	0	1	0	1	0	1	0
Scydmaeninae	43	28	6	0	11	4	30	18
Euaesthetinae	10	4	3	0	4	0	9	3
Solieriinae	1	1	–	–	–	–	1	1
Leptotyphlinae	9	8	1	0	2	1	7	7
Pseudopsinae	1	0	–	–	1	0	1	0
Paederinae	96	56	25	3	54	10	73	20
Staphylininae	128	76	34	1	77	9	102	38
Total	1083	669	216	27	483	84	821	395

Tot total, *Excl* exclusive

Table 4.2 Genera of Staphylinidae present in South America

	Total	Endemic
Argentina	302	25
Bolivia	207	10
Brazil	475	88
Chile	222	77
Colombia	232	8
Ecuador	231	24
French Guiana	118	6
Guyana	69	8
Paraguay	126	6
Peru	238	16
Suriname	59	2
Uruguay	39	0
Venezuela	192	7
TOTAL	846	408

number of genera are found in several subfamilies, e.g. Microsilphinae, Oxyporinae, etc., with only one genus.

The large land mass of South America embraces the highest number of staphylinid genera with 821 (48%) genera restricted to that region. Both Central America and the West Indies only reach 17% and 12%, respectively.

If only genera of South America are regarded, Brazil with the largest area has the highest number of genera (Table 4.2). Overall, a highly significant logarithmic relation between the size of the country's area (x) and the number of genera is found (genera number $= 121.3e^{0.07x}$, $R^2 = 0.40$, $p < 0.001$). However, regarding genera found exclusively in a given country, the relation reflects no significant result. The percentage of exclusive genera for Brazil is only 18%, but it is 34% for Chile. Thus, Chile has a special status among the countries of South America. It seems to have the highest percentage of endemic genera of South American countries.

4.3 Biodiversity and Geographical Pattern of Osoriinae

4.3.1 General Remarks to Osoriinae

The subfamily Osoriinae contains four tribes, i.e. Osoriini, Thoracophorini, Leptochirini and Eleusini (Fig. 4.1). Among these, both Leptochirini and Eleusini only have two and three genera, respectively, whereas both Osoriini and Thoracophorini are richer with 16 and 23 genera, respectively. In particular, in the tribe Thoracophorini, a remarkable morphological variety of genera is present. Osoriinae show ecologically a relationship to all kinds of forests or similar ecosystems (Irmler 2007, 2010a), where they mainly live under the bark, in decayed logs or in the litter layer. They inhabit all strata from the soil layer up to the rainforest canopy. Several genera, in particular among the tribe Osoriini, are living euedaphic in the soil and developed loss or reduction of eyes and wings. As a consequence of their relationship to forests, their elevation distribution correlates with the vertical distribution of forests from lowland rainforests via montane forests up to high cloud forests. A few species seem to live with ants or termites (Wasmann 1902; Bohác 1978). The size varies from approximately 1.5 mm to more than 14 mm.

4.3.2 Material and Methods

In the last 30 years, the Neotropical fauna of the subfamily Osoriinae was studied, and the genera were reviewed except for Leptochirini and Eleusini (e.g. Irmler 1981a, b, 2000, 2015a). For these reviews materials of all major museums were studied and the material identified to species level. In total 22,150 specimens were identified.

The data were transferred to a database that now included 6520 items concerning Osoriinae. For the tribes Eleusini and Leptochirini, only published data are included. The data were separated according to the countries belonging to the Neotropical region. Brazil as the largest country was separated into Amazonia legal including the states of Amazonas, Pará, Mato Grosso, Mato Grosso do Sul, Amapá, Roraima, Tocantins, Rondonia and Acre and the remaining states as southeastern Brazil. The species were counted, and the species composition of the different countries or regions was determined. Species similarity between countries was analysed

Fig. 4.1 Typical representatives of Osoriinae tribes (genus in brackets): (1) Thoracophorini (*Lispinus*), (2) Osoriini (*Holotrochus*), (3) Leptochirini (*Leptochirus*), (4) Eleusini (*Eleusis*)

using the Jaccard index with subsequent clustering. The Jaccard index calculates similarities on the basis of identical species compositions without regarding the number of specimens. High values between two countries indicate a high similarity of the fauna. The subsequent clustering was used to combine groups of countries with high similarity of the faunal composition. For the clustering method, unweighted average pair group and distance as percent similarity was performed using the program PAST (Hammer et al. 2014). The same program was used to perform the individual rarefaction method that calculates the relation between number of species and number of specimens. This test was made using the program PAST (Hammer et al. 2014). Distribution maps for the geographic analysis were made using the program ARCVIEW GIS 3.2 (ESRI 1999).

4.3.3 Pattern of Species Richness of Neotropical Countries

In total, 672 species and 44 genera of Osoriinae are known from the Neotropical region. Among the subfamily, Osoriini are more speciose than Thoracophorini, and South America has nearly twice as many species as Central America (Table 4.3). Among the tribe Osoriini, both the genera *Holotrochus* Erichson, 1840 and *Molosoma* Say, 1831 with 94 species and 106 species, respectively, are the most speciose genera of the subfamily. Most species are known only from South America, where more than 73% of the species are only recorded from this sub-continent. Among these, several genera are known from South America, but not from Central America or the West Indies, e.g. *Mimogonia* Coiffait, 1978, *Neolosus* Blackwelder, 1942,

Table 4.3 Number of osoriine species in South America, Central America and the West Indies

	Thoracophorini	Osoriini	All Osoriinae			
Region	Species	Species	Genera	Species	Endemic	Endemic (%)
South America	212	229	41	493	363	73.6
Central America	124	89	27	259	128	49.4
West Indies	57	44	16	106	40	37.7
Total	262	318	44	672		

Euctenopsia Bruch, 1942, *Verhaaghiella* Irmler, 2005, etc. In contrast, only one genus is known from Central America that was not yet recorded from South America and only one, i.e. *Antillosorius* Irmler, 2010, from the West Indies. Although the West Indies are known for their high rate of endemisms, only 38% of the osoriine species are endemic there.

The geographic distribution of species richness shows high numbers of species in the tropical countries from Brazil with 222 species along the Andean range with numbers ranging between approximately 100 and 170 species up to Mexico (Fig. 4.2). Subtropical countries such as Argentina have less than 50 species and the southernmost country, Chile, less than ten. In particular, the Andean countries Peru and Ecuador as well as Panama and Costa Rica have outstanding numbers of species compared to their size.

Besides the West Indian islands, with high rates of endemism for Osoriinae (Irmler 2015b), centres of endemisms are located in the montane areas of Central and South America. One is the Talamanca-Chiriqui region adjacent to the frontier between Costa Rica and Panama (Rosen 1985; Liebherr 1988). According to Irmler (2006, 2007), six of the 15 species of the genus *Tannea* Blackwelder, 1952 and three of the 16 species of the genus *Lispinus* Erichson, 1840 are endemic in this region. Another montane region is located in the Aragua mountains close to the Atlantic coast of Venezuela, where several Osoriinae can be regarded as endemic, e.g. *Osorius araguensis* Irmler, 2014 and *O. tschirnhausi* Irmler, 2010. The Atlantic rainforest of southern Brazil is a main centre of endemism (e.g. Ribeiro et al. 2011; Pardini et al. 2009) which is also true for Osoriinae. Endemic species are found in several genera, e.g. *Paratorchus* McColl, 1985 (Irmler 2015a), *Tannea* Blackwelder, 1952 (Irmler 2007) and *Holotrochus* Erichson, 1840. The genus *Glyptoma* Motschulsky, 1857 is most speciose in this region (Irmler 2015c). From the 19 Neotropical species of the genus *Glyptoma* Motschulsky, 1857, 7 species (37%) occur in the Atlantic rainforest region, but only four (21%) in the region of Amazonia legal.

However, the highest number of endemic species is found along the eastern slope of the Andean range in Peru and Ecuador (Irmler 2009a, 2012a, b). For this region, even several genera are regarded as endemic, e.g. *Anancosorius* Bernhauer, 1908, *Lispinuncus* Irmler, 2005 and *Neolosus* Blackwelder, 1942. Among the 238 species found in Peru and Ecuador, 61 (25%) species were only found there. Several genera are most speciose in the region, e.g. *Allotrochus* Fagel, 1955. Among the five Neotropical species, one is distributed from Mexico to Argentina, one is endemic to Cuba and three are endemic from Ecuador to Peru. The genus *Aneucamptus* Sharp, 1887 is represented with three species in the Neotropics: two are pantropic in the Neotropics, and one is endemic on the western slope of the Peruvian Andean range.

The relation of species richness to area can be seen in the West Indies (Fig. 4.3). The species richness is correlated with log area of the islands. Whereas from small islands such as Martinique only one osoriine species is recorded, large islands such as Cuba have 46 species. The species richness

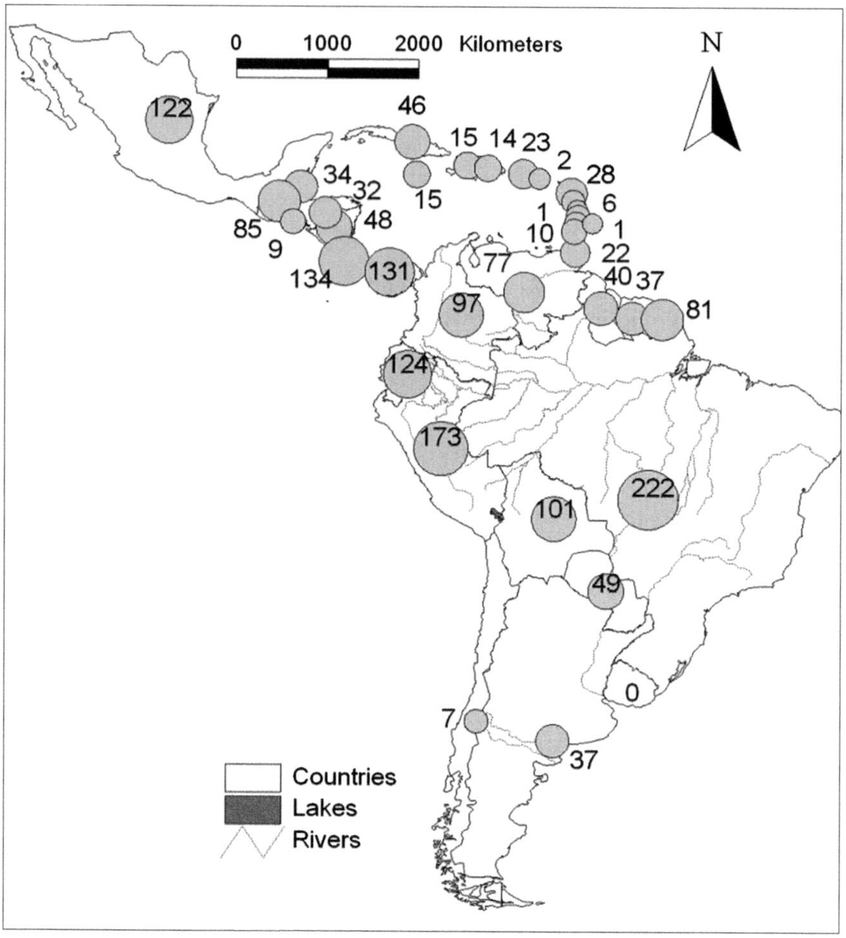

Fig. 4.2 Pattern of species richness (number in circles) of the subfamily Osoriinae for the Neotropical countries

increases with a rate of approximately 4.8 by the natural logarithm of area size. Using Wilson's (1988) formula $S = bA^z$, where S = number of species, A = island area and b = taxon area-specific constant, z is similarly high with 0.30 compared to the 0.28 calculated for ants by Wilson.

Certainly, species richness depends on the state of investigation which differs between the countries. Whereas small countries such as Costa Rica and Panama can be regarded as well investigated, large countries such as Brazil need more effort to come on the same state of investigation. The state of investigation for the total Neotropical region using the individual rarefaction method was studied separately for the tribes Osoriini and Thoracophorini (Fig. 4.3). The results show that for Thoracophorini nearly a steady state is found meaning that only few new species are expected by further sampling efforts. In contrast, no steady state is found for Osoriini which totals, however, much less specimens than Thoracophorini. It can be derived from the results that Thoracophorini are on average much more common than Osoriini and that much more species can be expected if more material is collected that contains Osoriini specimens. Regarding the species richness of genera, in particular, the species richness of the genera *Holotrochus* Erichson, 1839 and *Molosoma* Say, 1831 will increase.

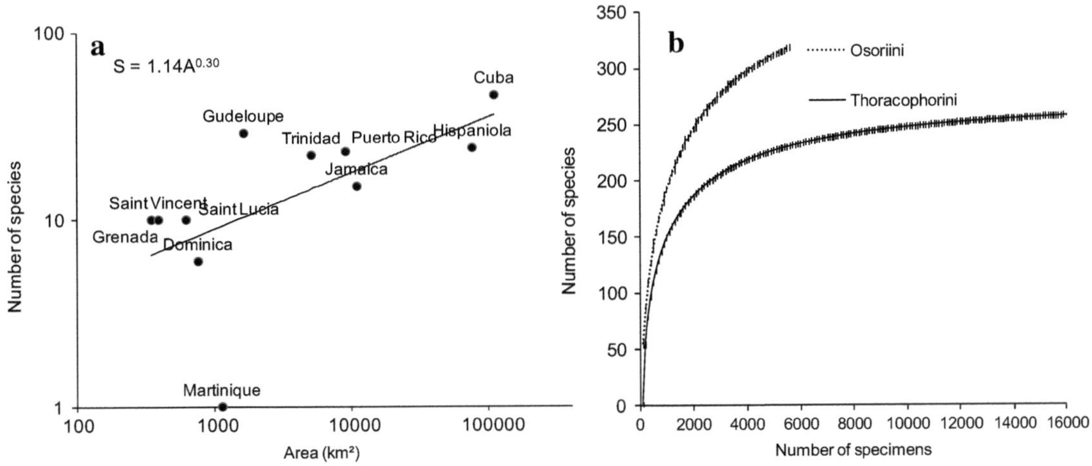

Fig. 4.3 (a) Relationship between area and species number of the West Indian islands; (b) rarefaction species richness for Thoracophorini and Osoriini; vertical lines indicate standard deviation. S = number of species. A = area

4.3.4 Species Similarity Between Neotropical Regions

Species similarity among countries using the Jaccard index with subsequent clustering is shown in Fig. 4.4. According to these results, Chile stands outside the remaining Neotropical countries and has no concurrent species with them. The southeastern countries Argentina, Paraguay and southeastern Brazil form one region of high similarity separated from the large area that mainly contains regions with lowland rainforest. The northwestern countries of South America are grouped together with the countries of the southern Central America. The last two groups form a cluster of higher similarity together with the northern Central American countries, i.e. Mexico, Honduras and Belize. The West Indian islands make up the last group of high species similarity. All these groups are characterized by a species similarity of approximately 20% or higher. Trinidad as the southernmost island plays an intermediate role between the South American mainland and the West Indies.

Figure 4.5 gives the geographical regions with high similarities of the faunal composition that can be derived from the cluster of similarity indices. According to these results, the northwestern Andean region has the highest number with 320 species followed by the lowland Amazonian rainforest with 204 species. On the continental mainland, the number of combining species between adjacent regions varies between 50 and 88 which range approximately between 15% and 18% depending on the adjacent regions. If combining species between not adjacent regions are regarded, the number decreases to approximately 8% which means that about 8% of the species can be considered to be distributed in the total tropical area of the Neotropics. The number of combining species is much lower between the mainland and the West Indian islands. It ranges between 3% and 10%. It is interesting to note that the number and the ratio of combining species are higher in the northern bridge between northern Central America and Cuba with 10% than at the southern bridge between both the northwest Andean region and Trinidad or between Trinidad and the southern West Indian islands with 3% and 4%, respectively. This relation might implicate that the northern bridge is closer and more often used for a faunal interchange.

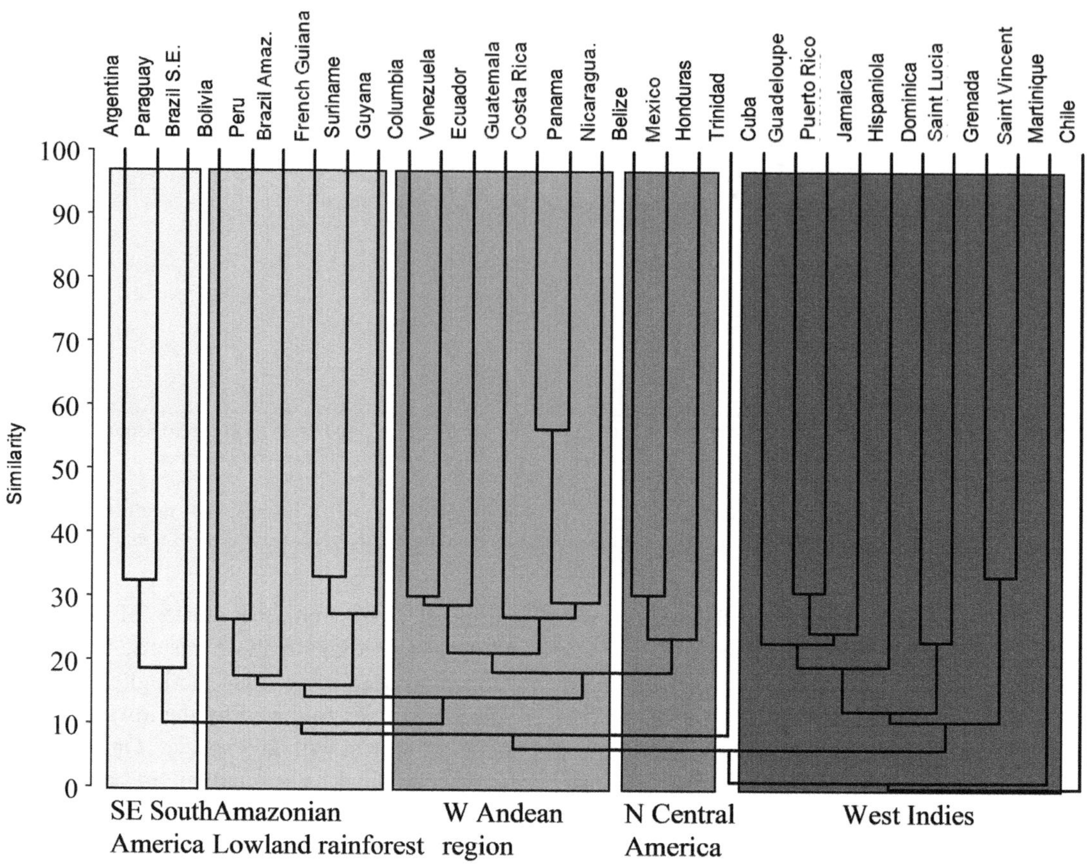

Fig. 4.4 Result of the cluster analysis with Jaccard indices for the Osoriinae subfamily between Neotropical countries regions; Brazil Amaz., Brazilian states of Amazonia legal; Brazil SE, southeastern Brazilian states not included to Amazonia legal

4.3.5 Transition Bridges and Distribution Patterns of Single Species and Species Groups

Within the Neotropical regions, several main zones exist with strong faunal exchange, e.g. the southern Central American bridge that combines northwestern South America with Central America (Morone 2014; Asiain et al. 2015) or the Yucatan-West Indian bridge (Nichols 1988; Peck 2005). According to Irmler (2015b), nine species of Osoriinae may have colonized Cuba via the Yucatan peninsula. The function of these two transition zones is reflected by the distribution of the two species *Lispinus laticollis* Erichson, 1840 and *Nacaeus dejectus* Sharp, 1887. Both species are frequent in Central America, but also occur in northern parts of South America. connected by the Isthmus of Panama (Fig. 4.6). They were also frequently found in Cuba and other Great Antillean islands which reflect a connection over the Yucatan bridge. Several more species represent such type of Circum-Caribbean distribution, e.g. *Holotrochus minor* Chevrolat and Fauvel, 1863 that is found in Central America, northern South America up to Ecuador as southernmost record and nearly the whole Caribbean region (Irmler 2016). It seems to be the northern sister species of a group of two other species found in the Guyanas and southern Brazil.

Fig. 4.5 Species richness of Osoriinae (number in circles) in the differentiated regions with number of combining species (number within arrows, percent in brackets)

Another important bridge is the southern Amazonian-northern Savanna bridge as transition zone between the eastern Andean region and the Atlantic rainforest region. Here, northern borders of Savanna species meet southern borders of lowland Amazonian rainforest species, and eastern borders of Andean species meet western borders of Atlantic rainforest species. This is exemplified by the genera *Pardirocephalus* Bruch, 1942 that finds its northern border in the Brazilian state of Mato Grosso and *Holotrochopsis* Fagel, 1959 with its southern border in the same region (Fig. 4.6). This region also functions as colonizing corridor for species from the Andean region such as for *Molosoma hanagarthi* Irmler, 2014 or the Atlantic rainforest

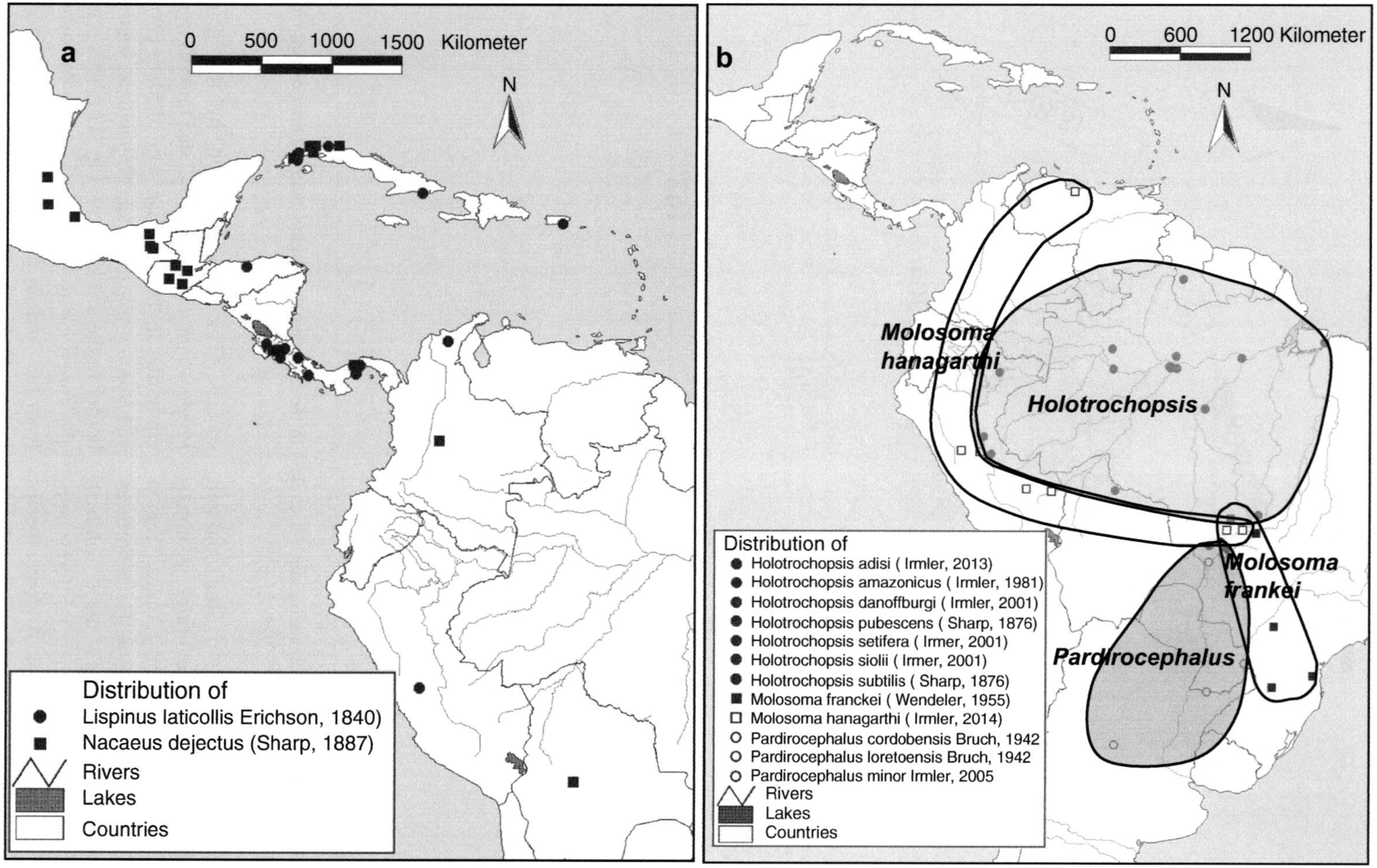

Fig. 4.6 Transition bridges with overlaps of species of Osoriinae distributions from different geographical regions; (**a**) Yucatan bridge, (**b**) southern Amazonian-Mato Grosso bridge

region such as for *Molosoma franckei* Wendeler, 1955.

Consequently, this region might be the bridge where several widely distributed species change from the southeastern region to the northwestern regions along the eastern side of the Andean slope or vice versa. For these species the large area of the Amazonian lowland rainforest seems to be a strong barrier. Among Osoriinae several species show a Circum-Amazonian distribution, e.g. *Holotrochus schubarti* Irmler, 1981 or *Nacaeus opacus* Fauvel, 1895 (Fig. 4.7).

Even species groups follow this type of distribution, e.g. the *Holotrochus simplex* group. Another example is represented by the genus *Dirocephalus* Silvestri, 1938. The genus is closely related to the northern South American savanna genera *Pardirocephalus* Bruch, 1942, *Euctenopsia* Bruch, 1942 and *Pselaphomimus* Bruch 1942 that seem to be inquilines in ant nests. Only the genus *Dirocephalus* shows a wide distribution along the eastern slope of the Andean range up to southern Central America (Costa Rica) and the Guyanas with several species, mostly endemic in restricted areas. In the Central Amazonian lowland rainforest, only one species occurs (Irmler 2005, 2009b). One genus, i.e. *Verhaghiella* Irmler, 2005, is endemic in the eastern Peruvian Andes. This shows that this bridge was open for a long time in the past and allowed the development of endemic species along this way.

Regarding additionally the results given by Irmler (2007, 2009a, 2012a, b) and Asian et al. (2015), the following distribution types can be differentiated: (1) northern Central America type with species distributed from Mexico to Costa Rica; (2) southern Central America-northwestern South America type with species distributed in the northern Andean region up to Nicaragua; (3) northern South America-southern Central America type with species distributed from Guyana and Venezuela up to Nicaragua and partly the southern West Indian islands; (4) lowland Amazonian rainforest type with species distributed from the eastern slope of the Andes to the mouth of the Amazon in the lowland rainforest; (5) southeastern Brazilian type with species in the Atlantic rainforest of Brazil and adjacent areas; (6) northern Savanna type with species distributed in the Savanna vegetation from southern Brazil, northern Argentina and Paraguay; (7) Circum-Amazonian type with species distributed from southern Central America along the eastern slope of the Andes up to southeastern Brazil; (8) pantropical type for the Neotropics with species occurring in the whole Neotropical region with tropical and subtropical rainforest; (9) pantropical type worldwide with species occurring in several tropical regions of the world, mainly invasive species (see following part); and (10) endemic type with species restricted to small areas, e.g. West Indian islands, Talamanca-Chiriqui region, Aragua mountains, small areas along the Andes from Colombia to Bolivia, Guyana, small areas of the Brazilian Atlantic rainforest and Chilean Pacific coast with subtropical forest.

4.3.6 Connection to Non-Neotropical Regions

Among the Neotropical Osoriinae fauna species are found that colonized from other continents, but Neotropical species also invaded other continents. Irmler (1981b) graphed the worldwide distribution of *Mimogonus fumator* Fauvel, 1889 that has been recorded from various locations of the Indo-Pacific and Australian region and also from the West Indian islands and Central America. Meanwhile, new records are known from several Central American countries that indicate a wide distribution there. Moreover, Assing (2012) recently published a new record from Israel, Wadi Kelt, St. George Monastery, moist spots on sandy soils, which shows that the species is exported worldwide to countries with warm climate. The origin of the species is unknown. It may be in South-East Asia. For the Neotropics, it is certainly an invasive species.

A similar history can be assumed for *Nacaeus impressicollis* (Motschulsky, 1857). It was originally described from the Indo-Pacific region, too. Later, Blackwelder (1943) described it as

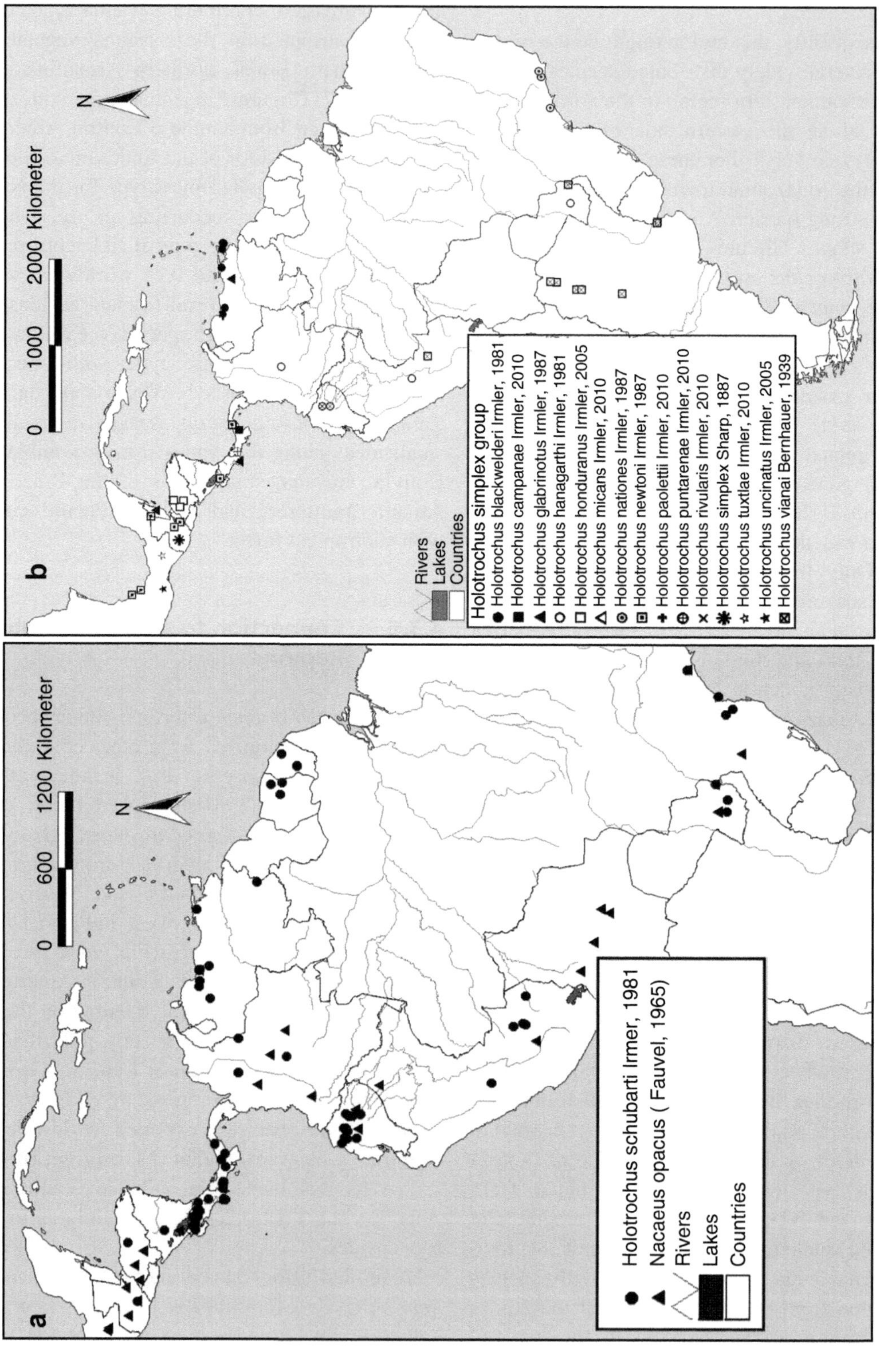

Fig. 4.7 Circum-Amazonian distribution of species from **(a)** Osoriini and Thoracophorini tribe and **(b)** Holotrochus simplex species group

Pseudolispinodes irregularis from the West Indian islands. The species is also known from Japan, where other related species occur. Actually, it is recorded from various Central and South American countries and can be also considered as an invasive species in the Neotropics. The same status can be given to *Nacaeus planellus* Sharp, 1887.

Examples of the export of Neotropical Osoriinae to other continents are also known. Outerelo et al. (2010) described *Holotrochus hispanicus* Outerelo, Gamarra, Urbaneja, Castañera and Monzó, 2010 from the Spanish province of Valencia which, however, is conspecific with *Holotrochus acromyrmicis* Bernhauer, 1921 and was also recently recorded from the Botanical Garden of Monaco (Lemair and Raffaldi 2015). The *Mimogonia europaea* Coiffait, 1978 is another example of an exported Neotropical species. The genus was firstly described on this species found in southern Portugal. Later, Irmler (e.g. 1981b, 2010b, 2013) described 18 new species from South America, but no records were given from Europe since the first description. This shows that the genus is widely distributed in the tropical and subtropical regions of South America with several species and records in Europe likely based on invasive species.

Among the tribe Osoriini, 47% of Neotropical genera are endemic (Irmler 2015a). Besides genera found in different tropical faunal kingdoms, the Neotropical fauna shows closer relations to the Paleotropic as well as to the Australian fauna.

Examples for the generic relationship with Africa are the genera *Mimogonus* Fauvel, 1903 and *Holotrochopsis* Fagel, 1955 (Fig. 4.8). As can be seen from Fig. 4.8, the number of species and their distributional pattern differ

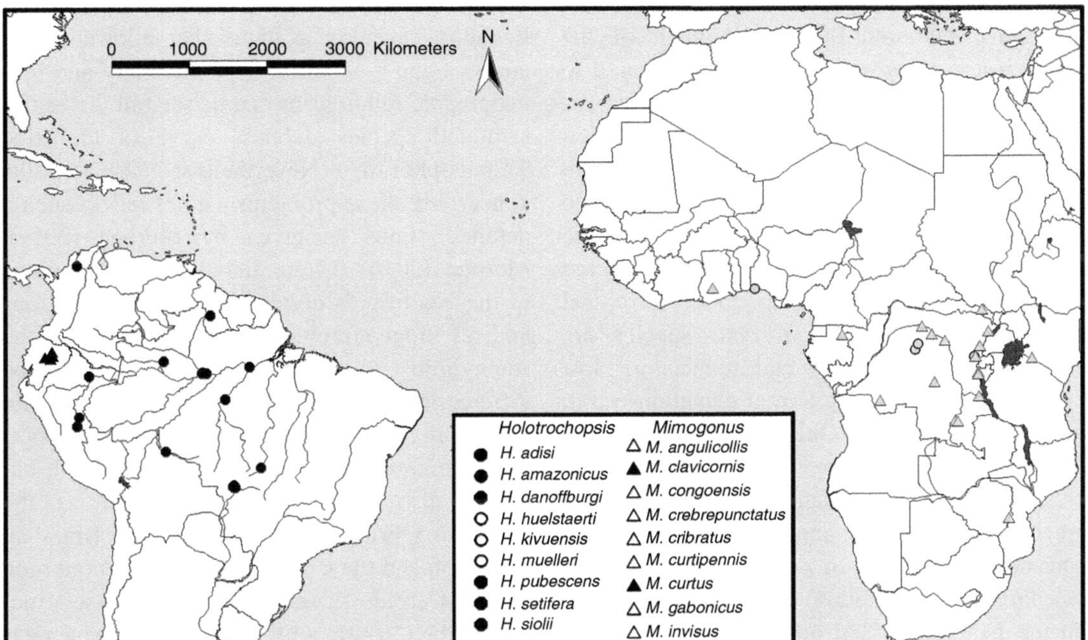

Fig. 4.8 Distribution of the genera *Holotrochopsis* and *Mimogonus* (Osoriinae) in South America and Africa excluding *Mimogonus fumator* Fauvel, 1889; black items, Neotropic; grey items, Paleotropic

fundamentally between South America and Africa. Excluding the pantropical *M. fumator* (Fauvel, 1889), only two species of *Mimogonus* Fauvel, 1903 occur in South America, but 13 species occur in Africa. Moreover, the two Neotropical species are restricted to a small area of the eastern slope of the Ecuadorian Andes. In contrast, the genus is widely distributed in the tropical montane and lowland rainforests of Africa. The genus *Holotrochopsis* Fagel, 1955 is speciose in South America with six species and poor in Africa with three species. In both continents the genus inhabits lowland rainforests. The distribution of the two genera may be explained by the different geneses of the areas. *Mimogonus* Fauvel, 1903 distribution in South America may be a relict of a formerly wider distribution. Both Neotropical species are edaphic insects with lost wings and reduced eyes and may have survived unfavourable conditions in deep soil layers. In contrast, the large Amazon basin with lowland rainforest may have favoured the development of the genus *Holotrochopsis* Fagel, 1955 compared to the smaller lowland rainforest area in Africa.

Another example is the genus *Paratorchus* McColl, 1985 that occurs with seven species in western Central and South America from Mexico to Chile, one species in the Brazilian Atlantic rainforest and 35 species in New Zealand (McColl 1982). Similarly as the Neotropical *Mimogonus* species, *Paratorchus* species are edaphic with lost wings and reduced or lost eyes. In the Andes, they live at elevations up to more than 3000 m. Only in Chile and the Brazilian Atlantic rainforest they inhabit lowland or submontane subtropical zones. Most species are endemic in very small areas that can be considered as relicts. In particular, the Chilean *P. chilensis* (Irmler 2005) inhabits a small area on the Isla Chiloe and adjacent parts, isolated from the other congeneric species, which is specific for the *Nothofagus* centre defined by Müller (1973). Moreover, considering the investigations of Harrison (2004) and Garcia (2012), the species must have survived the glacier period close to the glacier frontier. It can be supposed that the genus is part of the South American Transition Zone which shows relations to the Austral region (Urtubey et al. 2010; Morone 2015).

According to McColl (1982) the species occur on both islands, but single species are restricted to small areas, except two species with wide distribution, and to indigenous forests, where they live in the litter and humus layers from the lowland up to approximately 1200 m elevation. Thus, the ecology in New Zealand corresponds with the observations in South America that in colder climates lowlands or montane zones are inhabited, whereas under tropical climates high montane zones are their main habitat.

4.4 Conclusion

The discovery of the Neotropical staphylinid fauna is still insufficient in wide areas, e.g. in the large Amazon basin or parts of South Brazil. However, the extensive collection in the last decades, e.g. in Costa Rica, Panama and Ecuador, revealed a status that allows a first analysis and evaluation of biodiversity and biogeographic relations as can be seen in the nearly saturated species richness curve of the tribe Thoracophorini. Nevertheless, the results concerning these problems cannot reflect such a detailed status as given by Morone (2014). Morone (2014) differentiated three subregions in the Neotropics divided into eight dominions and 53 biogeographic provinces. Actually, the staphylinid distribution pattern can be differentiated on the dominion levels, but not on the province levels. The actual results support the differentiation given by Morone (2014), but reflect also additional results such as the connecting bridge between the South Brazilian dominion and the Chacoan and Parana dominion via the Cerrado. It seems that the diverse structure of the Cerrado with forest and savanna areas is appropriate for an exchange of faunal elements (Da Silva and Bates 2002). Moreover, modern agricultural management and intensification of trade traffic have already and will more intensively promote invasive species in the future. This can be seen already by many staphylinid species on pastures and agricultural fields which colonized from outside South America or are exported to other tropical countries. Therefore, a fast taxonomic inventory combining also several invertebrate groups is needed to follow this process and to find dangerous developments early enough to regulate this inexorable process.

Acknowledgements We heartily thank Dr. Alfred Newton, Chicago, for his support to find the correct number of species and genera of the Neotropical region.

References

Asenjo A, Irmler U, Klimaszewski J et al (2013) A complete checklist with new records and geographical distribution of the rove beetles (Coleoptera, Staphylinidae) of Brazil. Insecta Mundi 277:419

Asiain J, Hernandez J, Irmler U (2015) New national and state records of Neotropical Staphylinidae (Insecta: Coleoptera). Zootaxa 3974:76–92

Assing V (2012) On the Staphylinidae of Israel. Linzer Biol Beitr 44:351–363

Bates HW (1864) The naturalist on the river Amazon. A record of adventures, habits of animals, sketches of Brazilian and Indian life, and aspects of nature under the equator, during eleven years of travel. J. Murray, London, p 466

Blackwelder R (1943) Monograph of the West Indian beetles of the family Staphylinidae. Bull US Natl Mus 182:658

Blackwelder R (1944) Checklist of the coleopterous insects of Mexico, Central America, the West Indies, and South America – Staphylinidae. Bull US Natl Mus 185:100–168

Bohác J (1978) Description of the larva and pupa of *Thoracophorus brevicristatus* (Coleoptera, Staphylinidae). Acta Entomol Bohemoslov 75:394–399

Chani-Posse MR, Thayer MK (2008) Staphylinidae. In: Claps L, Debandi G, Roig-Juñent S (eds) Biodiversidad de Artrópodos Argentinos, vol 2. Sociedad Entomológica Argentina Ediciones, Mendoza, pp 471–494

Da Silva JMC, Bates JM (2002) Biogeographic patterns and conservation in the South American Cerrado: a tropical savanna hotspot. Bioscience 52:225–232

de Laporte FL (1834–1840) Études entomlogique ou description d'insectes nouveaux, et observation sur las synonymie. Méquignon, Paris, p 159

Erichson WF (1839-1840) Genera et species Staphylinorum Insectorum Coleopterorum Familiae. Morin, Berlin, p 994

ESRI (1999) ArcView GIS software program. Environmental Systems Research Institute

Garcia JL (2012) Late Pleistocene ice fluctuations and glacial geomorphology of the Archépelago de Chiloé, southern Chile. Geogr Ann Ser A 94:459–479

Gravenhorst ILC (1806) Monographia coleopterorum micropterorum. H. Dietrich, Göttingen, p 236

Hammer Ø, Harper DAT, Rayan PD (2014) PAST: paleontological statistics software package for education and data analysis. Paleontol Electron 4:1–19

Harrison S (2004) The Pleistocene glaciations of Chile. Dev Q Sci 2:89–103

Irmler U (1981a) Descriptions of new Neotropical *Holotrochus* and a key to the species of the genus (Coleoptera: Staphylinidae). Coleopt Bull 35:379–397

Irmler U (1981b) Neue Arten der Gattung *Mimogonia* Coiffait (1978) aus der Neotropis (Coleoptera, Staphylinidae). Entomol Bl 77:143–152

Irmler U (2000) The Neotropical species of the genus *Allotrochus* Fagel, 1955. Bull Inst Roy Sci Nat Belg, Entomol 70:247–250

Irmler U (2005) Review of the genus *Dirocephalus* Silvestri, 1938 and related genera in the Neotropical region (Coleoptera: Staphylinidae: Osoriinae). Bull Inst Roy Sci Nat Belg, Entomol 75:103–118

Irmler U (2006) The genus *Lispinus* (Coleoptera: Staphylinidae: Osoriinae) in Costa Rica – a key, a new species, ecological and biogeographical remarks. Brenesia 66:1–13

Irmler U (2007) The *Tannea* (Coleoptera: Staphylinidae: Osoriinae) species of Costa Rica – new species and records, a key, ecological, and geographical remarks. Brenesia 68:69–85

Irmler U (2009a) New species and records of the genus *Lispinus* with a key to the species from Peru (Coleoptera: Staphylinidae: Osoriinae). Zootaxa 2263:42–58

Irmler U (2009b) Two new species of the Neotropical *Dirocephalus* complex of genera (Coleoptera: Staphylinidae: Osoriinae). Koleop Rundsch 79:59–63

Irmler U (2010a) Two new species of the genus *Thoracophorus* Motschulsky, 1837 (Coleoptera: Staphylinidae, Osoriinae) with remarks on ecology of the genus in the Neotropical region. Psyche 138518:1–6

Irmler U (2010b) New species of the genus *Mimogonus* and *Mimogonia* (Coleoptera: Staphylinidae: Osoriinae) from the Neotropical region. Acta Entomol Mus Nat Prag 50:483–294

Irmler U (2012a) The *Lispinus* Erichson, 1840 species of Ecuador with description of a new species and a new synonymy (Coleoptera: Staphylinidae: Osoriinae). Koleop Rundsch 82:219–233

Irmler U (2012b) The *Tannea* Blackwelder species of Ecuador with description of new species (Coleoptera: Staphylinidae: Osoriinae). Neotrop Entomol 41:214–222

Irmler U (2013) New species of the genus *Mimogonia* and *Holotrochus* from South America (Coleoptera: Staphylinidae: Osoriinae). Acta Entomol Mus Nat Prag 53:155–176

Irmler U (2015a) New Neotropical genera and species of the tribe Osoriini (Coleoptera: Staphylinidae: Osoriinae). Contrib Entomol 65:173–196

Irmler U (2015b) Osoriinae of Cuba with description of new species and an identification key (Coleoptera: Staphylinidae). Acta Entomol Mus Nat Prag 55:145–172

Irmler U (2015c) The neotropical genus *Glyptoma* Erichson, 1839 with descriptions of new species and a key to the species (Coleoptera: Staphylinidae: Osoriinae). Contrip Entomol 66:297–325

Irmler U (2016) Four new species of the genus *Holotrochus* Erichson, 1839 from South America (Staphylinidae: Osoriinae). Entomofauna 37:85–100

Lemair J-M, Raffaldi J (2015) Deux nouveaux coléptères introduits dans les jardins de Monaco: *Coiffaitiella benjamini* (Marquet, 1875) (Coleoptera Curculionoidea Raymondionymidae) et *Holotrochus acromyrmicis* Bernhauer, 1920 (Coleoptera Staphylinidae Osoriinae). Le Coléoptérist 18:165–166

Liebherr JK (1988) Biogeographic pattern of West Indian Platynus carabid beetles (Coleoptera). In: Liebherr JK (ed) Zoogeography of Caribbean insects. Cornell University Press, London, pp 121–152

Linné C (1758) Systema naturae per regna tria naturae, secundum classes, ordines, genera, species, cum characteribus, diufferentiis, synoymis, locis. Holm, Laurentius, p 823

Lynch-Arribalzaga F (1884) Estafilinos de Buenos Aires. Bol Acad Nac Cienc 7:392

McColl HP (1982) Osoriinae (Insecta: Coleoptera: Staphylinidae). Fauna of New Zealand, Wellington, p 89

Morone JJ (2014) Biogeographical regionalization of the Neotropical region. Zootaxa 3782:110

Morone JJ (2015) Biogeographical regionalization of the Andean region. Zootaxa 3936:207–236

Müller P (1973) The dispersal centres of terrestrial vertebrates in the Neotropical realm: a study in the evolution of the Neotropical biota and its native landscapes. Junk, The Hague, p 244

Navarrete-Heredia JL, Newton AF, Thayer MK et al (2002) Illustrated guide to the genera of Staphylinidae (Coleoptera) of Mexico. Universidade de Guadalajara, Guadalajara, p 401

Newton AF (2015) Beetles (Coleoptera) of Peru: a survey of the families. Staphylinidae Latreille, 1802. BioOne. https://doi.org/10.2317/kent-88-02-283-304.1

Newton AF, Gutiérrez Chacón C, Chandler DS (2005) Checklist of the Staphylinidae (Coleoptera) of Colombia. Biota Colomb 6:1–72

Nichols SW (1988) Kaleidoscopic biography of West Indian Scaritinae (Coleoptera: Carabidae). In: Liebherr JK (ed) Zoogeography of Caribbean insects. Cornell University Press, London, pp 71–115

Olivier GA (1795) Entomologie,ou histoire naturelle des insectes, avec leurs caractères génériques et spécifiques, leur description, leur synonymie et leur figure enluminée. Coléoptères, vol 3. Lanneau, Paris

Outerelo R, Gamarra P, Urbaneja A et al (2010) *Holotrochus hispanicus* nov. sp. (Coleoptera, Staphylinidae, Osoriinae) de Valencia, España y su curioso fenómeno de tanatosis. Bol Real Soc Esp Hist Nat (Sec Biol) 104:97–105

Pardini R, Faria D, Accacio GM et al (2009) The challenge of maintaining Atlantic forest biodiversity: a multi-taxa conservation assessment of specialist and generalist species in an agro-forestry mosaic in southern Bahia. Biol Conserv 142:1178–1190

Peck SB (2005) A checklist of the beetles of Cuba with data on distribution and bionomics (Insecta: Coleoptera). Arthropod Fla Neighbor Land Areas 18:1–241

Peck SB (2016) The beetles of the Lesser Antilles (Insecta, Coleoptera): diversity and distributions. Insecta Mundi 0460:1–360

Perty M (1830) Delectus Animalium Articulatorum quae in itinere per Brasiliam annis MDCCCXVII-MDCCCXX jussu et auspiciis Maximiliaani Josephi I. Bavariae regis Augustissimi peracto collegerunt Dr. J.B. de Spix et Dr. C.F.Ph. de Martius. Lipsius, München

Puthz V (1998) Bibilographie der Publikationen von Alexander Bierig (1884-1963). Philippia 8:209–215

Ribeiro MC, Martensen AC, Metzger JP et al (2011) The Brazilian Atlantic forest: a shrinking biodiversity hotspot. In: Zachos FE, Habel JC (eds) Biodiversity hotspots. Distribution and protection of conservation priority area. Springer, Berlin, pp 405–434

Rosen DE (1985) Geological hierarchies and biogeographic congruence in the Caribbean. Ann Mo Bot Gard 72:636–659

Sharp D (1876) Contributions to the Staphylinidae of the Amazon valley. Transactions of the Entomological Society, London, pp 34–424

Sharp D (1883-1887) Fam. Staphylinidae. In: Biologia Centrali Americana: Insecta Coleoptera, vol 1. Taylor & Francis, London, pp 145–824

Thayer MK, Newton AF 2005 Distribution of austral species of Staphylinoidea: genera and species of Staphylinidae, Silphidae, Leiodidae, and Agyrtidae occurring in Australia, New Zealand, Chile and/or southern Argentina, or South Africa (also Madagascar) and neighboring islands. Field Museum of Natural History, Chicago. 313 pp. Online only, PDF available from URL: http://archive.fieldmuseum.org/peet_staph/db_1c.html

Urtubey E, Stuessy TF, Tremetsberger K et al (2010) The South American biogeographic transition zone: an analysis from Asteraceae. Taxon 59:505–509

von Wied M (1825) Reise nach Brasilien in den Jahren 1815 bis 1817. Kaulfuß und Kramer, Wien

Wasmann E (1902) Species novae insectorum Termitophilorum ex America Meridionali. Tijdschr Entomol 45:95–107

Wilson EO (1988) The biogeography of the West Indian ants (Hymenoptera: Formicidae). In: Liebherr JK (ed) Zoogeography of Caribbean insects. Cornell University Press, London, pp 214–230

5 Canada's Adventive Rove Beetle (Coleoptera, Staphylinidae) Fauna: A Long-Term Case Study on the Detection, Origin, Introduction Pathways, and Dynamic Distribution of Non-native Beetles

Jan Klimaszewski and Adam J. Brunke

Abstract

The family Staphylinidae, or rove beetles, consists of more than 62,290 described species worldwide, of which 1682 species have been recently recorded from Canada. One hundred and fifty-three of these species, in 73 genera and 13 subfamilies, are confirmed here as adventive; they constitute about 9% of the Canadian fauna and mostly originate from the western Palaearctic region. The highest number of adventive species is found in the Atlantic Provinces, Quebec, Ontario, and coastal British Columbia, which are areas with a long history of trade with Europe and Asia. Important historical pathways have been organic matter associated with livestock and soil used as dry ballast in ships destined for North America from Europe. Over Canada's trade history, moss and soil, likely imported with plant stock, have become more important. Unlike other beetles, relatively few staphylinids associated with woody organic matter have become established in North America as adventive. Although it is difficult to reconstruct precise introduction timelines, it is clear that adventive rove beetle species have recently and will continue to become established in North America. The combination of high diversity, inadequate taxonomic knowledge, and incomplete sequence reference libraries poses challenges to the accurate and rapid detection of adventive rove beetles in Canada.

5.1 Taxonomy, Origin, and Distribution

In Canada, 153 species in 73 genera and 13 subfamilies are considered as adventive, constituting about 9% of the Canadian fauna (Fig. 5.1). The majority of Canadian adventive species belong to two subfamilies: Staphylininae, with 51 species, and Aleocharinae, with 51 species. The genera with the most adventive species in Canada are *Philonthus* (16 spp.), *Atheta* (8 spp.), *Aleochara* (8 spp.), and *Quedius* (7 spp.).

The origin of nearly all 153 adventive rove beetles in Canada is West Palaearctic, especially Central Europe (Klimaszewski et al. 2013). This is mainly due to the climatic, biotic, and landscape similarities between this region and the

J. Klimaszewski (✉)
Natural Resources Canada, Canadian Forest Service, Laurentian Forestry Centre, Québec, QC, Canada
e-mail: jan.klimaszewski@canada.ca

A. J. Brunke
Canadian National Collection of Insects, Arachnids and Nematodes, Agriculture and Agri-Food Canada, Ottawa, ON, Canada

O. Betz et al. (eds.), *Biology of Rove Beetles (Staphylinidae)*,
https://doi.org/10.1007/978-3-319-70257-5_5

Fig. 5.1 Different adventive species of Staphylinidae recorded from Canada. Dates in parentheses represent approximate dates of first detection

northern Nearctic and to some 500 years of transport of goods from Europe to North America. Exceptions such as *Anotylus insignitus* (Gravenhorst), possibly originating from the Neotropical region, as well as *Rugilus ceylanensis* (Kraatz) and *Philonthus rectangulus* Sharp, originating from the Oriental region (Klimaszewski et al. 2013), are distinctly thermophilic: they are restricted to compost piles or farmland and avoid the northernmost areas of Canada. About 35 of the adventive rove beetle species recorded from Canada have been transported to other continents as well, and at least 22 of them are now considered cosmopolitan (Klimaszewski et al. 2013). Although populations of adventive staphylinids in western Canada are physically closer to those of the eastern Palaearctic region, their origin is certainly western Palaearctic, or via introduction to eastern North America, as eastern Palaearctic species have generally not become established in North America. However, an increasing trade with China may have already resulted in the introduction of additional undetected species to western North America, which has received far less sampling effort recently, compared to the east. The number of separate introduction events to North America for each species is difficult to ascertain. Based on the dates of first detection and on distribution patterns, at least 32 adventive rove beetles were likely introduced separately on the east and west coasts, but it is unknown whether these introductions were direct (from the native range) or whether they represent translocations from one part of North America to another Klimaszewski et al. (2013).

The number of adventive staphylinid species varies considerably among Canadian regions (Table 5.1). The Atlantic Provinces, Quebec, Ontario, and British Columbia (Majka and Klimaszewski 2008a, b; Klimaszewski et al. 2013) have the greatest number of adventive species. In terms of percentage of species represented by non-native rove beetles, the faunas of Prince Edward Island, Newfoundland, and Nova Scotia are the most strongly dominated by adventives (Table 5.1). The Prairie Provinces of Alberta, Saskatchewan, and Manitoba and the northern territories and regions (Yukon, Northwest Territories, Nunavut, and Labrador) have been the least invaded by adventives (Klimaszewski et al. 2013). It is anticipated that the north will become more extensively colonized by adventive species in the wake of climate warming and increased access to northern ports via ships as the ice-free season increases.

5.2 Detection of Adventive Species

The date of first detection of an adventive species in Canada only roughly approximates the actual time of entry. As the sampling of insects in Canada only began in the late nineteenth century, with the exception of a few localized efforts in the early to mid-nineteenth century, the earliest invaders were collected for the first time long (perhaps centuries) after their introduction (Klimaszewski et al. 2013). For many seventeenth- to nineteenth-century detections, the presence of a species is reported in historical publications based on specimens bearing no date of collection (e.g., Gravenhorst 1802). A few of these cases are North American species that were later discovered to be identical to previously described Palearctic species. The detection dates used herein are based either on this literature or on the oldest specimen known in collections as reported by Klimaszewski et al. (2013) and updated from the faunistic literature. In an attempt to reduce collection bias as much as possible, the period from 1620 to 1850 is treated as one unit in Fig. 5.2, and dates of first detection in North America (Klimaszewski et al. 2013) rather than Canada are used in the present work.

Table 5.1 Summary of the native and adventive staphylinid beetle fauna of Canada by province and territory (from Klimaszewski et al. 2013)

	YK	NT	NU	BC	AB	SK	MB	ON	QC	NB	NS	PE	NF	LB
Total species	265	209	26	774	440	223	390	861	740	625	420	86	317	148
Native species	254	204	26	697	399	193	356	749	635	539	341	53	251	134
Adventive species	11	5	0	77	41	31	34	112	105	86	79	33	66	14
Percent adventives	4.2	2.4	0.0	9.9	9.3	13.5	8.7	13.0	14.2	13.8	18.8	38.4	20.8	9.5

The estimates of species richness are updated from Bousquet et al. (2013). Abbreviations of Canadian provinces and territories: *AB* Alberta; *BC* British Columbia; *LB* Labrador; *MB* Manitoba; *NB* New Brunswick; *NF* Newfoundland (island); *NS* Nova Scotia; *NT* Northwest Territories; *NU* Nunavut; *ON* Ontario; *PE* Prince Edward Island; *QC* Quebec; *SK* Saskatchewan; *YT* Yukon Territory

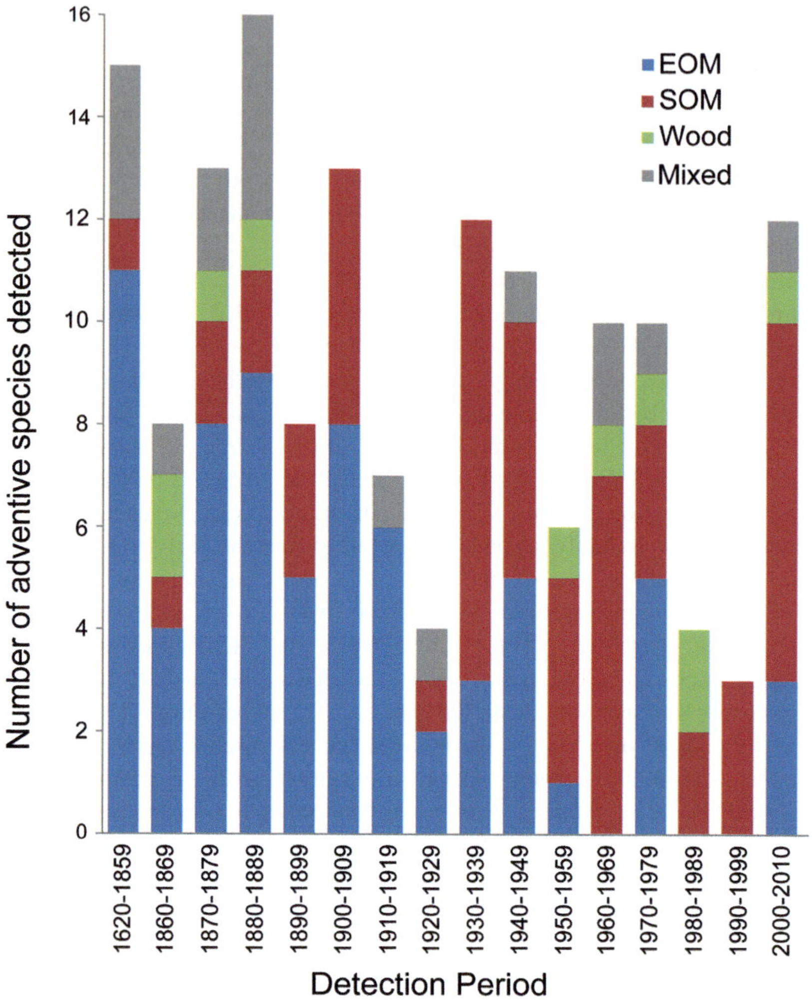

Fig. 5.2 Detection timeline of adventive Staphylinidae in Canada, based on the earliest date of detection in North America and categorized by general microhabitat preferences given in Table 5.2. EOM = ephemeral organic matter; SOM = stable organic matter; wood = bark or rotting woody matter; mixed = combining two or more categories. The intentional release of *Aleochara tristis* Gravenhorst is omitted

A complete list of confirmed adventive species in Canada is provided in Table 5.2. Some species reported to be adventive in North America, especially those with very early dates of detection, may actually be naturally Holarctic. Putatively adventive species known from the Russian Far East (mainly based on Schülke and Smetana 2015) were excluded from Table 5.2. Species reported from North America without supporting specimen data (e.g., *Lobrathium multipunctatum* (Gravenhorst)) were also excluded from Table 5.2. A list of these excluded species is provided in Table 5.3.

For 66 adventive staphylinid species, the first North American record is from Canada, thus suggesting that these species may have been directly introduced into Canada from their original range (Klimaszewski et al. 2013, 2015a, b). Sixty-eight species were first detected in the USA and may have spread to Canada (Klimaszewski et al. 2013). The point of first detection of the remaining 23 species is unknown as the older literature usually refers to the New World distribution as "North America." Of the 66 species that were possibly directly introduced into Canada, 19 species first became detected in the Maritime Provinces, 17 in Ontario, 16 in Quebec, and 11 in British Columbia. The most common points of entry were St. John's, Newfoundland (8 spp. introduced); the lower

Table 5.2 Adventive Staphylinidae recorded in Canada with date of first North American detection and microhabitat preferences

Species	Detection	Microhabitat	Category
Omaliinae			
Eusphalerum torquatum (Marsham)	1965	Broom flowers (Zanetti 2012)	SOM
Omalium rivulare (Paykull)	1878	Dung, compost, hay (Forbes et al. 2016)	EOM
Phyllodrepa floralis (Paykull)	1860	Compost, nests, flowers, (Zanetti 2012), dung (Forbes et al. 2016)	EOM
Xylodromus concinnus (Marsham)	1894	Dung, compost, nests (Zanetti 2012; Forbes et al. 2016)	EOM
Pselaphinae			
Euplectus karsteni (Reichenbach)	1894	Compost, dung (Wagner 1975)	EOM
Euplectus signatus (Reichenbach)	1933	Compost, dung (Wagner 1975)	EOM
Tachyporinae			
Cilea silphoides (Linnaeus)	1870	Compost (Assing and Schülke 2012)	EOM
Mycetoporus lepidus Gravenhorst	1953	Agricultural fields, litter (Majka and Klimaszewski 2008a)	SOM
Sepedophilus immaculatus (Stephens)	2010[a]	Forest litter, flood debris, and deadwood (Schülke 2011)	Wood/ SOM
Sepedophilus littoreus (Linnaeus)	1866	Moldy wood (Campbell 1976)	Wood
Sepedophilus marshami (Stephens)	1951	Moldy wood (Campbell 1976)	Wood
Sepedophilus testaceus (Fabricius)	1877	Moldy wood (Campbell 1976)	Wood
Tachinus corticinus Gravenhorst	1967	Litter, moss, compost, agricultural fields (Schülke 2012)	SOM/ EOM
Tachinus rufipes (Linnaeus)	1949	Litter, moss (Campbell 1973)	SOM
Tachinus subterraneus (Linnaeus)	1978	Compost (Klimaszewski et al. 2013)	EOM
Tachyporus dispar (Paykull)	1927	Litter, agricultural fields, compost (Campbell 1979)	SOM/ EOM
Tachyporus nitidulus (Fabricius)	1832	Litter, agricultural fields, compost (Campbell 1979)	SOM/ EOM
Tachyporus transversalis Gravenhorst	1963	Wetlands, especially moss (Campbell 1979)	SOM
Habrocerinae			
Habrocerus capillaricornis (Gravenhorst)	1931	Litter (Assing and Wunderle 1995)	SOM
Trichophyinae			
Trichophya pilicornis (Gyllenhal)	1897	Litter (Ashe and Newton 1993)	SOM
Aleocharinae			
Aleochara bilineata Gyllenhal	1870	Agricultural fields (Klimaszewski 1984)	EOM
Aleochara curtula (Goeze)	1906	Dung, carrion (Klimaszewski 1984)	EOM
Aleochara fumata Gravenhorst	1852	Forests, rotting fungi, carrion, litter, moss (Klimaszewski 1984)	SOM/ EOM
Aleochara lanuginosa Gravenhorst	1893	Dung, carrion (Klimaszewski 1984)	EOM
Aleochara lata Gravenhorst	1802	Dung, carrion (Klimaszewski 1984)	EOM
Aleochara morion (Gravenhorst)	1889	Dung, compost, carrion (Klimaszewski et al. 2002)	EOM
Aleochara tristis Gravenhorst	1966	Dung, compost (Klimaszewski 1984)	EOM
Aleochara villosa Mannerheim	1891	Bird nests, old hay, carrion (Webster et al. 2009)	EOM
Alevonota gracilenta (Erichson)	2009	Soil/edaphic (Assing and Wunderle 2008)	SOM
Aloconota cambrica (Wollaston)	1907	Forest litter (Klimaszewski and Winchester 2002)	SOM
Aloconota sulcifrons (Stephens)	1889	Dung, compost, litter, moss (Webster et al. 2009)	EOM/ SOM
Amischa analis (Gravenhorst)	1878	Litter, moss, soil (Majka and Klimaszewski 2008a)	SOM

(continued)

Table 5.2 (continued)

Species	Detection	Microhabitat	Category
Atheta (Chaetida) longicornis (Gravenhorst)	1977	Dung, compost (Klimaszewski et al. 2007)	EOM
Atheta (Datomicra) celata (Erichson)	1907	Mammal and bird nests (Majka et al. 2006)	EOM
Atheta (Datomicra) dadopora Thomson	1910	Dung, decaying fungi, litter (Klimaszewski et al. 2010)	EOM/ SOM
Atheta (Datomicra) nigra (Kraatz)	2012[a]	Dung (Klimaszewski et al. 2015b)	EOM
Atheta (Dimetrota) atramentaria (Gyllenhal)	1949	Compost, dung (Forbes et al. 2016)	EOM
Atheta (Dimetrota) subtilis (Scriba)	2001[a]	Rotting fungi (Klimaszewski et al. 2015b)	EOM
Atheta (Microdota) amicula (Stephens)	1906	Compost, dung, rotting fungi (Majka and Klimaszewski 2008a)	EOM
Atheta (Rhagocneme) subsinuata (Erichson)	2005	Compost (Klimaszewski et al. 2008a)	EOM
Autalia puncticollis Sharp	1907	Dung, carrion (Assing 1997)	EOM
Autalia rivularis (Gravenhorst)	1928	Dung, carrion (Assing 1997)	EOM
Callicerus obscurus Gravenhorst	1985	Soil/edaphic (Assing 2001)	SOM
Callicerus rigidicornis (Erichson)	2009	Soil/edaphic (Assing 2001)	SOM
Cordalia obscura (Gravenhorst)	1905	Compost, dung, carrion, bird nests (Horion 1967)	EOM
Crataraea suturalis (Mannerheim)	1833	Dung, compost, carrion (Klimaszewski et al. 2007)	EOM
Dalotia coriaria (Kraatz)	1873	Compost, carrion, litter, bark (Klimaszewski et al. 2007)	EOM/ SOM
Diglotta mersa Haliday	2004	Intertidal gravel (Klimaszewski et al. 2008b)	SOM
Dinaraea angustula (Gyllenhal)	1889	Litter, soil, compost (Webster et al. 2009)	EOM/ SOM
Drusilla canaliculata (Fabricius)	1906	Soil/edaphic, moss, litter (Assing 1994)	SOM
Falagria caesa Erichson	1913	Compost, dung (Hoebeke 1985)	EOM
Geostiba circellaris (Gravenhorst)	1949	Soil, litter (Assing 2005)	SOM
Gymnusa brevicollis (Paykull)	1889	Wetland, sphagnum (Klimaszewski 1979)	SOM
Halobrecta flavipes Thomson	1910	Seaweed, drift (Klimaszewski et al. 2008b)	EOM
Homalota plana (Gyllenhal)	1863	Under bark (Klimaszewski et al. 2007)	Wood
Ilyobates bennetti (Donisthorpe)	1981	Litter, soil, moss (Assing 1999)	SOM
Meotica exilis (Gravenhorst)	1998	Litter, moss (Majka and Klimaszewski 2008a)	SOM
Meotica pallens (Redtenbacher)	1899	Soil, flood debris (Muona 1991)	SOM
Mocyta fungi (Gravenhorst)	1894	Litter, moss, soil (Klimaszewski et al. 2015c)	SOM
Myrmecocephalus concinnus Erichson	1906	Litter (Hoebeke 1985)	SOM
Nehemitropia lividipennis (Mannerheim)	1863	Compost, dung (Klimaszewski et al. 2007)	EOM
Placusa incompleta Sjöberg	1968	Under bark (Klimaszewski et al. 2001)	Wood
Placusa tachyporoides (Waltl)	1889	Under bark (Klimaszewski et al. 2001)	Wood
Oligota chrysopyga Kraatz	1976[a]	Compost (Webster et al. 2016)	EOM
Oligota inflata (Mannerheim)	2005[a]	Compost (Klimaszewski et al. 2016)	EOM
Oligota parva Kraatz	1889	Compost, dung, hay (Majka and Klimaszewski 2008a)	EOM
Oligota pusillima Gravenhorst	1889[a]	Compost (Webster et al. 2016)	EOM
Oxypoda brachyptera (Stephens)	1977	Litter, soil (Assing 2012a)	SOM
Oxypoda opaca (Gravenhorst)	1936	Compost, carrion, hay (Hoebeke 1990)	EOM
Philhygra hygrotopora (Kraatz)	2004[a]	Riparian, moss, rotting seaweed (Webster et al. 2016)	SOM
Philhygra luridipennis (Mannerheim)	1997	Riparian, moss (Webster et al. 2012a)	SOM

(continued)

Table 5.2 (continued)

Species	Detection	Microhabitat	Category
Phloeocharinae			
Phloeocharis subtilissima Mannerheim	2001	Under bark (Majka and Klimaszewski 2004)	Wood
Oxytelinae			
Anotylus insecatus (Gravenhorst)	1914	Compost (Schülke 2012)	EOM
Anotylus insignitus (Gravenhorst)	1863	Compost, dung (Hammond 1976)	EOM
Anotylus nitidulus (Gravenhorst)	1889	Wetland, compost (Schülke 2012)	EOM/ SOM
Anotylus rugosus (Fabricius)	1670	Dung/compost (Schülke 2012)	EOM
Anotylus tetracarinatus (Block)	1877	Dung/compost (Schülke 2012)	EOM
Carpelimus gracilis (Mannerheim)	1889[a]	Riparian, compost (Webster et al. 2016)	EOM/ SOM
Carpelimus pusillus (Gravenhorst)	1871[a]	Riparian (Webster et al. 2016)	SOM
Coprophilus striatulus (Fabricius)	1974	Compost, dung, mammal nests (Majka and Klimaszewski 2008a)	EOM
Deleaster dichrous Gravenhorst	1934	Edaphic, damp areas (Klimaszewski et al. 2013)	SOM
Oxytelus laqueatus (Marsham)	1843	Dung/compost (Schülke 2012)	EOM
Oxytelus sculptus Gravenhorst	1670	Dung/compost (Schülke 2012)	EOM
Platystethus degener Mulsant and Rey	1957	Riparian (Newton et al. 2001)	SOM
Scydmaeninae			
Cephennium gallicum Ganglbauer	1988	Litter, under bark (Majka and Klimaszewski 2004)	Wood
Euaesthetinae			
Edaphus lederi Eppelsheim	1914[a]	Compost (Webster et al. 2016)	EOM
Steninae			
Stenus clavicornis (Scopoli)	1968	Moss, soil, edaphic (Puthz 2012)	SOM
Stenus fulvicornis Stephens	1962	Wetlands, moss (Puthz 2012)	SOM
Stenus lustrator Erichson	1975	Wetlands, moss (Puthz 2012)	SOM
Stenus melanopus (Marsham)	1965	Soil/edaphic, shorelines (Puthz 2012)	SOM
Paederinae			
Lathrobium fulvipenne (Gravenhorst)	1968	Edaphic, fields (Assing 2012b)	SOM
Lithocharis nigriceps Kraatz	1945	Compost (Assing 2012b)	EOM
Lithocharis ochracea (Gravenhorst)	1886	Compost, dung (Assing 2012b)	EOM
Medon fusculus (Mannerheim)	1959	Litter (Assing 2004)	SOM
Ochthephilum fracticorne (Paykull)	1968	Wet litter, moss (Assing 2009a)	SOM
Pseudomedon obsoletus (Nordmann)	1941	Compost, wet litter (Assing 2009b)	EOM/ SOM
Rugilus angustatus Geoffroy	1950	Wet litter (Assing 2012c)	SOM
Rugilus ceylanensis (Kraatz)	1934	Compost, dung, carrion (Assing 2012c)	EOM
Rugilus orbiculatus (Paykull)	1885	Compost, litter (Assing 2012c)	EOM
Rugilus rufipes Germar	1971	Compost, litter (Assing 2012c)	EOM/ SOM
Scopaeus minutus Erichson	2002	Soil, edaphic (Boháč 1985)	SOM
Sunius melanocephalus (Fabricius)	1924	Soil, edaphic (Assing 2008)	SOM
Staphylininae			
Bisnius cephalotes (Gravenhorst)	1860	Dung, compost (Forbes et al. 2016), bird nests (Smetana 1995)	EOM
Bisnius fimetarius (Gravenhorst)	1949	Dung, carrion (Smetana 1995)	EOM
Bisnius parcus (Sharp)	1892	Bird nests, carrion, dung (Smetana 1995)	EOM
Bisnius sordidus (Gravenhorst)	1844	Dung, compost (Smetana 1995)	EOM

(continued)

Table 5.2 (continued)

Species	Detection	Microhabitat	Category
Creophilus m. maxillosus (Linnaeus)	1929	Carrion, rarely compost, dung (Clarke 2011)	EOM
Gabrius appendiculatus Sharp	1978	Riparian, edaphic (Smetana 1995)	SOM
Gabrius astutoides (A. Strand)	1936	Riparian, moss, litter (Smetana 1995)	SOM
Gabrius nigritulus (Gravenhorst)	1870	Compost but also moss, litter (Smetana 1995)	EOM/ SOM
Gabronthus thermarum (Aubé)	1874	Compost, dung, hay (Smetana 1995)	EOM
Gauropterus fulgidus (Fabricius)	1870	Compost, dung, carrion (Smetana 1982)	EOM
Gyrohypnus angustatus (Stephens)	1912	Compost, dung, hay (Smetana 1982)	EOM
Gyrohypnus fracticornis (Müller)	1884	Compost, dung, hay (Smetana 1982)	EOM
Leptacinus batychrus (Gyllenhal)	1880	Compost, hay (Smetana 1982)	EOM
Leptacinus intermedius Donisthorpe	1903	Compost, dung, hay (Smetana 1982)	EOM
Leptacinus pusillus (Stephens)	1874	Compost, dung, hay (Smetana 1982)	EOM
Neobisnius lathrobioides (Baudi di Selve)	1940	Riparian (Webster et al. 2012b)	SOM
Neobisnius villosulus (Stephens)	1860	Riparian (Majka and Klimaszewski 2008b)	SOM
Ocypus aeneocephalus (De Geer)	1932	Edaphic (Kleeberg and Uhlig 2011)	SOM
Ocypus nitens (Schrank)	1944[a]	Edaphic, litter (Newton 1987)	SOM
Ontholestes murinus (Linnaeus)	1949	Dung (Smetana 1981)	EOM
Othius subuliformis Stephens	2000	Litter (Assing 2003)	SOM
Phacophallus parumpunctatus (Gyllenhal)	1874	Compost, dung, hay (Smetana 1982)	EOM
Philonthus carbonarius (Gravenhorst)	1905	Edaphic, litter (Smetana 1995)	SOM
Philonthus caucasicus Nordmann	1910	Dung, compost, hay (Smetana 1995)	EOM
Philonthus cognatus Stephens	1884	Edaphic (Smetana 1995)	SOM
Philonthus concinnus (Gravenhorst)	1909	Edaphic, fields (Smetana 1995)	SOM
Philonthus cruentatus (Gmelin)	1907	Compost, fungi, dung, carrion (Smetana 1995)	EOM
Philonthus debilis (Gravenhorst)	1957	Compost, hay (Smetana 1995)	EOM
Philonthus discoideus (Gravenhorst)	1844	Compost, dung (Smetana 1995)	EOM
Philonthus jurgans Tottenham	1881	Compost, dung, decaying fungi (Smetana 1995)	EOM
Philonthus longicornis Stephens	1840	Compost, dung (Smetana 1995; Forbes et al. 2016)	EOM
Philonthus politus (Linnaeus)	1670	Compost, dung, carrion (Smetana 1995)	EOM
Philonthus rectangulus Sharp	1907	Compost, dung, carrion (Smetana 1995)	EOM
Philonthus sanguinolentus (Gravenhorst)	1962	Dung, burrows, litter (Smetana 1995)	EOM/ SOM
Philonthus tenuicornis (Mulsant and Rey)	1979	Carrion, compost, dung, rotting fungi (Smetana 1995)	EOM
Philonthus umbratilis (Gravenhorst)	1860	Litter and compost, often near water (Smetana 1995)	EOM/ SOM
Philonthus varians (Paykull)	1889	Compost, dung, carrion (Smetana 1995)	EOM
Philonthus ventralis (Gravenhorst)	1802	Compost, dung, riparian litter (Smetana 1995)	EOM/ SOM
Quedius cinctus (Paykull)	1942	Compost, dung, fungi (Smetana 1971)	EOM
Quedius cruentus (Olivier)	1983	Woody debris, bark, litter, fungi (Brunke and Marshall 2011)	Wood
Quedius curtipennis Bernhauer	1934	Soil, edaphic, moss (Webster et al. 2012b)	SOM
Quedius fulgidus (Fabricius)	1834	Dung, hay, nests (Smetana 1971)	EOM
Quedius fuliginosus (Gravenhorst)	1996	Soil, edaphic, moss (Majka and Smetana 2007)	SOM
Quedius mesomelinus (Marsham)	1620	Dung, carrion, hay, fungi, compost, nests (Smetana 1971)	EOM

(continued)

Table 5.2 (continued)

Species	Detection	Microhabitat	Category
Quedius molochinus (Gravenhorst)	1949	Soil, edaphic (Majka and Smetana 2007)	SOM
Tasgius ater (Gravenhorst)	1802	Edaphic (Brunke et al. 2011)	SOM
Tasgius melanarius (Heer)	1935	Edaphic (Brunke et al. 2011)	SOM
Tasgius winkleri (Bernhauer)	1931	Edaphic (Brunke et al. 2011)	SOM
Xantholinus elegans (Olivier)	2007	Edaphic, agricultural fields (Assing 2012b)	SOM
Xantholinus linearis (Olivier)	1930	Edaphic, litter, fields, intercepted in moss (Smetana 1982)	SOM
Xantholinus longiventris Heer	1931	Edaphic, litter, fields (Smetana 1982)	SOM

[a]since Klimaszewski et al. (2013)
EOM ephemeral organic matter; *SOM* stable organic matter; *wood* bark or rotting woody matter

mainland of British Columbia (8 spp.); Montreal and vicinity, Quebec (8 spp.); Toronto and the western part of Lake Ontario, Ontario (3 spp.); and Quebec City and vicinity, Quebec (6 spp.) (Klimaszewski et al. 2013). Of the 68 species that appear to have spread to Canada from original points of introduction in the USA, the majority spread across the border into the lower mainland of BC or into southern Ontario and Quebec (Klimaszewski et al. 2013).

5.3 Methods of Introduction

Several introduction pathways into North America have been proposed for the adventive Canadian Staphylinidae including dry ballast, livestock bedding, wood, moss, and organic matter around the roots of plant stock (Lindroth 1957; Majka and Klimaszewski 2008a; Klimaszewski et al. 2013). Although it is nearly impossible to pinpoint the exact pathway for adventive staphylinids, we aimed to understand patterns of introduction pathways at a coarser level by using the earliest detection date available and ecological preferences for microhabitats. These microhabitats were broadly categorized into three groups: ephemeral, nutrient-rich organic matter (EOM), more stable organic matter (including soil) (SOM), and wood. Examples of EOM include compost or rotting exposed vegetation, dung, and carrion. Species regularly occurring in the waste piles of burrows or nests were included under EOM. Examples of SOM include leaf litter, moss, and soil; general soil surface-dwelling species were also included here. The wood category was reserved for those species specialized in living under the bark of trees or in rotting logs. A small percentage of adventive staphylinids are extremely eurytopic and were scored for multiple categories (11%). Potential introduction pathways for EOM-associated staphylinids include livestock bedding, manure, spoiled produce, and decaying livestock feed. SOM-associated staphylinids may have been accidentally introduced via dry ballast, moss or the soil, and other organic matter around roots or in pots of imported plants used in landscaping, urban forestry, the greenhouse industry, and agriculture. Woody debris-associated staphylinids may have entered with bark-bearing lumber or packing material, moldy woodchips, or rotting firewood.

An approximate introduction history of Canada's adventive Staphylinidae is summarized in Fig. 5.2. The earliest rove beetle introductions (1620–1850) into North America are known mainly from some of the first literature reporting on North American staphylinids and from seventeenth-century subfossils discovered in early human settlements (Bain and Prévost 2010). This cohort of 12 species is heavily dominated by those associated with EOM and typical of barnyard dung and decomposing livestock bedding (e.g., *Quedius mesomelinus*

Table 5.3 Staphylinidae omitted from the list of adventive species in Canada

Species	Reason
Omaliinae	
Acrolocha minuta (Olivier)	No specimen vouchers
Xylodromus depressus (Gravenhorst)	No specimen vouchers
Proteininae	
Proteinus atomarius Erichson	No specimen vouchers
Tachyporinae	
Bolitobius cingulatus Mannerheim	Possibly Holarctic, occurs in Far East Russia (Schülke and Smetana 2015)
Aleocharinae	
Atheta subrugosa Märkel & Kiesenwetter	No specimen vouchers
Atheta vestita (Gravenhorst)	Possibly amphi-Atlantic, occurs in Iceland and Greenland (Böcher 1988; Schülke and Smetana 2015)
Dochmonota rudiventris (Eppelsheim)	Possibly Holarctic, occurs in Far East Russia (Schülke and Smetana 2015)
Gyrophaena affinis (Sahlberg)	Possibly Holarctic, occurs in Far East Russia (Enushchenko and Semenov 2016)
Oxypoda operta Sjöberg	Possibly Holarctic, occurs in Far East Russia (Schülke and Smetana 2015)
Philhygra palustris Märkel & Kiesenwetter	Possibly Holarctic, occurs in Far East Russia (Schülke and Smetana 2015)
Oxytelinae	
Carpelimus bilineatus Stephens	No specimen vouchers, inadequate taxonomy
Carpelimus obesus (Kiesenwetter)	No specimen vouchers, inadequate taxonomy
Carpelimus subtilis (Erichson)	Possibly Holarctic, occurs in Far East Russia (Schülke and Smetana 2015)
Paederinae	
Lobrathium longiusculum (Gravenhorst)	Nearctic, listed as adventive by error in Bousquet et al. (2013)
Lobrathium multipunctum (Gravenhorst)	No specimen vouchers

(Marsham), *Philonthus politus* (Linnaeus), *Anotylus rugosus* (Fabricius), *Oxytelus sculptus* Gravenhorst, *Aleochara lata* Gravenhorst, and *Crataraea suturalis* (Mannerheim)). Some of the earlier introductions of SOM-associated staphylinids may have been via dry ballast carried by ships coming from Europe and later deposited on the shores of Maritime Canada and New England (Majka et al. 2008). Lindroth (1957) stated that this potentially important source of adventive species continued until just after World War I (WWI) and would have favored species typical of dry, vegetated waste earth (e.g., old fields), which was the source of most dry ballast. Staphylinids that were likely imported by dry ballast and also found in historical ballast source sites by Lindroth (1957) include *Drusilla canaliculata* (Fabricius), *Dinaraea angustula* (Gyllenhal), *Tachyporus nitidulus* (Fabricius), *Tasgius ater* (Gravenhorst), *Quedius curtipennis* Bernhauer, and *Philonthus concinnus* (Gravenhorst). Moisture-loving species and those detected much later than WWI are unlikely to have reached North America via ballast. Two hygrophilous species, *Neobisnius villosulus* (Stephens) and *Gymnusa brevicollis* (Paykull), were relatively early introductions into North America (late nineteenth century) and were most likely imported with moss. Two species, *Aleochara bilineata* Gyllenhal and *A. tristis* Gravenhorst, were intentionally introduced as biocontrol agents (Klimaszewski 1984; Soroka et al.

2002), but the former was present in North America far before this. It is unclear whether existing populations of the latter species are the result of later accidental introductions as they were not subsequently detected near release sites (Legner 1978).

5.4 Trends in Adventive Canadian Staphylinidae

Over time, the dominance of ephemeral organic matter (EOM)-associated staphylinids among introductions has shifted to a dominance of stable organic matter (SOM)-associated staphylinids, with some recent decades entirely lacking EOM species (Fig. 5.2). This shift may be associated with improved sanitary requirements for goods imported into North America. Unlike in other beetle groups, woody material appears to have played a minor and sporadic role as a pathway for adventive Staphylinidae in Canada. One reason could be that wood-associated staphylinids in the north temperate region typically occur under bark, which must be removed from wood upon import to Canada (CFIA 2011). Modern (post-1950) introductions appear to be predominantly associated with SOM, and many of these species are specialists of cool, permanently wet microhabitats [e.g., *Ochthephilum fracticorne* (Paykull), *Stenus lustrator* Erichson, *Tachyporus transversalis* Gravenhorst, *Philhygra hygrotopora* (Kraatz)]. The increasing rapidity of modern trade and the ability to climate control shipments may have aided the survival of these species, which would otherwise have desiccated. Other recently detected SOM-associated species are primarily soil surface-dwelling species (e.g., *Oxypoda brachyptera* (Stephens), *Quedius fuliginosus* (Gravenhorst), *Scopaeus minutus* (Erichson)) or soil crevice-dwelling species (e.g., *Alevonota gracilenta* (Erichson), two species of *Callicerus*) that may have been introduced via soil among the roots of plant stock destined for agriculture, urban forestry, landscaping, or floriculture, as in some click beetles (Elateridae) (Douglas 2011).

5.5 Future Directions

Detections of adventive species have not markedly slowed down over the centuries of Canada's international trade history, though we acknowledge that detections may occur much later than the establishment of a species. Of the 13 species newly detected in North America since the start of 2000, two are known only from the Yukon and two from Saskatchewan, regions that have been poorly sampled historically. It is likely that these species have been locally distributed for some time and that traces of their earlier presence were lost. One species, *Diglotta mersa* Haliday (present at least by 2004), is a flightless specialist of intertidal gravel along the Atlantic seacoast (Klimaszewski et al. 2008b), and it may have gone undetected in this infrequently collected microhabitat for many years. Both *Oligota inflata* Mannerheim (present at least by 2005) and *Scopaeus minutus* (present at least by 2002) belong to genera that are unrevised in the Nearctic region. This lack of taxonomic knowledge means that the large series of unidentified specimens of North American *Oligota* and *Scopaeus* in collections may eventually reveal more accurate introduction histories for these species and, potentially, may lead to the detection of additional adventive species. However, the majority of the most recent detection dates concern species in taxonomically well-known groups in the Nearctic region (e.g., *Sepedophilus*), or genera native only to the Old World (*Xantholinus*, *Alevonota*, *Callicerus*), which are therefore distinctive among the Canadian fauna. Several species were first detected in suburban and rural southern Ontario (*Xantholinus elegans* (Olivier), *Alevonota gracilenta*, two *Callicerus* species), which has a long history of insect collecting in both natural and anthropogenic areas (deposited mainly at the

University of Guelph and in the Canadian National Collection of Insects, Arachnids and Nematodes). Despite sampling artifacts, it is clear that at least some staphylinids have recently and will continue to become adventive in North America.

Achieving rapid and accurate detection of adventive rove beetles in Canada faces several challenges. Many staphylinids in North America belong to genera that have never been comprehensively revised in the Nearctic, making it difficult to identify and monitor native and non-native species for these groups or to determine whether species are adventive or naturally Holarctic; these include *Anotylus*, *Carpelimus*, *Oligota*, *Rugilus*, and *Scopaeus*. Regular sampling near Canadian ports such as St. John's (Newfoundland), Montreal and Quebec City (Quebec), Toronto (Ontario), and Victoria (British Columbia) may provide more detailed introduction timelines and reduce detection lag times. The classic Malaise trap survey protocol commonly used in Canada, while efficient for many flying insect groups, may not be sufficient to detect adventive staphylinids, unless preservative-filled trays are added underneath Malaise traps to create a flight intercept trap. Staphylinids typically drop downward upon impact with the trap and rarely fly upward to a Malaise trap head. Although staphylinids can be captured by Malaise traps, the sample captured in flight intercept traps is less taxonomically biased toward a few taxa and tends to contain far more individuals. The identification of samples from surveys is further challenged by the fact that only adult males are identifiable in many staphylinid groups. In such cases, adult females, larvae, and pupae are best identified by comparing their DNA with a comprehensive online molecular reference library. The Canadian National Collection of Insects, Arachnids and Nematodes (Ottawa, Canada) and the Laurentian Forestry Centre (Quebec City, Canada) have recently collaborated with the International Barcode of Life Project (University of Guelph, Guelph, Canada) to increase coverage of the barcode reference library used by the Barcode of Life Data System (BOLD, http://www.Boldsystems.org) for Canadian beetles (summarized in Bouchard et al. 2017). However, the BOLD database is far from being comprehensive for the diverse and poorly known Canadian Staphylinidae as less than 50% of the recorded species are included and even fewer are represented by multiple specimens (Bouchard et al. 2017). Improving the detection of adventive staphylinids, and insects in general, will require an integrated approach involving increased support for and implementation of taxonomic research, regular, efficient sampling in regions of high commercial trade, and the continued development of well-vouchered molecular reference libraries (e.g., BOLD).

References

Ashe JS, Newton AF (1993) Larvae of *Trichophya* and phylogeny of the tachyporine group of subfamilies (Coleoptera: Staphylinidae) with a review, new species and characterization of the Trichophyinae. Syst Entomol 18:267–286

Assing V (1994) Zur Kurzflügelkäferfauna xerothermer Flächen im südlichen Niedersachsen (Coleoptera: Staphylinidae). Göttinger Naturkundliche Schriften 3:7–31

Assing V (1997) Review of the Palaearctic species of *Autalia* Leach in Samouelle, 1819 (Coleoptera, Staphylinidae, Aleocharinae). Entomologische Blätter 93:69–85

Assing V (1999) A revision of *Ilyobates* Kraatz, 1856 (Coleoptera: Staphylinidae, Aleocharinae, Oxypodini). Beiträge zur Entomologie 49:295–342

Assing V (2001) A revision of *Callicerus* Gravenhorst, 1802, *Pseudosemiris* Machulka, 1935 and *Saphocallus* Sharp, 1888. Beiträge zur Entomologie 51:247–334

Assing V (2003) A revision of Othiini. XIII. Horizontal and vertical distribution of *Othius*, new species, and additional records (Coleoptera: Staphylinidae: Staphylininae). Entomol Probl 33:69–88

Assing V (2004) A revision of the *Medon* species of the Eastern Mediterranean and adjacent regions (Insecta: Coleoptera: Staphylinidae: Paederinae). Bonner Zoologische Beiträge 52:33–82

Assing V (2005) A revision of the species of *Geostiba* Thomson and *Tropimenelytron* Pace of the Eastern Mediterranean, the Caucasus, and adjacent regions (Coleoptera: Staphylinidae, Aleocharinae). Linzer Biologische Beiträge 37:903–1006

Assing V (2008) A revision of the *Sunius* species of the Western Palaearctic region and Middle Asia (Coleoptera: Staphylinidae: Paederinae). Linzer Biologische Beiträge 40:5–135

Assing V (2009a) On the western Palaearctic and middle Asian species of *Ochthephilum* Stephens, with notes on *Cryptobium koltzei* Eppelsheim (Coleoptera: Staphylinidae: Paederinae: Cryptobiina). Linzer Biologische Beiträge 41:397–426

Assing V (2009b) On the *Pseudomedon* species of the Palaearctic region (Coleoptera: Staphylinidae: Paederinae). Linzer Biologische Beiträge 41:1175–1189

Assing V (2012a) On the taxonomy and natural history of *Oxypoda brachyptera* and *O. tarda* (Coleoptera: Staphylinidae: Aleocharinae). Beiträge zur Entomologie 62:207–224

Assing V (2012b) Unterfamilie Trichophyinae Thomson, 1859 [pp 199–200]; Unterfamilie Habrocerinae Mulsant & Rey, 1877 [p 200]; Unterfamilie Paederinae Fleming, 1821 [pp 322–369, 380–383]; Tribus Xantholinini Erichson, 1839 [pp 507–526]. In: Assing V, Schülke M (eds) Freude–Harde–Lohse–Klausnitzer Die Käfer Mitteleuropas. Band 4. Staphylinidae (exklusive Aleocharinae, Pselaphinae, und Scydmaeninae), 2 Auflage. Spektrum Akademischer Verlag, Heidelberg, XII + 560 pp

Assing V (2012c) The *Rugilus* species of the Palaearctic and Oriental regions (Coleoptera: Staphylinidae: Paederinae). Stuttgarter Beiträge zur Naturkunde A, Neue Serie 5:115–190

Assing V, Schülke M (2012) Freude-Harde-Lohse-Klausnitzer-Die Käfer Mitteleuropas. Band 4. Staphylinidae (exklusive Aleocharinae, Pselaphinae und Scydmaeninae), 2 Auflage. Spektrum Akademischer Verlag, Heidelberg, XII + 560 pp

Assing V, Wunderle P (1995) A revision of the species of the subfamily Habrocerinae (Coleoptera: Staphylinidae) of the world. Revue suisse de Zoologie 102:307–359

Assing V, Wunderle P (2008) On the *Alevonota* species of the western Palaearctic region (Coleoptera: Staphylinidae: Aleocharinae: Athetini). Beiträge zur Entomologie 58:145–189

Bain A, Prévost M-A (2010) Environmental archeology and landscape transformation at the seventeen-century Ferryland Site, Newfoundland. Hist Archeol 44 (3):21–35

Böcher J (1988) The Coleoptera of Greenland. Meddelelser om Grønland, Bioscience 26. Commission for Scientific Research in Greenland, Copenhagen, Copenhagen, 100 pp

Boháč J (1985) Review of the subfamily Paederinae (Coleoptera, Staphylinidae) in Czechoslovakia. Acta Entomologica Bohemoslovaka 82:360–385

BOLD Barcode of life datasystems v 3. http://www.Boldsystems.org. Accessed 20 Oct 2016

Bouchard P, Smith ABT, Douglas H et al (2017) Biodiversity of Coleoptera. In: Footit B, Adler PH (eds) Insect biodiversity: science and society. Wiley-Blackwell, Oxford

Bousquet Y, Bouchard P, Davies AE et al (2013) Checklist of beetles (Coleoptera) of Canada and Alaska. Pensoft, Sofia-Moscow, 402 pp

Brunke AJ, Marshall SA (2011) Contributions to the faunistics and bionomics of Staphylinidae (Coleoptera) in northeastern North America: discoveries made through study of the University of Guelph Insect Collection, Ontario, Canada. ZooKeys 75:29–68

Brunke A, Newton A, Klimaszewski J et al (2011) Staphylinidae of eastern Canada and adjacent United States. Tribes and subtribes, and species of Staphylinina. Can J Arthropod Identif 12:1–110

Campbell JM (1973) A revision of the genus *Tachinus* (Coleoptera: Staphylinidae) of North and Central America. Mem Entomol Soc Can 90:1–137

Campbell M (1976) A revision of the genus *Sepedophilus* Gistel (Coleoptera: Staphylinidae) of America north of Mexico. Mem Entomol Soc Can 99:1–89

Campbell JM (1979) A revision of the genus *Tachyporus* Gravenhorst (Coleoptera: Staphylinidae) of North and Central America. Mem Entomol Soc Can 109:1–95

CFIA (Canadian Food Inspection Agency) (2011) Directive D-02-12: import requirements for non-processed wood and other non-propagative wood products, except solid wood packaging material, from all areas other than the continental United States. http://www.inspection.gc.ca/plants/plant-pests-invasive-species/directives/forestry/d-022/eng/1312383927173/1312384324372. Accessed 20 Oct 2016

Clarke D (2011) Testing the phylogenetic utility of morphological character systems, with a revision of *Creophilus* Leach (Coleoptera: Staphylinidae). Zool J Linn Soc 163:723–812

Douglas H (2011) New records of European wireworm pests and other click beetles (Coleoptera: Elateridae) in Canada and USA. J Entomol Soc Ont 142:11–17

Enushchenko IV, Semenov VB (2016) A review of the genus *Gyrophaena* Mannerheim 1830 (Coleoptera: Staphylinidae: Aleocharinae: Gyrophaenina) of the Caucasus and adjacent territories. Zootaxa 4126:301–337

Forbes V, Dugmore A, Olafsson E (2016) The life and death of barn beetles: faunas from manure and stored hay inside farm buildings in northern Iceland. Ecol Entomol 41:480–499

Gravenhorst JLC (1802) Coleoptera Microptera Brunsvicensia nec non exoticorum quotquot exstant in collectionibus entomologorum Brunsvicensium in genera familias et species distribuit. Carolus Reichard, Brunsuigae, lxvi + 206 pp

Hammond PM (1976) A review of the genus *Anotylus* C.G. Thomson (Coleoptera: Staphylinidae). Bull Br Entomol Soc 33:139–187

Hoebeke ER (1985) A revision of the rove beetle tribe Falagriini of America north of Mexico (Coleoptera: Staphylinidae: Aleocharinae). J N Y Entomol Soc 93:913–1018

Hoebeke ER (1990) First record of the Palaearctic species *Oxypoda opaca* (Gravenhorst) from North America (Coleoptera: Staphylinidae: Aleocharinae). J N Y Entomol Soc 97:448–454

Horion AD (1967) Faunistik der mitteleuropäischen Käfer. Band II. Staphylinidae. 3. Habrocerinae bis

Aleocharinae (Ohne Subtribus Athetae). P.C.W. Schmidt, Überlingen-Bodensee, xxiv + 419 pp

Kleeberg A, Uhlig M (2011) Die Staphylinina (Insecta: Coleoptera: Staphylinidae) in Mecklenburg-Vorpommern, 1847–2009: Erforschungsgeschichte, kommentierte, Artenliste, Verbreitung und Entwurf einer Roten Liste. Insecta, Zeitschrift für Entomologie und Naturschutz 13:5–137

Klimaszewski J (1979) A revision of the Gymnusini and Deinopsini of the world. Coleoptera: Staphylinidae, Aleocharinae. Agriculture Canada research branch Monograph No. 25, 169 pp

Klimaszewski J (1984) A revision of the genus *Aleochara* Gravenhorst of America north of Mexico (Coleoptera: Staphylinidae: Aleocharinae). Mem Entomol Soc Can 129:1–211

Klimaszewski J, Winchester N (2002) Aleocharine rove beetles (Coleoptera Staphylinidae) of the ancient Sitka spruce forest on Vancouver Island, British Columbia, Canada. Mémoires de la Société Royale Belge d'Entomologie 40:3–126

Klimaszewski J, Pelletier G, Germain C et al (2001) Diversity of *Placusa* (Coleoptera: Staphylinidae, Aleocharinae) in Canada, with descriptions of two new species. Can Entomol 133:1–47

Klimaszewski J, Pelletier G, Sweeney J (2002) Genus *Tinotus* (Coleoptera: Staphylinidae, Aleocharinae) from America north of Mexico: review of the types, distribution records, and key to species. Can Entomol 134:281–298

Klimaszewski J, Assing V, Majka CG et al (2007) Records of adventive aleocharine beetles (Coleoptera: Staphylinidae: Aleocharinae) found in Canada. Can Entomol 139:54–79

Klimaszewski J, Godin B, Pelletier G et al (2008a) Six new species and records of aleocharine beetles from Yukon and Alaska (Coleoptera: Staphylinidae: Aleocharinae). Can Entomol 140:265–291

Klimaszewski J, Webster RP, Assing V et al (2008b) *Diglotta mersa* (Haliday) and *Halobrecta flavipes* Thomson, two new species for the Canadian fauna (Coleoptera, Staphylinidae, Aleocharinae). ZooKeys 2:175–188

Klimaszewski J, Langor D, Majka CG et al (2010) Review of adventive species of Coleoptera (Insecta) recorded from eastern Canada, Pensoft Series Faunistica No. 94. Pensoft, Sofia, 272 pp

Klimaszewski J, Brunke A, Assing V et al (2013) Synopsis of adventive species of Coleoptera (Insecta) recorded from Canada. Part 2: Staphylinidae, Pensoft Series Faunistica No. 104. Pensoft, Sofia, 360 pp

Klimaszewski J, Godin B, Langor D et al (2015a) New distribution records for Canadian Aleocharinae (Coleoptera, Staphylinidae), and new synonymies for *Trichiusa*. ZooKeys 498:51–91

Klimaszewski J, Webster RP, Sikes D et al (2015b) A review of Canadian and Alaskan species of the genera *Clusiota* Casey and *Atheta* Thomson, subgenus *Microdota* Mulsant and Rey (Coleoptera, Staphylinidae, Aleocharinae). ZooKeys 524:103–136

Klimaszewski J, Webster R, Bourdon C et al (2015c) Review of Canadian species of the genus *Mocyta* Mulsant & Rey (Coleoptera, Staphylinidae, Aleocharinae), with the description of a new species and a new synonymy. ZooKeys 487:111–139

Klimaszewski J, Labrecque M, Bourdon C (2016) Twelve new species and fifty three new provincial distribution records of Aleocharinae rove beetles of Saskatchewan, Canada (Coleoptera, Staphylinidae). ZooKeys 610:45–112

Legner EF (1978) Muscidae—face fly. In: Clausen, C.P. (ed.). Introduced parasites and predators of arthropod pests and weeds: a world review, vol 480. USDA Agriculture Handbook, Washington, DC, pp 354–355, 545

Lindroth CH (1957) The faunal connections between Europe and North America. Wiley, New York, 344 pp

Majka CG, Klimaszewski J (2004) *Phloeocharis subtilissima* Mannerheim (Staphylinidae: Phloeocharinae) and *Cephennium gallicum* Ganglbauer (Scydmaenidae) new to North America: a case study in the introduction of exotic Coleoptera to the port of Halifax, with new records of other species. Zootaxa 781:1–15

Majka CG, Klimaszewski J (2008a) Introduced Staphylinidae (Coleoptera) in the maritime provinces of Canada. Can Entomol 140:48–72

Majka CG, Klimaszewski J (2008b) Adventive Staphylinidae (Coleoptera) of the maritime provinces of Canada: further contributions. ZooKeys 2:151–174

Majka CG, Smetana A (2007) New records of introduced species of *Quedius* Stephens, 1829 (Coleoptera: Staphylinidae) from the Maritime Provinces of Canada. Proc Entomol Soc Wash 109:427–434

Majka CG, Klimaszewski J, Lauff RF (2006) New Coleoptera records from owl nests in Nova Scotia, Canada. Zootaxa 1194:33–47

Majka CG, Klimaszewski J, Lauff RF (2008) The coastal rove beetles (Coleoptera, Staphylinidae) of Atlantic Canada: a survey and new records. ZooKeys 2:115–150

Muona J (1991) The North European and British species of the genus *Meotica* Mulsant & Rey (Coleoptera, Staphylinidae). Deutsche Entomologische Zeitschrift (N.F.) 38:225–246

Newton AF Jr (1987) Four *Staphylinus* (sensu lato) species new to North America, with notes on other introduced species (Coleoptera: Staphylinidae). Coleopt Bull 41:381–384

Newton A, Thayer MK, Ashe JS et al (2001) 22. Staphylinidae Latreille, 1802. In: Arnett RH, Thomas MC (eds) American beetles, vol 1. CRP Press, Boca Raton, pp 272–418

Puthz V (2012) Unterfamilie Steninae MacLeay, 1825. In: Assing V, Schülke M (eds) Freude–Harde–Lohse–Klausnitzer – Die Käfer Mitteleuropas. Band 4. Staphylinidae, 2 Auflage. Spektrum Akademischer Verlag, Heidelberg, XII + 560 pp

Schülke M (2011) Zur identität von *Sepedophilus immaculatus* Stephens (Coleoptera, Staphylinidae, Tachyporinae). Linzer biologische Beiträge 43:1609–1615

Schülke M (2012) Unterfamilie Tachyporinae MacLeay, 1825 [pp 130–199]; Unterfamilie Oxytelinae Fleming, 1821 [pp 207–266, 283–284]. In: Assing V, Schülke M (eds) Freude–Harde–Lohse–Klausnitzer – Die Käfer Mitteleuropas. Band 4. Staphylinidae, 2 Auflage. Spektrum Akademischer Verlag, Heidelberg, XII + 560 pp

Schülke M, Smetana A (2015) Staphylinidae. In: Löbl I, Löbl D (eds) Catalogue of Palaearctic Coleoptera. Brill, Leiden, pp 901–1134

Smetana A (1971) Revision of the tribe Quediini of North America north of Mexico (Coleoptera: Staphylinidae). Memoirs of the Entomological Society of Canada No. 79, vi + 303 pp

Smetana A (1981) *Ontholestes murinus* (Linné 1758) in North America (Coleoptera: Staphylinidae). Coleopt Bull 35:125–126

Smetana A (1982) Revision of the subfamily Xantholininae of America north of Mexico (Coleoptera: Staphylinidae). Memoirs of the Entomological Society of Canada No. 120, iv + 389 pp

Smetana A (1995) Rove beetles of the subtribe Philonthina of America north of Mexico (Coleoptera: Staphylinidae): classification, phylogeny and taxonomic revision, Memoirs on Entomology. Associated Publishers, Madison, 964 pp

Soroka JJ, Kuhlmann U, Floate KD et al (2002) *Delia radicum* (L.), cabbage maggot (Diptera: Anthomyiidae). In: Mason PG, Huber PG (eds) Biological control programmes in Canada, 1981–2000. CABI Publishing, Wallingford, pp 99–104

Wagner JA (1975) Review of the genera *Euplectus*, *Pycnoplectus*, *Leptoplectus*, and *Acolonia* (Coleoptera: Pselaphidae) including Nearctic species north of Mexico. Entomol Am 49:125–207

Webster RP, Klimaszewski J, Pelletier G et al (2009) New Staphylinidae (Coleoptera) records with new collection data from New Brunswick, Canada. I. Aleocharinae. ZooKeys 22:171–248

Webster RP, Klimaszewski J, Sweeney JD et al (2012a) New Staphylinidae (Coleoptera) records with new collection data from New Brunswick and an addition to the fauna of Quebec: Aleocharinae. ZooKeys 186:83–118

Webster RP, Smetana A, Sweeney JD, DeMerchant I (2012b) New Staphylinidae (Coleoptera) records with new collection data from New Brunswick and an addition to the fauna of Quebec: Staphylininae. ZooKeys 186:293–348

Webster RP, Klimaszewski J, Bourdon C et al (2016) Further contributions to the Aleocharinae (Coleoptera, Staphylinidae) fauna of New Brunswick and Canada including descriptions of 27 new species. ZooKeys 573:85–216

Zanetti A (2012) Unterfamilie Omaliinae MacLeay, 1825 [pp 49–117] – Unterfamilie Proteininae Erichson, 1839 [pp 117–123]. In: Assing V, Schülke M (eds) Freude–Harde–Lohse–Klausnitzer – Die Käfer Mitteleuropas. Band 4. Staphylinidae I. Zweite neubearbeitete Auflage. Spektrum Akademischer Verlag. Heidelberg, XII + 560 pp

Systematics, Natural History, and Evolution of the Saw-Lipped Rove Beetles (Euaesthetinae): Progress and Prospects for Future Research

6

Dave J. Clarke

Abstract

The rove beetle subfamily Euaesthetinae is reviewed and information on the systematics, ecology, and evolution presented. Key morphological features of adults and larvae are discussed, and the current state of morphology-based phylogenetics and paleontological research is reviewed. Natural history information is compiled for most genera, and general ecological trends are highlighted. Euaesthetinae are probably monophyletic but with a suprageneric taxonomic structure likely poorly reflected by the current classification. They are nearly globally distributed in most habitats, and collection data suggests that their ecological diversity is not yet fully known or confirmed. The southern hemisphere and high-elevation faunas globally comprise mostly flightless species restricted to ground litter of diverse habitats. A division into groups extending from the general ground litter into either soil (endogenous) or aboveground habitats (mediated by high-moisture microhabitats, typically dense bryophyte growths) is suggested. Although Euaesthetinae are generally found in mesic habitats, a group of seemingly "periaquatic" taxa are primarily found in Holarctic riparian and wetland sites. Probable "surface runners" and arboreal (foliage-dwelling) species form two other (overlapping) ecological groupings, and the occurrence of some species in vertebrate and ant nests requires further investigation. Biological inferences are drawn from several different morphological features of the group suggesting diverse life histories for these tiny beetles. Updated fossil information is provided, and this indicates needed taxonomic changes and suggests a greater extinct taxonomic diversity than previously known. The fossil record and ecology of the group suggest that euaesthetine lineages are resistant to extinction over geological time making these beetles well-suited to historical biogeographic studies.

6.1 Introduction

Euaesthetinae (Fig. 6.1), or "saw-lipped rove beetles" (on account of most included species having a serrate or denticulate labral margin), are a poorly known group of predatory ground-dwelling staphylinids occurring in forests and other habitats worldwide. Herman (2001) listed 724 species in 26 genera (762 spp. in Thayer 2005), with the current total standing at 1155 species (A. Newton unpub.). Although it is therefore a small group compared with some of the

D. J. Clarke (✉)
Department of Biological Sciences, University of Memphis, Memphis, TN, USA
e-mail: dclarke@fieldmuseum.org

O. Betz et al. (eds.), *Biology of Rove Beetles (Staphylinidae)*,
https://doi.org/10.1007/978-3-319-70257-5_6

larger subfamilies, some genera present notable radiations. As most species are cryptic, occurring primarily in the leaf litter and similar microhabitats, finding them requires specialized collecting techniques, and so they are only rarely encountered in the field. Unsurprisingly, this practical issue has contributed to a much slower rate of increase in biological and ecological knowledge compared to related groups like Steninae for which significant biological knowledge has accumulated (e.g., Betz 1998a; Chap. 11 of this book). To date there has not been a focused study on any aspect of the biology of Euaesthetinae, and there may be only a single literature report on any aspect of their life history (Remillet 1969; an *Octavius* Fauvel, 1873 specimen fed diplurans for ~1 month). Available information is scattered in the primary literature and consists of specimen-level collection data reported in the context of taxonomic studies, with some larger contributions presenting valuable in-depth generalizations about the ecology of regional faunas (e.g., Orousset 1988, 2012). The development of evolutionary research on this group has therefore been impacted by this general paucity of biological information and has been further affected by both a lack of synthetic systematic studies and a slower discovery of well-preserved fossils than for other staphylinid groups. Given their cryptic habits and undersampling, the extant global diversity of Euaesthetinae is undoubtedly much greater than currently known. Systematic research is just beginning to advance beyond alpha taxonomic work on (mainly) the larger widespread genera following preliminary work by Leschen and Newton (2003), Clarke and Grebennikov (2009), and others to develop sets of phylogenetically informative morphological characters. Paleontological research on Euaesthetinae (and the closely related group Steninae) is in its infancy, but some notable recent papers reported on the existence of morphologically derived (crown-group) taxa as early as the Early Cretaceous, ~135–100 Ma (Lefebvre et al. 2005; Clarke and Chatzimanolis 2009). The author's ongoing work on the comparative morphology of Euaesthetinae (especially of their mouthparts), inter- and infra-generic morphological diversity, new fossil discoveries, and the synthesis of information presented herein confirms that the group is much more biologically and ecologically diverse than previously thought.

6.1.1 Recognition of Adults and Larvae of Euaesthetinae

Most species of Euaesthetinae are minute, ~1–2.5 mm long, but with size outliers occurring in both directions, such as in some undescribed species of *Protopristus* Broun, 1909 <0.5 mm and the "giant" *Edaphus goliath* Puthz, 2007 of South America, reaching 5.1 mm (Puthz 2007a). They are generally slender or tubular in form though as a group Euaesthetinae are morphologically diverse in characters and habitus considering their minute size (Fig. 6.1; see also Orousset 1988 for habitus illustrations of five genera). Euaesthetinae are closely related to Steninae (see below), but extant taxa are distinguished from this megadiverse and comparatively homogenous group (and other staphylinid subfamilies) by the combination of several characters including the evenly denticulate margin of the labrum (Fig. 6.2c; *Stenaesthetus* Sharp, 1874, *Schatzmayrina* Koch, 1934, and several species of *Edaphus* Motschulsky, 1857 are exceptional in having a smooth or nearly smooth margin; e.g., Fig. 6.2f), the well-developed pronotosternal suture (Fig. 6.2d, arrow), the pointed apex of the ninth sternite (male) or second gonocoxite (female) (Fig. 6.2g–h), and the conspicuous macrosetae (usually a row of 3–4) along the posterolateral edge of the metacoxa (Fig. 6.2e, arrow). Other characteristics include the falcate or otherwise slender-curved mandibles (Fig. 6.2a) that are hidden beneath the labrum in repose, the clubbed antennae (Fig. 6.2a; both shared with Steninae), and reduced tarsal segmentation in most genera [4-4-4 in the majority, e.g., Fig. 6.2i; 5-5-4 in Stenaesthetini; 5-5-5 only in *Edaphosoma* Scheerpeltz, 1976 and *Nordenskioldia* Sahlberg, 1880 (Nordenskioldiini), and in *Fenderia* Hatch, 1957 and *Stictocranius* LeConte, 1866 (Stictocraniini)]. Although generally distinct in both habitus and characters, no adult

Fig. 6.1 Habitus photos of select Euaesthetinae (**a**) Euaesthetini: *Edaphus* sp. (**b**) Austroesthetini: *Austroesthetus* sp. (**c**) Stenaesthetini: "EuaAUS." (**d**) Euaesthetini: *Octavius* sp. Scale bars = 1 mm

character has yet been found that is both present in all species of Euaesthetinae and apparently unique to the group (see below).

Larvae have been described for about one third of euaesthetine genera and are known (but not yet described) for several others (Fig. 6.4). Larval descriptions of varying detail have been contributed by several authors (e.g., Kasule 1966; Newton 1990; Clarke and Grebennikov 2009), with the most complete larval description being for that of a New Zealand *Agnosthaetus* species (Clarke 2011). Like the adults, larvae are generally characteristic but cannot easily be distinguished from other subfamilies by any single character alone. They lack a separate labrum (this has fused to the head capsule forming an apically toothed nasale), have abdominal terga that form entire plates (undivided at middle, except, exceptionally, in at least some Australian *Edaphus*), and, if present at all, have an unarticulated (fixed) maxillary mala. All three characters are shared with Steninae. From that subfamily, euaesthetine larvae can be most easily distinguished by the combination of their generally much smaller size, three-segmented antennae that are shorter than the head, longest leg seta located on the tibia, and the absence or strong reduction of the maxillary mala. Further, when the mala is present it is positioned somewhat dorsally (rather than projecting mesially) such that it is difficult to observe in ventral view (see Kasule 1966: fig. 60; Grebennikov and Newton 2009; Clarke and Grebennikov 2009: figs. 7C–F, 8; Clarke 2011: figs. 41 and 44). Traditionally, larvae of Euaesthetinae and Steninae have also been distinguished from each other by the structure of the head capsule and ligula (with distinct neck and nuchal carina and narrow, acute, or finger-like ligula in Euaesthetinae; no neck or nuchal carina, and distinctly broad and bilobed ligula in Steninae; e.g., Kasule 1966; Newton 1990; Frank 1991; Leschen and Newton 2003). However, known *Edaphus* larvae lack a distinct neck and nuchal carina, and only *Euaesthetus* Gravenhorst, 1806, larvae actually have a somewhat developed neck (Newton 1990; Clarke and Grebennikov 2009: fig. 5C). Moreover, an undescribed litter-dwelling stenine genus endemic to southeastern Australia also has a distinct neck and nuchal carina (and short antennae), unlike the known larvae of other stenines (Clarke et al., in prep.), and it is now known that the structure of the ligula in both subfamilies is more variable than previously thought (see also Welch 1966: 250). Euaesthetine larvae are also very similar to those of the related subfamilies Staphylininae, Paederinae, Pseudopsinae, and Leptotyphlinae. From the first three of these, Euaesthetinae can be most easily distinguished by the lack of a

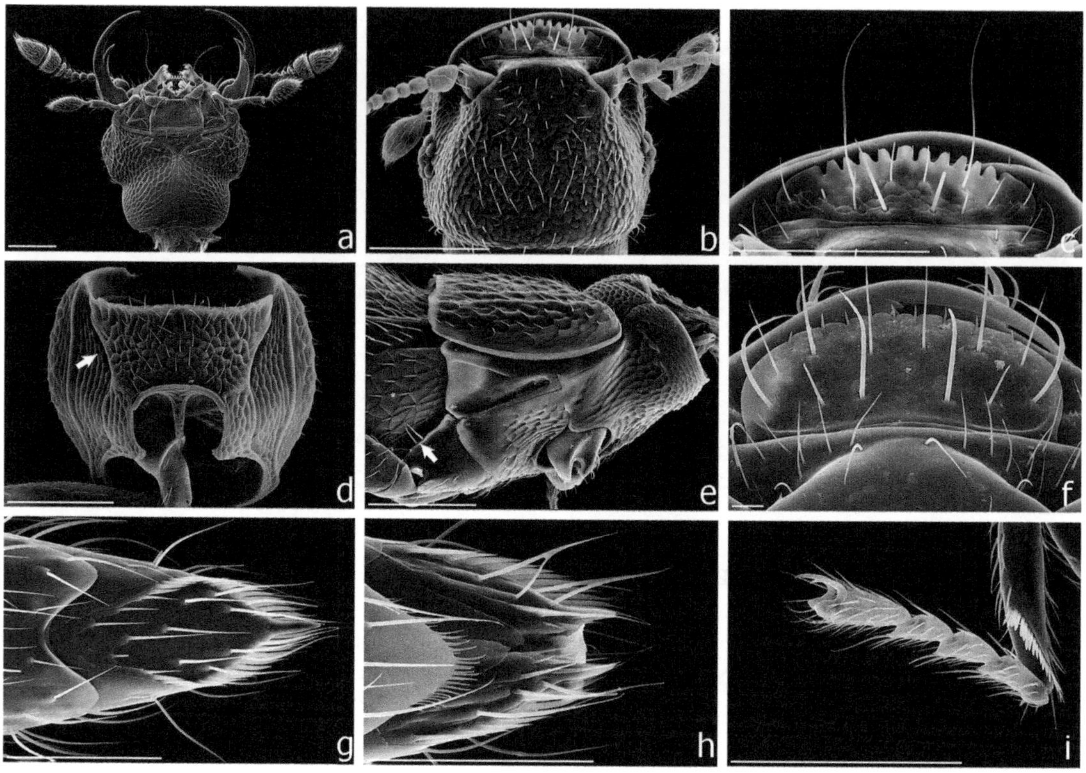

Fig. 6.2 Euaesthetinae, details of morphology (**a**) head (ventral), (**b**) head (dorsal), (**c**) labrum (dorsal), (**d**) prothorax (ventral), (**e**) pterothorax (right lateral), (**f**) labrum (dorsal), (**g**) male abdominal apex (ventral), (**h**) female abdominal apex (ventral), (**i**) metatarsus (lateral). (**a**) *Austroesthetus* sp., (**b–e**, **g–h**) *Chilioesthetus* sp., (**f**, **i**) *Edaphus* sp. Scale bars = 100 μm. See text for discussion of structures identified by arrows

distinct neck constriction, the presence of only one pair of parasclerites per abdominal segment, and the mesially open maxillary foramen (Grebennikov 2005). From leptotyphline larvae, Euaesthetinae can be distinguished by the lack of a subapical tooth along the inner mandibular margin, a dorsally instead of mesially projecting maxillary mala, and abdominal spiracles placed in membrane and not embedded in the terga (Grebennikov and Newton 2008).

6.2 Biodiversity and Systematics

6.2.1 Species Richness and Taxonomy

Although a small subfamily, the rate of species description for Euaesthetinae shows no sign of leveling off (Fig. 6.3) and demonstrates how much basic taxonomic work still remains in this group. The lower rate of taxonomic description for Euaesthetinae compared to that for its putative sister group Steninae reflects both the astonishing diversity in the genus *Stenus* Latreille, 1797, and also the greater taxonomic effort expended on that subfamily. Nearly 89% of the 1155 nominal species of Euaesthetinae are in the mainly pantropical genera *Edaphus* (593 spp.), *Octavius* (260 spp.), and *Stenaesthetus* (110 spp.), and the mainly Holarctic genus *Euaesthetus* (55 spp.). For these genera the rate of new species descriptions has generally been high and continues to be so (e.g., *Edaphus*: Puthz 2006a, b; *Octavius*: Puthz 2006c; *Stenaesthetus*: Puthz 2013a; *Euaesthetus*: Puthz 2014a). The rest of the 137 nominal species in the subfamily are scattered among the numerous species-poor and more or less geographically restricted genera

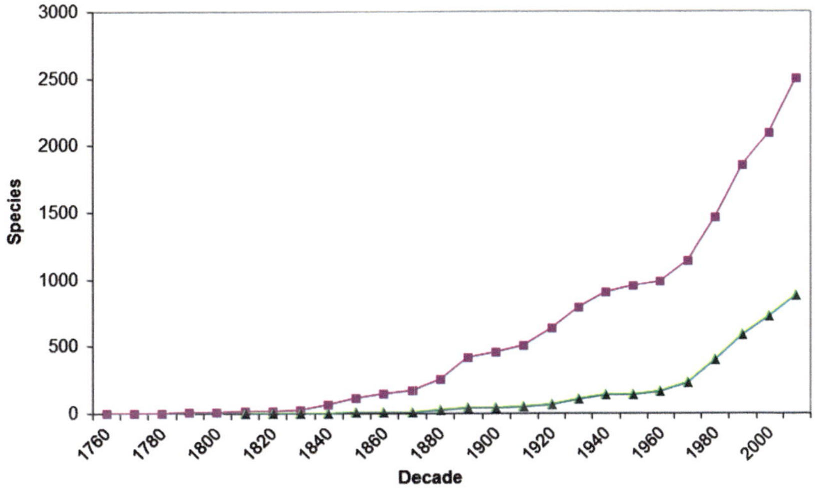

Fig. 6.3 Rate of species description for subfamilies Euaesthetinae (triangles) and Steninae (squares)

occurring primarily in temperate regions (Fig. 6.4). Similar diversity increases for these faunas have been slower, with most notable increases reported only recently and for only a few genera [e.g., South African *Octavius*, from 15 to 53 spp. (Puthz 2006c; Janák 2014); New Zealand *Agnosthaetus*, from 6 to 34 spp. (Clarke 2011); Chinese *Edaphosoma*, from 6 to 22 spp. (Puthz 2010a, b)]. The numbers of undescribed species reported in Fig. 6.4 are the result of my own in-progress surveys of the austral euaesthetine fauna (with substantial help from V. Puthz), and these further emphasize that the rate of species description for this subfamily will continue to increase. The difficulty with collecting euaesthetines (mostly via Berlese/Winkler extraction from sifted plant material) in combination with the still vast and ecologically diverse areas of unsurveyed habitat in just the austral areas alone helps to explain the trend in Fig. 6.3 and strongly suggests that the real diversity in any region is much greater than what is currently known (e.g., Puthz 1978). For example, in a review of Neotropical *Edaphus*, Puthz (2014b) indicates >240 new species of *Edaphus* known to him (most of these from the Oriental Region). In Chile, just one euaesthetine species has been described from the large and southernmost Magellanic Province of Chile (*Nothoesthetus australis* Sáiz, 1970), and the published figures of this species and also *N. obesus* Sáiz, 1970 (both known only from females), suggest that neither species are even congeneric with *N. coiffaiti* Sáiz, 1970, the type species. In Australia the lack of collecting effort is more pronounced. For example, fewer than 40 total specimens representing at least a dozen species in three genera have been collected from the cool temperate forests of Western Australia. These include the two most unusual Australian species of Austroesthetini (the only fully winged species of an austral-endemic genus—an undescribed species of *Austroesthetus* Oke, 1933; the only blind, flightless, and soil-dwelling species of *Chilioesthetus* Sáiz, 1968; see below).

As typical for Staphylinidae, alpha taxonomy of Euaesthetinae has historically been (and continues to be) entirely based on morphological characters, with differences between species largely anchored on male genital structures. It is common in this group for the only illustrations accompanying descriptions to be of the aedeagus, secondary sexual structures, and (less frequently) female genitalia (spermatheca). For some genera there are not yet any published habitus images, and much imaging work remains to help increase the profile of this interesting subfamily. Important aspects of future taxonomic work will include the development of electronic identification guides with comprehensive image documentation for all genera and major species

	Des. sp.	Und. sp.	Lar.	Biogeographic Regions WP	EP	NA	AF	OR	NE	PA	MA	SA	AU	NZ	CH	Wid. sp.
NORTHERN																
Nordenskioldiini																
Edaphosoma	22	?						22								0
Nordenskioldia	2	?			1	1										0
Stictocraniini																
Fenderia	6	?	D			6										0
Stictocranius	5	?				2		3								0
Euaesthetini																
Euaesthetus	55	?	D	2[5]	5[7]	21[6]		3[2]	12[5]							12
Ctenomastax	7	?		4[2]	1[1]		[1]									2
Euaesthetotyphlus	1	?		1												0
TROPICAL																
Stenaesthetini																
Stenaesthetus	110	?	U		[2]		19[2]	35[2]	22	[1]	28	2[2]				4
Euaesthetini																
Edaphus	593	~75	D	1	29	4	82	299	104	46	7	2	12		1	6
				[1]	[2]	[1]	[1]	[5]		[1]	[2]	[1]	[1]	i[1]		
Octavius	260	?	D	59	[1]		37	16[2]	26	9[1]	58	53	**(1)**			2
Tamotus	11	~1						**(1)**	11							0
Schatzmayrina	3	~1					1[1]	[1]	1		**(1)**	[1]				1
[Macroturellus]	1	-					1									0
{Phaenoctavius}	1	-							1							0

Fig. 6.4 Biodiversity and biogeographic data for Euaesthetinae

AUSTRAL																
Alzadaesthetini																
Alzadaesthetus	2	1	D												3	0
Austroesthetini																
Austroesthetus	4	13	D										17			0
Chilioesthetus	6	10	D										6		10	0
Kiwiaesthetus	7	5												12		0
Mesoaesthetus	2	17											19			0
Nothoesthetus	10	4	D												14	0
Tasmanosthetus	1	2											3			0
Gen1_CHI	-	2													2	0
Gen2_CHI	-	1	U												1	0
Euaesthetini																
{Coiffaitia}	6	~2									8					0
{Neocoiffaitia}	1	~1									2					0
Protopristus	3	~120	D										22	101		0
Gen1_SAF	{1}	4	U?									5				0
Gen2_SAF	{1}	1										2				0
Gen3_TAS	-	1											1			0
Stenaesthetini																
Agnosthaetus	34	0	D											34		0
[Tyrannomastax]	2	-									2					0
EuaAUS	-	2											2			0
TOTALS	1155	263	13	67	36	34	140	378	177	55	105	64	157	147	31	27

The 32 described and undescribed genera of Euaesthetinae are divided into a northern fauna, a tropical fauna, and an austral fauna. Numbers of described and known undescribed species are shown in their respective regions; total numbers of widespread species (two or more regions; right column) are shown inside square brackets within regions. *Euaesthetus:* 12 species in WP/EP (4 spp.), EP/OR (2 spp.), NA/NE (5 spp.), and NA/WP/EP (1 sp.); *Ctenomastax*: 2 species in WP/EP (1 sp.) and WP/AF (1 sp.); *Stenaesthetus:* 4 species in AF/SA (2 spp.), EP/OR (1 sp.), and EP/OR/PA (1 sp.); *Edaphus:* 6 species in EP/OR (2 spp.), OR/MA (1 sp.), OR/PA (1 sp.), AF/MA/AU (1 sp.), and WP/NE/OR/AU/NZ (1 sp.); *Octavius:* 2 species in EP/OR (1 sp.) and OR/PA (1 sp.). Column totals exclude bracketed numbers; the total for AU includes the ~75 undescribed Australian *Edaphus* omitted from that cell due to space. *Des. sp.* Described species, *Und. sp.* known undescribed species (~ estimated; others confirmed by dissection), *Lar.* larval records for genera (D = described; U = undescribed; U? = uncertain larval association), *WP* Western Palearctic, *EP* Eastern Palearctic, *NA* North America, *AF* Afrotropical, *OR* Oriental, *NE* Neotropical, *PA* Pacific, *MA* Madagascar region, *SA* Southern Africa, *AU* Australia, *NZ* New Zealand, *CH* Chile and Argentina, *Wid. sp.* widespread species, *i* introduced species, [*Genus*] likely a synonym of another genus, {*Genus*} belongs in another subfamily, **(1)** new records for that region, {1} described species currently misplaced in another genus, numbers in square brackets, e.g., [1], are the total number of widespread species in the region

Fig. 6.4 (continued)

groups. At present, no molecular taxonomic or phylogeographic studies have been performed, but these would likely reveal many more species than are currently known from morphology alone. Figure 6.4 summarizes current estimates of known undescribed species for the austral fauna, details of which are discussed later in this section.

The generic-level taxonomy of extant Euaesthetinae is still in flux (Fig. 6.4), with three included genera misplaced in Euaesthetinae (*Coiffaitia* Kistner and Shower, 1965, and *Neocoiffaitia* Orousset, 1988 probably belonging in Solieriinae; and *Phaenoctavius* Pace, 1986 in Oxytelinae, this genus likely being a synonym of *Carpelimus* Leach, 1819) and at least two currently valid genera that should probably be placed in synonymy with others, pending detailed study of type material. For example, the genus *Tyrannomastax* Orousset, 1988 has derived mouthparts similar to those of *Stenus* (as discussed by Leschen and Newton 2003) but in nearly all important respects has the characters of *Stenaesthetus* including several unique or diagnostic characters such as the filamentous antennal structure, prothoracic structures, form of the elytral epipleural carina, ventral abdominal carinae, and genital structure. From the figures in the original description, the African monotypic genus *Macroturellus* Orousset, 1987 is clearly a highly derived member of a primarily African and Oriental group of *Octavius* species comprising distinctive forms like *O. batesi* (Sharp, 1876) and *O. bicolor* (Cameron, 1938), both originally described in the separate genera *Turellus* Sharp, 1876 and *Doletica* Cameron, 1938, respectively, reflecting the shared unique form of these species. Most notably, *M. pulcher* Orousset, 1987, *O. bicolor*, and several related species share among other characters a unique pronounced lateral flanging of the prothorax and strongly tapering abdomen as well as longitudinally carinate elytra (e.g., Orousset 1987: fig. 11; Kistner 1961a: figs. 9–16), similar to those of Pseudopsinae. The current taxonomic status of World Euaesthetinae, as summarized in Fig. 6.4, reveals that a significant number of new but as yet undescribed genera are known in the austral region but none are known for other regions. This perhaps reflects the well-known bias toward northern hemisphere faunas in both collecting and taxonomic effort but may also reflect different evolutionary histories for austral and other regions. Six putative new genera in three tribes are known for the austral region and await description and proper phylogenetic study. Two undescribed genera have been identified among minute South African species described in *Octavius* (Puthz 2006c; "Gen1_SAF" for *O. angusticollis* Puthz, 2006, and four undescribed species; "Gen2_SAF" for *O. bacillus* Puthz, 1986, and one undescribed species) that may actually be closely related to *Protopristus* of Australia and New Zealand. Five more species described in the same paper (*O. caecigenus* Puthz, 2006; *O. longesulcatus* Puthz, 2006; *O. unocellus* Puthz, 2006; *O. brevisulcatus* Puthz, 2006; and *O. inoptatus* Puthz, 2006) plus one other (*O. atomus* Puthz, 1986) and at least three more undescribed ones belong in either *Tasmanosthetus* Puthz, 1978 or a new genus placed near *Tasmanosthetus* and *Nothoesthetus* Sáiz, 1970, in Austroesthetini; unlike *Tasmanosthetus*, these species have abdominal parasclerites (hence their inclusion here within Euaesthetini in Fig. 6.4). At least one Tasmanian species and possibly others from the Australian mainland are probably related to *Protopristus* but may require a new genus ("Gen3_TAS") because they lack the key character of that genus, the derived ligula tooth (Puthz 1978; Newton 1985), as well as having differently structured pharyngeal characters (Clarke, in prep.). Two other new genera from Chile are more enigmatic because of their unusual combinations of seemingly primitive and derived characters. A new genus of Stenaesthetini in Australia has been previously reported in the literature (Puthz 1978) and has been since referred to by the tag name "EuaAUS" (Clarke and Grebennikov 2009), which hereafter is used to refer this genus. The phylogenetic placement of all these species within the context of a global phylogenetic analysis of Euaesthetinae is a fundamental goal of future phylogenetic work on Euaesthetinae, especially since the South African taxa imply new Gondwanan area connections that, as far as I am aware, have not been previously recognized for this subfamily. Ongoing morphological phylogenetic studies by the author are addressing the generic-level classification of Euaesthetinae with future taxonomic changes likely resulting in a reduction in the number of currently valid described genera in the subfamily, with an eventual total of ~30 genera for Euaesthetinae when all new genera are described and other taxonomic actions implemented.

6.2.2 Phylogeny and Suprageneric Classification

Euaesthetinae, Steninae, and Megalopsidiinae have been placed together in the "stenine group" (Hansen 1997; Leschen and Newton 2003) within the "staphylinine group" of subfamilies (Lawrence and Newton 1982). Monophyly of Euaesthetinae has been previously questioned on account of there being no clear ubiquitous synapomorphies for the group (e.g., Thayer 2005) and only weak support for it in the first phylogenetic analysis of the stenine group (Leschen and Newton 2003). In a later much-expanded analysis of adult and larval characters, Clarke and Grebennikov (2009) recovered the first strong support for monophyly of Euaesthetinae, including 19 hypothesized adult and larval synapomorphies. In an analysis of the staphylinine group of subfamilies, Grebennikov and Newton (2009) also consistently recovered Euaesthetinae as a monophyletic group in analyses based on adult morphology, larval morphology, and 18S rDNA data, as did McKenna et al. (2015) in a phylogenetic analysis of Staphyliniformia using 28S rDNA and *CAD* sequences. However, none of these studies included all genera, nor representatives from all tribes of Euaesthetinae, and likely did not include the most basal lineages within the "euaesthetine subgroup" (Clarke and Chatzimanolis 2009: Euaesthetinae + Steninae), such as the enigmatic northern temperate genera *Nordenskioldia* and *Ctenomastax* Kraatz, 1870, or the now much better known Oriental genus *Edaphosoma* (Puthz 2010a). As well as including such taxa, future and more rigorous tests of euaesthetine monophyly will also need to be based on analyses that include multiple diverse species sampled from each genus (especially *Octavius*, *Edaphus*, and *Stenaesthetus*) as well as molecular data sampled for a wider range of genera and genes (Clarke in prep.). Only few molecular phylogenetic studies have included Euaesthetinae, and most of these included too few genera to draw any suitable conclusions; a clear priority for future phylogenetics within this group will be to expand the taxon sampling of molecular phylogenies. Previous morphological phylogenies (Clarke and Grebennikov 2009; Grebennikov and Newton 2009), augmented by ongoing surveys of character diversity within Euaesthetinae and Steninae (Clarke unpublished), indicate that the most promising synapomorphies for the subfamily include:

In adults:

(1) The presence of differentiated setae at the apex of antennomere X
(2) The denticulate apical margin of the labrum (though this is not uniform within the subfamily)
(3) A line of macrosetae on the posterolateral margin of the metacoxa
(4) The mesal edge of the gonocoxite and apex of male sternite IX produced into a spine

In larvae:

(5) The markedly reduced or apparently absent maxillary mala
(6) The dorsad (rather than mesad) orientation of the mala
(7) The reduced cardo, which is much narrower than the base of the stipes [though a more significant feature of the euaesthetine larval cardo, including *Euaesthetus* (Newton 1990: fig. 38B.4) is the postero-oblique displacement of its apical edge from the basal edge of the stipes; see also illustrations in Clarke and Grebennikov 2009]
(8) The stipes markedly narrowed distad
(9) The longest leg seta located on the tibia

Only the adult labrum and larval cardo characters might be restricted to Euaesthetinae (some Paederinae and Leptotyphlinae adults have teeth along the labral edge, but these apparently do not form a serrated edge as they do in Euaesthetinae), but few possible synapomorphies are likely uniform within the subfamily, and several of them occur in other subfamilies.

The current suprageneric classification of Euaesthetinae consists of a system of six tribes (Scheerpeltz 1974) and has been recognized as in need of critical phylogenetic review (Newton 1985; Clarke and Grebennikov 2009). An overview of the salient historical milestones in the development of the higher classification of Euaesthetinae, as well as the problems with it, was given by Clarke and Grebennikov (2009). The six tribes represent artificial groupings based on just a few superficial but commonly used characters (Newton 1985) including tarsal formula, presence/absence of wings, and abdominal "margination" (presence/absence of parasclerites), which vary both within and among genera. Clarke and Grebennikov (2009) recovered only weak support for the monophyly of one of these tribes (Stenaesthetini) and discussed the problems associated with using tarsal formula and abdominal margination as characters for distinguishing suprageneric taxa: the tarsal formula 4-4-4, for example, likely represents independent and often only partial instances of basal tarsomere fusion, whereas abdominal "margination" obfuscates several different character states (e.g., one vs. two pairs of parasclerites per segment; parasclerites on some segments but not others), an issue that is further complicated by the uncertain morphological origin and therefore homology of the individual lateral sclerites (see Naomi 2014). The extent of the evolutionary lability of "margination" within Euaesthetinae is therefore not yet certain; in Steninae this issue is clearer, with margination varying not only within *Stenus* (e.g., within the Australian fauna; within even individual species elsewhere) but also within a new genus of only three species (Clarke et al., in prep.)! In addition, the current system of tribes did not include all the known genera when originally proposed and until recently was not widely adopted nor critically evaluated by others (Newton and Thayer 1992). Although it is maintained here for organizational reasons, the numerous exceptions to the character combinations used to define the tribes make the system not particularly useful for either identification or for drawing general biological or ecological conclusions. A major goal of the author's ongoing phylogenetic work is to establish a new, holistic suprageneric classification system for Euaesthetinae based on a robust phylogenetic hypothesis for the euaesthetine subgroup.

6.2.2.1 Austroesthetini

In Australia, this tribe includes species generally lacking abdominal parasclerites, though there is considerable variability in the precise lateral structure of the abdomen in different genera: parasclerites are completely absent in *Mesoaesthetus*, but the terga and sterna are separated by a fine suture; there is a single parasclerite on segment III only in *Austroesthetus* and *Chilioesthetus*, with segments IV–VI tubular; and parasclerites are absent entirely in *Tasmanosthetus*, with each tergum and sternum of segments III–VI fused into a solid ring. In New Zealand, this tribe is represented by the single genus *Kiwiaesthetus*, and in Chile/Argentina the genera *Chilioesthetus* and *Nothoesthetus* (see also Fig. 6.4)—all three genera have different states for abdominal margination. Until recently, the tribe was thought to comprise only wingless—or at least flightless—species (Scheerpeltz 1974), but at least the female of one undescribed *Austroesthetus* species (from Western Australia) has abbreviated wings beneath full-length elytra (the species is possibly sexually wing dimorphic or sexually flight dimorphic), and other species are now known to be micropterous or variously brachypterous instead of wingless (Clarke, unpublished). Recent phylogenetic work on Euaesthetinae included all the then-known genera of Austroesthetini and concluded that the tribe is most likely polyphyletic (Clarke and Grebennikov 2009).

6.2.2.2 Euaesthetini

This tribe consists of a morphologically heterogeneous assemblage of species ranging from fully winged robust species of *Ctenomastax* and *Edaphus* to the minute elongate-slender, eyeless, and wingless species of *Octavius* and *Protopristus*. The tribe is widely understood to be non-monophyletic (e.g., Clarke and Grebennikov 2009) and includes genera with quite different structural features, including degree of metacoxal separation, type of abdominal margination (number, form, and presence of parasclerites), and presence and form of various head, antennal, and mouthpart structures. Ongoing comparative morphological studies of particularly internal head and mouthpart characters (but also others) indicate the group is more than likely polyphyletic, with groups of genera dispersed throughout the phylogeny of the group.

6.2.2.3 Stenaesthetini

Of the six tribes recognized in the current higher classification of Euaesthetinae, Stenaesthetini is the only one that may be monophyletic. Clarke and Grebennikov (2009) recovered a weakly supported clade comprising EuaAUS, *Stenaesthetus* (including *Gerhardia* Kistner, 1960, now a synonym), and *Agnosthaetus* based on an analysis of adult morphological characters. This clade was suggested to be supported by a single uniquely optimized synapomorphy—a carinate groove on the basomesal surface of the mesocoxa that receives the intermesocoxal process. Subsequent extensive dissection efforts, however, indicate instead that this character may be too variable within genera, and similar conditions are seen elsewhere in the subfamily (casting doubt on its phylogenetic value). The main character tying these genera together is the 5-5-4 tarsal formula (unique within Euaesthetinae), and although a classic composite character ("tarsal formula") in beetle classification at all taxonomic levels, proper phylogenetic treatment of it would preclude interpreting the state "5-5-4" as a unique synapomorphy for Stenaesthetini (see discussion of characters 93–95 in Clarke and Grebennikov 2009: 393), at least when considering it as the numeric character "number of tarsomeres". Further comparative study of this character could, however, shed light on whether this reduction in metatarsomere number may be different from that seen in Euaesthetini and Austroesthetini, since the arrangement of "landmark" setae on the basal metatarsomere in Stenaesthetini is dissimilar from that in these other tribes (and differs within it). If Stenaesthetini does represent a monophyletic group, its presence in Australia—represented only by EuaAUS, a species-poor and range-restricted taxon—seems anomalous. It may be relictual, considering the widespread occurrence of *Stenaesthetus* in the tropics (but its apparent absence from Australia and New Guinea), as well as the occurrence of the relatively diverse genus *Agnosthaetus* (34 spp.) in neighboring New Zealand (Clarke 2011).

6.2.2.4 Alzadaesthetini, Stictocraniini, and Nordenskioldiini

Of the three remaining tribes, Alzadaesthetini is monotypic, comprising only the genus *Alzadaesthetus* Kistner, 1961, with two species, and both Stictocraniini (=Fenderiini) and Nordenskioldiini each have two genera. Neither *Alzadaesthetus* nor Stictocraniini were found to be monophyletic by Clarke and Grebennikov (2009), and the two genera in the latter tribe (*Edaphosoma* and *Nordenskioldia*) have not yet been studied phylogenetically.

6.3 Ecology and Biology

Information on the ecology and biology of Euaesthetinae is largely buried in the primary taxonomic literature. The following review gathers much of this and summarizes unpublished label data for thousands of specimen-level collection records held in a database, most of which at the time of writing are for southern hemisphere taxa. Although *Edaphus* is arguably the most diverse genus of Euaesthetinae, space prevents a complete synthesis of its ecology and biology here.

6.3.1 General Ecological Patterns

Euaesthetinae are broadly characterized as litter- or soil-dwelling rove beetles that occur worldwide from near polar environments to the tropics, where they are most diverse, and from sea level to high-elevation mountains. They are primarily found in forested regions in all major biomes but may also be found in other diverse vegetation types. Predominant ecological communities include tropical rainforest, boreal and temperate forests, montane and alpine habitats, shrublands, and wetlands. In Australia, where Euaesthetinae are generically most diverse, they are mainly found in cool temperate rainforest dominated by *Nothofagus* Blume, 1851 and *Eucalyptus* L'Heritier, 1879 species but can also be found in savannah, dry sclerophyll forest and scrubland, as well as alpine grasslands and meadows; in New Zealand they are found in all forest types as well as indigenous alpine grasslands (Clarke 2011); in South Africa they are most diverse in podocarp and Afromontane forests; and in Chile they are more or less restricted to cool temperate rainforests.

At the global/regional scale, the elevational distribution of many genera changes with decreasing latitude. In the Holarctic, the genus *Euaesthetus* extends south into both the Oriental Region and the Neotropical Region, but in these regions species are seemingly restricted to high-elevation mountains. Several southern hemisphere genera show this pattern also (e.g., *Austroesthetus*, *Kiwiaesthetus*, *Mesoaesthetus*), becoming increasingly restricted to alpine or high-elevation habitats from southern to northern limits of their ranges. This suggests that many of the genera are cold-tolerant and well-adapted to high altitude environments.

The general ground-dwelling Euaesthetinae can be divided into different "ecological groups" of genera that have microscale distributions extending beyond the general leaf litter and into either above- or belowground microhabitats (with little overlap). The first group comprises genera that are also found in aboveground situations on tree trunks, large boulders, and fallen logs, in association with bryophyte communities that grow on these substrates (e.g., *Austroesthetus*, *Mesoaesthetus*, *Tasmanosthetus*, EuaAUS, *Alzadaesthetus*, *Stenaesthetus* in part). A second group comprises genera that are also found, perhaps even more abundantly than in litter, in the soil beneath the surface litter (e.g., *Chilioesthetus*, *Euaesthetotyphlus*, *Octavius*, *Protopristus*, *Tasmanosthetus*, *Stenaesthetus* in part), and this group could be considered a truly endogeous fauna. Results of recent soil sampling by entomologists and collectors working in Australia and New Zealand suggest that soil may be the preferred microhabitat of the immature stages of at least some genera. This hypothesis may help explain the comparative rarity of euaesthetine larvae in collections and the still unknown larvae of several otherwise well-collected genera.

Another useful ecological group of genera can be recognized, though not comprising a set of genera mutually exclusive from the first two. Some genera can be identified as having a stronger association with riparian and general wetland-type habitats than others (*Euaesthetus*, *Nordenskioldia*, *Schatzmayrina*, possibly *Ctenomastax*). This periaquatic group contrasts with the remaining genera of the subfamily that are found in more diverse (but also generally moist) habitats, though some of the genera in this third group likely also occur in both above- and below-ground microhabitats. Yet another ecological group could recognize "surface runners," those genera that, based on ecomorphological criteria, are thought to be primarily found, or have been collected, in open situations such as on river banks (e.g., *Ctenomastax*, *Schatzmayrina*, some *Octavius*) versus those cryptic taxa that usually remain hidden within the litter or soil. A subset of these "surface runners" might also be recognized as a distinct group that may be arboreal, walking on exposed plant surfaces. This subset probably would include genera like *Tamotus* Schaufuss, 1872, some species of *Stenaesthetus* and *Octavius* (e.g., Puthz 1977: *O. dybasi* Puthz, 1977; Palau), and likely some species of *Edaphus*.

6.3.2 Tribe Nordenskioldiini

6.3.2.1 *Edaphosoma* (India, Nepal, China)

The genus *Edaphosoma* is evidently a specialist alpine genus found nearly exclusively in high mountainous areas (Puthz 2010a). The Nepalese species *E. janetscheki* Scheerpeltz, 1976 is recorded from 4800 to 5250 m (Scheerpeltz 1976) and *E. nepalensis* Puthz, 1979 from 5000 m. Chinese species are known from 650 to 4350 m, with nearly all collections occurring over 2000 m. Specimens have been collected from *Salix* L., *Rhododendron* L., oak and alder scrub and woodland, from bamboo and generally subalpine to alpine habitats dominated by grass heath and moss. Most collections have come from sifted debris, including grass, dead wood, mushrooms, and pine needles.

6.3.2.2 *Nordenskioldia* (Siberia, Russia; British Columbia, Canada)

This genus, known only from type material, seems to be a cold-tolerant, subarctic group and one of the most northerly distributed taxa in Euaesthetinae. The three known specimens of *N. glacialis* Sahlberg, 1880 were collected under or among stream-edge stones (an unusual collecting situation for Euaesthetinae); those of *N. columbiana* Puthz, 1974 from among stream-edge *Salix* litter at ~1600 m. *Nordenskioldia* may prefer wet riparian microhabitats. Fieldwork in Yoho National Park in 2008 by the author failed to recollect *N. columbiana* from the type locality as well as numerous other sites throughout Yoho National Park, Banff National Park, and Kootenay National Park. The habitat at the type locality is coniferous forest with sparse shrubby understory but with thicker broadleaf growth near streams. Leaf litter is rare, with dense and thick moss mats forming a uniform and predominant ground cover. Future collecting efforts should more deliberately target riparian areas and, especially, should include Berlese processing of partly submerged moss and other semiaquatic debris (A. Newton, pers. comm.) and should also focus on direct searching of stream-edge gravels.

6.3.3 *Tribe Stictocraniini* (=*Fenderiini*[1])

6.3.3.1 *Fenderia* (USA) and *Stictocranius* (USA, China)

Fenderia is a locally endemic genus restricted to the Pacific North West states (California, Oregon, Washington), occurring mainly in coastal areas (Puthz 2003). Most collections have been from old-growth coniferous or mixed conifer-hardwood forest from low to mid elevations, 40–1234 m, where species are commonly collected by Berlese-processing leaf litter samples; most published collection records are from Douglas fir and redwood duff. Larvae are known, also from leaf litter samples; at Mary's Peak (Oregon), a long series of both adults and larvae have been collected from around a huge decaying stump in old-growth *Abies procera* Rehder forest.

Stictocranius is commonly found in montane Appalachian forests of eastern North America (*Stictocranius puncticeps* LeConte, 1866) with a second species, *S. mariae* (Hatch, 1957), found in montane areas of the western USA (where the types were collected from grass sod) and British Columbia, Canada. Originally thought to be primarily found in the Appalachian Mountains (Puthz 1974), *S. puncticeps* is now known to be

[1] This name has now been placed in synonymy with Stictocraniini, a previously overlooked but now resurrected valid name for this tribe (see Newton 2017).

widespread in forested areas of the eastern USA and Canada (Ontario), where most collections have been from Berlese-processed leaf litter samples. This species seems to be commonly collected from deep litter deposits and occurs in both hardwood and coniferous forests. In China, the three species have been found from 1600 to 2000 m, with one collection from a mixed deciduous forest with bamboo and small meadows. Larvae are unknown.

6.3.4 Tribe Alzadaesthetini

6.3.4.1 *Alzadaesthetus* (Chile/Argentina)

Comparatively little material with collecting data is available for *Alzadaesthetus*. Specimens have been collected at low to mid elevations (10–1300 m) in mixed *Nothofagus*-conifer forest and *Saxegothaea* Lindley, 1851-dominated Valdivian rainforest with dense *Chusquea* Kunth, 1822 (bamboo) understory, with most taken in leaf litter samples or by pyrethrum knockdown. A few series of several specimens have also been collected from processed aboveground moss samples, suggesting this genus may be common outside of the general leaf litter. Larvae have been collected in Berlese-processed leaf litter samples. No ecological differences between the two described species are evident, but the different morphologies of *A. chilensis* Kistner, 1961 and *A. furcillatus* Sáiz, 1972 suggest they may be distinct (see Sect. 6.3.8.2). For example, *Alzadaesthetus* is one of the few euaesthetine genera with species differing in abdominal margination, and the lack of shared derived characters suggests included species may not be congeneric (Clarke and Grebennikov 2009). Also, a unique feature of *A. furcillatus* (and a related undescribed species) is the densely papillate underside of the elytra; these species are (almost) wingless, and these structures have no obvious function, but the morphological differences between *A. furcillatus* and *A. chilensis* suggest that ecological differences exist.

6.3.5 Tribe Austroesthetini

6.3.5.1 *Austroesthetus* (Australia)

This genus is found primarily in cool temperate rainforest dominated by *Nothofagus* and *Eucalyptus* species but can also be found in both wet and dry sclerophyll forests, in coastal scrub and sand heath communities, in forest of all successional stages, and in both riparian and ridge-top communities. The genus occurs from sea level to >1600 m; most of the lowland records are from Victoria and Tasmania, whereas nearly all of the higher elevation collections (>1300 m) are from mountains in Queensland and New South Wales. Species are most commonly collected from general forest litter but are also frequently found in moss growing on a variety of substrates (living trees, old logs, rocks, and ground). Common collections from moss strongly indicate that at the microhabitat scale the most important environmental variable is high moisture content, since most of these collections were from aboveground mosses. One collection from wet moss at a seepage area near the crest of a mountain indicates that mossy microhabitats allow species to exist in drier, more exposed areas, including outside the general forest floor environment. Specimens have also been commonly collected from old logs (with or without fungi, via pyrethrum knockdown), from under bark of *Nothofagus* and *Eucalyptus* logs, and (rarely) from sifted woody materials. One collection from moss with ants is probably an incidental, rather than a myrmecophilous, association. Larvae have been taken from both litter and under bark, so there is some evidence that the life cycle can be completed in both aboveground microhabitats and the general ground litter.

6.3.5.2 *Chilioesthetus* (Australia, Chile)

Chilioesthetus is relatively rare in Australia, known from less than 50 specimens. Species are found primarily in cool temperate *Eucalyptus*-dominated rainforest and sclerophyll forest in Western Australia or subtropical montane rainforest in Queensland. In Western Australia, species have been found from near sea level in

Walpole National Park to 783 m in Stirling Range National Park. In contrast, nearly all Queensland collections have been from elevations above 1000 m (up to 1260 m). Collection records indicate a more restricted range of microhabitats for Australian *Chilioesthetus* than the related *Austroesthetus*. Most specimens have been collected via Berlese processing of general leaf and log litter, and unlike *Austroesthetus*, no specimens have been taken from moss nor via pyrethrum knockdown, whereas soil washing has produced several specimens (soil samples; 18 cm depth). This suggests a generic difference in microhabitat preference in Australia, with *Chilioesthetus* species possibly preferring soil microhabitats rather than surface leaf litter. The generally smaller size and reduced eyes of Australian *Chilioesthetus* species (compared to *Austroesthetus* and Chilean *Chilioesthetus*; completely blind in an undescribed Western Australian species) may support this inference.

Chilean *Chilioesthetus* are more common and have been collected from near sea level to 1250 m in mixed *Nothofagus*-conifer forest and Valdivian rainforest, often with *Chusquea*-dominant understory. Most specimens have been collected by Berlese-processing mixed *Nothofagus*-podocarp leaf litter samples. However, several collecting events from fungi (bracket fungi and soft fungi; fogging fungusy logs), from debris under logs, and from fine debris under bark of *Nothofagus* logs (Puthz 2008a) are unusual and suggest a broader range of microhabitats for this genus in Chile; larvae were also associated with these specimens collected from logs, as well as others from leaf litter. Adults of most species of *Chilioesthetus* are characterized by coarse, imbricate surface sculpturing with prominent lateral pronotal and elytral grooves, and many species (mostly Chilean, but also some Australian) are commonly observed to have a waxy encrustation on especially the dorsal surfaces (does not completely dissolve in KOH), and it is possible that this is secreted from foveae or pores near the grooves (see also *Octavius*) and may be related to their preference for moist habitats.

6.3.5.3 *Kiwiaesthetus* (New Zealand)

Kiwiaesthetus species are primarily found in all cool temperate forest types in New Zealand, including *Nothofagus* and podocarp-broadleaf forest, but also occur in subtropical kauri (*Agathis* Salisbury, 1807)-dominated forest in the far north of the country (Puthz 2008b). However, several species are also common in tussock grassland communities and in alpine areas above the tree line. Although species can be found from near sea level to ~1900 m, the genus seems to be most common at high elevations, with nearly all records from North Island [*K. whitehorni* (Puthz, 2008); one undescribed species] found in high montane forest and alpine scrub and tussock communities. The majority of collections have been nearly equally from moss and diverse types of forest and ground litter. Specimens have been hand collected from among the bases of tussock grasses and from under plant mats. Like other genera, the association of *Kiwiaesthetus* with moss indicates moisture is the most important microhabitat variable, which may also explain the ease of collecting specimens from the bases of alpine grasses. Only few specimens have been taken via pyrethrum knockdown (from old logs) and from soil. Larvae are unknown, and given how common this genus is in collections (and its apparent abundance), the immatures must have unusual seasonality or else occur in an as-yet unknown habitat, possibly deep in the soil.

6.3.5.4 *Mesoaesthetus* (Australia)

Mesoaesthetus species are found primarily in cool temperate rainforest dominated by *Eucalyptus* and *Nothofagus* species but can also be found in wet sclerophyll forest, scrub, and open woodlands. Species have been found from near sea level to ~1500 m; in Tasmania species are found primarily in lowland to montane habitats (~50–1100 m), whereas in the northern mainland part of their range (Victoria), they have been found almost exclusively above 1200 m. The majority of collections have been from diverse leaf litter types, including fern, woody, and fungusy debris. Species are also commonly found in moss growing on a variety of substrates (living trees, old logs, ground). Like

Austroesthetus and *Kiwiaesthetus*, the association of *Mesoaesthetus* with moss indicates moisture as the most important microhabitat variable. Only few specimens have been taken via pyrethrum knockdown (from old logs) and from soil. As one of the most abundant genera in collections, it is unusual that larvae have not yet been found, indicating an unusual microhabitat for the immatures (probably soil).

6.3.5.5 *Nothoesthetus* and Undescribed Genera (Chile)

The relatively few available *Nothoesthetus* specimens with ecological data have been taken largely from *Nothofagus* and mixed evergreen-conifer forest, Valdivian rainforest, and scattered forest remnants from 500 to 1080 m. This genus also occurs in forests with dense *Chusquea* understory, and all specimens, including larvae, have been taken from Berlese-processed leaf litter samples. The species of *Nothoesthetus* are generally distinguishable only by their genitalia, and these are remarkably diverse (e.g., Puthz 2012a: figs. 2 and 4). The undescribed genus "Gen1_CHI" is known from two undescribed species, one from a series collected in the Aysen Province from sifted moss growing on logs and the other from litter. "Gen2_CHI," perhaps the most morphologically isolated euaesthetine taxon, has been collected in litter from *Cupressus* L.-*Eucalyptus* forest in Arauco Province.

6.3.5.6 *Tasmanosthetus* (Tasmania, Australia)

Tasmanosthetus species are found primarily in cool temperate *Nothofagus* rainforest, as well as *Eucalyptus* and wet sclerophyll forest, and scrubland. The genus occurs from near sea level to montane habitats (>950 m). Species have been commonly collected from forest litter of diverse types (fern, broadleaf, pine) but are also frequently found in moss growing on a variety of substrates (living trees, logs, ground, and rocks). Although common collections from moss (including one from wet *Sphagnum* Linnaeus, 1753 moss at a forest seep) indicate that *Tasmanosthetus* species prefer moist microhabitats, they evidently occur only at ground level. A few specimens have also been collected from soil suggesting that, like *Chilioesthetus* and *Protopristus*, this genus also occupies both the litter layer and the edaphic zone. The minute size and reduced eyes (blind in some) of all *Tasmanosthetus* species are consistent with a soil-dwelling lifestyle.

6.3.6 Tribe Euaesthetini

6.3.6.1 The Rare Genera (*Ctenomastax, Euaesthetotyphlus, Tamotus, Macroturellus*)

Several Euaesthetini genera are poorly known, collected only once or very rarely (and most not recently) such that little about their biology can be confirmed. An exception is *Euaesthetotyphlus*, known only from type material of the only known species, *E. almajensis* Coiffait and Decu, 1970. The series of 29 specimens (most of which are now presumed lost) were found at 200 m elevation at the base of a slope near a river in the limestone region of western Almajului Mountains, Romania, and collected by washing a sample of calcareous soil—described as being relatively cool (15°) at 5–15 cm depth and moist despite a prevailing drought (Coiffait and Decu 1970). Genera like *Protopristus* (Australia, New Zealand) and many *Octavius* species (e.g., from the Western Palearctic, South Africa, and Madagascar) are also collected in this way and are similarly adapted to subterranean life, being minute, flightless, and (many of them) nearly eyeless.

The holotype and only known specimen of *Macroturellus* (Cameroon) was probably collected at light (Orousset 1987), and there are no other hints about its biology, except that published figures suggest that it must be unusual given the highly modified structure of the labrum and front of the head (including the subgeniculate antennae) combined with the unusual combination of reduced eyes but presumably functional wings (Orousset 1987: figs. 1–3, 6, and 12).

The odd genus *Ctenomastax* is the only one largely restricted to the Mediterranean Region (North Africa, southern Europe) and is among the most distinctive genera of Euaesthetinae (e.g., Orousset 1990a: 131)—in several respects resembling *Stenus* (Steninae). Puthz (1988a) lists an elevation range of 1000–1500 m for *C. kiesenwetteri* Kraatz, 1870 (no other data are published), and compiled the ecological information then known: specimens have been found at the edges of a pond and collected from shoreline wrack (of fresh and salt waters, the latter unusual for Euaesthetinae) and under stones on clay soil. These records suggest that *Ctenomastax* species (like most Steninae) should be found most commonly in riparian habitats and immediately adjacent to bodies of water. As typical for riparian insects, specimens have also come to light and have been captured in flight, and it has been suggested that on clay-limestone plains they can be found in cracks in the ground (see Puthz 1988a). Orousset (1990a) describes the sand dune desert habitat (with palms) in the vicinity of where *C. mirei* Orousset, 1990 was collected, which demonstrates *Ctenomastax* can survive in arid regions, likely doing so by dispersing between localized bodies of water (species are winged); he also notes a corollary to this in that the holotype and paratype showed no detectable variation despite the great distance separating their respective localities.

The genus *Tamotus* is widespread throughout Central America, South America, and the Caribbean but relatively rarely collected. Most collections are singletons or small series from flight traps set in lowland tropical rainforest. Apart from *T. similis* Puthz, 2002, with the broadest elevational range of 200–1400 m, the other 10 species for which data are available seem to occur in lowland forest (100–400 m). A few records are from Berlese-processed forest litter, and Puthz (2007b) reports one record of *T. cariniceps* (*lapsus* for *T. carinifrons* Puthz, 1986) from an emergence trap with floats on aquatic macrophytes. With the frequency of flight trap records and paucity of litter records, these observations are notable since they suggest an arboreal habitat for *Tamotus* (see below, Sect. 6.3.8.4).

6.3.6.2 *Edaphus* (Summarized Mainly for the Australian Fauna)

The genus *Edaphus* is incredibly speciose, morphologically diverse, and likely more ecologically diverse than presently understood. A comprehensive overview of the ecological knowledge of this genus will be provided elsewhere, though information for the Australian fauna is provided here since this has recently been compiled in connection with an in progress revision of the fauna. Australian *Edaphus* species can be found in a diversity of vegetation types. Most collections have been from *Nothofagus* and *Eucalyptus* rainforest and others from subtropical forest, wet sclerophyll forest, and scrubland and woodland habitats. Species can be found from near sea level to ~1600 m. Most collections have been from various Berlese-processed forest litter and other substrates, but, unlike Austroesthetini and Stenaesthetini: EuaAUS, Australian *Edaphus* seem to avoid moss; no specimens have yet been collected from that microhabitat, except for the syntypes of *E. melculus* (Oke, 1933) that were collected "in moss with ants on stone" (Puthz 1978). In this case, the occurrence in moss was probably incidental to the association with ants: Oke (1933) reports finding on several occasions specimens of *E. melculus* var. *camponoti* (Oke, 1933) (=*E. termitophilus* Bernhauer, 1916: Puthz 1978) in nests of *Camponotus* Mayr, 1861 ants, including an instance of more than 50 specimens in one nest. Puthz (1978) reports a female of *E. termitophilus* collected with an *Aphaenogaster longiceps* (Smith, 1858) host and notes the very close resemblance shared by *E. melculus* and *E. termitophilus*—which also suggests the biology of these species is similar. Both are probably myrmecophilous, and, though not verified, the name of the latter species suggests that the type series was associated with termites. If an association of *Edaphus* with ants and/or termites could be confirmed, it would be a novel biological discovery for Euaesthetinae (not recorded for Euaesthetinae according to a recent review of myrmecophily in Staphylinidae by Parker 2016), though not unprecedented for the larger euaesthetine subgroup (Euaesthetinae + Steninae: some species of *Stenus*). *Edaphus* specimens have also been collected from old logs with fungi (via pyrethrum knockdown) and directly from fungi, rotting fruit, stream-edge flood debris, and under the bark of rotting logs. Most Australian species are fully winged, and many specimens have therefore also been taken in flight intercept traps. Larvae have been collected from Berlese-processed leaf, bark, and log litter samples.

Within *Edaphus*, there is a wide range of morphological variation in both overall body form and in the appearance and modification of specific structures, with many species groups being characterized by unique morphological characters. Among the more enigmatic of these characters and warranting special mention are the so-called "atrium pockets" of the Neotropical *ventralis* species group (Puthz 2006a: figs. 15–18). In several species of this group, these pockets comprise a cuticular shield that apparently encloses each of the spiracles of the first one to few abdominal segments in both sexes. It is as yet unclear what the function of these unique structures could be, but it is possible they function as some kind of regulatory structure, perhaps related to an unusual ecological association or microhabitat for these beetles. While many unique structures in *Edaphus* and other genera seem to be related to sexual selection, structures such as the atrial pockets have no obvious function and seem to indicate the evolution of unusual, but as-yet undiscovered biologies for many groups.

6.3.6.3 *Euaesthetus* (Holarctic)

This genus, the most northerly distributed of all Euaesthetinae, occurs into the far north of Europe, Russia, and Canada and as far south as Thailand in the Old World (Puthz 1994) and Panama in the New World (Puthz 2001a). At these southern range extremes, *Euaesthetus* beetles are restricted to high mountains (1500 m in the former, 1127–2900 m in the latter) and found in elfin cloud forest, grassland (large tussocks), montane mesophilous forest (with *Quercus* L., *Cupressus*), cloud forest, and mixed conifer-hardwood forest. Specimens have been sifted from tussock bases, extracted from Berlese-processed leaf and log litter, root mats, and flood debris. Most *Euaesthetus* species are found in North America, with many species distributed widely throughout. Puthz (2014a) provided collection data for the North American species, detailing the wide collecting circumstances (low and high elevation) in which these beetles have been found. In general they are mainly in wetland ecosystems (marshes, bogs, swamps) and riparian vegetation. They can be collected most abundantly from wet debris, moss, reeds, and litter in close proximity to streams and in dried river beds and ponds. Species have also been frequently found from litter sifted from the nests of various mammal species.

6.3.6.4 *Octavius* (Europe, Africa, Madagascar, Oriental and Neotropical Regions, Australia)

The genus *Octavius* comprises at least three or four morphological species groups (Clarke in prep.). The group for which there is by far the most available information includes flightless and frequently blind species from Europe, Madagascar, and South Africa (note, other species groups also have blind and/or flightless species). Orousset (2012) has written extensively about the ecology and biology of the French fauna and notes that although the environments of the Mediterranean and more northern mainland parts of France are quite different, the wingless *Octavius* there show great ecological tolerance, being broadly distributed in different habitats and elevations in both areas. In this region, this group of *Octavius* is mainly collected by soil washing and litter sifting. As similarly summarized for the Madagascan fauna by Orousset (1988), this group divides into those that live primarily in the soil (minute, all anophthalmous) and those in the surface leaf litter (larger, microphthalmous). They are found there primarily in dense mountain forests, from 900 to 2650 m, being most abundant from 1000 to 2000 m. In addition to different montane forest types, many species also occur at very high-elevation habitats, in various types of mountain scrub, and extending near to summits in turf-like communities where specimens have been sifted from dense mats of herbs and grasses growing on rocks. They have also been collected by Berlese-processing plant debris and soil samples taken from 0 to 10 cm depths. A few species there have been found in the litter accumulations of epiphytes. In South Africa the rich fauna of this group is found in similarly diverse habitats and collecting situations, similar as well to the austral euaesthetine fauna in general. They have been found from 10 to 1800 m (most records from above ~1100 m), mostly from podocarp and Afromontane forest, but they are also commonly collected from other forest types including coastal forest, various degraded forests, and mountain/alpine marshes and fynbos. Specimens have been collected from diverse litter types and also frequently from bryophytes. South African *Octavius* are generally very coarsely sculptured, and a waxy encrustation covering specimens is frequently observed (see also *Chilioesthetus*). This may be excreted from cuticular pores, which are sometimes visible in clean specimens under SEM examination. This possible excretion is yet to be studied and may relate to the prevalence of these beetles in moist habitats.

The Neotropical *Octavius* fauna comprises only ~27 species (e.g., Puthz 2001b), most of which form part of a distinctive group of minute, slender species, usually winged and with large eyes (also found throughout the Oriental and African tropics). This group is common at high elevations (>2000 m) but can be found throughout tropical lowlands, and specimens have been collected from all types of litter and in flight traps and associated with rotten wood. The species *O. panamensis* has been collected from the rubbish heaps of *Atta* ant nests and from thatch from a snapping ant nest, though it is unclear whether this is merely a facultative association. In Australia, this species group is so far known only from a single species from Christmas Island and from four localities on mainland Australia (*O. biroi* Puthz, 1977). Specimens have been found near sea level on Christmas Island, in rainforest near Mt. Tozer and on the Iron Range (Queensland), and in open *Eucalyptus* forest at 1490 m in Kosciusko National Park (New South Wales). Most collection records are from Berlese-processed forest litter, but specimens have also been found in flood debris, under bark, in fungi, and by malaise trapping.

By far the most enigmatic *Octavius* species belong to a group comprising species resembling *O. bicolor* and *O. flavescens* (Kistner, 1961). Several of these species are so unusual in their morphology that they were described in separate genera by different authors, some being placed in different staphylinid subfamilies (e.g., Piestinae: *Doletica bicolor* Cameron, 1938; Micropeplinae: *Nepalopeplus himalayicus* Coiffait, 1982). Sharp (1876), in describing his new genus *Turellus* (in Piestini), states of *T. batesi* Sharp, 1876 that it is "perhaps the most interesting of the Staphylinidae discovered by Mr. Bates, and ... the insect is one of the most anomalous of the Staphylinidae..." This group is comprised of mainly winged but also some flightless and blind species (e.g., *O. anophthalmus* Puthz, 1991) and is found throughout Africa and the Oriental Region (including in the high mountains of Africa and Nepal); only a single species (*O. batesi*) has so far been described from the Neotropical Region. Although found in largely similar collecting situations as other *Octavius* (including records from mammalian nests; e.g., Kistner 1961a), the peculiar morphology of species in this group, including the evolution of several novel structures and unusual character combinations, strongly suggests an as-yet unknown and possibly unique biology for these species.

6.3.6.5 *Protopristus* (Australia, New Zealand, Subantarctic Islands)

In Australia, *Protopristus* species are found in cool temperate rainforest dominated by *Nothofagus* and *Eucalyptus* species but can also be found in a variety of other habitat types (see *Austroesthetus*) from near sea level to 1650 m. Nearly all of the available Australian *Protopristus* material has been collected by Berlese processing of forest leaf litter samples, though recent soil sampling has also proven to be productive (in both Australia and New Zealand) and strongly indicates that this genus is common in both surface litter and subterranean microhabitats. However, the maximum depth below surface that this genus (and other soil-dwelling genera) might be found is not known. Some specimens (including larvae) have been taken from moss in Tasmania, but this is an otherwise rare microhabitat for Australian *Protopristus*. Larvae have also been collected from leaf litter; however, much more soil sampling has occurred in New Zealand than Australia, and this method has produced nearly all of the numerous larval specimens known from there (Clarke in prep.). New Zealand *Protopristus* have been collected from sea level to 1700 m, in litter from diverse forest types and also from alpine tussock grasslands and coastal megaherb communities on the Subantarctic Islands. In contrast to the Australian fauna, much New Zealand material has been collected from both moss and by soil washing, the latter likely reflecting greater soil sampling effort in New Zealand.

6.3.6.6 *Schatzmayrina* (Africa, Asia, South America)

Schatzmayrina is a small genus of only three species. One of these, *S. oxyclypea* Koch, 1934, is among the most widely distributed of euaesthetines; it occurs throughout Africa and Asia (Puthz 2007b) but it is unclear whether this distribution is entirely natural or partly human-assisted. The genus has been recorded from low to high elevations (150–1660 m in Southeast Asia; 800 m in Congo; 500–1240 m in South Africa) in diverse vegetation types. According to Kistner's (1961a: 30) translation of C. Koch's original description, Egyptian *S. oxyclypea* records were "in fields under stones," and the species "swarms over dirt on the shores of the Nile." Although stream-edge associations are known or suspected for other genera, it would be highly unusual to observe both large numbers of any Euaesthetinae at one time but also that they would be actively moving on open ground (such as can be observed for many *Stenus* species). *Schatzmayrina* does seem to prefer riparian habitats or otherwise high-moisture microhabitats away from rivers. Since species are fully winged, specimens have been collected at light and in flight intercept traps. Most records have been from forest (e.g., dense gallery forest) or from generally wet places including swamps and bogs. In Brazil, *S. braziliana* Puthz, 2007 has been found in white-water inundated forest where it was collected with an arboreal "photo eclector" during rising waters (Puthz 2007b). Other *Schatzmayrina* records are from Afromontane forest, from mixed exotic plantings, and from open, highly disturbed, and drier habitats such as dry forest, recently burned vegetation, forest edge or ecotones, savannah, and grasslands (Kistner 1961a, 1962). Specimens have been collected by sifting plant debris near watercourses, from diverse debris under *Cyperus auricomus* Clarke, and by Berlese-processed general litter. However, the genus has also been commonly collected under stones and from soil (including deep soil), and from samples taken from the base of hollow trees. In South Africa, the author has collected an unidentified *Schatzmayrina* species (likely *S. oxyclypea*) together with *Edaphus* sp. and *Octavius* sp. from fungi [probably *Lenzites elegans* (Spreng.)] growing prolifically on the stumps of recently felled trees, and this fits with the seemingly moisture-loving habits of this group. Of further potential interest are the several records suggesting *Schatzmayrina* may be a loosely associated nest inhabitant of small African mammals (Kistner 1961a). Specimens have been sifted from debris from several different rodent nests, but the association may be merely incidental since these nests were also in areas otherwise frequented by the beetles (from burrows in grasses; nests made of grasses on swampy soil).

6.3.7 Tribe Stenaesthetini

6.3.7.1 "EuaAUS" (Australia: Victoria)

This undescribed genus is found in cool temperate *Nothofagus* and *Eucalyptus* rainforest and wet sclerophyll forest from ~250 to 560 m elevation. The more common of the two known species has been collected primarily from Berlese-processed leaf litter but has also been found abundantly (via Berlese processing and pyrethrum fogging) in moss growing on the ground, on trees, and on old logs both in contact with the ground and not. It would seem that this taxon, like several other austral genera, is also able to exist outside the general forest floor by utilizing the high-moisture microhabitats provided by dense bryophyte growth. The other species is known from a single specimen collected much further north in the Grampian Ranges. This region is much drier than southern Victoria, and so a collection from there is unusual. Larvae remain unknown despite much collecting in relevant areas but will most likely be found in the soil, a microhabitat that has not yet been extensively targeted where this genus occurs.

6.3.7.2 *Agnosthaetus* (New Zealand)

A recent revision of this genus (Clarke 2011) included compilations of detailed ecological data gathered entirely from specimen labels. This information demonstrates a number of key ecological characteristics of the fauna: (1) several species seem to be "alpine specialists" (e.g., *A. lanceolatus* Clarke, 2011; *A. ecarinatus* Clarke, 2011; and other species found primarily in the Southern Alps). (2) Most species appear not to be restricted to specific forest types (e.g., *Nothofagus* versus podocarp-broadleaf forest, early succession versus mature old-growth forest, etc.). For many species, the distribution data suggest that they can be found in a wide range of general vegetation types ranging from forest through shrubland and tussock grassland (the three major vegetation types in New Zealand). (3) With the exception of primarily alpine species, other well-sampled species appear to be broadly distributed with respect to elevation; as a whole, *Agnosthaetus* beetles have been collected from sea level to >1600 m in the mountains. (4) Within major vegetation types, individual species are associated with a broad spectrum of different but related microhabitats, including general forest leaf litter, moss, wet debris, decaying vegetation, dead wood, and rarely soil and fungi. The general habitat requirement seems to be moist litter or vegetation, rather than a specific vegetation type—such habitats, e.g., moss, can be found both in "sheltered" forested environments through more open early successional forest, and in scrubland and even highly "exposed" grassland communities, such as high-elevation tussock grasslands. Similar to several other genera, some species have been collected via pyrethrum fogging of old logs and substrates with bryophytes, indicating that these beetles can also extend beyond the ground layer. Larvae have been collected from litter, moss, and soil (Clarke 2011).

6.3.7.3 *Stenaesthetus* (Africa, Madagascar, Oriental Region, Neotropical Region)

The genus *Stenaesthetus* is distributed in tropical areas of Africa, Asia, and South and Central America, with extensions into temperate areas of South Africa and mountainous areas of northern India, Nepal, and China. Although this broad distribution entails that *Stenaesthetus* species are found in diverse environments and habitat types, the group appears to be generally orophilic—preferring high-elevation montane and alpine habitats. The majority (~3/4) of species are flightless (brachypterous or apterous), showing morphological features typically correlated with the loss of flight and adaptation to endogeous life (e.g., eye reduction, depigmentation, increasing fusion of thoracic sclerites); nearly all of them are also restricted in distribution (most are likely narrow-range endemics). Flightless *Stenaesthetus* species occur in sheltered, mainly ground-layer microhabitats; as an ecomorphological group, they therefore seem to form a tropical ecological analog of the austral fauna. Although ecological data are otherwise scant for *Stenaesthetus*, several papers treating the faunas of the Oriental Region (Puthz 2013a), Madagascar (Orousset 1988), Africa (Puthz 2011), and the Neotropical Region (Orousset 1990b) were important in their compilation of basic ecological data for a broad cross section of *Stenaesthetus* diversity.

Approximately one third of described *Stenaesthetus* species are Oriental wingless species in the *quadrisulcatus* species group (Puthz 2013a). This group comprises distinctly orophilic species, almost all of them having been collected from mountain ranges, most from elevations exceeding 1400 m, and up to 4150 m in Nepal (*S. quadrisulcatus* Cameron, 1930). Most species from this group have been collected in general forest litter (leaf litter, grasses, ferns, and dead wood) from diverse forest types dominated by *Quercus*, *Rhododendron* L., *Abies* Miller, *Tsuga* Carrière, bamboo, or pine and from riparian and non-riparian habitats. Numerous collections from outside of the general forest litter suggest that these flightless species are also common in a broad range of other microhabitats, including in dead wood, under bark, on fungi, and under stones and logs. Although available data demonstrate that species from the *quadrisulcatus* group can occur in aboveground microhabitats (in dead logs not entirely contacting the ground, in aboveground bryophytes), none have been collected from the soil, and so the only indications that this group may also be soil-dwelling are from specimens taken under stones and logs.

Madagascan *Stenaesthetus* are also flightless and so far as known occur only in forested and alpine habitats on the eastern (wetter) side of the island (Orousset 1988). Species in the *rugosus* species group were reported as restricted to the northern part of the island and at low to mid elevations, from 80 to 1200 m; in contrast to the *quadrisulcatus* group, nearly all specimens were collected by soil washing and from sifted moss growing on soil. The *vadoni* species group, seemingly more orophilic and widespread, was recorded from 110 to 2000 m (most records from >1000 m), with species also collected by soil washing and from sifted moss (and general litter). Some species in this group therefore extend above the tree line and have been collected from moss growing on rocks. The species *S. haribe* Janák, 1996 and *S. miskovai* Janák, 1996 (*vadoni* group) were collected from thick moss growth and humus accumulations from around tree roots in a wet forest, 1150–1300 m. Janák (1996) also described *S. dunayi* (*rugosus* group) from a disparate locality in central Madagascar, collected at >1800 m from moss and litter on an old stump. This validates Orousset's (1988) caution that apparent ecological differences between species groups may only reflect collecting effort, and it seems likely that species in the *quadrisulcatus* species group may yet be found in soil samples.

In eastern Africa (Ethiopia, Kenya, Tanzania) a group of wingless species superficially similar to the abovementioned species groups has also been collected so far only at high-elevation sites. Three described species are known from mountains in Ethiopia, 2100–3230 m, where they were collected from soil under embedded stones near a stream and under embedded rocks and roots of ferns and grass mats (Puthz 1988b). Two similar species from Tanzania (Puthz 2012b) and one from Kenya (Puthz 1986) were found from 1050 to 1350 m by sifting forest litter. The species *S. leleupi* (Kistner, 1967), collected in forest litter from sites at 300 to 800 m elevation, also morphologically related to these species, is seemingly anomalous in its distribution; it is currently the only known species of this wingless "type" (e.g., a group of several wingless species, many originally described in the now-sunk genus *Aulacosthaetus* Bernhauer, 1939) from South Africa and is seemingly closely related (morphologically) to Madagascan and possibly Brazilian wingless species in the *mrazi* species group (see below). This is also the only *Stenaesthetus* species for which larvae have been collected (a single specimen, associated with a large series of adults, awaits description by the author).

The *mrazi* species group includes ten described wingless South American species collected in forests (e.g., Atlantic forest) south of the Brazilian Plateau, from 50 to 1100 m (Orousset 1990b). Although ecological/collection data are unavailable for most of these, the collection of *S. carinipennis* Puthz, 2011 from an abandoned cocoa plantation (Puthz 2011) suggests that at least some species may be able to withstand major disturbance (deforestation etc.) or at least retain a stronghold for some time following deforestation.

Nearly one quarter of described *Stenaesthetus* species are winged and fully flight capable, and most of these belong to either the "*Gerhardia*" species group (e.g., see Puthz 2011) in Africa and the Oriental Region or the *illatus* species group (Orousset 1988) in the Neotropical Region. The former group mostly consists of species sharing similar derived male genital segment and aedeagal characters (see below, Sect. 6.3.8.3), and several of these species also have broad distributions within the regions they occur in. For example, *S. sunioides* Sharp, 1874, the type species, extends from Pakistan east through Asia to Japan and Indonesia (Puthz 2013a) and from 200 to 2400 m; it has been suggested that the current range of this species (among the largest of Euaesthetinae) has been facilitated by human transport and disturbance since it is found frequently in rice fields and other disturbed habitats (Kistner 1962; Puthz 2013a), and it is clearly able to tolerate a wide range of environmental conditions. The three other Oriental species of this group have been found in mid- to high-elevation sites, from 200 to 1800 m. Most specimens have been collected from sifted litter from a variety of vegetation types (e.g., teak forest, bamboo, and grasslands) with some records of *S. conflictatus* Puthz, 1995 from elephant dung likely being incidental. African species of the *Gerhardia* species group have been collected from primarily high-elevation sites, from 150 to 2850 m (most with elevation data from above ~1300 m). Specimens have been collected from diverse vegetation types, mainly wet or humid habitats including rainforest, forest remnants, secondary and plantation forests, bamboo forest, gallery forest, general periaquatic and riparian habitats, and wetlands, but also in semi-arid habitats including sclerophyllous forest and savanna. These species have also been frequently collected from sifted litter of various types, as well as from soil and rotten wood, but have also been found in microhabitats more unusual for Euaesthetinae [under bark, from dead and live flowers, on mushrooms, and from the nests of a few different mammals, also at light (Kistner 1962; Puthz 2013a)].

The *illatus* group in the Neotropical Region is known from throughout the Amazonian basin and the Brazilian Plateau, reaching as far as the Andes in the northwest and occurring from 200 to 1630 m (though elevation data is scant for the known species). All species for which ecological data are available have been collected from rainforests, though *S. castaneus* Orousset, 1990 has been collected from a diverse range of wet and dry habitats including gallery forest (with bamboo), *Eucalyptus* plantation forest, bromeliad groves, Cactaceae scrub, and dry forest. Most *illatus*-group specimens have been collected by sifting leaf litter and dead wood, but some have also been found in rotten palm flowers, beaten from foliage (see below, Sect. 6.3.8.4), and taken at light (Orousset 1990b).

6.3.7.4 *Tyrannomastax* (Madagascar)

The genus *Tyrannomastax* is known only from the type series of the two described species (Orousset 1988), both of which were collected in sifted forest litter from southeastern Madagascar. The genus differs from *Stenaesthetus* only in the remarkably derived labial apparatus, in which several structures are modified or somewhat exaggerated compared to *Stenaesthetus*, and the two species are overall morphologically close to the *vadoni* species group. The labium appears to be modified to function as a prehensile prey-capture apparatus, and it has been suggested that if not homologous to the apparatus found in Steninae: *Stenus* (Leschen and Newton 2003; Betz and Kölsch 2004), then it is at least similar in its function. The anatomy and morphology of the labium are certainly different from that of *Stenus* (Clarke in prep.) and are fundamentally identical to that in *Stenaesthetus*. The modified labial palps are rigid, elongated, subconnate, and laterally extended, with the apical palpomere directed ventrad. This modification, combined with the ventrally curving, elongated, and subsclerotized (yet flexible) "paraglossae," seems to combine to form a claw-like apparatus (e.g., see Orousset 1988: fig. 415) that may serve to restrain captured prey (perhaps by pinning it to the ground) in order to draw it closer to the mouthparts. It is notable as well that the mentum is equipped with a pair of long ventrally projecting spines that may also assist in this function. Moreover, detailed examination of numerous *Stenaesthetus* species by the author has revealed a range of similar yet clearly nonhomologous modifications of the mouthparts, suggesting different morphological solutions to the functional problems associated with prey capture and/or feeding strategies within this group.

6.3.8 Biology and Morphology

In contrast to the wealth of biological information that has accumulated for Steninae (Chap. 11, this book), no studies of the biology or life history of any euaesthetine species have yet been made. The generalized falcate structure of the mandibles of Euaesthetinae suggests they are predators of other micro-arthropods, as in most other subfamilies placed in the "staphylinine group," whose included taxa are characterized by extraoral digestion (Lawrence and Newton 1982; Grebennikov and Newton 2009). Predation and preferred prey have not yet been directly observed for Euaesthetinae—though both predatory behavior and prey are well-studied for Steninae (e.g., Betz 1998b; Leschen and Newton 2003) and they have similarly structured mandibles and maxillae. The one exception may be a report by Remillet (1969), who kept a specimen of *Octavius massatensis* Coiffait, 1959 alive for a month feeding it diplurans. Notable intergeneric differences in the mandibles as well as other mouthpart structures further suggest either different predatory strategies or different prey in the different genera, perhaps partly resulting from ecological partitioning of, or adaptation to, alternative prey resources.

It is typical to find Euaesthetinae as singletons or small series in samples. However, the occasional finds of large numbers in litter samples might indicate a general (and possibly temporary) clustering within populations. Clusters of individuals on suitable microhabitats or substrates would be consistent with sexual selection and sexual conflict-type behaviors occurring in those species. This may be a general explanation for the morphologies discussed below. There is otherwise no other available information on the life history or biology of Euaesthetinae; only morphological variation allows for some inference. For genera in which larvae are known, they seem to occur in the same microhabitats as adults; however, this might not be true for common genera like *Kiwiaesthetus* and *Mesoaesthetus*, whose larvae remain unknown despite the abundance of adults in collections, and which thus likely occur in different habitats from the adults.

6.3.8.1 Morphological Traits Linked with Subterranean Life

In describing the high degree of adaptation of French *Octavius* to a subsurface soil-dwelling lifestyle, Orousset (2012) describes all stages of these beetles as developing in the soil, where they migrate throughout the rhizosphere depending on temperature and humidity. It is unclear whether his comments are based on observation or inference from field experience, but he describes excessive moisture as likely promoting vertical movement toward the surface and drought or freezing temperatures promoting movement in the other direction, where they likely would take refuge in deep crevices and pockets or partly decomposing roots and other plant structures.

Various morphological characters seem to be linked to the association with soil habitats and likely other aspects of their as-yet unknown biologies as well. Wing loss or reduction, generally associated with reduced pterothoracic volume, fusions of the elytra with each other and with the pterothorax (and to varying degrees), and other changes, is nearly ubiquitous in the austral-endemic genera and many other flightless groups, with many species also being blind (or with a reduced number of ommatidia) and showing varying degrees of depigmentation. These species also tend to have shorter, more robust legs. The association with leaf litter and soil habitats is therefore strongly reflected in several aspects of the morphology of numerous unrelated groups of Euaesthetinae and represents a major example of convergence involving a syndrome of interrelated and likely non-independent morphological changes.

6.3.8.2 Dimorphism Linked to Mating and Sexual Selection?

Sexual dimorphism involving nongenital or genital segment characters is a significant feature of many Euaesthetinae and suggests general behaviors and life history traits correlated with these characters. The most prominent category of dimorphic morphologies includes (usually male-specific) secondary sexual characters of the abdomen. Interspecifically variable modifications of the male abdominal sternites are known for many genera of Euaesthetinae. For example, males of most or all species of *Agnosthaetus*, *Alzadaesthetus*, *Edaphosoma*, *Kiwiaesthetus*, and *Mesoaesthetus* and some *Tamotus* and *Edaphus* species show varying degrees of (always) species-specific modification of one or more abdominal sternites (cuticular structures or modifications; often with modified setae). The degree of modification is typically highly interspecifically variable, but qualitative patterns of correlated variation (with other structures) among genera are perplexingly inconsistent and lead to interesting functional questions. For example, in *Agnosthaetus* and *Edaphosoma*, there are abdominal dimorphisms as well as strong interspecific differences in both the median lobe shape and internal sac structures (Puthz 2010a; Clarke 2011). But male sternites modified to a similar degree in *Mesoaesthetus* and *Kiwiaesthetus* species are not matched by such divergent median lobe shapes and internal sac structures (e.g., Puthz 2008b). The reverse pattern is seen in other groups. For example, the male abdomen of EuaAUS species is not modified, but the aedeagus is a highly modified lanceolate structure with reduced parameres and elaborate internal sac structures, while the female gonocoxites are characterized by strong secondary sclerotization. *Nothoesthetus* and *Protopristus* species show a similar pattern. Males of *Protopristus* species almost never exhibit secondary sexual abdominal modifications but likewise have complicated aedeagal morphology, including particularly complex internal sac structures, and the females likewise have heavily sclerotized internal vaginal structures (possibly related to, or in place of, the spermatheca). Puthz (2010a, 2013b) illustrates for males of several *Edaphosoma* species extraordinary forked processes arising from the apex of sternite VI, as well as substantial genitalic differences between species. Males of both described *Alzadaesthetus* species have secondary sexual characters on the abdomen, but these are completely different in each species: *A. furcillatus* and one new species both have an exaggerated explanate protuberance on the ventral posterior margin of segment VI (Sáiz 1972: fig. 8). Curiously, this structure is apically concave and lined with appressed, sub-tuberculate, and transversely ribbed spines. The function of this structure must presumably relate to mating, but the apical tuberculate surface also resembles a stridulatory file (though there is no obvious plectrum-like structure; this could be the hind legs). *Alzadaesthetus chilensis* males have instead a seemingly only minor modification, and to the apex of sternite IV (Kistner 1961b: fig. L), and the two species have dissimilar male genitalia. Secondary sexual abdominal modifications are not limited to the ventral side: Puthz (1990: figs. 3 and 11) illustrates remarkable modifications to the male tergites of *Edaphus nitidifrons* Puthz, 1990 and *E. sumatrensis* Schaufuss, 1887, which involve cuticular and setal developments.

Further notable examples of sexual dimorphism involve other structures or body regions, including the antennae, mouthparts, and legs. In *Edaphus*, males often have exaggerated (usually elongate) apical antennomeres compared to females (e.g., *E. ventralis* Puthz, 2006, *E. ventricula* Puthz, 2006; see Puthz 2006a), and males of other species show a modified eighth antennomere. The Australian taxon EuaAUS is notable for pronounced sexual labrum (and to a lesser degree, mandibular) dimorphism that may relate to intersexual dietary specialization. Similar modifications have been described for *Agnosthaetus* also, including perhaps the only examples of Euaesthetinae (*A. newtoni* Clarke, 2011 and *A. thayerae* Clarke, 2011) with more than one tooth along the inner mandibular edge and only in males (Clarke 2011). These derived labral morphologies are notable in providing examples of dimorphisms in which both sexes have a modified structure, rather than simply the male possessing some structure, or exaggerated structure, that is lacking or poorly developed in the female. Sexually dimorphic characters of the legs are common. Males of several *Octavius* and *Protopristus* species have modified metafemora (swollen and occasionally with a tooth on the inner face), and in many species, also in *Fenderia*, males have tenent setae on the ventral side of the tarsi (e.g., Orousset 2012: fig. 20; Clarke and Grebennikov 2009: fig. 13K), possibly used to grasp the female during mating.

Female-specific dimorphic characters not related to genitalia or the genital segment are extremely rare in Euaesthetinae and are perhaps the most interesting examples of sexual dimorphism. One locally endemic undescribed species of *Kiwiaesthetus* in northwest Nelson, New Zealand, is notable for an enlarged mesothoracic process in the females. This structure, with a concavity on each side resembling a socket, is unique to this species, and nothing like it is known anywhere else in Euaesthetinae suggesting it is involved in some novel biological function. Other examples of female dimorphisms include apparent sexual wing dimorphism and associated sexual eye dimorphism in some undescribed Australian *Edaphus* (and possibly some *Austroesthetus*) species that indicate sedentary females and dispersive males.

These and many other modifications in Euaesthetinae likely relate to mating and have different sensory and mechanical functions. The very different morphologies and locations involved in dimorphism suggest a diverse set of life history attributes and mating strategies within Euaesthetinae that are yet to be studied. Together these intergeneric morphological patterns indicate that biological differences likely exist between genera (or groups of genera) as well.

6.3.8.3 Complex Genital Structures

Complex genital structures (usually only of males) occur in most species of Euaesthetinae, suggesting that mating is characterized by sexual selection or sexual conflict-type behaviors (Eberhard 1985). Two categories of male sexual structures in Euaesthetinae are most important to emphasize. The first consists of male aedeagus structures that interact with the female during copulation (mainly elaborate internal sac structures) and are the most pronounced and widespread examples of genital complexity in Euaesthetinae. Nearly all genera show diverse internal sac structures, but both the presence and absence of derived internal sac structures likely indicate biological differences among taxa. The wide range of structural diversity seen in the internal sac suggests highly nuanced biological functions likely associated with sexual competition or sexual selection and suggests that these latter two mechanisms are prevalent in the mating biology of many Euaesthetinae. An unusual example of complex genital morphology is seen in several African *Stenaesthetus* species (Puthz 1995). These species have an extremely elongate flagellum that in repose is coiled within the median lobe. During mating, presumably this structure engages with the comparably elongate spermatheca in the female, as has been described for an aleocharine staphylinid with similar structures (Gack and Peschke 2005). Interestingly, the novel internal modifications of the ninth tergite and sternite in these *Stenaesthetus* species may be used analogously to how the elytra is used in the aleocharine—a behavior termed "shouldering" by those authors is used to ensure retraction of the flagellum without entanglement, and presumably a similar biomechanical problem occurs in these *Stenaesthetus* species.

The second category of male genital characters includes two internal structures that do not physically interact with the female during copulation and are found in males of many species of *Edaphus*, *Schatzmayrina*, *Stenaesthetus*, and *Tamotus*. Puthz (1973: figs. 24, 28–29) has described two internal structures, which he named the "double-trumpet" and "vesica seminalis." These are best developed in *Tamotus*, where they can occupy a substantial space in the abdomen. The "double-trumpet" structure is seen in scattered species in the aforementioned genera and is thought to represent a sperm pump or be otherwise involved in sperm transfer. Weakly coiled longitudinal muscles surrounding this structure (and resembling an extended corkscrew) have been observed and support the notion that the two "trumpet" ends of the structure can contract toward each other. The "vesica seminalis" is found mainly in *Tamotus* but also in *Edaphus* (e.g., Puthz 2010b) and may function as a temporary sperm reservoir. Ultimately, the functions of these remarkable structures in Euaesthetinae remain unknown, and detailed studies are required; there may be, for example, as-yet unrecognized correlations between the differential presence, form, or size of these internal structures and that of other genitalic or apparently morphologically unlinked structures.

6.3.8.4 Arboreal Habitats for Some Euaesthetinae?

Several collections of South American *Stenaesthetus* have been directly from foliage. This ecological association is apparently unusual for Euaesthetinae but may be more widespread than currently appreciated, at least for the winged fauna. The tarsi of these and related *Stenaesthetus* species are clothed with explanate setae, including several elongate spatulate setae near the apices of the tarsomeres, which is consistent with the idea that these beetles walk on flat surfaces. Other cases of this ecomorphological pattern are known: the tarsal morphology of *Tamotus*, combined with collection records (see above), suggests this genus may be primarily an arboreal one—*Tamotus* is unique within Euaesthetinae in having strongly bilobed penultimate tarsomeres (Puthz 1973: figs. 18–20) with dense tenent setae, features again typical of beetles that crawl on smooth plant surfaces.

6.4 The Fossil Record and Evolution of Euaesthetinae

6.4.1 Cenozoic Fossil Record

The fossil record for Euaesthetinae is scant compared with other staphylinid subfamilies and is represented by only six described amber specimens in four nominal genera from three tribes and an additional unassigned compression fossil (reviewed in Clarke and Chatzimanolis 2009; Chap. 3 of this book). Considering the extraordinary diversity of extant *Edaphus*, it is notable that no fossil species of that genus has yet been discovered. One silicified fossil extracted from calcareous nodules from the Barstow Formation, California (~middle Miocene), and now preserved in the US National Museum (USNM-MO 561993), was recently identified as potentially belonging to this genus or to Pselaphinae (A. Newton pers. comm.; Fig. 6.5a, b). Although the author's preliminary study of this specimen in situ could not improve the confidence in the identification (originally described as *Carpelimus* sp.; Palmer 1957), the observable details of the dorsal side are consistent with *Edaphus* but are insufficiently preserved to allow conclusive determination. The form of the abdomen is reminiscent of *Edaphus* (Fig. 6.5a), and Palmer's description of the ventral characters is also largely consistent with this genus. But the "many closely spaced spines" on the posterior face of the mesocoxa would be unusual for *Edaphus*, and further study using sophisticated imaging may be necessary to confirm its placement.

The genera in Baltic amber (Eocene; ~44 Ma) include *Euaesthetus*, *Octavius* (Euaesthetini), and *Stenaesthetus* (Stenaesthetini), but the specimens were not sufficiently well preserved to allow formal naming (Puthz 2006d, 2008c). Only the first two generic records are from within the current distributional limits of these genera; *Stenaesthetus* is now largely restricted to the tropics. However, although all of the recorded species show clear affinities with extant species, they also either belong to species groups not now known from the Western Palearctic or comparable temperate regions elsewhere or are most similar to species far disjunct from the Palearctic. Being similar to *O. securifer* Puthz, 2006 (Puthz 2008c), the Baltic amber *Octavius* specimen seems to belong in a species group now largely restricted to (or at least with greatest diversity in) South Africa, rather than that in which the European species belong. Moreover, the same is true for another recently discovered undescribed *Octavius* species from Baltic amber (Fig. 6.5c). This specimen, while having a seemingly more primitive antennal

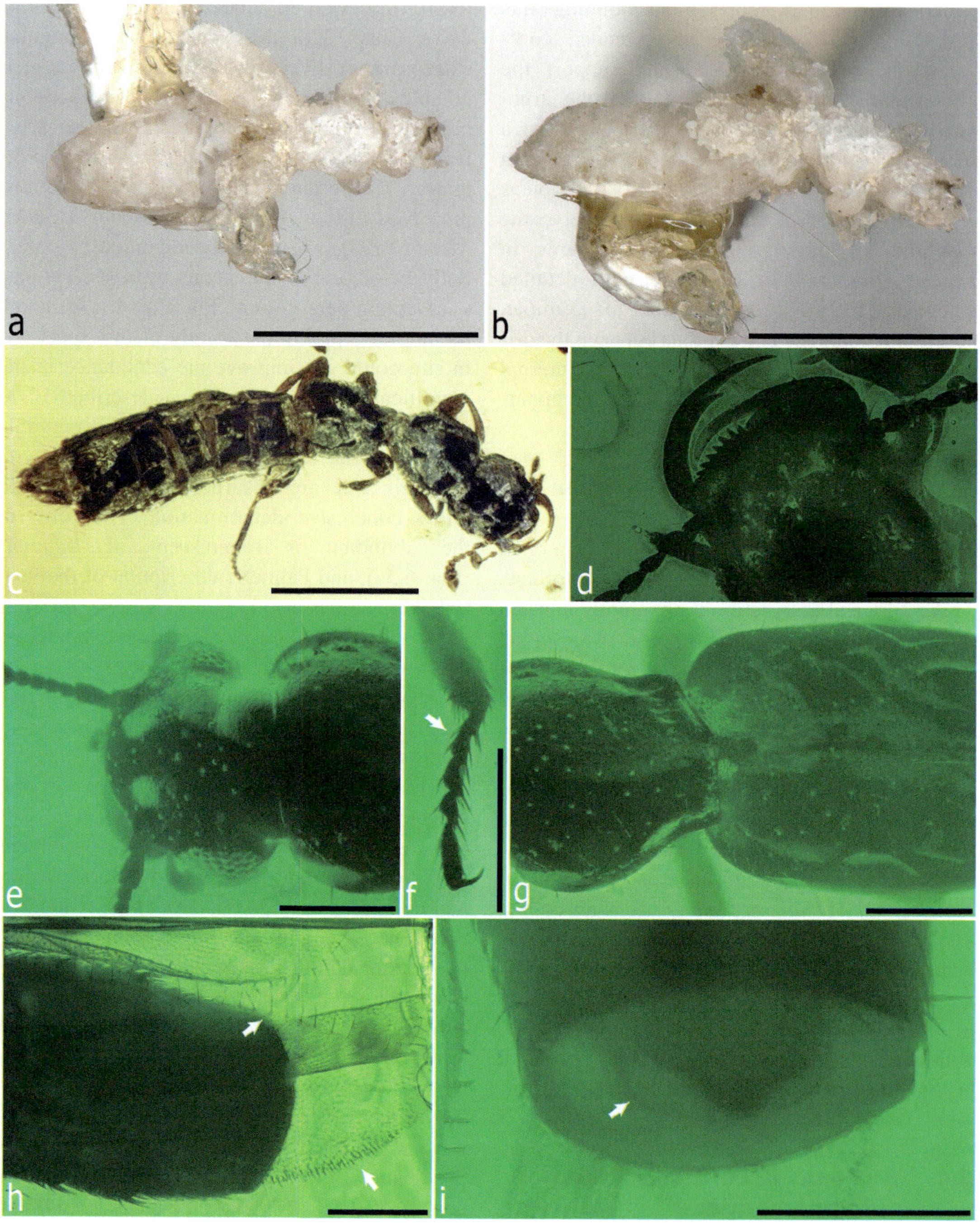

Fig. 6.5 Some fossil Euaesthetinae: (**a**) specimen #USNM-MO 561993 ("*Carpelimus* sp." = ?*Edaphus*; dorsal); (**b**) same (dorsolateral); (**c**) *Octavius* sp. (Baltic amber); (**d**) Euaesthetinae, gen. nov. (Cretaceous Burmese amber); (**e–i**) holotype of *Nordenskioldia pentatarsus* (Lefebvre et al., 2005); (**e**) head and front of prothorax (dorsal); (**f**) right metatarsus (dorsal oblique); (**g**) pronotum and elytra; (**h**) detail of wings; (**i**) telescoped abdominal apex. Scale bars in (**a–b**) = 1000 μm, (**c–i**) = 15 μm. See text for discussion of structures identified by arrows

club that is rare within the genus, belongs to a largely pantropical species group along with species like *O. neotropicus* Puthz, 1977 and *O. biroi*. This new fossil does not "fit" with most of the rest of the blind or nearly blind Palearctic *Octavius* fauna (belonging in a different species group). It has relatively large eyes typical of the tropical species group in which it likely belongs and may provide further indication of the paleoclimate of the Baltic region during the Eocene (Grimaldi and Engel 2005).

6.4.2 Cretaceous Fossil Record

Mesozoic fossils substantiate the view that crown-group Euaesthetinae—those having the putative synapomorphies listed above, notably the serrate labral edge—appeared as early as the Early Cretaceous, 125–135 Ma (e.g., Fig. 6.5d) and that these fossils reflect a pattern of scattered extant higher taxa in Staphylinoidea occurring already by the early Mesozoic (Lefebvre et al. 2005). The first and oldest recorded species, described from Lebanese amber in the extinct genus *Libanoeuaesthetus* Lefebvre et al., 2005 was subsequently transferred to the extant genus *Nordenskioldia* (Puthz 2008c). A detailed comparative morphological study of this fossil is currently under way, and although Puthz (2008c) correctly pointed out the superficial (albeit fairly extensive) original description of this species, it can now be confirmed that it does not belong in *Nordenskioldia*. It lacks the diagnostic characters of that genus: deep dorsal tentorial pits on the vertex, median pronotal impressions, and basal arcuate abdominal ridges (Fig. 6.5e, g; Clarke, in prep.). New fluorescent confocal imaging of this fossil has also revealed other characters potentially informative for its eventual phylogenetic placement. For example, among other details not previously studied or in need or re-evaluation, the tarsal formula may be 4-4-4, not 5-5-5 as in *Nordenskioldia* (e.g., see Lefebvre et al. 2005: fig. 3C, who illustrated a distinct tarsomere articulation at the position of the arrow in Fig. 6.5f but also illustrated only a line bisecting the basal pro- and mesotarsomere—i.e., no clear articulation). The pronotum also has basolateral impressions and a basal line of foveae and the pterothorax an elongate scutellum (Fig. 6.5g). Fossils with visible wings can be particularly valuable since characters from this structure may be visible. Although the original description of *Nordenskioldia pentatarsus* (Lefebvre et al., 2005) illustrated elongate setulae along the trailing wing edge, Fig. 6.5h also documents the presence of short setulae along the leading edge (Fig. 6.5h, top and bottom arrows, respectively), and the configuration of these and other wing structures varies among winged euaesthetine genera. The apical abdominal structure identifies the holotype as female based on the visible angulate eighth sternite, which has the form characteristic of females of at least Austroesthetini, Stenaesthetini, and Alzadaesthetini, and a few genera from other tribes, and protrudes slightly apicad from tergite VIII (marked by the arrow in Fig. 6.5i). It is notable that the type of antennal club of *N. pentatarsus* is most similar to that of *Austroesthetus*; indeed this specimen bears a strong overall resemblance to the only winged (currently undescribed) species of *Austroesthetus* from Western Australia, though unlike that genus it has a completely margined abdomen.

The description of *Octavius electrospinosus* Clarke and Chatzimanolis, 2009, introduced the second oldest euaesthetine from Burmese amber (~100 Ma), and based on observable characters that fossil seemed to be best placed in the extant genus *Octavius*. This conclusion has subsequently been "corroborated" by a recent phylogenetic analysis of Euaesthetinae and Steninae that placed this species as the sister group of a Neotropical *Octavius* species (Zyla et al. 2017). With this and the Baltic amber fossils discussed above, the *Octavius* lineage now has a confirmed paleontological record spanning ~50 Ma. Another undescribed euaesthetine fossil from Burmese amber either is an additional fossil species of this genus or is morphologically close (Clarke et al. unpub.).

6.4.3 Extinction Resilience and the "Environmental Buffer Effect"

It is becoming increasingly clear that Cretaceous "euaesthetine subgroup" diversity comprised a mix of both extinct and extant lineages. Recently discovered Burmese amber fossils are proving to be diverse, with at least three new genera of Euaesthetinae (e.g., Fig. 6.5d) and one of Steninae now known (Zyla et al. 2017; Clarke et al. in prep.). The hypothesized existence of crown-group Euaesthetinae and still-extant euaesthetine genera in the Lower Cretaceous is significant for at least two reasons. First, it demonstrates the great antiquity of these derived lineages, highlighting the long-term persistence of morphological taxa. Second, it places their minimum age, and by extension all other basally subtending nodes, into deep geological time. The existence of "cloistered habitats" (Stanley 1984), such as the consistently mesic habitats where euaesthetines occur (Sect. 6.3, above), led Clarke and Chatzimanolis (2009) to propose the hypothesis that the continuous presence of mesic habitats over geological time may explain the apparent morphological stasis in these lineages, but likely also other lineages as well (e.g., Cai et al. 2014: Olisthaerinae, Jurassic; Clarke et al. in prep.: ?*Stenus*, Burmese amber). Generally, this hypothesis can be labeled the "environmental buffer effect."

Assuming that observable ecological associations were similar in the Cretaceous ("uniformitarianism"; e.g., Gould 1965), it is straightforward to understand how Euaesthetinae and other groups may have been buffered from extinction through geological time. Using the New Zealand fauna as an example, within genera the distributions of individual species collectively span diverse geological settings, climatic regimes, and vegetation types (Clarke 2011). If these lineages are "paleoaustral" (sensu Fleming 1963), having drifted with New Zealand since the breakup of Gondwana, it is likely that they were little affected by the extreme and cyclical climatic/environmental changes of the Tertiary period and particularly those of the Pliocene and Pleistocene periods that resulted in widespread extinction of many plant and animal groups and drastic changes in composition of the regional biota in New Zealand (see Mildenhall 1980; McGlone 1985; McGlone et al. 2001; Worthy et al. 2007). This resilience to extinction may be a virtue of Euaesthetinae and other litter-dwelling staphylinids when considering target groups for biogeographic analysis.

6.5 Conclusions

Taxonomic work on the saw-lipped rove beetles has revealed a great diversity at the species level, but distinct lineages also still remain to be further described and explored. More collecting in association with monographic work on the austral fauna in particular is needed to better understand this fauna and its connection to northern temperate and tropical faunas, as well as the biogeographic structure of individual groups. Most genera in the austral region are relatively species-poor, but the genus *Protopristus* has radiated into a diverse group, likely comparable in richness to some of the tropical and northern genera, which include most of the diversity of Euaesthetinae. An increasing database of specimen-level collection data is beginning to reveal much new insight into the biological and ecological diversity of Euaesthetinae. This information suggests that there are distinct differences among genera and groups of taxa in microhabitat preferences but that the larger habitat within which many species and genera occur is free to vary considerably. Moreover, ecological data associated with specimen records indicates that general collection methods used to find Euaesthetinae may not be able to provide sufficiently resolved information on the biologies and ecological requirements of most taxa, suggesting new efforts to directly locate in the field and rear specimens in captivity could provide critical and novel data on the natural history of Euaesthetinae. The increasingly better known fossil record and future dedicated ecomorphological and biological studies will eventually expand the kinds of evolutionary inferences that can be made about the group, offering insight into general phylogenetic and biogeographic problems. The Cretaceous fauna from Myanmar was diverse, comprising both extant and extinct genera, while the Tertiary record preserves fossils so far assignable only to extant taxa.

Acknowledgments Funding for this work was provided by the US National Science Foundation, PEET Grant #DEB-0118749 (to Margaret Thayer and Al Newton); The University of Illinois, at Chicago; a Lester Armour Graduate Fellowship (Field Museum of Natural History); an Ernst Mayr Travel Award in Animal Systematics (Harvard University/Museum of Comparative Zoology); a Grant-in-Aid of Research (Sigm Xi); a Chicago Consular Corps Scholarship (University of Illinois, at Chicago); and a 21st Anniversary Research Fund Grant (Entomological Society of New Zealand). I thank the editors for inviting me to contribute this chapter. Volker Puthz, John Nunn, Chenyang Cai and Diying Huang, Dany Azar, and all of the curators and collection managers of institutions from which I have sourced materials for this contribution are gratefully acknowledged. Yvonne Matos and Richard Bloomfield assisted with specimen databasing. I thank Al Newton and Margaret Thayer, especially, for access to literature and unpublished taxonomic catalogs and databases used in this work; Shuhei Yamamoto for permission to use the Baltic amber fossil *Octavius* photo; Mark Florence (USNM) for supplying images of the possible *Edaphus* fossil; and Volker Puthz and the editors for critically reading the manuscript. I thank Director Omar Skalli and Lab Coordinator Lauren Thompson (Integrated Microscopy Center, University of Memphis) for their assistance with fluorescent microscopy and Betty Strack (Field Museum of Natural History) for assistance with scanning electron microscopy. Parts of this chapter are based on my doctoral dissertation, earlier drafts of which were generously read by Margaret Thayer and Lawrence Heaney (Field Museum of Natural History) and Roberta Mason-Gamer, Mary Ashley, and Roy Plotnick (University of Illinois, at Chicago). I also acknowledge funding from NSF to Duane McKenna (DEB1355169) and thank Seunggwan Shin, Stephanie Haddad, and Cristian Beza-Beza (University of Memphis) for useful discussion that improved earlier draft versions of parts of this manuscript.

References

Bernhauer M (1916) Results of Dr E Mjöberg's Swedish scientific expeditions to Australia 1910–1913: 7 Staphyliniden. Arkiv för Zoologi 10(5):1–7

Betz O (1998a) A behavioural inventory of adult *Stenus* species (Coleoptera: Staphylinidae). J Nat Hist 33:1691–1712

Betz O (1998b) Comparative studies on the predatory behaviour of *Stenus* spp (Coleoptera: Staphylinidae): the significance of its specialized labial apparatus. J Zool (Lond) 244:527–544

Betz O, Kölsch G (2004) The role of adhesion in prey capture and predator defence in arthropods. Arthropod Struct Dev 33(1):3–30

Cai C, Beattie R, Huang D (2014) Jurassic olisthaerine rove beetles (Coleoptera: Staphylinidae): 165 million years of morphological and probably behavioral stasis. Gondwana Res 28(1):425–431

Clarke DJ (2011) A revision of the New Zealand-endemic rove beetle genus *Agnosthaetus* Bernhauer (Coleoptera: Staphylinidae). Coleopt Soc Monogr 10:1–118

Clarke DJ, Chatzimanolis S (2009) Antiquity and long-term morphological stasis in a group of rove beetles (Coleoptera: Staphylinidae): description of the oldest *Octavius* species from Cretaceous Burmese amber and a review of the "Euaesthetine subgroup" fossil record. Cretac Res 30(6):1426–1434

Clarke DJ, Grebennikov VV (2009) Monophyly of Euaesthetinae (Coleoptera: Staphylinidae): phylogenetic evidence from adults and larvae, review of austral genera, and new larval descriptions. Syst Entomol 34:346–397

Coiffait H, Decu VG (1970) Recherches sur les Coléoptères endogés des Carpates roumaines III Staphylinidae – Euaesthetinae: *Euaesthetotyphlus almajensis* n. gen. n. sp. Annales de Spéléologie 25:377–382

Eberhard WG (1985) Sexual selection and animal genitalia. Harvard University Press, Cambridge

Fleming CA (1963) The nomenclature of biogeographic elements in the New Zealand biota. Trans R Soc NZ 88:13–22

Frank JH (1991) Staphylinidae (Staphylinoidea). In: Stehr FW (ed) Immature insects, vol 2. Kendall/Hunt Publishing, Dubuque, pp 341–352

Gack C, Peschke K (2005) 'Shouldering' exaggerated genitalia: a unique behavioural adaptation for the retraction of the elongate intromittant organ by the male rove beetle (*Aleochara tristis* Gravenhorst). Biol J Linn Soc 84(2):307–312

Gould SJ (1965) Is uniformitarianism necessary? Am J Sci 263:223–228

Grebennikov VV (2005) Older-instar larvae of Pseudopsinae (Coleoptera: Staphylinidae): morphological description of three genera and phylogenetic placement of the subfamily. Eur J Entomol 102 (4):713–724

Grebennikov VV, Newton AF (2008) Minute larvae of Leptotyphlinae (Coleoptera: Staphylinidae): description of three genera with discussion on the monophyly and phylogenetic position of the subfamily as inferred from larval morphology. Zootaxa 1817:49–58

Grebennikov VV, Newton AF (2009) Good-bye Scydmaenidae, or why the ant-like stone beetles should become Staphylinidae sensu latissimo (Coleoptera). Eur J Entomol 106:275–301

Grimaldi DA, Engel MS (2005) Evolution of the insects. Cambridge University Press, New York

Hansen M (1997) Phylogeny and classification of the staphyliniform beetle families (Coleoptera). Biologiske Skrifter, Det Kongelige Danske Videnskabernes Selskab 48:1–339

Herman LH (2001) Catalog of the Staphylinidae (Insecta: Coleoptera) 1758 to the end of the second millennium IV Staphylinine Group (Part 1) Euaesthetinae, Leptotyphlinae, Megalopsidiinae, Oxyporinae, Pseudopsinae, Solieriinae, Steninae. Bull Am Mus Nat Hist 265:1807–2440

Janák J (1996) Zwei neue madagassische *Stenaesthetus*-Arten (Coleoptera, Staphylinidae, Euaesthetinae). Entomofauna: Zeitschrift für Entomologie 17 (20):337–341

Janák J (2014) New species and subspecies of *Octavius* from South Africa, with a key and additional distribution records (Coleoptera: Staphylinidae: Euaesthetinae). Acta Entomologica Musei Nationalis Prague 54(1):195–231

Kasule FK (1966) The subfamilies of the larvae of Staphylinidae (Coleoptera) with keys to the larvae of the British genera of Steninae and Proteininae. Trans R Entomol Soc Lond 118:261–283

Kistner DH (1961a) Euaesthetinae (Coleoptera Polyphaga), Fam. Staphylinidae. Exploration du Parc National de la Garamba Mission H de Saeger, en collaboration avec P Baert, G Demoulin, I Denisoff, J Martin, M Micha, A Noirfalise, P Schoemaker, G Troupin et J Verschuren (1949–1952) Fascicule 21 (2):11–40. Institut des Parcs Nationaux du Congo et du Ruanda-Urundi, Bruxelles

Kistner DH (1961b) A new genus and species of Euaesthetinae from Chile (Coleoptera: Staphylinidae). Pan Pac Entomol 37:216–220

Kistner DH (1962) A revision of the Old World species of the genus *Stenaesthetus* Sharp and a review of the Euaesthetinae of Angola (Coleoptera: Staphylinidae). Publicações Culturais da Companhia de Diamantes de Angola 54:93–107

Lawrence JF, Newton AF (1982) Evolution and classification of beetles. Annu Rev Ecol Syst 13:261–290

Lefebvre F, Vincent B, Azar D, Nel A (2005) The oldest beetle of the Euaesthetinae (Staphylinidae) from Early Cretaceous Lebanese amber. Cretac Res 26:207–211

Leschen RAB, Newton AF (2003) Larval description, adult feeding behavior, and phylogenetic placement of *Megalopinus* (Coleoptera: Staphylinidae). Coleopt Bull 57(4):469–493

McGlone MS (1985) Plant biogeography and the late Cenozoic history of New Zealand. NZ J Bot 23:723–749

McGlone MS, Duncan RP, Heenan PB (2001) Endemism, species selection and the origin and distribution of the vascular plant flora of New Zealand. J Biogeogr 28:199–216

McKenna DD, Farrell BD, Caterino MS, Farnum CW, Hawks DC, Maddison DR, Seago AE, Short AEZ, Newton AF, Thayer MK (2015) Phylogeny and evolution of Staphyliniformia and Scarabaeiformia: forest litter as a stepping stone for diversification of nonphytophagous beetles. Syst Entomol 40(1):35–60

Mildenhall DC (1980) New Zealand late Cretaceous and Cenozoic plant biogeography: a contribution. Palaeogeogr Palaeoclimatol Palaeoecol 31:197–233

Naomi S-I (2014) The lateral sclerites of the pregenital abdominal segments in Coleoptera (Arthropoda: Hexapoda). Jpn J Syst Entomol 20(2):319–335

Newton AF Jr (1985) South Temperate Staphylinoidea (Coleoptera): their potential for biogeographic analysis of austral disjunctions. In: Ball GE (ed) Taxonomy, phylogeny and zoogeography of beetles and ants, Series Entomologica, vol 33. W. Junk, Dordrecht, pp 180–220

Newton AF (1990) Insecta: Coleoptera: Staphylinidae adults and larvae. In: Dindal DL (ed) Soil biology guide. Wiley, New York, pp 1137–1174

Newton AF (2017) Nomenclatural and taxonomic changes in Staphyliniformia (Coleoptera). Insecta Mundi 0595:1–52

Newton AF, Thayer MK (1992) Current classification and family-group names in Staphyliniformia (Coleoptera). Fieldiana: Zoology (NS) 67:1–92

Oke C (1933) Australian Staphylinidae. Proc R Soc Victoria (NS) 45:101

Orousset J (1987) Un nouveau genre d'Euaesthetinae africain: *Macroturellus pulcher* n. gen., n. sp. Bulletin de la Société Entomologique de France 91 (7–8):219–227

Orousset J (1988) Faune de Madagascar, 71. Insectes, Coléoptéres: Staphylinidae, Euaesthetinae. Muséum National d'Histoire Naturelle, Paris

Orousset J (1990a) Note sur le genre *Ctenomastax* Kraatz (Coleoptera, Staphylinidae). Revue Française d'Entomologie (N.S.) 12(3):131–133

Orousset J (1990b) Révision des Euaesthetinae néotropicaux du genre *Stenaesthetus* Sharp (Coleoptera, Staphylinidae). In: Berti N (ed) Miscellanées sur les Staphylins, Mémoires du Muséum National d'Histoire Naturelle (A: Zoologie), vol 147, pp 9–55

Orousset J (2012) Catalogue commenté et illustré des espèces du genre *Octavius* Fauvel, 1873, de la faune de France (Coleoptera, Staphylinidae, Euaesthetinae). Bulletin de la Société entomologique de France 117 (2):133–166

Palmer AR (1957) Miocene arthropods from the Mojave Desert California. Geol Surv Prof Pap 294-G:237–280

Parker J (2016) Myrmecophily in beetles (Coleoptera): evolutionary patterns and biological mechanisms. Myrmecol News 22:65–108

Puthz V (1973) On some Neotropical Euaesthetinae (Coleoptera, Staphylinidae). Stud Neotropical Fauna 8(1):51–73

Puthz V (1974) A new revision of the Nearctic *Edaphus*-species and remarks on other North American Euaesthetinae (Coleoptera, Staphylinidae). Rev Suisse Zool 81(4):911–932

Puthz V (1977) Die Gattung *Octavius* Fauvel (Coleoptera: Staphylinidae) weltweit verbreitet! Opusc Zool (Bp) 14:105–126

Puthz V (1978) Revision of the Australian Euaesthetinae (Coleoptera: Staphylinidae). Mem Natl Mus Vic 39:117–133

Puthz V (1986) Drei neue *Stenaesthetus*-Arten aus Afrika (Coleoptera, Staphylinidae). 48. Beitrag zur Kenntnis der Euaesthetinen. Zeitschrift der Arbeitsgemeinschaft Österreichischer Entomologen 37 (3–4):117–120

Puthz V (1988a) Bemerkungen über die Gattung *Ctenomastax* Kraatz (Coleoptera, Staphylinidae). 53. Beitrag zur Kenntnis der Euaesthetinen. Entomologische Blätter für Biologie und Systematik der Käfer 84(1–2):51–60

Puthz V (1988b) Three new *Stenaesthetus* species from Ethiopia (Insecta, Coleoptera: Staphylinidae). 55th contribution to the knowledge of Euaesthetinae. Reichenbachia 26(3):9–13

Puthz V (1990) Die Gattung *Edaphus* Motschulsky in Sumatra (Coleoptera, Staphylinidae). 65. Beitrag zur Kenntnis der Euaesthetinen. Rev Suisse Zool 97 (4):815–825

Puthz V (1994) Beiträge zur Kenntnis der Euaesthetinen LXXIII. Bemerkungen über die Altweltlichen *Euaesthetus*-Arten (Staphylinidae, Coleoptera). Philippia: Abhandlungen und Berichte aus dem Naturkundemuseum im Ottoneum zu Kassel 6 (5):389–396

Puthz V (1995) Sexualität mit dem Lasso? – oder: Revalidierung der Gattung *Gerhardia* Kistner, 1960 (Coleoptera, Staphylinidae). 74. Beitrag zur Kenntnis der Euaesthetinen. Entomologische Blätter für Biologie und Systematik der Käfer 91(1–2):119–125

Puthz V (2001a) Neotropical *Euaesthetus* Gravenhorst (Coleoptera: Staphylinidae). 82th contribution to the knowledge of Euaesthetinae. Dugesiana 8(1):29–36

Puthz V (2001b) Neotropische Arten der Gattung *Octavius* Fauvel, 1873 (Coleoptera, Staphylinidae). 81. Beitrag zur Kenntnis der Euaesthetinen. Zeitschrift der Arbeitsgemeinschaft Österreichischer Entomologen 53:11–30

Puthz V (2003) Revision of the genus *Fenderia* Hatch, 1957 (Coleoptera, Staphylinidae). In: Cuccodoro G, Leschen RAB (eds) Systematics of Coleoptera: papers celebrating the retirement of Ivan Löbl, Memoirs on Entomology, International, vol 17, pp 637–648

Puthz V (2006a) The ventralis group of *Edaphus*: neotropical euaesthetines with special abdominal characters (Coleoptera, Staphylinidae). Dugesiana 13 (2):77–90

Puthz V (2006b) Description of new species and notes on some Neotropical Euaesthetinae (Coleoptera: Staphylinidae). Dugesiana 13(1):25–38

Puthz V (2006c) New southern African species and distribution records of the genus *Octavius* Fauvel (Coleoptera: Staphylinidae). Ann Transv Mus 43:1–28

Puthz V (2006d) Die erste Euaesthetinae aus dem Baltischen Bernstein (Coleoptera, Staphylinidae). 90. Beitra zur Kenntnis der Euaesthetinen. Entomologische Blätter 101:127–128

Puthz V (2007a) New Neotropical *Edaphus*-species (Coleoptera: Staphylinidae). Amazoniana 19 (3/4):97–130

Puthz V (2007b) First record of the euaesthetine genus *Schatzmayrina* Koch in the Neotropics (Coleoptera: Staphylinidae). Mitteilungen des Internationalen Entomologischen Vereins 32(3/4):83–88

Puthz V (2008a) Three new species of *Chilioesthetus* Saíz (Coleoptera: Staphylinidae) (98th Contribution to the knowledge of Euaesthetinae). Dugesiana 15(1):21–25

Puthz V (2008b) *Kiwiaesthetus*, a new genus of Euaesthetinae from New Zealand (Coleoptera, Staphylinidae). 100th contribution to the knowledge of Euaesthetinae. Zeitschrift der Arbeitsgemeinschaft Österreichischer Entomologen 60:59–69

Puthz V (2008c) Über Euaesthetinen aus dem Bernstein (Coleoptera, Staphylinidae). 99 Beitrag zur Kenntnis der Euasethetinen. Entomologische Blätter 103 (104):59–62

Puthz V (2010a) Die Gattung *Edaphosoma* Scheerpeltz, 1976 in China (Coleoptera, Staphylinidae). Entomologische Blätter 106:289–306

Puthz V (2010b) Zwei neue, bemerkenswerte *Edaphus*-Arten (Coleoptera, Staphylinidae). Mitteilungen des Internationalen Entomologischen Vereins 35 (3/4):159–163

Puthz V (2011) Afrikanische *Stenaesthetus*-Arten (Coleoptera: Euaesthetinae). 109. Beitrag zur Kenntnis der Euaesthetinen. Linzer Biologische Beiträge 43 (1):793–812

Puthz V (2012a) Zwei neue Arten der Gattung *Nothoesthetus* Saíz aus Chile (Coleoptera, Staphylinidae). Mitteilungen des Internationalen Entomologischen Vereins 37(3):159–162

Puthz V (2012b) Two new *Stenaesthetus* species from Tanzania (Coleoptera, Staphylinidae: Euaesthetinae). Annales Historico-Naturales Musei Nationalis Hungarici 104:127–130

Puthz V (2013a) Revision der orientalischen *Stenaesthetus*-Arten (Coleoptera: Euaesthetinae) 114. Beitrag zur Kenntnis der Euaesthetinen. Linzer Biologische Beiträge 45(2):2077–2113

Puthz V (2013b) Eine weitere neue *Edaphosoma*-Art aus China. Mitteilungen des Internationalen Entomologischen Vereins 38:45–48

Puthz V (2014a) Nordamerikanische Arten der Gattung *Euaesthetus* Gravenhorst (Coleoptera, Staphylinidae). Linzer Biologische Beiträge 46(1):845–876

Puthz V (2014b) Review of the new world *Edaphus* Motschulsky (Coleoptera, Staphylinidae) 120th contribution to the knowledge of Euaesthetinae. Linzer Biologische Beiträge 46(1):799–827

Remillet M (1969) Observations biologiques sur plusieurs Coléoptères hypogés de France. Annales de Spéléologie 24(1):183–186

Sáiz F (1972) Nuevos Euaesthetinae de Chile (Col. Staphylinidae) (III contribución). Anales del Museo de Historia Natural de Valparaíso 5:173–187

Scheerpeltz O (1974) Coleoptera: Staphylinidae (exclus Subfam. Paederinae, except pars min). In: Hanström B, Brinck P, Rudebeck G (eds) South African animal life: results of the Lund University Expedition in 1950–1951, vol 15. Swedish Natural Science Research Council, Stockholm, pp 43–394

Scheerpeltz O (1976) Wissenschaftliche Ergebnisse der von Prof. H. Janetschek im Jahre 1961 in das Mt.-Everest-Gebiet Nepals unternommenen Studienreise (Col. Staph…) (157. Beitrag zur Kenntnis der paläarktischen Staphyliniden, gleichzeitig 12. Beitrag zur…orientalischen…). Khumbu Himal, Ergebnisse des Forschungsunternehmens Nepal Himalaya 5 (Zoologie, Wirbellose):3–75

Sharp D (1876) Contributions to an insect fauna of the Amazon Valley. Coleoptera—Staphylinidae. Trans Entomol Soc Lond 1876:27–424

Stanley SM (1984) Does bradytely exist? In: Eldredge N, Stanley SM (eds) Living fossils. Springer Verlag, New York, pp 278–280

Thayer MK (2005) 11. Staphylinoidea 11.7. Staphylinidae Latreille, 1802. In: Beutel RG, RAB L (eds) Coleoptera, Vol. 1 morphology and systematics (Archostemata, Adephaga, Myxophaga, Polyphaga partim). Handbook of zoology Vol. IV, Arthropoda: Insecta, Part 38. (Kristensen NP, Beutel RG, eds). De Gruyter, Berlin, New York, pp 296–344

Welch RC (1966) A description of the pupa and third instar larva of *Stenus canaliculatus* Gyll. (Col., Staphylinidae). Entomol Month Mag 101:246–250

Worthy TH, Tennyson AJD, Jones C, McNamara JA, Douglas BJ (2007) Miocene waterfowl and other birds from central Otago, New Zealand. J Syst Paleontol 5:1–39

Zyla D, Yamamoto S, Wolf-Schwenninger K, Solodovnikov A (2017) Cretaceous origin of the unique prey-capture apparatus in mega-diverse genus: stem lineage of Steninae rove beetles discovered in Burmese amber. Sci Rep 7:45904. https://doi.org/10.1038/srep45904

Part II

Ecology, Conservation and Biotic Interactions

Effect of Environmental Conditions on Distribution Patterns of Rove Beetles

7

Ulrich Irmler and Erhard Lipkow

Abstract

This chapter aims to give a review about the impact of environmental conditions on Staphylinidae. Densities in diverse ecosystems from northern to tropical and from marine to alpine ecosystems are listed. As most Staphylinidae belong to the soil fauna, soil conditions are of main interest. Thus, life forms of soil-dwelling species are described. The effects of the parameter moisture, soil pH, acidity, and salinity on staphylinid occurrence are taken into consideration. Furthermore, the staphylinid faunas of main contrasting ecosystem type are reviewed: forests vs. agricultural fields, coasts vs. montane, and alpine ecosystems. Finally, since many rove beetle species have an affinity to nests, fungi, etc., the importance of microhabitats for Staphylinidae is described. Here, the special ecology of dung-dwelling and fungi-dwelling species is given such as the impact of the ephemeral food resource on the development.

7.1 Introduction

According to Herman (2001), more than 62,290 species with 3418 genera were described in the family Staphylinidae at the end of the second millennium. On that note, Staphylinidae is one of the most species-rich insect families in the world. This high number alone indicates that the Staphylinidae are widely distributed and occur in various habitats, zones, and ecosystems. It is certainly the family of Coleoptera that provides the highest variety of ecological groups. Bohac (1999) differentiated 5 classes of life forms with more than 14 ecological groups. Representatives of the family are found among euedaphic species in deep soil layers, on the soil surface, on the vegetation up to the canopy of forests, and from marine habitats in the eulittoral zone of the sea up to high mountain zones (Thayer 2005). Moreover, the variety of feeding habits is astonishing. Many staphylinids are nonspecialized predators; others prey specific insects or mites; there are many fungus feeders, pollen feeders, and humus feeders but no leaf feeders except few species of genera such as of the genus *Himalusa* (Klimaszewski et al. 2010). Numerous species live associated with social insects such as ants, wasps, or termites, where they feed directly on their hosts, are fed by them, or feed on remnants of their food or waste. Others live in the nests of mammals and birds; often they live either in their holes in the soil, in the large nests of

U. Irmler (✉)
Department of Applied Ecology, Institute for Ecosystem Research, University of Kiel, Kiel, Germany
e-mail: uirmler@ecology.uni-kiel.de

E. Lipkow
Kiel, Germany
e-mail: erlipkow@googlemail.com

O. Betz et al. (eds.), *Biology of Rove Beetles (Staphylinidae)*,
https://doi.org/10.1007/978-3-319-70257-5_7

raptors, or in episitic behavior directly on their skin. Thus, many staphylinid species are adapted to microhabitats such as the under bark habitat, large fungi (*Basidiomycetes*), and carcass. Most species of Staphylinidae have a well-developed flight ability which enables them to move large distances and seek microhabitats with short-term food resources. They are able to find heterogeneously distributed microhabitats such as spatially and temporally restricted dung heaps. Due to the isolated distribution and unpredictable occurrence of dung heaps, dung-inhabiting staphylinids must have a high mobility, e.g., great flight ability, highly sensitive chemical organs, and short larval development.

The following chapter tries to provide a rough overview of the ecological role of rove beetles in different ecosystem layers, zones, and microhabitats.

7.2 Soils and Staphylinids

7.2.1 Densities on Soil Floors

Investigations publishing staphylinid densities are rare. In general, staphylinids are collected by trapping methods, such as pitfall traps, flight intercept traps, etc., or are collected directly by hand. Bohac (1999) published data on staphylinid densities. Table 7.1 combines his data with our own investigations in Germany and in tropical Brazil as well as data from other references. According to these data, ecosystems of temperate zones have higher densities of staphylinids, on average, than tropical ecosystems. Temperate zones and wet ecosystems, e.g., wet alder forests or wet pastures, have higher densities than dry or oligotrophic ecosystems such as peat bogs. High densities are also recorded at sites with high organic detritus, such as wrack at beach sites. However, these sites also show extremely high variances of densities (Ruiz-Delgado et al. 2014). Low densities are recorded at agricultural fields in both temperate and tropical zones and in high alpine zones, but not in northern birch forests. According to these density data, staphylinids are one of the most frequent groups among the macrofauna on the soil floor. Densities are similar to spider densities (Palmgren and Biström 1979; Irmler 1995).

7.2.2 Life Forms and Functional Groups

The affinity to the soil habitat developed in various ways. Bohac (1999) differentiated the life forms Epigeobios and Geobios. In the following classification of life forms, the classic separation in epedaphic, hemiedaphic, and euedaphic species is maintained. Although many groups match the descriptions of Bohac (1999), another arrangement results from the specific focus to the soil.

The epedaphic rove beetles are characterized by large eyes and by moderately long legs. They can be separated into three groups: (1) species running on the soil or litter surface which forage randomly for food (Figs. 7.1 and 7.2), (2) optically oriented species walking slowly on the litter surface or on bare soil surface, and (3) litter-dwelling species living in the litter layer which can be regarded as an intermediate life form between the surface dwellers and the hemiedaphic species.

(1) The species running on the soil or litter surface have moderately well-developed eyes; their femurs are thick due to the well-developed leg muscles used for long and fast running events (Fig. 7.1). The eye-to-head-length ratio ranges between 0.28 and 0.33 for the few investigated species; the meso-femur width-to-length ratios are between 0.40 and 0.50. In general, they are black, dark brown, or colorful. Species that fit into this group look like *Ocypus* and *Philonthus* in temperate zones or *Xenopygus* and *Xanthopygus* in tropical zones. The running behavior is also shown by the high amount of catches in pitfall traps compared to their density (indicator for running activity) (Fig. 7.2). The species exhibit a slightly clumped dispersion pattern (Fig. 7.3).
(2) The optically orienting species have thick and large eyes because they have to fix their eyes on their prey. Legs are long, but femurs are thin; they walk slowly since no long

Table 7.1 Densities (ind. m^{-2}) of Staphylinidae in different ecosystems of temperate and tropical zones

Temperate zone	Mean or range	SD	Country	Reference
Agricultural field	29	(11)	Germany	Own investigation
Dry pasture	84	(33)	Germany	Own investigation
Wet pasture	169	(28)	Germany	Own investigation
Oak forest (sandy soils)	58	(36)	Germany	Irmler (1995)
Oak forest	50–170		Russia	Bohac (1999)
Spruce forest (sandy soils)	66	(33)	Germany	Irmler (1995)
Spruce forest	99–187		Russia	Bohac (1999)
Pine forest	75–118		Russia	Bohac (1999)
Beech forest (sandy soils)	101	(38)	Germany	Irmler (1995)
Beech forest (loamy soils)	100	(51)	Germany	Irmler (1995)
Alder brook	113	(32)	Germany	Irmler (1995)
Alder forest	675–783		Russia	Bohac (1999)
Alder forest	350–470		Czech Rep.	Bohac (1999)
Montane spruce forest	30–110		Slovakia	Bohac (1999)
Heath land	78–110		Russia	Bohac (1999)
Peat bog	85–198		Russia	Bohac (1999)
Peat bog	5–68		Czech Rep.	Bohac (1999)
Peat bog	10–160		Germany	Bohac (1999)
Coastal ecosystems				
Salt marsh (upper)	16.7	(8.3)	Germany	Own invest.
Beach, sandy	825	(403)	Denmark	Larsen (1936)
Beach (wrack)	80–120		Spain	Ruiz-Delgado et al. (2014)
Beach (wrack)	10–420		Brazil	Ruiz-Delgado et al. (2014)
Northern and alpine zone				
Birch forest	120		Finland	Palmgren and Biström (1979)
Alpine zone (Caucasus)	3–10		Russia	Onipchenko (2004)
Alpine zone (3000–3600 m)	19–58		China	Tan et al. (2013)
Tropical zone				
Varzea forest (Amazonas)	74	(19)	Brazil	Irmler (1978)
Varzea forest (Amazonas)	50	(10)	Brazil	Irmler (1978)
Blackwater forest (Amazonas)	73	(54)	Brazil	Irmler (1978)
Agricultural field (Mato Grosso)	13	(19)	Brazil	Own investigation
Cerrado forest (Mato Grosso)	5	(7)	Brazil	Own investigation
Pasture (Mato Grosso)	15	(22)	Brazil	Own investigation

SD Standard deviation

distances have to be passed. The eye-to-head-length ratio ranges between 0.33 and 0.93; meso-femur width-to-length ratios are between 0.16 and 0.35. Typical species of this group are found in the genus *Stenus*, *Rugilus*, or some *Quedius*. In contrast to the former group, the number of catches in pitfall traps is usually lower than their densities because they run less and also spot the traps better (Fig. 7.2). The species are, similar to the preceding life form, dark or colorful. In regard to the dispersion patterns, the optically orienting species show near random values (Fig. 7.3).

(3) The last group of epedaphic species lives in leaf litter. The eyes are less developed and are distinctly reduced in comparison with the preceding two groups, as they live in a darker habitat. Eye-to-head-length ratio ranges between 0.18 and 0.24, which is even smaller than the randomly running species. The meso-femur width-to-length ratios are

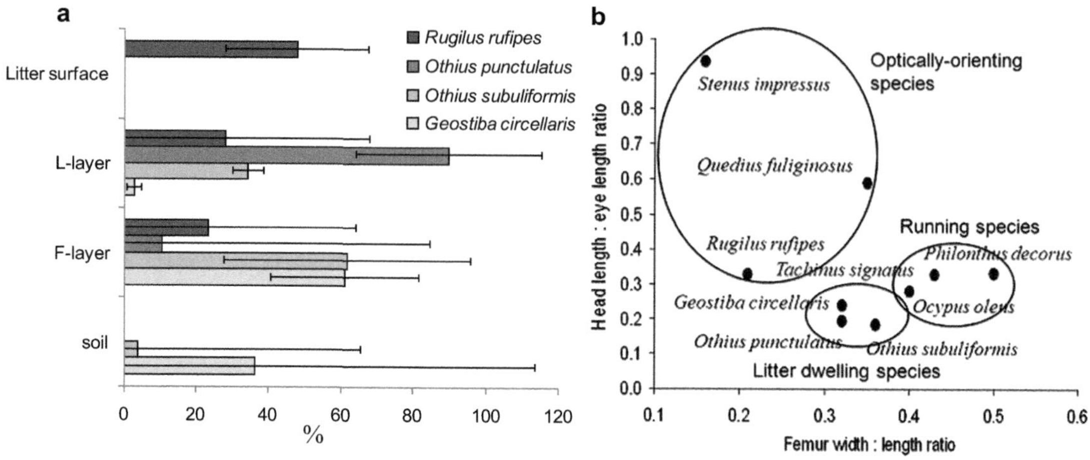

Fig. 7.1 (**a**) Vertical distribution of staphylinids in the litter layer of a northern German beech forest, (**b**) head-to-eye-length ratio vs. meso-femur width-to-length ratio for different life forms of the beach forest floor

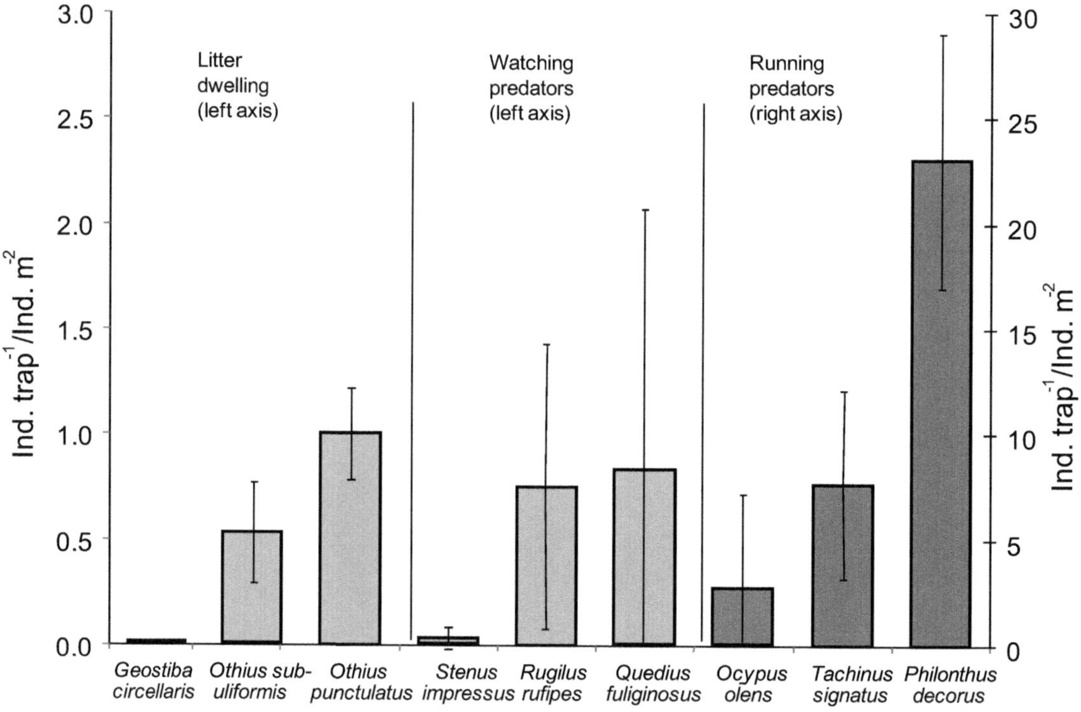

Fig. 7.2 Relation (ind. trap^{-1}/ind. m^{-2}) between activity density (ind. trap^{-1}) and density (ind. m^{-2}) to indicate the running activity of rove beetles

moderately high, ranging between 0.32 and 0.36. They can be named as "sliders" because walking among the narrowly packed foliage is impossible. The sliding movement is also shown by the dorsoventral depressed body and legs which are narrowly attached to the body. They more or less swim among the foliage. Regarding their vertical distribution, they live deeper in the L- or F-layer of the litter than the soil surface dwellers (Fig. 7.1).

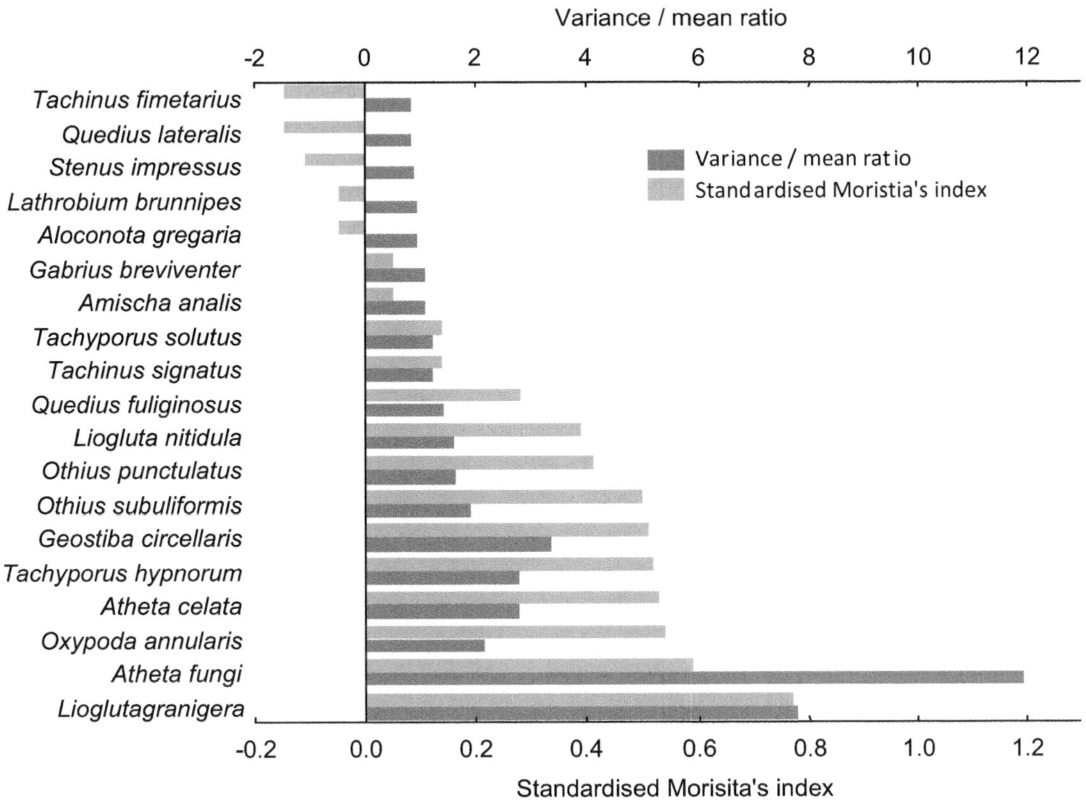

Fig. 7.3 Dispersion indices of different edaphic staphylinid species in a northern German beech forest; Standardized Morisita's index: clumped patterns above zero, uniform patterns below zero; values near zero indicate random dispersions, based on 18 samples in a northern German beech forest

Because resistance by the foliage layers is high, they move slowly (Fig. 7.3). Similar to the optically orienting group, sliders are caught in pitfall traps less often than species with high running activity (Fig. 7.2). In contrast to the two preceding life forms, litter dwellers are usually paler, light brown, or pale reddish because they rarely have direct contact to sunlight. Species of this group are found in the genus *Othius* in temperate zones or *Diochus* in tropical zones.

Dwellers of deep litter layers such as *Geostiba circellaris* also belong to the hemiedaphic life forms. This can be seen by their strongly reduced eyes, small size, short legs, and reduced elytra and hind wings. Within the vertical distribution, they live mainly in the F-layer or even in the mineral soil layer (Fig. 7.1). Running activity is extremely reduced due to the dense packages of litter foliage (Fig. 7.2). Comparing the three investigated species on the vertical gradient, *Othius punctulatus* is ca. 10–15 mm long. *O. subuliformis*, which lives in an intermediate layer, is ca. 4.8–5.5 mm long. *Geostiba circellaris*, which lives in the deepest layer of litter, is ca. 2.2–2.9 mm long. This indicates that the species become smaller the deeper they live in the litter layers. Regarding the dispersion patterns, litter-dwelling species are slightly clumped (Fig. 7.3). However, cumulative dispersion patterns with a tendency to aggregation are also found, e.g., *Atheta fungi*, *Liogluta granigera*, and *Oxypoda annularis*, which might be caused by the fact that some food resources are distributed heterogeneously in the litter, such as

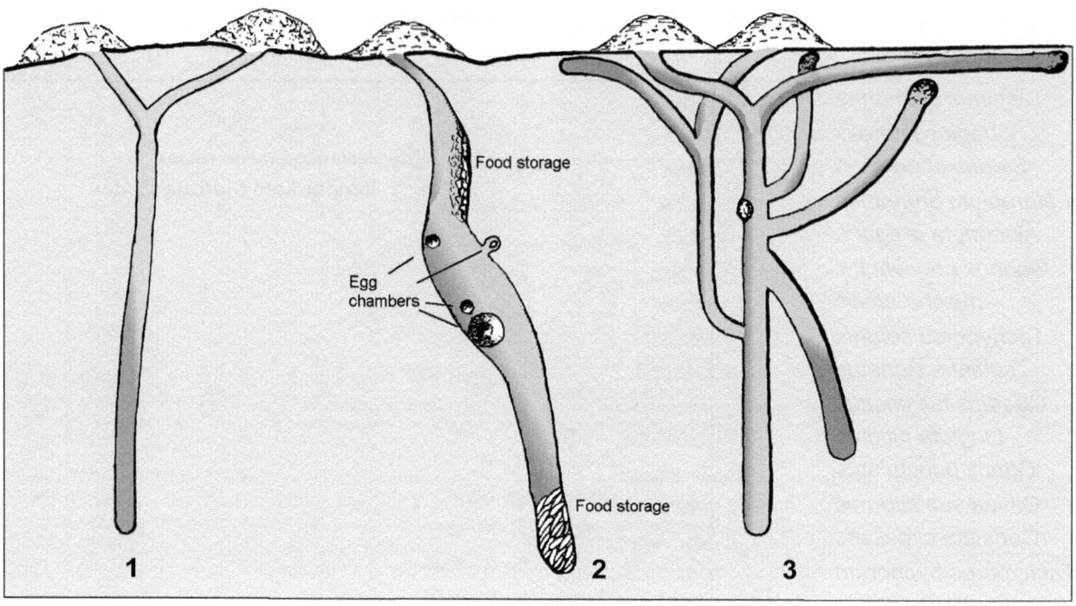

Fig. 7.4 Permanent holes of *Bledius arenarius* (1), *Bledius tricornis* (2), and *Bledius bicornis* (3) (according to Larsen 1936, modified)

specific fungi in the litter layer (Reise and Weidemann 1975).

The hemiedaphic species live in the mineral soil and are able to change the soil structure by burrowing in the substrate. They can be separated into two types: (4) species digging permanent holes in the soil surface and (5) species digging in the soil without producing permanent holes.

(4) Species of this group are represented mainly by the genus *Bledius*. They have broad pro-tibiae and large eyes. First investigations were made by Larsen (1936) on Danish beaches and Wadden sea sites. Larsen (1936) found three different types of permanent holes for the algae-feeding *Bledius* species: (1) simple unbranched holes, (2) more complicated holes with repositories and breeding chambers, and (3) branched hole systems without specific rooms (Fig. 7.4). Larson (1936) emphasized that *Bledius* species burrow their holes by carrying sand particles with their mandibles to deposit them on the adjacent soil surface, such as ants, or press them into the walls to stabilize them. The large tibiae are used for counter bearing against the hole walls and not for digging; the narrowed pronotum is necessary to bend the forebody during this activity.

(5) The species of this group exhibit intermediate morphological characters between the species digging permanent holes and the euedaphic species. Due to their small size, on average smaller than 2 mm, they usually belong to the soil mesofauna. Species of this life form are found in the genera of Aleocharinae and Oxytelinae, e.g., *Diglotta*, *Phytosus*, *Geostiba*, *Taxiera*, *Hydrosmectina*, *Meotica*, *Carpelimus*, and *Thinobius*. They often have long mandibles, such as some *Bledius*, that they use to carry sand particles during burrowing activities. Eyes, antennae, elytra, and hind wings might be reduced, depending on the stability of their habitat. Legs are usually longer than in the euedaphic group. For the species examined, length of mesotiba vs. total length was 0.12 ± 0.02. In unstable habitats, e.g., river margins, wings and antennae are fully developed, because

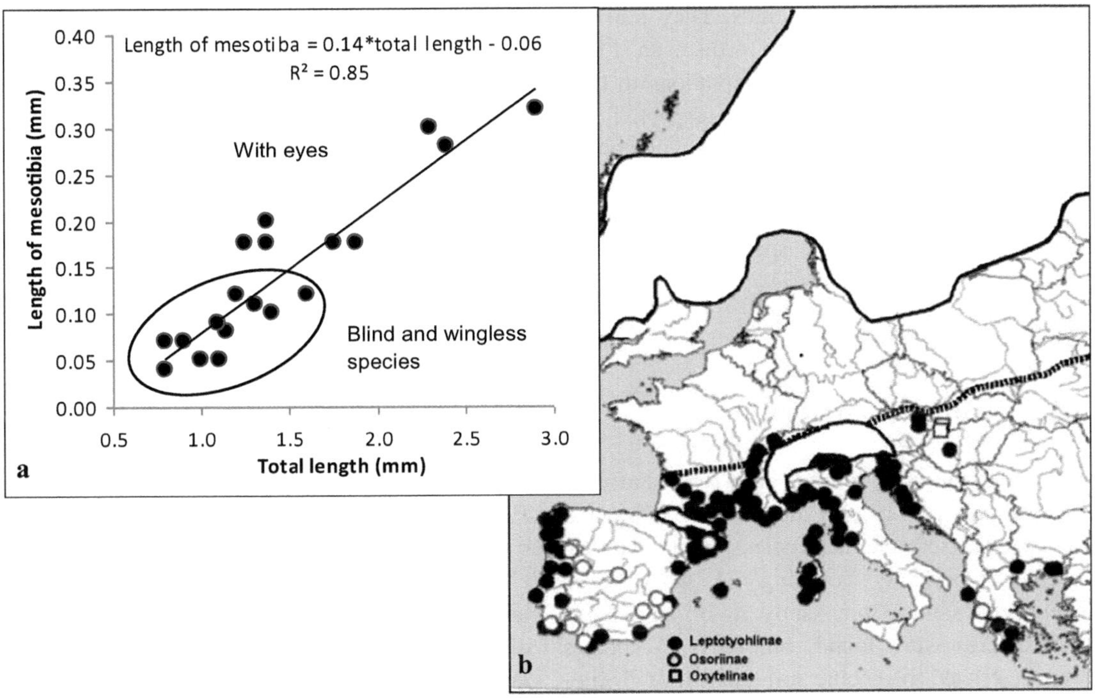

Fig. 7.5 (**a**) Relation between total body length and length of mesotibia for hemiedaphic species (with eyes, with or without wings) and euedaphic species (blind and without wings); data for euedaphic species measured according to graphs given by Coiffait (1972); (**b**) distribution of blind and wingless Staphylinidae in Western Europe (*white*, covered by glaciers in last glacier period; *dotted line*, southern border of permafrost soils)

they have to leave their deep soil habitat during unpredictable flooding events. In stable habitats, such as woods, eyes and wings are usually reduced, e.g., *Geostiba circellaris*.

The last life form of the soil-dwelling species is represented by the euedaphic species that live in deep soil layers and rarely move to the soil surface (6) or in deep holes in the soil made by mammals or other great animals (7).

(6) Typically, euedaphic species are extremely small and belong to the soil mesofauna, because they live in soil gaps, small holes, and similar soil structures. They cannot change the soil structure by burrowing. Although the deep soil is an extreme habitat for staphylinids, species in many subfamilies are found that live euedaphicly in the soil, e.g., Oxytelinae (blind and wingless species of *Thinobius*), Leptotyphlinae (nearly all species), Osoriinae (*Geomitopsis*, *Lusitanopsis*, *Heterocylindropsis*, *Rhabdopsis*, and others), and Aleocharinae (blind and wingless species of *Hydrosmectina*). The eyes are extremely reduced or totally absent. In contrast to the blind cave species, legs are also reduced to adapt to the narrow habitat between the soil particles. The ratio mesotibia length vs. total length is 0.07 ± 0.02, which means that legs are significantly shorter than for the hemiedaphic species (*t*-test: T, 5.7; $p < 0.001$). The morphological adaptations result in a wormlike-shaped body which is necessary to move in the narrow habitat. Species of this group are absent from regions that were covered by glaciers or permafrost soils during the glacial period. In Europe, they are only found south of the Alps (Fig. 7.5).

(7) Species living in the large holes (caverns) of soil-living mammals can hardly count as soil-dwelling. Although they live in soil caverns, their affinity to soils is lower than to the nest

habitat of their host animals. They must be placed to the inquiline life forms even if they also have morphological adaptations to their subterranean habitat in some respect. Some species have reduced eyes and a pale coloration, e.g., *Lathrobium pallidum* Nordmann, 1837; legs and antennae might be elongated, such as in cavernicole species, e.g., *Bisnius scribae* (Fauvel 1867) and *Rheochara spadicea* (Erichson 1837).

7.2.3 Soil Parameters

Although many staphylinids have a close affinity to soils, only few species show a narrow demand on specific soil characteristics. In a study in Northern Germany counting 65 woody and open ecosystems from sandy to organic soils, 265 species were found, but only 14 species were restricted to specific soil conditions (Irmler and Gürlich 2007). Only two species were found in sandy habitats, eight species in systems with high organic matter, and four species in ecosystems with low organic matter. Most species were distributed over a wide gradient of various soils (Table 7.2). This was also evident by the low eigenvalue in an ordination analysis, which was only 0.48 in spite of the high variance in soil parameters. The wide range of various soil conditions used by rove beetles is also emphasized by many other investigations that studied the relation between staphylinids and habitat conditions, including various soils (Vogel and Uhlig 1982; Steinmetzger and Tietze 1982; Rose 2001). Rose (2001) also investigated a wide range of habitat and soil conditions in coastal ecosystems but found a very low differentiation along the parameter gradients, with an eigenvalue of only 0.41. In his study, litter type and canopy cover were the most important factors, while soil pH and moisture and salt content were less important.

In spite of the overall wide range of soils used by staphylinids, few species are restricted to specific soil conditions. In a study on coastal staphylinids along the sand–loam–gravel gradient, *Bledius defensus* and *Oxytelus insecatus* were found in very short ranges of sand–silt–clay mixture (Irmler 2012). *Bledius defensus* was only found in a mixture of approximately 50% sand to 45% silt/clay and *Oxytelus insecatus* in a mixture of 30–60% sand and 35–45% silt/clay. Regarding the sand-shingle gradient, *Cafius xantholoma* preferred a higher percentage of shingle, whereas *Polystomota grisea*, *P. punctatella*, and *Phytosus spinifer* preferred sandy beaches (Fig. 7.6).

According to many investigations, moisture is one of the most important parameters controlling the distribution of staphylinid beetles (Irmler 1993; Ottesen 1996). The moisture of the soil on agrarian fields was a crucial factor for *Philonthus cognatus* when selecting sites for overwintering (Holland et al. 2007). Many species are known to be restricted to moist or wet lake or creek edges. Krogerus (1948) studied the distribution of staphylinid beetles in a wet–dry gradient and found 4 species restricted to wet conditions (80%–90% soil moisture), 5 species to moist conditions (50%–90%) with preference of high moisture, and 13 species that avoid moist conditions. In our own investigations of a wet–dry gradient which included a total of 122 species in forests and 79 species in grassland, 9 species preferred dry conditions (<30% mean yearly soil moisture) and 8 species wet conditions (>50% mean yearly soil moisture) (Table 7.3). Most species demanded intermediate moisture conditions and had a wide tolerance in regard to the high standard deviation of the weighted mean. However, species of both dry and wet habitats had a low standard variation, which indicates that they were restricted to either dry or wet conditions with a short range. For most species, the moisture demands corresponded to forest and grassland habitats. In both grassland and forests, moisture demands are significantly higher for four species and lower for one species.

Larsen (1936) and Krogerus (1948) studied the soil moisture demands under experimental conditions with six coastal and four bank staphylinid beetles. According to their experiments, the results support the field distribution. Living in dry dune habitats, both *Bledius pygmaeus* und

Table 7.2 Results of Spearman correlation analyses (R-values) between staphylinid species and soil parameters (Modified according to Irmler and Gürlich 2007)

Species	Sand (%)	pH	Corg (%)	Tree cover (%)	Ecological groups of species on/of
Zyras humeralis	0.35	0.31	−0.37		Alkalic, sandy, humus-poor soils
Atheta euryptera	0.32	0.38			Alkalic, sandy soils
Ocypus brunnipes		0.45			Alkalic soils
Oxypoda opaca		−0.29			Acidic soils
Ilyobates bennettii			0.43		Humus-rich soils
Lathrobium brunnipes			0.35		
Aleochara bilineata			−0.39		Humus-poor soils
Plataraea brunnea			−0.29		
Anthobium atrocephalum			0.31	0.42	Humus-rich forests
Philonthus decorus			0.34	0.57	
Platydracus fulvipes			0.34	0.32	
Quedius fuliginosus			0.46	0.39	
Tachinus pallipes			0.32	0.33	
Zyras limbatus			−0.35	−0.33	Humus-poor soils of open habitats

Corg: organic carbon

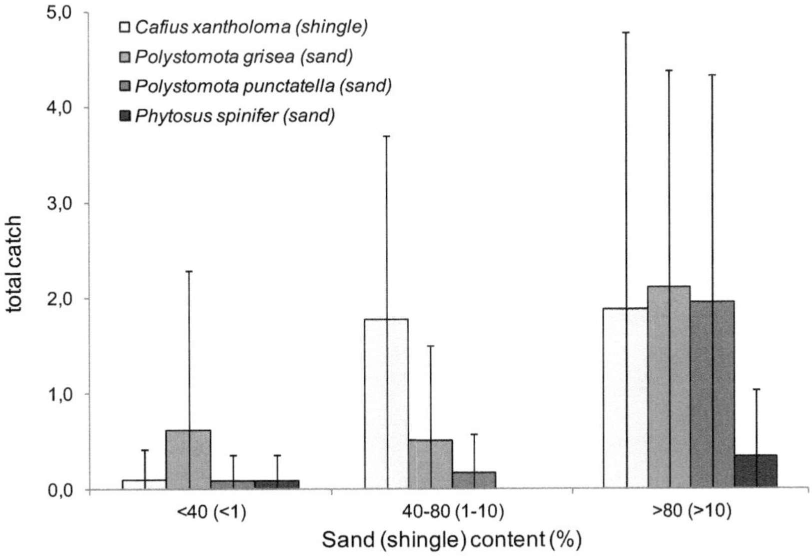

Fig. 7.6 Total catch of four staphylinid species at different beach types of the Baltic Sea (total $n = 40$, based on data of Irmler 2012)

B. opacus selected dry conditions of 2.5–7% and 3–10% moisture in the experiment and the field, respectively. In contrast, *Bledius diota* demand soil moisture at a minimum of 21%. This also corresponds to the field investigation where *Bledius diota* was not found at sites with moisture of 17 % and lower. The bank species investigated by Krogerus (1948) reflected both corresponding results between field and experiment (e.g., *Paederus riparius*, *Oxytelus rugosus*) and contrasting results between experiment and field (e.g., *Stenus cicindeloides*, *Philonthus quisquilarius*). There may be various reasons for the discrepancy between soil moisture demands in the field and the physiological demands. It might be referred either to moisture demands of eggs or larvae or a combination of effects between temperature and moisture.

Table 7.3 Weighted mean with standard deviation of soil moisture (%) for species in northern German deciduous forests and grassland indicating soil moisture demands and tolerance

	Deciduous forest		Grassland	
Species	W. mean	SD	W. mean	SD
Tachyporus nitidulus	–	–	10.9	0.2
Xantholinus linearis	–	–	10.9	0.2
Oligota pusillima	–	–	10.3	0.6
Oxypoda sericea	–	–	10.8	4.1
Tachinus corticinus	–	–	15.7	13.2
Tachyporus dispar	–	–	22.6	18.6
Oxypoda exoleta	–	–	23.7	19.0
Geostiba circellaris	29.7	11.1	–	–
Philonthus cognatus	31.0	4.2	37.3	18.7
Philonthus laminatus	31.6	1.8	–	–
Oxypoda annularis	32.4	4.1	–	–
Lathrobium brunnipes	32.6	11.3	50.5	0.0
Tachinus rufipes	33.0	5.2	–	–
Stenus impressus	33.9	6.7	–	–
Othius subuliformis	34.3	10.3	–	–
Othius punctulatus	34.5	12.0	–	–
Philonthus decorus	34.8	10.4	–	–
Quedius fuliginosus	35.2	10.5	50.5	0.0
Rugilus rufipes	36.7	11.2	49.6	6.0
Stenus clavicornis	32.8	2.0	37.3	18.7
Stenus juno	38.0	6.7	43.9	14.8
Gabrius subnigritulus	–	–	38.9	18.2
Anotylus rugosus	44.2	13.3	46.6	12.1
Stenus bimaculatus	45.8	2.0	50.5	3.4
Carpelimus elongatulus	46.7	5.7	50.5	0.0
Anthobium atrocephalum	47.3	14.0	–	–
Anthobium unicolor	47.6	13.3	–	–
Atheta aquatica	58.6	6.6	–	–
Olophrum piceum	59.5	7.3	–	–
Atheta volans	–	–	50.5	0.0
Stenus cicindeloides	–	–	50.5	0.0
Stenus similis	–	–	50.5	0.0
Myllaena intermedia	55.2	9.1	–	–
Atheta elongatula	56.1	9.1	47.7	10.1
Ocalea picata	57.9	8.7	50.5	0.0

W. mean Weighted mean, *SD* standard deviation

According to Bong et al. (2013), *Paederus fuscipes* develops better under higher temperatures than under cold conditions. However, they need high moisture conditions for their development at higher temperatures. Eggs need high water uptake for their development, which may differ from the water demands of adults, e.g., for *Ocypus olens* (Lincoln 1961).

Soil moisture shows a high variation throughout seasons, with high values during winter or spring or during rainy seasons. Moreover, the groundwater table varies in the same way. Irmler (2009) investigated the time lag between the moisture changes over a period of 7 years and the reaction of staphylinid beetles. He found time lags of 0–2 months between strong rainfall

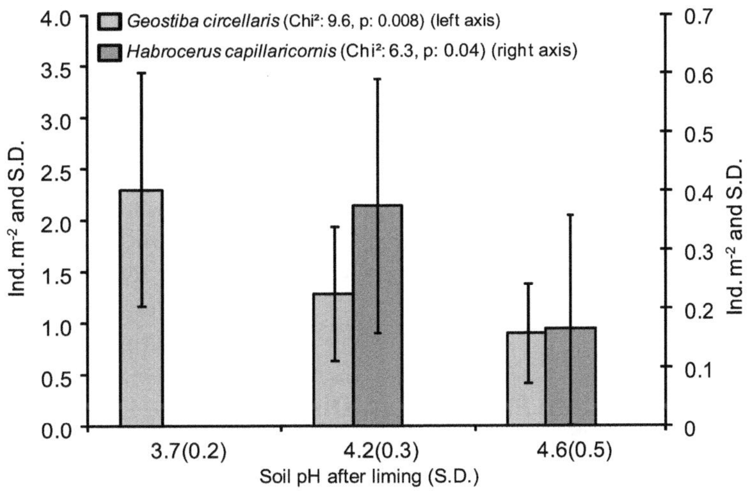

Fig. 7.7 Effect of liming on the density of *Geostiba circellaris* and *Habrocerus capillaricornis* in a northern German beech forest ($n = 9$ in each plot), with results of Kruskal–Wallis ANOVA between the three liming groups including Chi^2 and p values

events and the reaction of *Anthobium atrocephalum* populations. The species reaction to high rainfall events was positive. In an alder wood, time lag was both 3 and 4 months between groundwater table changes and the reactions of *Ocalea picata* and *Myllaena intermedia* populations, respectively. The negative correlation coefficients indicate that under high groundwater tables with backwater conditions, the species retreat from the wettest sites to avoid submersion.

The acidity of substrate also plays a role in the distribution of staphylinid species. Typical species of acid bogs, e.g., *Acylophorus wagenschieberi* and *Stenus kiesenwetteri*, may be restricted to that habitat, not only by the wet conditions but mainly by the acid substrate. The two European *Acylophorus* species prefer moist conditions with contrasting soil acidity: *A. wagenschieberi* on acid bogs (Staniec 2005) and *A. glaberrimus* alkaline mire conditions at pond edges. According to Gryuntal (2009), *Geostiba circellaris* prefers acid forests on moder litter, whereas *Tachinus marginellus* avoids acidic conditions. Preference experiments using Staphylinidae in an acid–alkaline gradient are not available. However, liming experiments in forests may show the reaction of staphylinid beetles under changing soil pH (Fig. 7.7). With an increase of soil pH in a northern German liming experiment in a beech forest, the densities of both *Geostiba circellaris* and *Habrocerus capillaricornis* decreased and increased, respectively. However, among the 64 species recorded, there were only these 2 species that reacted to the pH increase, which means that rove beetles rarely react to slightly changing soil acidity.

According to Frank and Ahn (2011), 392 staphylinid species were confined to coastal habitats worldwide. Moore and Legner (1976) mentioned ecological details but gave no information on the dependence on or the tolerance to salinity. Larsen (1936) tested the preferences of various rove beetles in her thesis about the burrowing beetles of the coast. According to her preference experiments, the species that do not live directly at the sea margin, such as *Carpelimus despectus* and *Bledius opacus*, avoided salinity greater than 5‰. Others, e.g., *Bledius longulus*, *B. arenarius*, and *Bledius tricornis*, showed a wide range, from 0 to 20‰. Only the species directly living under submersion conditions, e.g., *B. spectabilis* and *B. diota*, preferred salinities between 30‰ and 60‰. Larsen (1936) assumed that the species preferred the sites with high salinity in order to avoid the parasitoid ichneumonids. Topp and Ring (1988) also made experimental studies with intertidal rove beetles, including studies on the influence of the salinity. They studied the two species

Liparocephalus cordicollis Le Conte and *Diaulota densissima* Casey from British Colombia. Both species survive a submergence of more than 2 weeks under marine water. *L. cordicollis* has an osmotic regulation and can stabilize their body weight at a range between 2‰ and 45‰. Respiration was constant within the salinity range of 2–30‰, but at higher levels of 45‰, oxygen consumption increased.

7.3 Ecosystem Preferences, Diversity, and Occurrence

7.3.1 Temperate and Mediterranean Forests

Forests are one of the most diverse ecosystems, which is due to the habitat diversity offered by trees. Living trees with their diverse structure offer different climate conditions in the vertical layers and the bark; dead trees, in addition, offer still more habitats due to the succession of decaying wood. The group of Staphylinidae is regarded to be an essential group for the evaluation of forest biodiversity (Parmain et al. 2015). Besides the alpha diversity at a site, forests differ in their tree composition and soil conditions, e.g., moist vs. dry forests or sandy vs. loamy forests. The sand–moisture gradient was primarily accountable for the ordination of forest ecosystems in Northern Germany using the staphylinid composition (Irmler 1993).

The ascertained species richness mainly depends on the sampling effort, the different collecting methods, and the duration of the investigation. In Europe, total species richness of forests was 42 species, with 12.3 ± 4.5 species in each of 10 forests in Norway (Andersen et al. 1990), and 205 species, with 48 ± 30.6 in each of 12 forests in Northern Germany (Irmler 1993). In the 12 German forests, a lowest number of species was found in the extremely wet alder forests (17 species); the highest numbers were found in the moderately moist alder forest (121 species). Dry forests on sandy or loamy soils exhibited intermediate species richness (24–83 species). A total of 132 species, with 40.2 ± 24.3 species in each of 9 forest sites, were found on sandy soils with pine forests in Central Germany (Steinmetzger and Tietze 1982). Bohac (1999) published values ranging between 12 and 52 species for diverse Russian forests and 23–81 species for Czech forests. In beech forests of Central Germany, the species richness varied between 91 (Karlsruhe) (Friebe 1982), 87 (Berlin) (Weigmann et al. 1989), and 117 (Göttingen) (Hartmann 1976). Few investigations pertain to the biodiversity of Mediterranean forests. Zanetti (2011) recorded 80 species out of 1200 specimens from a Sardinian *Quercus ilex* forest area, which is less than in the Central European region.

A survey on rove beetles living on dead wood in three different deciduous and spruce forests of Northern Germany recorded 170 species (Irmler et al. 1997). Age of dead wood was the major factor controlling the species richness of rove beetles (Fig. 7.8). The species richness increased with increasing age and was highest on the soil surface. Density of rove beetles living on dead wood ranged between 5 and 137 ind. m^{-2} of wood surface; on the soil surface, it ranged between 27 and 120 ind. m^{-2} (125 and 190 ind. m^{-2} including larvae) (Irmler 2009). The species composition was more similar between the different ages of decay than between the forests standing on different soil conditions and having different tree compositions. Some species were exclusively found on dead branches of beech wood; other species were additionally found on alder wood or on beech, alder, and spruce wood (Table 7.4). Whereas young branches had species exclusively found there, stumps had no exclusive species and shared many species with the surrounding soil surface.

Seasonality in temperate forests is controlled by the temperature. Most species have their highest activity in spring and early summer (Kasule 1968; Hartmann 1979; Friebe 1982). Little knowledge is available about the long-term fluctuations. Irmler (2009) published a 7-year study and found a negative cross-correlation between the population density and temperature for the six species *Oxypoda*

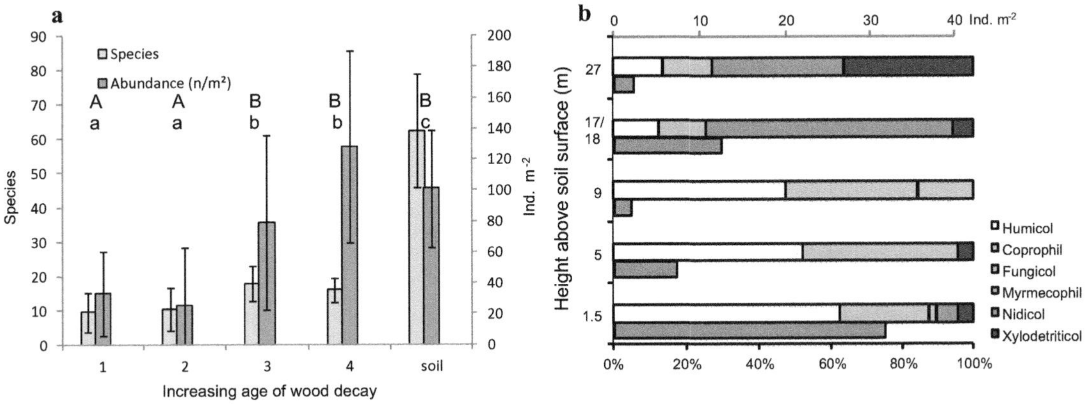

Fig. 7.8 (**a**) Number of species and abundance of rove beetles on dead wood of different age of decay compared to the soil; (1) bark closely fixed to the log, partly with fungi; (2) bark loosely fixed to the log, already with moss; (3) the original structure of the wood is still visible, but the wood is already rotten; (4) the structure of the wood is amorphous, but single wood pieces are still existent; different capitals indicate significant differences (small for species, large for abundance); (**b**) vertical distribution of total number of specimens and percentages of ecological groups in northern German beech and alder forests; humicol species include phytodetriticol species (Modified Irmler et al. 1997; Irmler 1998)

annularis, *Geostiba circellaris*, *Atheta fungi*, *Othius subuliformis*, *Anthobium atrocephalum*, and *Anthobium unicolor*, which means that the populations of these species decreased in years with a low mean temperature. Reaction time ranged from a few months to 1 year, depending on the number of generations. Large species, e.g., *Othius subuliformis*, reacted slower than smaller species, e.g., *Atheta fungi*. However, several species responded positively to high amounts of litter fall, such as *Oxypoda annularis*, *Geostiba circellaris*, *Atheta fungi*, and *Othius punctulatus*. As Collembola densities also corresponded with litter fall (Irmler 2006), the reaction of rove beetles on litter fall fluctuations may be referred to either their collembolan food, e.g., *Othius punctulatus*, or to a higher amount of fungi or beech nuts.

Another specific character of forests is the distinct vertical structuring in different layers, i.e., several layers of soil, soil vegetation, bushes, and different tree layers. Little knowledge is available about the staphylinid fauna of the canopy layer. In Northern Germany, the vertical distribution of staphylinids was investigated in beech and alder forests using flight intercept traps at different heights from 1.5 to 27 m above the soil surface (Irmler 1998). The total number decreased with an increase in height (Fig. 7.8). Lowest numbers were found at lower and upper borders of the canopy at 9 m and 27 m. Mainly species living in the litter layer or in rotting plant matter were found close to the soil surface at a height of 1.5 m. Among these groups, only a few species occurred that primarily lived in the forest leaf litter; most species live in rotting plant litter and switch between agricultural and forest ecosystems in the course of the year, e.g., *Tachinus rufipes*, *Tachyporus* spp., and others (Lipkow 1966). They generally use the forest resources after harvest and for overwintering. Mainly species living in the nests of birds or on dead wood were found in the canopy layer. Many species near the soil surface layer must be regarded as species unspecific for forests, because they use resources of different ecosystems. In contrast, the species of the canopy are specific for the forest ecosystem because they use resources that are only found there. They were caught by the flight intercept traps because their resources are heterogeneously distributed within the forests. Coprophilous species were recorded over the whole vertical gradient in high percentages. They are extremely active flyers because their resources are also distributed quite heterogeneously. None of the

Table 7.4 Density of rove beetle species (ind. m^{-2}) exclusively found on dead wood in three northern German forests ($n = 50$ emergence traps filled with dead wood); density refers to surface area of wood; significant values due to Bonferroni corrected Kruskal–Wallis ANOVA with $p < 0.05$ italic (Modified according to Irmler et al. 1997)

	Beech				Alder				Spruce			
	Young log/stump		Old log/stump		Young log/stump		Old log/stump		Young log/stump		Old log/stump	
	M	SD	M	SD	M	SD	M	SD	M	SD	M	SD
Anomognathus cuspidatus	2.9	7.2	1.1	3.2	0.3	0.4	0.1	0.2	.	.	.	.
Atheta celata	0.1	0.1	*1.0*	*1.6*	0.3	0.5	.	.	.	.	.	.
Atheta coriaria	3.6	11.4	*23.1*	*45.5*	.	.	.	.	.	.	.	.
Baptolinus affinis	0.6	2.0	1.0	2.9	.	.	.	.	.	.	.	.
Bolitochara obliqua	*0.3*	*0.4*	.	.	.	.	.	.	.	.	.	.
Dinaraea aequata	0.4	0.6	*1.0*	*2.9*	0.5	0.5	0.2	0.5	.	.	.	.
Leptusa pulchella	0.5	1.3	*2.1*	*3.1*	.	.	0.5	1.5	.	.	.	.
Lordithion exoletus	0.1	0.2	*0.8*	*1.6*	.	.	.	.	.	.	.	.
Lordithion trinotatus	0.1	0.1	*7.7*	*7.2*	.	.	.	.	.	.	1.7	2.4
Phloeocharis subtilissima	0.1	0.1	0.9	1.8	0.8	1.3	1.3	2.2	5.9	8.4	3.4	4.8
Phloeonomus punctipennis	2.1	5.2	*3.8*	*10.6*	0.4	0.8	0.5	1.5	0.4	0.6	.	.
Phloeopora angustiformis	0.1	0.2	0.4	1.1	.	.	.	.	.	.	.	.
Phoeopora teres	.	.	0.3	1.0	.	.	.	.	.	.	.	.
Quedius invreae	1.5	4.1	*6.0*	*9.9*	.	.	2.1	3.9	2.0	1.7	3.0	0.7
Quedius maurus	0.5	1.2	3.8	7.6	.	.	1.9	5.9	.	.	2.5	3.5
Quedius xanthopus	0.1	0.3	*6.9*	*5.4*	0.5	0.9	0.7	2.3	.	.	*6.4*	*5.5*

M Arithmetic mean, *SD* standard deviation

species can be regarded as being specific for the forest ecosystem. They are found in a variety of ecosystems in which feces are available.

Many of the coprophilous and humicolous/phytodetriticolous species that occurred in the investigated vertical gradient were also recorded by aeroplanes at extreme heights (Weidel 2010): *Anotylus rugosus* and *A. tetracarinatus* were recorded at 300 m and *Philonthus cognatus* and *Xantholinus longiventris* at 1500 m.

7.3.2 Agricultural Fields

In contrast to the stable forest ecosystems, arable land is extremely unstable due to yearly agricultural management such as tilling, the change of agricultural crops, etc. Staphylinidae are mainly predators and are thus considered to be beneficial insects on agricultural land. In particular, *Tachyporus* species, e.g., *Tachyporus hypnorum*, are effective predators on cereal aphids (Vickerman and Sunderland 1975; Coombes and Sotherton 1986). In spite of their beneficial effects, densities of staphylinids under intensive farming conditions are distinctly lower than in forests or other agricultural systems (Table 7.1). Although the staphylinid fauna of agricultural land is strongly influenced by the surrounding habitats from where species invade the fields yearly (Topp 1977; Sotherton 1985), an autochthonous staphylinid fauna must exist, as can be seen by the dynamics of larvae and adults (Fig. 7.9).

Several staphylinid species are able to overwinter in arable fields and reproduce there in early spring (Gilgenberg 1986; Basedow 1990). They produce larvae that live in late spring. According to Basedow (1990), the rate of reproduction varies between crops and intensities but may be even higher under intensive agriculture, such as for *Lathrobium fulvipenne*, if moisture conditions favor the living conditions. Other species, e.g., *Tachyporus hypnorum* or *Tachinus rufipes* (Lipkow 1966), mainly overwinter at field margins or in adjacent forests and must

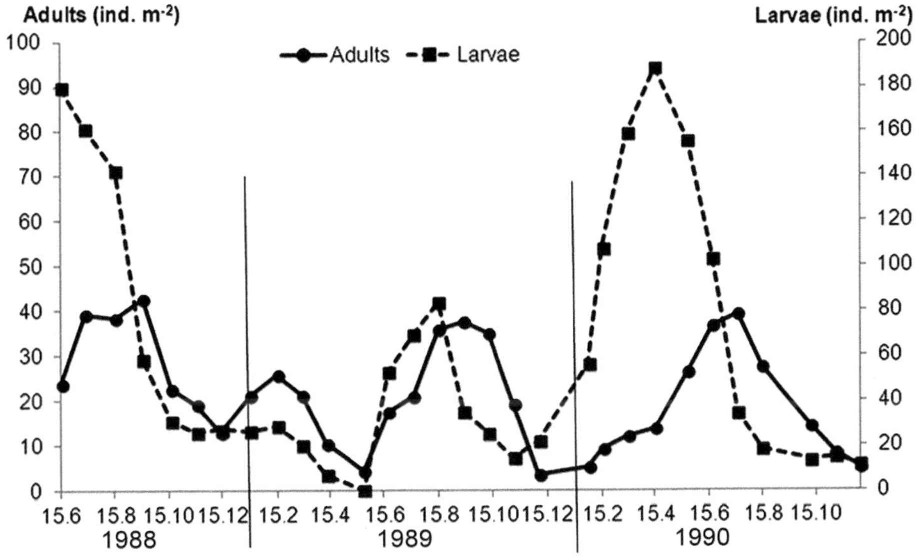

Fig. 7.9 Dynamics of adults and larvae on an agricultural field in Northern Germany based on fluctuating mean of 3 months (monthly samples of four replicates over 3 years gained by heat extraction)

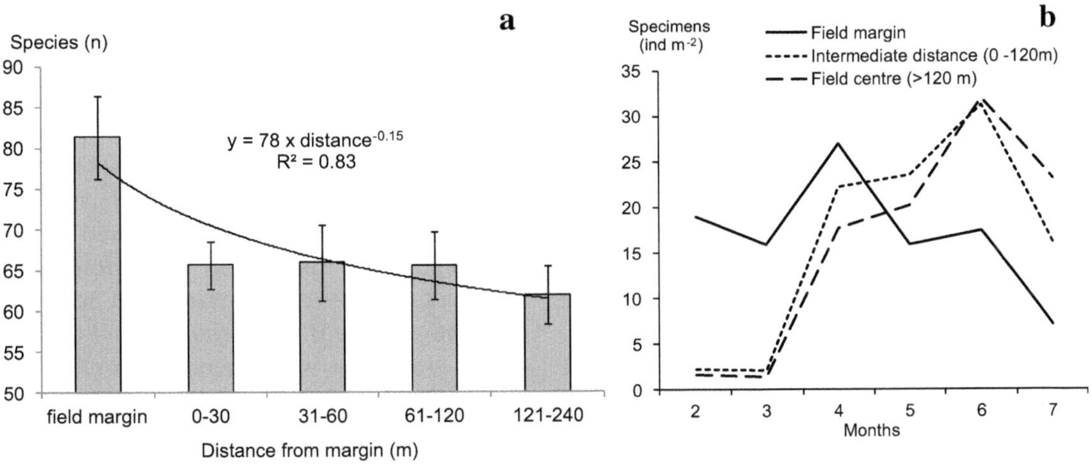

Fig. 7.10 (**a**) Species richness calculated as sample rarefaction for ten pitfall traps in a northern German agricultural field; (**b**) seasonal occurrence of *Tachyporus hypnorum* in different areas of an arable field ($n = 122$ pitfall traps, year 2001) in Northern Germany

actively fly into the field center (Fig. 7.10). They are active in the summer time when activity has already decreased in the field margins.

Although field margins usually have higher numbers of staphylinid species (Fig. 7.10), after the change from intensive to organic farming in long-term investigations, no significant increase of species or specimens was found during the succession (Andersen and Eltun 2000; Schröter 2010). According to Schröter (2010), 11 species profited from the organic practices on arable fields, e.g., *Tachyporus obtusus* and *Tachinus fimetarius*, and the same number of species profited from conventional farming, e.g., *Philonthus rotundicollis* and *Lathrobium fulvipenne*. Several of the specific species of

Table 7.5 List of dominant Staphylinidae restricted to coastal habitats in Northern Europe

Eulittoral		Supralittoral		
Rocky	Sandy	Silt mudflat	Sandy beach	Shingle beach
Micralymma marinum	*Diglotta mersa*	*Bledius tricornis*	*Bledius spectabilis*	*Polystomota grisea*
	Diglotta sinuaticollis	*Bledius frisius*	*Phytosus balticus*	*Cafius xantholoma*
	Bledius subniger	*Brundinia marina*	*Phytosus spinifer*	*Omalium riparium*
			Polystomota punctatella	
			Atheta vestita	
			Omalium riparium	

arable fields show a wide distribution. Due to the similar conditions under agriculture, they are dominant on most European agricultural fields such as *Xantholinus linearis*, *X. longiventris*, *Tachyporus hypnorum*, *Philonthus cognatus*, and *Anotylus rugosus* (Bohac et al. 1999; Andersen and Eltun 2000; Gilgenberg 1986; Lupi et al. 2006; Schröter 2010). Moreover, European species are invaders in North America, e.g., *Gyrohypnus angustatus*, *Tachinus corticinus*, *Sepedophilus marshami*, etc. (Levesque and Levesque 1995, see also Chap. 5 in this book).

7.3.3 Coasts

The 392 species specialized on coastal habitats (Frank and Ahn 2011) have to be divided into several ecological groups (Table 7.5). Few species occur under intertidal conditions and live under submersion conditions two times per day. Among these species, some live on rocky cliff coasts such as *Micralymma marinum* (Thayer 1985) and others in sandy mudflats (Topp 1975) or in tropical mangroves (Frank and Ahn 2011) such as *Bryothinusa sakishimana* or *Linoglossa murphyi* (Sawada 1991). According to Topp (1975), the two species *Diglotta mersa* and *D. sinuaticollis* use the same resources of a small area of aerated sand a few centimeters below the sand surface at a distance of 60 m from the mean high tide level. They overwinter in the dune zone above the mean high tide level and inhabit their eulittoral habitat again by hydrochory in each summer.

Most of the coastal species, however, live in the supralittoral zone under fewer, but unpredicted, submersion conditions. These species are either restricted to salt grassland in northern or in southern regions (e.g., Heydemann 1962; Irmler and Heller 2002) or to beaches that have to be divided into sandy beaches and shingle beaches (Irmler 2012) (Table 7.5).

On salt marshes covered by grassland, the species richness increased with increasing elevation and concurrently with decreasing tidal floods (Fig. 7.11). The number of specialized species, e.g., *Brundinia marina*, decreases with an increase in elevation. Other species such as *Bledius tricornis* have their maximum at intermediate elevations (60 cm above mean high tide level; MHT). The rise in species richness with increasing elevation is mainly referred to the high number of species invading from fresh grassland sites to high elevated salt marshes, e.g., *Amischa analis* and *Oxypoda brachyptera*.

The wave and wind energy at beaches are decisive factors for the occurrence of coastal staphylinid species (Irmler 2012). At sandy beaches with moderate wave and wind impact, the species composition differs significantly from wind- and wave-exposed shingle beaches (Table 7.5). Wrack is another important habitat on beaches (Ruiz-Delgado et al. 2014). According to Ruiz-Delgado et al. (2015), the common staphylinid species in SW Spain, e.g., *Carpelimus rivularis*, *Cafius xantholoma*, and *Remus sericeus*, were not correlated with temperature, moisture, or carbohydrate concentration. The wide distribution of coastal staphylinid species as documented by

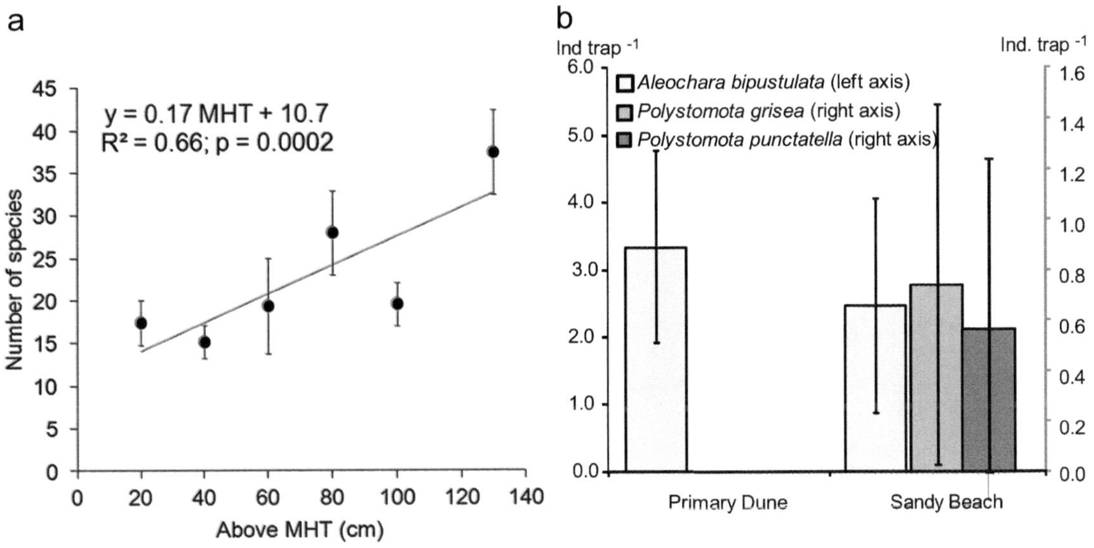

Fig. 7.11 (**a**) Vertical increase of species number (species trap^{-1} year^{-1}) of Staphylinidae in northern German saline grassland at the North Sea; (**b**) distribution of three rove beetles in the beech dune gradient; MHT, Mean high tide level (according to Irmler and Heller 2002 and Irmler 2012)

Frank and Ahn (2011) indicates that coastal species have a wide range of ecological demands on the nearshore conditions.

Many of the species found on sandy beaches are restricted to that habitat and do not even occur at adjacent sites such as primary dunes (Fig. 7.11). However, species of the dune habitat can invade the sandy beaches, e.g., *Aleochara bilineata*. Thus, the habitat of the specialized beach species is very narrow, mostly no wider than 10–20 m in the vertical direction. For the intertidal species *Diglotta brasiliensis*, which lives under similar conditions as the European *Diglotta submarina* and *D. mersa*, the coastal habitat zone is only 4–5 m wide (Da Rosa et al. 2008).

7.3.4 Montane and Alpine Habitats

In his fundamental work on the distribution of alpine and subarctic Coleoptera, Holdhaus (1954) mentioned several staphylinid species of the high alpine zone of the European high mountains, e.g., *Stenus hoelzeli*, *Leptusa winkleri*, and *Chilopora holdhausi* for the Southern Alps. Several species have a boreo–alpine distribution because they occur in Scandinavia and the Alps, e.g., *Atheta depressicollis*, *Oxypoda tirolensis*, and *O. nigricornis*. Many of the alpine staphylinid species are endemic in extremely small areas; some live only on the top of one mountain such as many *Leptusa*, *Geostiba*, and *Alpinia* species. Zerche (2006) reported that the species of the genus *Ophthalmoniphetodes* (Omaliinae: Coryphiini) live under snow patches in early spring. They are flightless and occur in numerous endemic species in alpine and subalpine zones of the Balkan Peninsula. As one of the best investigated countries regarding Staphylinidae, 44 endemic and subendemic species have been recorded in Austria (Paill and Kahlen 2009). Along the altitude zones, the highest number of species is found in a range from 1500 to 2500 m above NN in different alpine habitats, e.g., alpine grassland and alpine dwarf shrub heath (Fig. 7.12). The high number of endemic species in the mountains of Southeastern Europe is derived from the isolation during the glacial period.

In the Dolomite Alps, Schatz (2008) differentiated four assemblages of Staphylinidae on alpine grassland: calcareous boulder grassland, shady rock face, alpine calcareous grassland, and

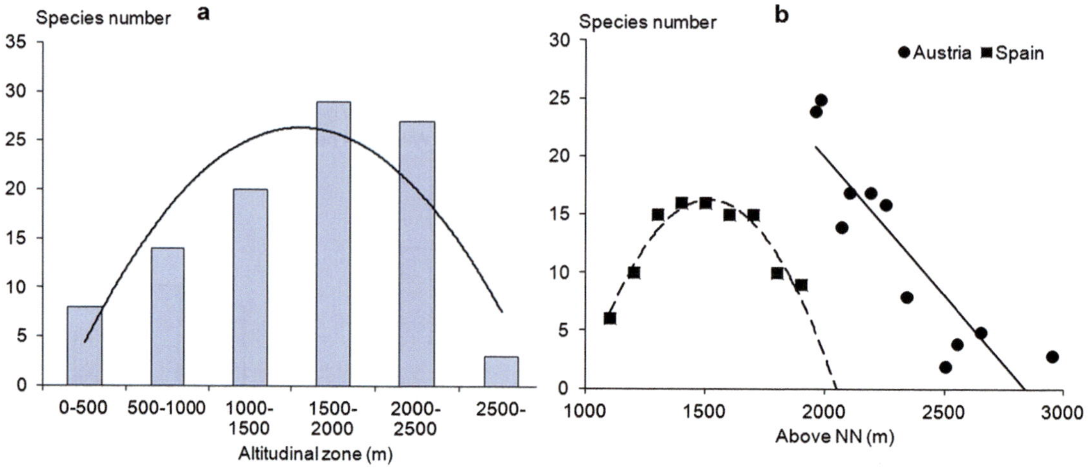

Fig. 7.12 (**a**) Vertical distribution of endemic and subendemic Staphylinidae in Austria according to Paill and Kahlen (2009); elevation range of species is considered; (**b**) vertical distribution of species richness of Staphylinidae gradients of the Central Alps (Austria) and in Sierra de Guadarrama (Central Spain; only the subfamily Staphylininae was considered)

volcanic boulder grassland. Typical species are, for example, *Ocypus alpestris*, *Philonthus frigidus*, and *Philonthus montivagus*. In the central Alps, De Zordo (1979) reported four assemblages at an altitude of 1900–3100 m: valley grassland, dwarf shrub heath, lichen heath, and alpine grassland with typical species, e.g., *Tachyporus macropterus*, *Omalium ferrugineum*, *Atheta leonardi*, and *Coryphium gredleri*. According to this investigation, the species richness decreased corresponding to the altitude gradient (Fig. 7.12). Fernandez et al. (2010) investigated the staphylinid fauna of the Central Spain Sierra de Guadarrama and found the highest species richness in an intermediate altitude of 1500 m. Several species, e.g., *Quedius latinus* and *Ocypus olens*, were restricted to this intermediate altitude. In Norway, alpine Staphylinidae assemblages were composed of species with affinities to dry and humid meadows as well as mires (Ottesen 1996). They showed niche segregation along the humidity, altitude, and season gradients. For example, characteristic spring species were *Stenus carbonarius* in wet habitats at low altitudes (~1300 m) and *Bryoporus rugipennis* in dry habitats at high altitudes (~1500 m); characteristic summer species were *Olophrum boreale* in humid habitats at high altitudes and *Stenus ludyi* in dry habitats at low altitudes. Only a few groups exhibited slightly separated niche dimensions, e.g., *Boreophilus henningianus*, *Arpedium quadrum*, and *Psephidonus longipes* (all in autumn at intermediate soil moistures and high latitudes). In contrast to their overall well-developed flight ability, alpine species displayed a low dispersion potential. Half of the 21 species found in the foreland of a Norwegian alpine glacier were found in sites that were more than 63 years old (Bråten et al. 2012). Sites with *Tachinus elongatus* and *Mycetoporus erichsonanus* had to have a minimum age of 200 years.

Little is known about Staphylinidae for the tropical mountain region. Some species at high altitudes of the Andean mountains are known that belong mainly to the Aleocharinae subfamily, e.g., *Atheta chimborazicola* at 4400 m or *Atheta atacazomontis* at 3890 m in the high montane paramos of Ecuador (Pace 2008). In regard to the leaf litter of montane forests of the Colombian Andes, two altitude zones were differentiated by staphylinid assemblages: one below approximately 1500 m and the other above 1500 m elevation (Gutièrrez Chacòn and Ulloa Chacòn 2006). High tropical mountains also seem to have a high rate of endemic staphylinid species, e.g., 39% are reported for the African Kilimanjaro (Hemp and Winter 1999).

7.4 Microhabitats

7.4.1 Dung-Inhabiting Staphylinidae

Coprobiontic Staphylinidae, e.g., *Platysthetus arenarius*, feed and reproduce in dung. Coprophilous Staphylinidae are not dependent on dung as a food source or reproduction habitat, e.g., *Philonthus* species are found in diverse decaying organic matter such as rotting plants, carrion, and fungi which contain *Diptera* larvae. In dung, *Philonthus*, *Tachinus*, *Megarthrus*, *Anotylus*, *Platystethus*, *Autalia*, *Aleochara*, *Atheta*, and *Oxypoda* are common and frequent European staphylinid genera (Koskela 1972). Staphylinidae are considered to be the most important predators in dung. Due to the diversity of species and high population densities, *Philonthus* species are the most important predators on eggs, larvae, and adults of Staphylinidae and other Coleoptera in dung, e.g., Scarabaeidae and Hydrophilidae.

Some of the dung-inhabiting Staphylinidae are considered invasive species in America, e.g., the European species *Philonthus rectangulus* and *P. longicornis* in Argentina and Chile. Larvae of the Staphylinidae have been also found in the dung of cows, horses, and/or sheep (Table 7.6).

Table 7.6 Number of common larvae of Staphylinidae (ind. l^{-100}) living in dung of different farm animals in Northern Germany (according to Lipkow 2011)

Species	Dung of		
	Cattle	Horse	Sheep
Philonthus splendens F.	13.3	6.8	–
Philonthus cruentus Gmelin	–	–	8.7
Philonthus marginatus Stroem	7.0	9.8	7.8
Philonthus varians Payk.	4.7	2.0	1.7
Gyrohypnus angustatus Steph.	–	2.0	–
Oxytelus laqueatus Marsh.	5.5	2.9	1.7
Platystethus arenarius Fourc.	7.8	2.4	1.7

7.4.1.1 Development and Overwintering

Three types of egg-laying behavior are differentiated for dung-inhabiting Staphylinidae (Fig. 7.13): (A) *Philonthus* species lay individual eggs under the cow pats between dung and soil, 1–4 cm from the cow pat margin; (B) *Platystethus arenarius* females build egg chambers inside the dung to deposit 10–20 eggs; (C) *Tachinus lignorum*, *T. rufipes*, and *T. laticollis* females move the tip of their abdomen with the fixed egg over the substratum (soil) to stick soil particles to the shell for camouflage to protect eggs against predators.

The time of larvae development from L1 to adulthood depends on the temperature and differs among species. Dung-inhabiting staphylinid species have a shorter development time than species that do not live in dung (Fig. 7.14). The development time of the investigated dung-inhabiting species (*Philonthus marginatus*, *P. varians*, *P. carbonarius*, *P. splendens*, and *P. cruentatus*) ranged from 18 to 25 days. The non-dung-inhabiting species (*Tachinus rufipes*, *P. rubripennis*, *P. cognatus*, *P. decorus*, *Tachyporus hypnorum*, *T. dispar*, *Atheta lividipennis*, *Anotylus rugosus*, *Stenus comma*, *Acylophorus wagenschieberi*, and *Drusilla canaliculatus*) have a longer development time, ranging from 21 to 49 days. The two regressions are significantly different according to the *t*-test ($t = 12.04, p < 0.01$).

Larvae of dung-inhabiting Staphylinidae leave the dung for pupation. Pupation occurs under the dung heap after a chamber is built near the dung.

Dung-inhabiting Staphylinidae usually leave the dung for overwintering in forests, in hedges, and at meadow margins, e.g., *Philonthus splendens*, *P. marginatus*, *Tachinus laticollis*, *T. marginellus*, *Oxytelus laqueatus*, and *Autalia rivularis* (Renken 1956).

7.4.1.2 Settlement and Food Uptake

Diptera are the first insects that arrive at fresh dung for food uptake and reproduction. Shortly afterward, Hydrophilidae reach the dung and

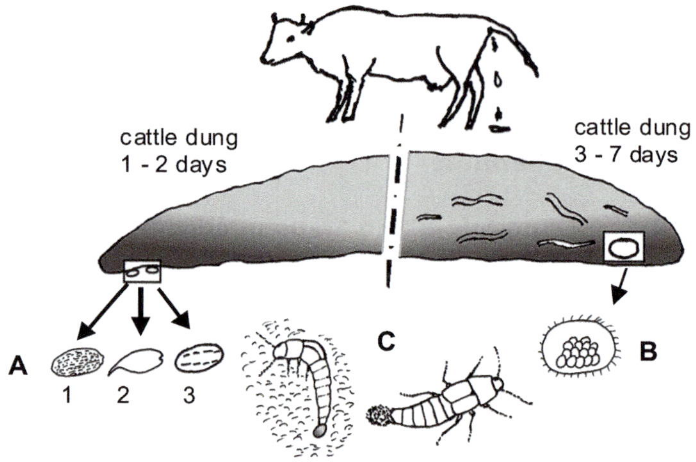

Fig. 7.13 Types of egg deposits in dung-inhabiting Staphylinidae: (**a**), eggs laid separately beneath the dung by *Philonthus splendens* (1), *P. marginatus* (2), and *P. varians* (3); (**b**), eggs laid in clusters into egg chambers in the dung by *Platystethus arenarius*; (**c**), eggs camouflaged with soil particles by *Tachinus* species (According to Lipkow 2011)

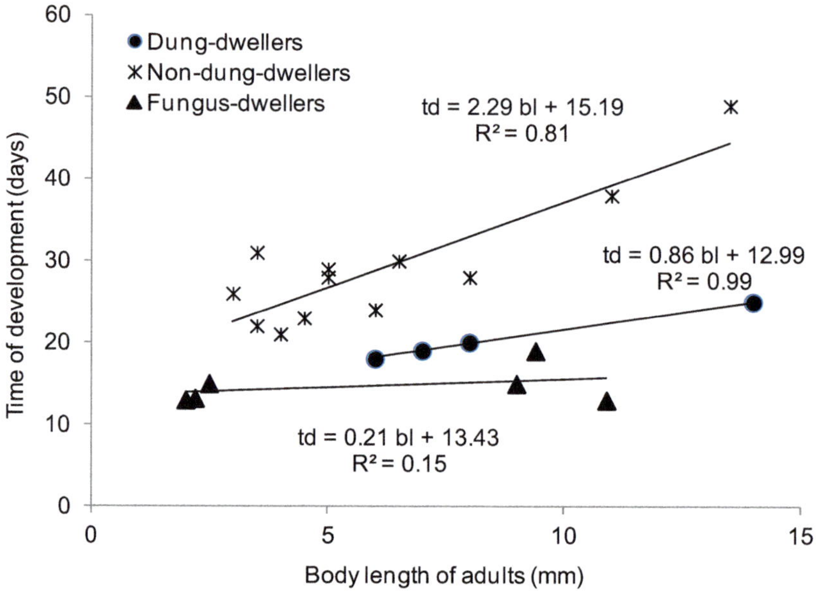

Fig. 7.14 Correlations between body length (bl) of adults and time of development (td) from egg to adult of dung-inhabiting, fungus-inhabiting, and soil surface-dwelling species under ~20 °C temperature conditions; data based on own research and from Staniec and Pietrykowska-Tudruj (2007), Eghtedar (1970), Wildschut et al. (1981), Lipkow (1966), Topp (1971), Weinreich (1968), Schminke (1978), Ashe (1981), Andresen (1984), Henley and Goodrich (1994), and Leschen and Allen (1988)

swim inside. In general, dung-inhabiting Staphylinidae arrive a few hours to a few days later. The time of arrival differs for species and depends on the dung conditions. According to observations in the field, most of the coprophilous Staphylinidae preferred 3–6-day-old dung: *Philonthus marginatus* preferred fresh dung; *P. splendens*, *Oxytelus laqueatus*, and *Platystethus arenarius* were found in fresh and 3–6-day-old dung; *P. fimetarius*, *P. albipes*, and *Anotyles tetracarinatus* preferred dung older than 7 days (Lipkow 2011). Most of the predatory Staphylinidae use tunnels of other dung beetles, e.g., *Sphaeridium*, *Cercyon* (Hydrophilidae), *Aphodius* (Aphodiidae), and larvae of *Diptera*, to enter and move in the dung.

Although *Oxytelus laqueatus* and *Platystethus arenarius* are considered to be coprophagous (Schlüter 1988), they are also found in other decaying matter. At present, a coprophagous feeding behavior is not verified. In contrast to the carnivorous *Philonthus* and *Tachinus* species, Oxytelinae do not feed on sliced larvae of *Tenebrio molitor* in the laboratory. The predator *Ontholestes murinus* waits near the dung heap to catch flies by running and grasping with the mandibles after landing. Petrenko (2013) observed that dung beetles are caught by *O. murinus* after they raise their hard elytra and open their vulnerable soft abdominal tergites. After catching the prey, the predatory rove beetles immediately disappear with the prey into the dung tunnels. *Philonthus* usually lands next to the dung. After landing, *Philonthus splendens* either disappears under the dung pat, e.g., for egg-laying, or roam on the pat surface. *P. splendens* can only walk on dried surfaces of cow pats. While walking on the pat, *P. splendens* searches tunnels made by the dung beetle *Sphaeridium*. After finding an entrance, *P. splendens* puts its head into the tunnel and either enters it or pulls the head back and continues searching for other tunnel entrances. After successful preying, *P. splendens* leaves the pat. Most of its prey, e.g., adults and larvae of dung beetles, small Staphylinidae, or larvae of flies, is found inside the dung. When the larvae of flies have disappeared from the dung for pupating in the soil, *Philonthus* species also disappear.

In Panama, the large Xantholini *Eulissus chalybaeus* (14–17 mm) and *E. rutilus* were observed to prey on the 5–7-mm-long Scarabaeinae *Canthon angustatus*, *C. lamprinus*, and the large *Dichotomius satanas*. Predation was probably inside the tunnels of *D. satanas* (Young 2011). The staphylinid *Leistotrophus versicolor* was observed roaming the neighborhood of a human dung pile and attacked the dung beetle *Canthidium cupreum* by grasping it with the mandibles (Noriega and Navarrete-Heredia 2013). *Philonthus*, *Tachinus lignorum*, and *T. rufipes* larvae have preoral digestion. They catch their prey with their mandibles, infuse digestion fluid into the food, and absorb the pulp (Lipkow 2011).

Philonthus and *Ontholestes* species have a specific mandible brush that may function as a filter to separate liquid and solid components of the food.

7.4.1.3 Dung-Inhabiting Staphylinidae and Biological Control of Pests

Predatory Staphylinidae, e.g., *Philonthus* species, are antagonists of the dung-inhabiting larvae of the horn fly *Haematobia irritans*, a pest of cattle (Cabrera Walsh and Chani-Posse 2003), and have already been used for biological control in Texas (Hunter et al. 1991). *Philonthus* species in Argentina have been documented as predators of dung-breeding flies that are assumed to be an invasive European species (Chani-Posse 2004).

Aleochara species are also antagonists of dung-breeding flies. Adults hunt on eggs and larvae of flies, whereas larvae are ectoparasitoids of fly pupae. The first larval instars actively search for host pupae, chew an opening, enter, and begin to feed. Subsequently, the entrance hole is clogged with fecal material. Pupation occurs either inside the host pupae or outside in cocoons. South African *Aleochara* species were introduced to Australia to control the buffalo fly *Haematobia irritans exiqua* (Wright and Müller 1989). Coprophilous Sarcophagidae (*Diptera*) are the most frequent hosts of *Aleochara verberans* in Argentina (Walsh and Chiani-Posse 2003).

7.4.1.4 Dung Preferences

Recent investigations show that Staphylinidae have no preference for dung of specific mammals (Lipkow and Irmler 2016).

In contrast to the many investigations of the relation of Staphylinidae to dung of domesticated mammals, e.g., cows and horses, few investigations have been made with the dung of wild mammals, e.g., *Anotyles* sp. preferred dung of Sika deer in coniferous forests in Southwest Japan (Yamamotu et al. 2014).

Many of the dung-inhabiting Staphylinidae inhabit pastures and woods (e.g., *Oxytelus*

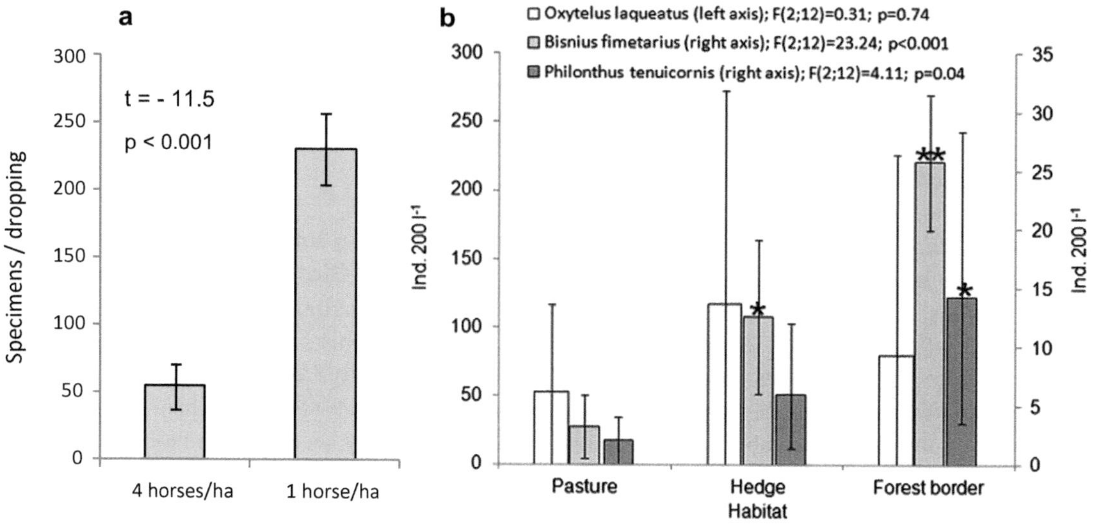

Fig 7.15 (**a**) Number of specimens on horse drops on pastures with different number of horses with results of *t*-test between means; (**b**) number (ind. l^{-200}) of three staphylinid species in horse dung; asterisks indicate significant differences with * $p < 0.05$ and ** $p < 0.01$

laqueatus) in equal abundance, but few species prefer woods, the border of woods, or hedges, e.g., *Tachinus pallipes*, *T. humeralis*, *T. proximus*, *Rugilus rufipes*, *Quedius scintillatus*, *Q. cinctus*, *Bisnius fimetarius*, or *Philonthus tenuicornis* (Fig. 7.15).

Abundance of Staphylinidae correlates with the density of dung droppings (Fig. 7.15). If only a few dung droppings were available on a pasture, more specimens of dung-inhabiting staphylinids were found on the individual droppings (Lipkow 2011).

7.4.2 Associations Between Staphylinidae and Fungi

7.4.2.1 General Traits

A general overview of life history and morphology of the mouthparts of fungus-dwelling Staphylinidae is published by Lipkow and Betz (2005). Associations with fungi, including mycophagy (fungus-feeding), are found in many subfamilies of Staphylinidae (Newton 1984). A clear distinction between saprophagy and mycophagy is difficult because both feeding items are mixed in wood and leaf decay. Fungi are probably a secondary feeding item because Staphylinidae, in general, are primarily predators (Thayer 2005). Shifts in feeding preferences are found between tribes and genera as well as in the same genus (e.g., *Sepidophilus*). Normally *Tachinus rufipes* (L) is a predator but can switch to feed exclusively on yeast (Protoascomycetidae) in laboratory conditions.

Interactions with fungi occur in each of the four subfamily groups in the form of mycophagy or of predation on other fungus-dwelling organisms (Scheerpeltz and Höfler 1948; Benick 1952; Thayer 2005). Fungivorous species are found in numerous subfamilies, e.g., Micropeplinae, Neophoninae, Habrocerinae, and Aleocharinae. In Aleocharinae larvae and adults of *Gyrophaena* Mannerheim, *Phanerota* Casey, and *Eumicrota* Casey obligatorily feed on spores of the hymenium of *Agaricales* and *Polyporales* (Andreesen 1984). Larvae and adults of *Placusa* Erichson and some species of *Homalota* Mannerheim feed on subcortical spores and hyphae (Ashe 1993). *Meronera* Sharp feed on surface hyphae. *Stictalia* Casey, *Pseudatheta* Cameron (Ashe 1993), *Pagla* Blackwelder, and *Polylobus* Solier are considered spore feeders (Betz et al. 2003), and *Oxypoda* Mannerheim feed on spores of agaricoid *Basidiomycetes* (Henneberg 2004). In the Tachyporinae subfamily *Sepedophilus* Gistel,

Tachinus Gravenhorst, and *Coproporus* Kraatz are fungus feeders. In the Scaphidiinae and Oxyporinae subfamilies, e.g., *Scaphium* Kirby, *Scaphisoma* Leach, *Cyparium* Erichson, *Oxyporus* Fabricus, *Baeocera* Erichson, and *Scaphobaeocera* Csiki, respectively, feed on fungi (Betz et al. 2003).

A more general description of fungus-feeding and host associations is published by Schigel (2012). DNA gut content analysis of abundant Staphylinidae shows that the mycobiota in posterior gut extracts was dominated by Saccharomycetales and Sordariomycetes (Stefani et al. 2016).

Fleshy, short-lived mushrooms (*Agaricales*) are ephemeral microhabitats, similar to dung heaps, and can be used only for a few days. The beetles respond to these short-time food resources by colonizing the mushroom immediately after the opening of the pileus and by rapid larval development (Fig. 7.14). The development time of the fungus-dwelling species (*Phanerota fasciata*, *Gyrophaena joyioides*, *G. gentilis*, *Oxyporus stygicus*, *O. vittatus*, and *O. major*) from egg to adult ranges from 13 to 19 days and is almost independent from the size of the species. The difference from the soil-dwelling species is significant using the *t*-test (significance between the two regressions: $t = 39.3$, $p < 0.001$). Because pupation of both ecological groups occurs in the soil, the development of the pupae lasts nearly an equal amount of time (fungus dwellers, 9.7 ± 0.8 days; soil dwellers, 10.4 ± 1.8 days).

Hymenomycetales (macrofungi) with hymenial fruiting bodies, belonging to the *Basidiomycetes*, have been the focus of most investigations. Hence, special adaptations of fungus-breeding beetles are expected to be seen in specific life-history traits, such as short developmental time. Life-history traits in fungus-dwelling staphylinids have not been studied in detail. In long-living tree fungi (*Aphyllophorales*), the density of Staphylinidae is lower than in fleshy fungi (Krasutski 2010).

Little is known about the association between Staphylinidae and fungi in deeper soil layers and in decaying plant material, dung, or carrion (i.e., Myxomycetes, *Ascomycetes*, Fungi imperfecti).

Fungivorous Staphylinidae (e.g., *Oxyporus*, *Gyrophaena*, *Sepedophilus*) and some predatory or parasitoid Staphylinidae, e.g., *Bolitobius* and *Aleochara*, prefer young fresh mushrooms containing spores. In contrast, decaying mushrooms mainly attract predaceous Staphylinidae, e.g., *Tachinus*, *Philonthus*, *Ontholestes*, and *Atheta*. Predaceous Staphylinidae (e.g., *Aleochara*) found in mushrooms feed on the fungivorous larvae of flies (e.g., Anthomyiidae).

7.4.2.2 Life History

Gyrophaena

Most species of the Gyrophaenina prefer several species of mushrooms and are not restricted to feeding on single fungus species. When the preferred mushrooms are not available, less preferred mushrooms are used (Ashe 1984; Andreesen 1984). The Central European species *Gyrophaena joyioides*, *G. gentilis*, *G. affinis*, *G. fasciata*, and *G. nana* feed on a wide spectrum of fungi, whereas the feeding spectrum of *G. minima* and *G. bihamata* is narrower. In the laboratory, *G. joyioides* fed on fungal "tissue," if spores of the preferred fungus were not available (Andreesen 1984). According to Henneberg (2004), several species of *Gyrophaena* might coexist on the same host mushroom. He concludes that interspecific competition between different *Gyrophaena* species is probably more important for the larvae than for the adults. In dry summers with a low density of mushrooms, thousands of adults of *Gyrophaena* might live on a single "fruiting body." Despite the high density, their individual distances might be more or less uniform (Fig. 7.16). At high population densities, e.g., 25 ind. cm^{-2}, gyrophaenine beetles show aggressive behavior; they defend their small territory. Most gyrophaenine beetles leave a mushroom before the decaying process starts. It seems that the sex ratio of the species in the mushrooms is balanced. In Central Europe, *Gyrophaena* species have 2–3 generations per year.

Mating of *G. joyioides* Wüsthoff was observed at the base of the mushroom cup and lasted about 20 min. *Gyrophaena* species prefer

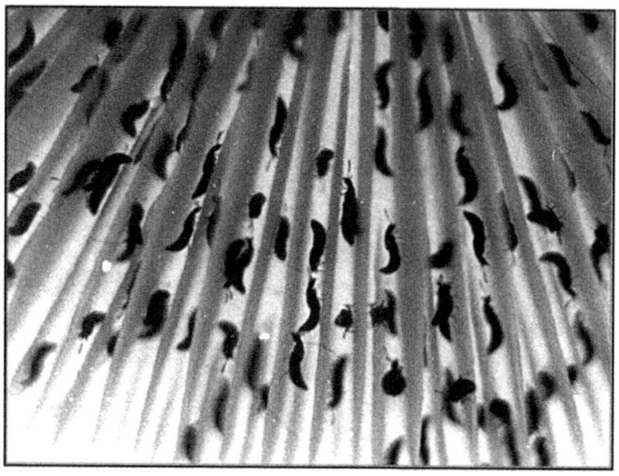

Fig. 7.16 Adults of *Gyrophaena joyoides* in the interlamellar space of a fruiting body of *Megacollybia platyphylla* (*Agaricales*) (Modified according to Andreessen 1984)

young fruiting bodies of *Agaricales* for oviposition and deposit their eggs in the hymenium when sporulation begins. The placement of egg deposition differs between *Gyrophaena* species and range between the distal part (*G. affinis*) to the proximal basis of the gill (*G. joyioides*). *G. minima*, *G. affinis*, and *G. joyioides* deposit individual eggs, whereas *G. joyioides* also produce clusters of three eggs. The eggs are covered with material from the periphery of the gills. *G. gentilis* is ovoviviparous. The first instar completely develops in the egg in the female body. Viviparity in *G. gentilis* seems to be an adaptation to the rapid decay of fungi as hosts (Andreesen 1984).

The development of the ovarioles and the maturation of the eggs seem to be induced after the female comes in contact with the mushroom (Henneberg 2004). The larvae feed almost continuously during the day and night. The development of the larvae of *G. joyioides* can be finished in just 1–2 days at 20 °C.

Eumicrota beetles, which prefer persistent polyporous fungi, have a longer larval development than *Gyrophaena* beetles, which prefer fleshy gilled mushrooms. The females of *Eumicrota* construct egg chambers at the ventral side of the hymenium of mushrooms. Mating takes place adjacent to the egg chamber. After oviposition, the female takes up the egg with the mandibles and places it in the egg chamber cleaned beforehand. The number of eggs per chamber varies between 5 and 24. Females of *Eumicrota* remain 5–6 days in the egg chamber. They groom the eggs and repel intruders such as older larvae, males, and other females (Ashe 1987).

Oxyporus

Some species of the New World Oxporinae prefer specific mushrooms out of a wide range of "acceptable" hosts. The larvae seem to be specialized to one or two species of mushrooms (Hanley and Goodrich 1994). Because of the large head with its long and crossing mandibles, entomologists assumed that *Oxyporus* species are predaceous; however, larvae and adults feed exclusively on the fungal tissue of *Agaricales* (Lipkow 1997). Fungal material chewed by the mouthparts is saturated with digestive fluid. Both larvae and adults digest their food preorally. Larvae begin feeding just minutes after leaving the egg. In *O. occipetales* Fauvel, both the first and the second instars last about 1 day, whereas the third instars need about 6 days (Goodrich and Hanley 1995). The third instars dig burrows into the soil to build pupal chambers. The pupal stage lasts 6 days.

Females of *O. japonicus* construct egg chambers in the stipe or cap of mushrooms (*Agaricales*). They pile up material of chewed fungus at the opening of the chambers after oviposition and cover the eggs with bits of chewed fungus. The young larvae presumably feed on this material. Females remain in the egg chamber and repel conspecific females and predaceous insects. Most fruit bodies contain only one egg chamber (Setsuda 1994). *O. germanus* have been found on various fleshy fungi (Hwang et al. 2002).

References

Andersen A, Eltun R (2000) Long-term developments in the carabid and staphylinid (Col., Carabidae and Staphylinidae) fauna during conversion from conventional to biological farming. J Appl Ecol 124:51–56

Andersen T, Ligaard S, Pedersen T et al (1990) Pitfall catches of Carabidae and Staphylinidae (Coleoptera) in a temporarily protected forest area on the Eidanger peninsula, Telemark, Norway. Fauna Norv Ser B 37:13–22

Andreesen B (1984) Studien zur Ökologie und Biologie pilzbewohnender Kurzflügelkäfer (Coleoptera: Staphylinidae). Unpublished Diploma thesis, University of Kiel, Germany, 64p

Ashe JS (1981) Studies on the life history and habits of *Phanerota fasciata* Say (Coleoptera, Staphylinidae, Aleocharinae) with notes on the mushroom as a habitat and descriptions of the immature stages. Coleopt Bull 3:81–96

Ashe JS (1984) Major features of the evolution of relationships between gyrophaenine staphylinid beetles (Coleoptera: Staphylinidae) and fresh mushrooms. In: Wheeler Q, Blackwell M (eds) Fungus-insect relationships: perspectives in ecology and evolution. Columbia University Press, New York, pp 227–255

Ashe JS (1987) Egg chamber production, egg protection, and clutch size among fungivorous beetles of the genus *Eumicrota* (Coleoptera: Staphylinidae) and their evolutionary implications. Zool J Linn Soc 90:255–273

Ashe J (1993) Mouthpart modifications correlated with fungivory among aleocharine staphylinids (Coleoptera: Staphylinidae: Aleocharinae). In: Schaefer CW, Leschen R (eds) Functional morphology of insect feeding. Thomas Say Publications in Entomology, Lanham, pp 105–130

Basedow T (1990) Jährliche Vermehrungsraten von Carabiden und Staphyliniden bei unterschiedlicher Intensität des Ackerbaus. Zool Beitr N F 33:459–477

Benick L (1952) Pilzkäfer und Käferpilze. Ökologische und statistische Untersuchungen. Acta Fenn 70:250

Betz O, Thayer MK, Newton AF (2003) Comparative morphology and evolutionary pathways of the mouthparts in spore-feeding Staphylinoidea (Coleoptera). Acta Zool 64:179–238

Bohac J (1999) Staphylinid beetles as bioindicators. Agric Ecosyst Environ 74:357–372

Bohac J, Jedlicka P, Frouz J (1999) Changes in communities of staphylinid beetles (Coleoptera, Staphylinidae) during secondary succession in abandoned fields. In: Tajovsky K, Pizl V (eds) Soil zoology in Central-Europe. ISB AS CR, Budejovice, pp 19–25

Bong L-J, Neoh K-B, Jaal Z et al (2013) Influence of temperature on survival and water relations of *Paederus fuscipes* (Coleoptera: Staphylinidae). J Med Entomol 50:1003–1013

Bråten AT, Flo D, Hågvar O et al (2012) Primary succession of surface active beetles and spiders in an alpine glacies foreland, Central South Norway. Arct Antarct Alp Res 44:2–15

Cabrera-Walsh G, Chiani-Posse M (2003) Abundance and seasonal distribution of predatory coprophilous Argentine rove beetles (Coleoptera: Staphylinidae), and a discussion of their effect on the community of dung breeding flies. Coleopt Bull 57:43–50

Chani-Posse MR (2004) Eight Argentinean species of dung-inhabiting *Philonthus* Stephens (Coleoptera: Staphylinidae). Stud Neotropical Fauna Environ 39:212–232

Coiffait H (1972) Coléoptères Staphylinidae de la region Paléarctique occidentale. Nouv Rev Entomol 2:651

Coombes DS, Sotherton NW (1986) The dispersal and distribution of polyphagous predatory Coleoptera in cereals. Ann Appl Biol 108:461–474

Da Rosa LC, Borzone CA, Caron E (2008) Occorencia de *Diglotta brasilienis* (Coleoptera: Staphylinidae: Aleocharinae) em duas praias estuarinas Baía de Paranguá, sul do Brasil. Rev Bras Zootec 25:563–565

De Zordo I (1979) Ökologische Untersuchungen an Wirbellosen des centralalpinen Hochgebirges (Obergurgel, Tirol). III. Lebenszyklen und Zönotik von Coleopteren. Veröff Univ Innsbruck 118:132

Eghtedar E (1970) Zur Biologie und Ökologie der Staphyliniden *Philonthus fuscipennis* Mannh.und *Oxytelus rugosus* Grav. Pedobiologia 10:169–179

Fernández V, Gamarra P, Outerelo R et al (2010) Distribución de stafilíninos necrófilos (Coleoptera, Staphylinidae, Staphylininae) a lo largo de un gradiente altitudinal en la Sierra de Guadarrama, España. Bol R Soc Esp Hist Nat 104:61–86

Frank JH, Ahn K-J (2011) Coastal Staphylinidae (Coleoptera): a worldwide checklist, biogeography and natural history. Zookeys 107:1–96

Friebe B (1982) Die Makroarthropodenfauna eines Buchenwaldbodens unter besonderer Berücksichtigung der Coleoptera. Dissertation, Universität Karlsruhe, p 141

Gilgenberg A (1986) Die Verteilungsstruktur der Carabiden- und Staphylinidenfauna verschieden bewirtschafteter landwirtschaftlicher Flächen sowie eines Waldes. Dissertation, Universität Bonn, p 262

Goodrich MA, Hanley RS (1995) Biology, development and larval characters of *Oxyporus major* (Coleoptera: Staphylinidae). Entomol News 106:161–168

Gryuntal SY (2009) Soil mesofauna of taiga burozems. Eurasian Soil Sci 42:1374–1381

Gutièrrez Chacòn C, Ulloa Chacòn P (2006) Composición de Estafilinido (Coleoptera: Staphylinidae) asociados a hojarasca en la Cordilliera oriental de Colombia. Fol Entomol Mex 45:69–81

Hanley RS, Goodrich MA (1994) Natural history, development, and immature stages of *Ocypus stygicus* Say (Coleoptera, Staphylinidae). Coleopt Bull 48:213–225

Hartmann P (1976) Struktur und Dynamik von Staphyliniden-Populationen in Buchenwäldern des Solling. Verh Ges Ökol 5:75–81

Hartmann P (1979) Biologisch-Ökologische Untersuchungen an Staphylinidenpopulationen verschiedener Ökosysteme des Solling. Dissertation, Universität Göttingen, p 173

Hemp C, Winter JC (1999) Ethnozoologische Feldforschung am Kilimanjaro. Arthropoda. Bayreuther Forum Ökol 64:167–199

Henneberg L (2004) Biologie der Interaktion zwischen Koleopteren und agaricoiden Basidiomyceten. Ph.D. thesis, University of Marburg, Germany, 363p

Herman L (2001) Catalogue of the Staphylinidae (Insecta: Coleoptera). 1758 to the end of the second millennium. I. Introduction, history, biographical sketches, and omaliine group. Bull Am Mus Nat Hist 265:649

Heydemann B (1962) Der Einfluss des Deichbaus an der Nordsee auf Larven und Imagines von Carabiden und Staphyliniden. Ber 9. Wandervers Dtsch Entomol 45:237–273

Holdhaus K (1954) Die Spuren der Eiszeit in der Tierwelt Europas. Abh Zool-Bot Gesell Wien 18:493

Holland JM, Thomas CFG, Birkett T et al (2007) Spatio-temporal distribution and emergence in arable fields in relation to soil moisture. Bull Entomol Res 97:89–100

Hunter JS, Fincher GT, Bay DE et al (1991) Seasonal distribution and dial flight activity of Staphylinidae (Coleoptera) in open and wooded pastures in East Central Texas Staphylinidae associated with dung. J Kansas Entomol Soc 64:163–173

Hwang WS, Hanley R, Ahn KJ (2002) Immature stages of *Oxporus germanus* Sharp (Coleoptera: Staphylinidae: Oxyporinae). J Kansas Entomol Soc 75:214–222

Irmler U (1978) Die Struktur der Carabiden- und Staphylinidengesellschaften in zentralamazonischen Überschwemmungswäldern. Amazoniana 6:301–326

Irmler U (1993) Die Kurzflügelkäfer (Staphylinidae) des Bodens schleswig-holsteinischer Wälder. Verh Westdtsch Entomol Tag, pp 69–77

Irmler U (1995) Die Stellung der Bodenfauna im Stoffhaushalt schleswig-holsteinischer Wälder. Faun-Ökol Mitt 18:199

Irmler U (1998) Die vertikale Verteilung flugaktiver Käfer (Coleoptera) in drei Wäldern Norddeutschlands. Faun-Ökol Mitt 7:387–404

Irmler U (2006) Climatic and litter fall effects on collembolan and oribatid mite species and communities in a beech wood based on a 7 years investigation. Eur J Soil Biol 42:51–62

Irmler U (2009) Gradiental changes and temporal fluctuations of rove beetles (Coleoptera: Staphylinidae) in northern German woodland. Faun-Ökol Mitt 9:1–15

Irmler U (2012) Effects of habitat and human activities on species richness and assemblages of Staphylinidae (Coleoptera) in the Baltic Sea coast. Psyche 2012:1–12

Irmler U, Gürlich S (2007) What do rove beetles (Coleoptera, Staphylinidae) indicate for site conditions. Faun-Ökol Mitt 8:439–455

Irmler U, Heller K (2002) Zonierung der Staphylinidae in einem Salzgrünland der schleswig-holsteinischen Nordseeküste. Faun-Ökol Mitt 8:219–229

Irmler U, Heller K, Warning J (1997) Kurzflügelkäfer (Staphylinidae) aus Totholz schleswig-holsteinischer Wälder. Faun-Ökol Mitt 7:307–318

Kasule K (1968) Field studies on the life-histories of some British Staphylinidae. Trans Soc Brit Entomol 1:49–80

Klimaszewski J, Pace R, Center TD, Couture J (2010) A remarkable new species of *Himalusa* Pace from Thailand (Coleoptera, Staphylinidae, Aleocharinae): phytophagous aleocharine beetle with potential for bio-control of skunkvine-related weeds in the United States. Zookeys 35:1–12

Koskela H (1972) Habitat selection of dung-inhabiting Staphylinids (Coleoptera) in relation to the age of dung. Ann Zool Fenn 9:156–171

Krasutski BV (2010) Coleoptera associated with tree fungus *Trichaptum biforme* (Fr. in Klotzsch) (Basidiomycetes; Aphyllophorales) in the forests of the Urals and the Trans-Ural area. Entomol Rev 90:679–688

Krogerus H (1948) Ökologische Untersuchungen über Uferinsekten. Acta Zool Fenn 53:157

Larsen EB (1936) Biologische Studien über die tunnelgrabenden Käfer auf Skallingen. Videnskabelinge Meddelelser fra Dansk naturhistorisk Forening 109:231

Leschen RAB, Allen RT (1988) Immature stages, life histories and feeding mechanisms of three Oxyporus spp. (Coleoptera: Staphylinidae: Oxyporinae). Coleopt Bull 42:321–333

Levesque C, Levesque G-Y (1995) Abundance, diversity and dispersal power of rove beetles (Coleoptera: Staphylinidae) in a Raspberry plantation and adjacent sites in eastern Canada. J Kansas Entomol Soc 68:355–370

Lincoln DCR (1961) The oxygen and water requirements of the egg of *Ocypus olens* Müller (Staphylinidae, Coleoptera). J Insect Physiol 7:265–272

Lipkow E (1966) Biologisch-ökologische Untersuchungen über *Tachyporus*-Arten und *Tachinus rufipes* (Col., Staphyl.) Pedobiologia 6:140–177

Lipkow E (1997) Zur Biologie, Fortpflanzung, Wirtswahl und Konkurrenzvermeidung von *Oxyporus* Arten (Coleoptera: Staphylinidae). Faun–Ökol Mitt 7:297–305

Lipkow E (2011) Observations to the Life history with dung-inhabiting Staphylinidae (Coleoptera). Faun–Ökol Mitt 9:225–246

Lipkow E, Betz O (2005) Staphylinidae and fungi. Faun–Ökol Mitt 8:383–411

Lipkow E, Irmler U (2016) Habitat choice experiments with dung-inhabiting beetles (Coleoptera: Staphylinidae, Hydrophilidae, Scarabaeidae). Faun-Ökol Mitt 9:471–481

Lupi D, Colombo M, Zanetti A (2006) The rove beetles (Coleoptera Staphylinidae) of three horticultural farm in Lombardy (Northern Italy). Boll Zool agr Bachic 38:143–165

Moore I, Legner EF (1976) Intertidal rove beetles (Coleoptera: Staphylinidae). In: Cheng L (ed) Marine insects. North-Holland Companay, Oxford, pp 521–565

Newton AF (1984) Mycophagy in Staphylinoidea (Coleoptera). In: Wheeler QD, Blackwell M (eds) Fungus-insect relationships. Columbia University Press, New York, pp 302–353

Noriega JA, Navarrete-Heredia JL (2013) Quantification of predation on the dung beetle *Canthidium cupreum* (Col., Scarabaeidae) by *Leistotrophus versicolor* (Col., Staphylinidae). Coleopt Bull 67:190–1993

Onipchenko VG (2004) Alpine ecosystems in the northwest Caucasus. Springer, Dordrecht, p 410

Ottesen PS (1996) Niche segregation of terrestrial alpine beetles (Coleoptera) in relation to environmental gradients and phenology. J Biogeogr 23:353–369

Pace R (2008) New records of Aleocharinae from Ecuador and Peru with description of new species, new subgenera and new genera (Coleoptera, Staphylinidae). Biodiversity of South America, I. Membr Biodivers 1:225–398

Paill W, Kahlen M (2009) Coleoptera (Käfer). In: Rabitsch W, Essl F (eds) Endemiten – Kostbarkeiten in Österreichs Pflanzen- und Tierwelt. Naturwiss Ver und Umweltbundesamt GbmH, Wien, pp 627–783

Palmgren P, Biström O (1979) Populations of Araneae (Arachnoidea) and Staphylinidae (Coleoptera) on the floor of a primeval forest in Mäntyharju, southern Finland. Ann Zool Fenn 16:177–182

Parmain G, Bouget C, Müller J et al (2015) Can rove beetles (Staphylinidae) be excluded in studies focusing on saproxylic beetles in central European beech forests? Bull Entomol Res 105:101–109

Petrenko AA (2013) About hunting methods of predatory rove beetles using *Ontholestes murinus* (Coleoptera: Staphylinidae) as an example. Kharkov Entomol Soc Gaz 21:9–11 (in Russian)

Reise K, Weidemann G (1975) Dispersion of predatory forest floor arthropods. Pedobiologia 15:106–128

Renken W (1956) Untersuchungen über Winterlager von Insekten. Z Morph Ökol Tiere 45:34–106

Rose A (2001) Räumliche und zeitliche Verteilungsmuster von Kurzflügelkäfern (Coleoptera, Staphylinidae) auf Nordsee-Düneninseln unterschiedlicher Sukzessionsstadien. Arch Zool Pub 5:220

Ruiz-Delgado C, Reyes-Martínez J, Sánchez-Moyano JE et al (2015) Distribution patterns of supralittoral arthropods: wrack deposits as a source of food and refuge on exposed sandy beaches (SW Spain). Hydrobiologia 742:205–219

Ruiz-Delgado C, Vierheller Vieira J, Gomes Veloso V et al (2014) The role of wrack deposits for supralittoral arthropods: an example using Atlantic sandy beaches of Brazil and Spain. Estuar Coast Shelf Sci 136:61–71

Sawada K (1991) On new genera and species of intertidal Aleocharinae (Coleoptera: Staphylinidae) and Goniacerinae (Pselpahinae) from Singapore and Japan. Raffles Bull Zool 39:141–152

Schatz I (2008) Kurzflügelkäfer (Coleoptera: Staphylinidae) im Naturpark Schlern – Rosengarten (Südtirol, Italien). Gredleriana 8:377–410

Scheerpeltz O, Höfler K (1948) Käfer und Käferpilze. Verlag Jugend und Volk, Wien, p 351

Schigel DS (2012) Fungivory and host associations of Coleoptera: a bibliography and review of research approaches. Mycology 3:258–272

Schlüter D (1988) Spezielle Beiträge zur Biologie von *Platystethus arenarius Fourc*. Diploma thesis, Freie Universität Berlin

Schminke G (1978) Einfluß von Temperatur und Photoperiode auf Entwicklung und Diapause einiger Staphylinidae. Pedobiologia 18:1–21

Schröter L (2010) Lauf- und Kurzflügelkäfer während der Umstellung zum ökologischen Landbau auf Hof Ritzerau. Faun-Ökol Mitt 36:144

Setsuda KI (1994) Construction of the egg chamber and protection of the eggs by female *Oxyporus japonicus* Sharp (Coleoptera: Staphylinidae). Jpn J Entomol 62:803–809

Sotherton NW (1985) The distribution and abundance of predatory Coleoptera overwintering in field boundaries. Ann Appl Biol 106:423–429

Staniec B (2005) A description of the developmental stages of *Acylophorus wagenschieberi* Kiesenwetter, 1850 (Coleoptera, Staphylinidae), with comments on its biology, egg parasite and distribution in Polandy. Dtsch Entomol Z 52:97–113

Staniec B, Pietrykowska-Tudruj E (2007) Developmental stages of Philonthus rubripennis Stephens (Col. Staphylinidae) with comments on its biology. Dtsch Entom Z 54:95–113

Stefani FOP, Klimaszewski J, Morency MJ, et al (2016) Fungal community in the gut of rove beetles (Coleoptera: Staphylinidae) from the Canadian boreal forest reveals possible endosymbiontic interactions for dietary needs.In: Fungal ecology. Elsevier, Amsterdam

Steinmetzger K, Tietze F (1982) Ein Beitrag zur Faunistik und Habitatbindung der Staphyliniden in Kiefernforstgesellschaften der Dübener Heide (Insecta, Coleoptera). Faun Abh Dresden 9:61–77

Tan B, Fuzhong W, Wanqin Y et al (2013) Seasonal dynamics of soil fauna in the subalpine and alpine forests of west Sichuan at different altitudes. Acta Ecol Sin 33:12–22

Thayer MK (1985) *Micralymma marinum* (Stroem) in North America: biological notes and new distributional records (Cleoptera: Staphylinidae). Psyche 92:49–55

Thayer MK (2005) Staphylinidae. In: Kristensen NP, Beutel RG, Leschen R (eds) Handbook of zoology. vol IV. De Gruyter, Berlin, 40pp

Topp W (1971) Zur Biologie und Larvalmorphologie von Atheta sordida Marsh. Ann Ent Fenn 37: 85–89

Topp W (1975) Zur Besiedlung einer neu enstandenen Insel. Untersuchungen am Hohen Knechtsand. Zool Jb Syst 102:215–240

Topp W (1977) Einfluss des Strukturmosaiks einer Agrarlandschaft auf die Ausbreitung der Staphyliniden (Col.) Pedobiologia 17:43–50

Topp W, Ring RA (1988) Adaptations of Coleoptera to the marine environment. II. Observations on rove beetles (Staphylinidae) from rocky shores. Can J Zool 66:2469–2474

Vickerman GP, Sunderland KD (1975) Arthropods in cereal crops: nocturnal activity, vertical distribution and aphid predation. J Appl Ecol 12:755–766

Vogel J, Uhlig M (1982) Zur Staphylinidenfauna zweier Leipziger Stadtparks. Faun Abh Staatl Mus Tierk Dresden 9:195–204

Walsh GC, Chani-Posse M (2003) Abundance and seasonal distribution of predatory coprophilous argentine rove beetles (Col., Staphyl.) and their effects on dung breeding flies. Coleopt Bull 57:43–50

Weidel H (2010) Das Aeroplankton in der Norddeutschen Tiefebene über Schleswig-Holstein. Faun-Ökol Mitt 9:111–129

Weigmann G, Kratz W, Heck M et al (1989) Ballungsraumnahe Waldökosysteme. Teilprojekt 1.5 Bodenbiologische Dynamik immissionsbelasteter Forsten. UBA, Berlin

Weinreich E (1968) Über den Klebfangapparat der Imagines von *Stenus* Latr. (Coleoptera, Staphylinidae) mit einem Beitrag zur Kenntnis der Jugendstadien dieser Gattung. Zeitschrift für Morphologie der Tiere 62:162–210

Wildschut MA, Heessen HJL, Brunsting AMH (1981) Duration of the developmental stages and timing of the end of the reproductive season of Pterostichus oblongopunctatus (Fabricius) (Col., Carabidae) and Philonthus decorus (Gravenhorst) (Col., Staphylinidae). Neth J Zool 32:49–62

Wright EJ, Müller P (1989) Laboratory studies of host finding, acceptance and suitability of the dung-breeding fly *Haematobia thirouxi* potans (Dipt. Muscidcae) by *Aleochara* species (Col. Staph.) Entomophaga 34:61–71

Yamamotu S, Ikeda K, Kamitaini S (2014) Species diversity and community structure of rove beetles (Col., Staphyl.) attracted to dung of sika deer in coniferous forests of Southwest Japan. Entomol Sci 17:52–58

Young OP (2011) Staphylinid predation on large dung beetles (Coleoptera: Staphylinidae, Scarabaeinae) in Panama. Coleopt Bull 65:227–229

Zanetti A (2011) Contribution to the knowledge of Staphylinidae from southern Sardinia (Coleoptera). Conserv Habitat Invert 5:331–352

Zerche L (2006) Monographie der paläarktischen Coryphiini. Suppl. 3: Revision der Gattung *Ophthalmoniphetodes* Zerche – tertiäre Relikte im Schnee (Coleoptera, Staphylinidae, Omaliinae). Nova Suppl Entomol 19:222

A Review of Nearctic Rove Beetles (Staphylinidae) Specialized on the Burrows and Nests of Vertebrates

8

Adam J. Brunke and Joel Buffam

Abstract

Ecosystem engineers, such as social insects, burrowing mammals, and beavers, have a large physical impact on their environment and create new niches for a diversity of organisms. New species or entire lineages of commensal insects have evolved in response to these new microhabitats, and the hyperdiverse rove beetles (Staphylinidae) are no exception. A well-known assemblage of commensal rove beetles in Central Europe is closely associated with the nests and burrows of mammals and birds. Despite similar temperate biomes and an even greater diversity of suitable host vertebrates in the Nearctic region, this fauna remains poorly known, and its documentation is scattered in collections and the primary literature. This review provides an overview of the rove beetles putatively specialized for life in burrows and nests. A total of 46 species from 7 of the 25 Nearctic subfamilies are known thus far, with most in the Aleocharinae and Staphylininae. Hosts for Nearctic Staphylinidae include groundhog, prairie dogs, ground squirrels, pocket gophers, mountain beaver, gopher tortoise, woodrats, North American beaver and muskrat, and several species of birds. Numerous potential vertebrate hosts remain overlooked, and most known hosts are rarely sampled and then only within a small portion of their distribution. Many new species likely remain to be discovered in these microhabitats.

A. J. Brunke (✉) • J. Buffam
Canadian National Collection of Insects, Arachnids and Nematodes, Agriculture and Agri-Food Canada, Ottawa, ON, Canada
e-mail: adam.brunke@canada.ca

8.1 Ecosystem Engineers and Their Impact on Insect Speciation

Organisms that create new habitat or microhabitat opportunities for other species through substantial modification of their environment are considered "ecosystem engineers" (Jones et al. 1997; Reichman and Seabloom 2002). A well-known example is the North American beaver, which creates lentic environments via damming, positively impacting populations of wetland fauna such as fish, birds, reptiles, and invertebrates (Rosell et al. 2005). In the hyperdiverse invertebrates, ecosystem engineering has resulted in the evolution of species or entire lineages of commensals, specialized on these new opportunities. Examples include a group of *Drosophila* flies developing primarily in beaver-harvested wood (Spieth 1979), beetle species developing only in the dung of burrowing mammals (Connoir et al. 2014), and an enormous suite of invertebrates tightly associated with the

O. Betz et al. (eds.), *Biology of Rove Beetles (Staphylinidae)*,
https://doi.org/10.1007/978-3-319-70257-5_8

colony activities of social insects (e.g., Parker 2016). As the largest family of insects with over 62,000 described species (Newton, unpublished catalog), the rove beetles (Staphylinidae) would be expected to have evolved a diverse range of commensal relationships with ecosystem engineers. The myriad lineages associated with the nest-building activities and creation of "refuse piles" by social insects are probably the best known staphylinid commensals and constitute some of the largest (Marlowe et al. 2015) and most bizarre (Parker 2016) species of the entire family. Another relatively well-known group of commensal staphylinids is associated with burrows and nests of vertebrates ("nidicolous" species) in Central Europe, particularly birds, moles, and ground squirrels (Hicks 1959; Assing and Schülke 2012). The most specialized type of vertebrate-staphylinid relationship is exhibited by *Myotyphlus* Fauvel of Australia (Jenkins-Shaw and Solodovnikov 2017) and the "*Amblyopinus* group of genera" in the Neotropical region (Ashe and Timm 1988), both Staphylininae, Amblyopinina. These remarkable beetles crawl upon their rodent and marsupial hosts unharmed and are thought to prey upon their nest parasites in a mutualistic relationship (Ashe and Timm 1988). Despite a diverse fauna of ecosystem engineering vertebrates in North America, including North American beaver, groundhog, pocket gophers, prairie dogs (Banfield 1974), and gopher tortoise (Jones and Franz 1990), very little is known about their associated staphylinid faunas. Rove beetles, unlike other beetle families such as the Histeridae or Scarabaeidae, are often ignored and are never the focus of surveys in North American vertebrate nests (e.g., Kovarik et al. 2008). Some remarkable nidicolous staphylinids have been described from these microhabitats (e.g., Smetana 1971a; Campbell 1979), but focused sampling has not been conducted across a broad range of hosts and distributions, and the known nidicolous species likely represent a small proportion of the true diversity. What little is known is generally fragmented across the primary taxonomic literature and has never before been reviewed. Here, we provide an overview of the Nearctic rove beetles that are putative specialist inhabitants of the burrows and nests of mammals and birds (hereafter "specialized nidicoles"). The purpose of this review is to assemble basic data on specialist nidicoles in the Nearctic, bring greater attention to these under-collected microhabitats in North America, identify major gaps in knowledge (poorly sampled hosts and regions), and provide a foundation for future evolutionary studies of this phenomenon.

8.2 Methods

8.2.1 Literature Survey for Specialized Nidicoles

Primary taxonomic and faunistic literature on the Nearctic Staphylinidae was surveyed in detail and supplemented by taxon accounts provided by Newton et al. (2001). While a diverse fauna of Staphylinidae has been reported from the nests and burrows of mammals and birds, this review focuses on a much smaller subset of species that are putative specialists of these microhabitats. Specialist nidicoles are often collected outside of burrows or nests during the spring in flight traps (e.g., pan or flight intercept traps) (Brunke, pers. obs.), and these records were not counted against a species' status as a specialist. Common names and taxonomy of vertebrate hosts follow that of the IUCN Red List (IUCN 2017). In several cases, material in the Canadian National Collection of Insects, Arachnids and Nematodes (Ottawa, Canada) was examined for additional host records; previously unpublished hosts are marked in Table 8.1. Collecting techniques for sampling vertebrate burrows and nests are not reviewed here, but the reader is referred to the following papers: Majka et al. (2006), Powell et al. (2017), and Smetana (1971a, 1995). Nidicolous species adventive in the Nearctic were not considered in this review as they are already well-represented in the European literature.

Table 8.1 Staphylinidae putatively specialized on the burrows and nests of vertebrates

Species	Host	Reference
Aleocharinae		
Aleochara costanoan Klimaszewski	"Gopher" [=pocket gopher]	Klimaszewski (1984)
Aleochara laramiensis (Casey)	Prairie dog, "gopher" [=pocket gopher]	Klimaszewski (1984)
Aleochara lobata Klimaszewski	Prairie dog	Klimaszewski (1984)
Aleochara ocularis Klimaszewski	Groundhog, red fox	Klimaszewski (1984)
Aleochara rubripennis (Casey)	Groundhog, ground squirrel	Klimaszewski (1984)
Aleodorus partitus (LeConte)	North American beaver	Hoebeke (1985)
Atheta (Dimetrota) alesi Klimaszewski & Brunke	Groundhog	Brunke et al. (2012)
Atheta (Dimetrota) bubo Klimaszewski & Webster	Great horned owl	Webster et al. (2016)
Atheta (Dimetrota) makepeacei Klimaszewski & Webster	Barred owl	Webster et al. (2016)
Atheta richardsoni Klimaszewski & Larson	Richardson's ground squirrel, black-tailed prairie dog	Klimaszewski et al. (2016)
Haploglossa barberi (Fenyes)	Bank swallow	Klimaszewski and Ashe (1991)
Haploglossa nebulosa (Casey)	"Rodent nest in tree hollow"	Klimaszewski and Ashe (1991)
Hylota ochracea Casey	Boreal, barred, and northern saw-whet owl nests	Klimaszewski et al. (2016)
Myrmecocephalus gatinaeuensis Hoebeke	North American beaver	Hoebeke (1985)
Leptotyphlinae		
Cubanotyphlus largo Frank	Key Largo woodrat	Frank and Thomas (1984)
Micropeplinae		
Micropeplus browni Campbell	North American beaver, "animal nest under log"	Campbell (1968)
Micropeplus neotomae Campbell	Dusky-footed woodrat	Campbell (1968)
Oxytelinae		
Anotylus neotomae (Hatch)	Woodrat	Hatch (1957)
Coprophilus castoris Campbell	North American beaver	Campbell (1979)
Oxytelus, undescribed species	Prairie dog	[a]museum specimens (CNC)
Paederinae		
Acrostilicus hospes Hubbard	Gopher tortoise	Hubbard (1896)
Staphylininae		
Bisnius howdeni Smetana	Gophers (*Geomys*)	Smetana (1995)
Bisnius lautus (Casey)	Groundhog, Belding's and Richardson's[a] ground squirrels, gopher, prairie dog	Smetana (1995)
Bisnius palmi Smetana	Tree swallow, great horned owl, squirrel	Smetana (1995)
Bisnius pugetensis Hatch	Groundhog, gopher (*Thomomys*), fox	Smetana (1995)
Gabrius hippisleyae Smetana	Mouse nest, squirrel midden	Smetana (1995)
Gabrius vindex Smetana	North American beaver	Smetana (1995)
Hemiquedius castoris, Brunke and Smetana	North American beaver, muskrat	Brunke et al. (2017)
Heterothops marmotae Smetana	Groundhog	Smetana (1971b)

(continued)

Table 8.1 (continued)

Species	Host	Reference
Linohesperus emarginatus Smetana	Woodrat	Smetana (1982)
Linohesperus similis Smetana	Woodrat	Smetana (1982)
Philonthus gopheri Hubbard	Gopher tortoise	Smetana (1995)
Philonthus testudo Smetana	Gopher tortoise	Smetana (1995)
Quedius (Megaquedius) explanatus LeConte	"Gopher" [=pocket gopher]	Smetana (1971a)
Quedius (Megaquedius) manatobensis (Casey)	Northern pocket gopher	Smetana (1971a)
Quedius (Megaquedius) syphax Smetana	Unknown	
Quedius (Megaquedius) validus Smetana	Unknown	
Quedius (Microsaurus) campbelli Smetana	North American beaver, muskrat	Smetana (1971a)
Quedius (Microsaurus) compransor Fall	Baird's, plains[a], and Central Texas[a] pocket gophers	Smetana (1981)
Quedius (Microsaurus) nidicola Smetana	Mountain beaver	Smetana (1971a)
Quedius (Microsaurus) pullmani Hatch	Belding's and Franklin's ground squirrels, prairie dogs, gopher (*Thomomys*), American badger[a]	Smetana (1971a)
Quedius (Microsaurus) spelaeus aplodontiae Smetana	Mountain beaver	Smetana (1971a)
Tachyporinae		
Bolitopunctus punctatissimus Campbell	Woodrat	Campbell (1993)
Mycetoporus nidicola Campbell	Woodrat	Campbell (1991)
Tachinus beckeri Campbell	Columbian ground squirrel	Campbell (1988)
Tachinus smetanai Campbell	"Gopher" [=pocket gopher]	Campbell (1973)

[a]Specimens in Canadian National Collection of Insects, Arthropods and Nematodes (Ottawa, Canada)

8.2.2 Specimen Imaging and Illustrations

Line illustrations were performed in Adobe Illustrator CC. Photomontage was accomplished using a motorized Nikon SMZ25 microscope and NIS Elements BR v4.5. Photos were processed in Adobe Photoshop CC.

8.3 Diversity of Specialized Nidicolous Staphylinidae

A review of the taxonomic and faunistic literature on Nearctic Staphylinidae (see Table 8.1 and References) resulted in a total of 46 staphylinid species putatively categorized as specialized nidicoles (Table 8.1). Most species were in the Aleocharinae and Staphylininae. Specialized nidicoles are known from 7 of the 25 staphylinid subfamilies recorded from the Nearctic (Newton et al. 2001) (Table 8.1). As can be expected, subfamilies with life histories specialized on macrofungi (Oxyporinae) or subcortical microhabitats (Piestinae) are not represented, but, surprisingly, species of diverse groups such as the Omaliinae, Pselaphinae, and Scydmaeninae have yet to be reported exclusively from burrows and nests in the Nearctic region. The genera with the most nidicolous species thus far are *Quedius* Stephens (9 spp.), *Aleochara* Gravenhorst (5 spp.), *Bisnius* Stephens (4 spp.), and *Atheta* Thomson (4 spp.). Entirely nidicolous lineages in the Nearctic include *Haploglossa* Kraatz (Aleocharinae), *Acrostilicus* Hubbard (Paederinae), and the *Quedius* subgenus *Megaquedius* Casey (Staphylininae). Nearctic vertebrates with a nidicolous staphylinid fauna, mostly rodents, generally create two types of structures: burrows and accumulations of decaying plant debris. The marmots, ground squirrels, and prairie dogs (tribe Marmotini) host many nidicolous species in common within a given region and were therefore grouped together. With the exception of some shared species between pocket gopher and marmotine squirrel assemblages (e.g., *Quedius pullmani* Hatch), specialist nidicoles were specific to host type (Table 8.1, Figs. 8.1, 8.2, 8.3, 8.4, 8.5, 8.6) as outlined below. Hosts with the richest specialist assemblages were the marmotine squirrels, with 13 specialist species and the pocket gophers (Family Geomyidae) with 10 species. Although phylogenetically unrelated, beavers and muskrats create similar lodges (Banfield 1974), were found to have identical nidicoles, and were therefore also treated together.

8.4 Burrowers

8.4.1 Groundhogs, Ground Squirrels, and Prairie Dogs

Burrows created by rodents of the tribe Marmotini are known to host specialized nidicoles from staphylinid subfamilies

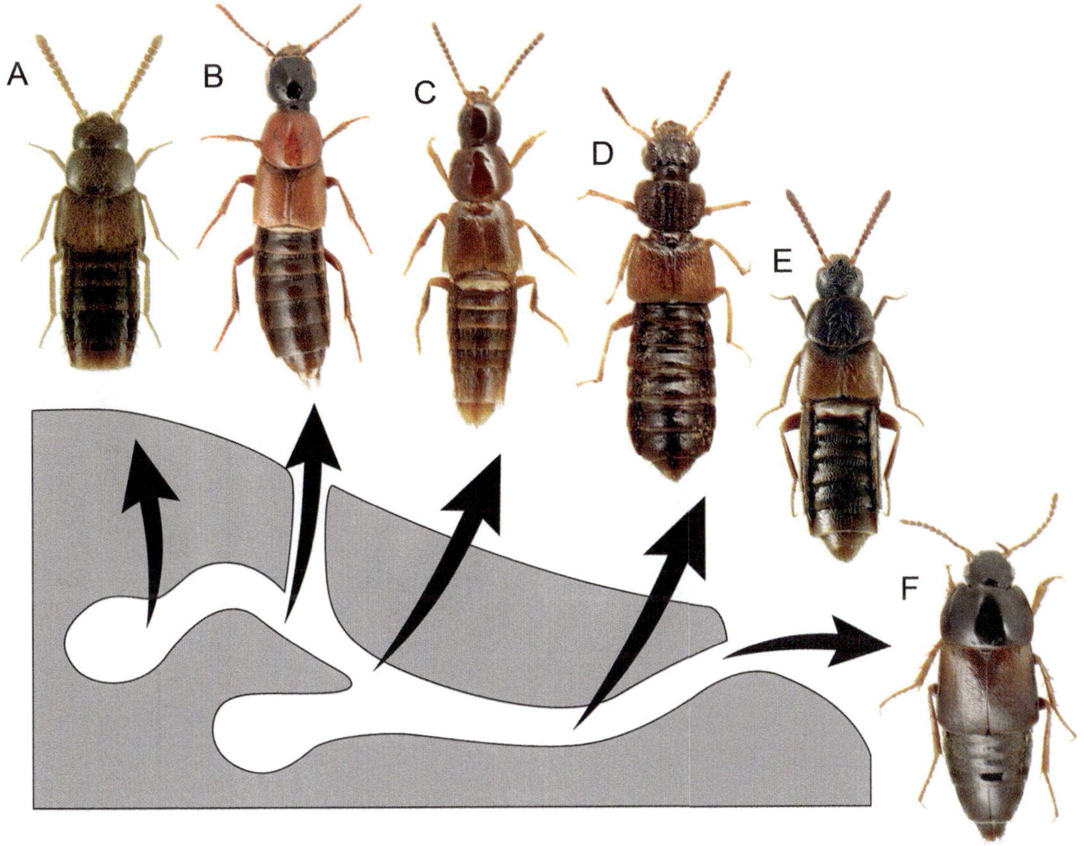

Fig. 8.1 Nearctic rove beetles putatively known as specialized inhabitants of groundhog, ground squirrel, and prairie dog (Sciuridae, Xerinae, Marmotini) burrows: *Atheta richardsoni* Klimaszewski and Larson (**a**); *Bisnius lautus* (Casey) (**b**); *Heterothops marmotae* Smetana (**c**); *Oxytelus*, undescribed species (**d**); *Aleochara rubripennis* (Casey) (**e**); and *Tachinus beckeri* Campbell (**f**). Groundhog burrow illustration adapted from Schwartz and Schwartz (2016)

Fig. 8.2 Nearctic rove beetles putatively known as specialized inhabitants of pocket gopher (Geomyidae) burrows: *Aleochara costanoan* Klimaszewski (**a**), *Quedius* (*Megaquedius*) sp. (**b**), *Bisnius howdeni* Smetana (**c**), *Quedius* (*Microsaurus*) *compransor* Fall (**d**), *Tachinus smetanai* Campbell (**e**). Burrow illustration adapted from Hickman (1977)

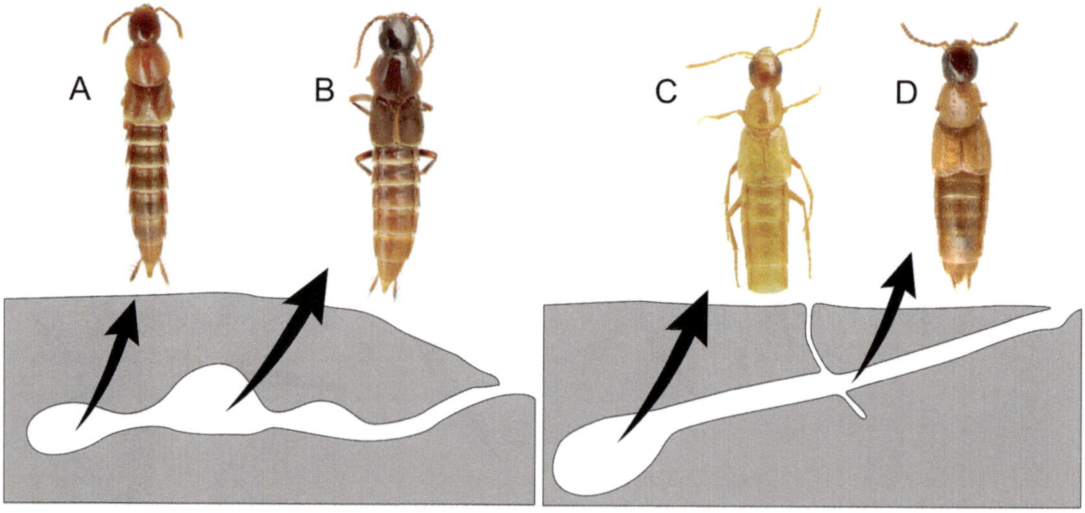

Fig. 8.3 Nearctic rove beetles putatively known as specialized inhabitants of mountain beaver (*Aplodontia rufa* (Rafinesque)) (**a**, **b**) and gopher tortoise (*Gopherus polyphemus* Daudin) (**c**, **d**) burrows: *Quedius* (*Microsaurus*) *nidicola* Smetana (**a**), *Quedius* (*Microsaurus*) *spelaeus aplodontiae* Smetana (**b**), *Philonthus gopheri* Hubbard (**c**), *Philonthus testudo* Smetana (**d**). Burrow illustrations adapted from Link (2004) (left) and Jones and Franz (1990) (right)

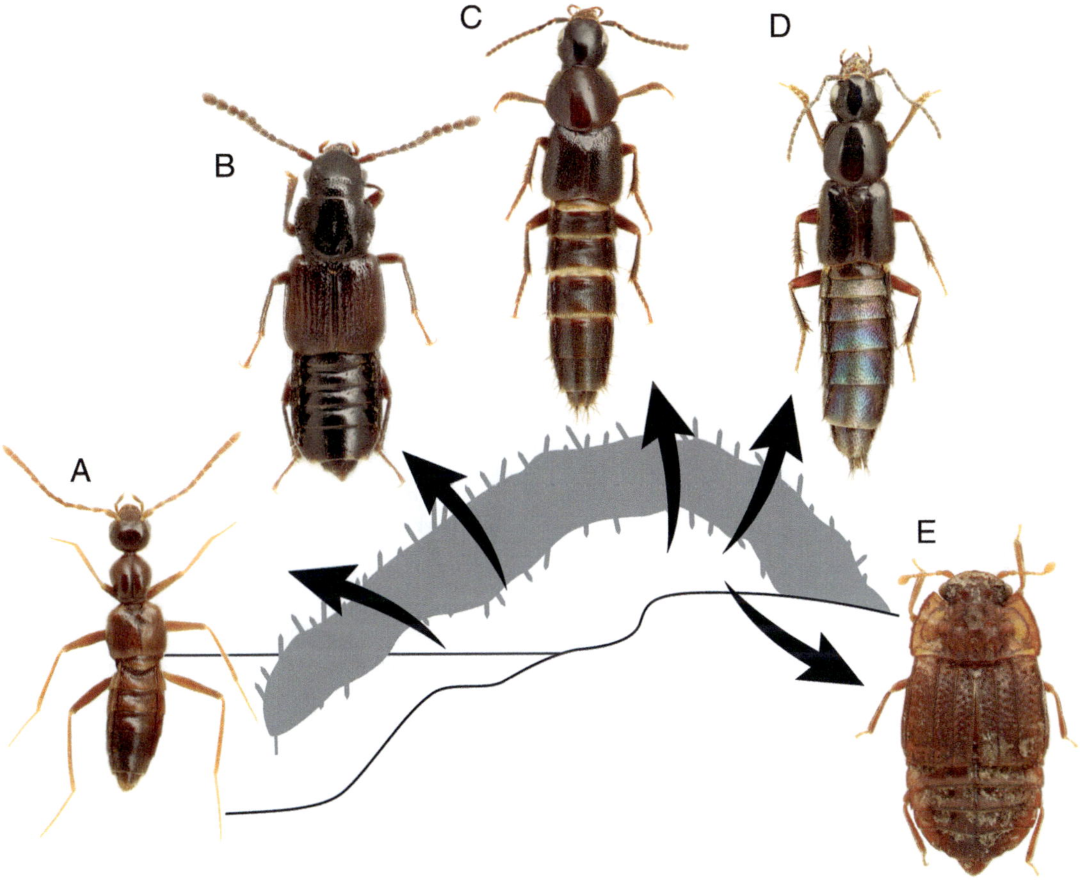

Fig. 8.4 Nearctic rove beetles putatively known as specialized inhabitants of North American beaver (*Castor canadensis* Kuhl) and muskrat (*Ondatra zibethicus* (L.)) lodges: *Myrmecocephalus gatinaeuensis* Hoebeke (**a**); *Coprophilus castoris* Campbell (**b**); *Quedius* (*Microsaurus*) *campbelli* Smetana (**c**); *Hemiquedius castoris*, Brunke and Smetana (**d**); *Micropeplus browni* Campbell (**e**). Lodge illustration adapted from Link (2004)

Aleocharinae, Staphylininae, Oxytelinae, and Tachyporinae (Table 8.1, Fig. 8.1). Currently, only the nidicolous fauna of the groundhog (*Marmota monax* (L.)) could be considered reasonably well-known (e.g., Smetana 1995) but only in the northeastern part of its broad distribution in the lowlands of North America (Banfield 1974). Beetles can be found in and around the burrow entrance during spring (Smetana 1995), and a few specimens will appear in flight traps placed nearby (Brunke, pers. obs.). Most of the nidicoles in marmotine burrows are probably associated with their host's dung as their closest, free-living relatives are typical members of dung communities elsewhere (*Aleochara*, *Atheta*, *Bisnius*, *Oxytelus*, *Tachinus*), except *Heterothops marmotae* Smetana (Fig. 8.1c), which may be more closely associated with the nest material itself as related species are typically in decaying plant matter (e.g., *H. fusculus* (LeConte)) (Smetana 1971a). All *Aleochara*, as far as known, are parasitoids of higher Diptera (Klimaszewski 1984) and may be parasitoids of nidicolous flies such as those that predictably develop within *Marmota* burrows (Griffiths 1997). However, no host records yet exist for any species of the nidicolous *Aleochara* subgenera *Echochara* Casey or *Calochara* Casey (Klimaszewski 1984). An analogous nidicole assemblage composed of *Aleochara*,

Fig. 8.5 Nearctic rove beetles putatively known as specialized inhabitants of woodrat (*Neotoma* spp.) nest piles: *Linohesperus emarginatus* Smetana (**a**), *Mycetoporus nidicola* Campbell (**b**), *Anotylus neotomae* (Hatch) (**c**), *Micropeplus neotomae* Campbell (**d**)

Bisnius, and *Heterothops* species has evolved in Central Europe with another member of Marmotini, the European ground squirrel (Smetana 1995).

8.4.2 Pocket Gophers

Unlike the marmotines, pocket gophers of the family Geomyidae are primarily fossorial and create an extensive closed-burrow system where entrances are plugged when not in use (Banfield 1974). Presently, specialized nidicolous staphylinids are known from Aleocharinae, Staphylininae, and Tachyporinae (Table 8.1, Fig. 8.2). The assemblage contains some of the largest nidicolous rove beetles in the Nearctic, such as *Quedius* (*Microsaurus*) *compransor* Fall and species in the subgenus *Quedius* (*Megaquedius*) (Fig. 8.2b, d). Currently both are known only from pocket gopher burrows but may also live undetected in marmotine burrows. These large species may be predators of the relatively large dung-feeding histerid and scarabaeid beetles in the same community (e.g., Connoir et al. 2014). Most nidicolous beetles living with pocket gophers are known only from this microhabitat (Connoir et al. 2014), possibly due to the generally closed nature of the burrow system. Pocket gopher specialists are not typically associated with a specific gopher species (Connoir et al. 2014), and, based on the fragmentary data at hand, neither are the staphylinids. Rather, distinct regions of the distribution of Geomyidae may be characterized by distinct nidicolous beetles, such as *Quedius* (*Megaquedius*) *manitobensis* (Casey) with the northern pocket gopher in the north and an undescribed *Quedius* (near *Q. compransor*) (examined by the first author) with extreme

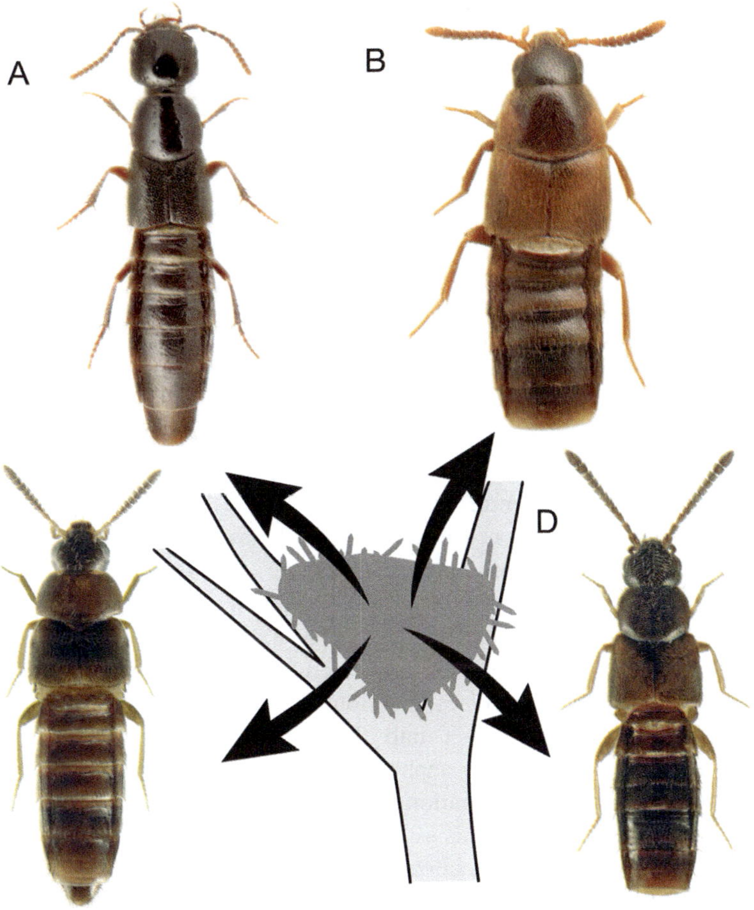

Fig. 8.6 Nearctic rove beetles putatively known as specialized inhabitants of bird nests: *Bisnius palmi* Smetana (**a**), *Haploglossa barberi* (Fenyes) (**b**), *Hylota ochracea* Casey (**c**), *Atheta (Dimetrota) bubo* Klimaszewski and Webster (**d**)

eastern populations of the plains pocket gopher in Indiana (Powell et al. 2017).

8.4.3 Mountain Beaver

The mountain beaver is the sole member of family Aplodontiidae and the sister group to the extant squirrels (Piaggio et al. 2013). Although it is reasonably common within its range in the coastal Western Lowlands and Cascades, south to the northern part of the Sierra Nevada (Banfield 1974), little is known about the commensal insects that may occupy its burrow. It tends to create burrows near water, especially streams, forming a main nest area and fecal chamber (Banfield 1974). The large species *Quedius spelaeus aplodontiae* Smetana was collected from the main nest, and the much smaller *Quedius nidicola* Smetana was most often in the fecal chamber (Smetana 1971a) (Table 8.1, Fig. 8.3, a–b). As very few burrow samples are known and the mountain beaver occurs from lower mountain valleys to nearly 3000 m (Banfield 1974), the majority of this nidicolous assemblage is probably still unknown.

8.4.4 Gopher Tortoise

The gopher tortoise is a burrowing reptile that occurs in the dry and sandy longleaf pine ecosystem of the southeastern United States (Florida, Georgia, South Carolina, Alabama, Mississippi, and Louisiana) (Jones and Dorr 2004). Its burrows are a refuge from xeric conditions for small mammals, reptiles, and amphibians but also for invertebrates (Jones and Franz 1990). Tortoise dung, which accumulates at the blind end of the burrow, provides a resource for many invertebrates including beetles (Jackson and Milstrey 1989). Hubbard (1896) reported two staphylinids from these burrows with their larvae (*Philonthus gopheri* Hubbard (Fig. 8.3c) and *Acrostilicus hospes* Hubbard). *Acrostilicus*, still with a single species, was vaguely described by Hubbard (1896) and stated to resemble *Stilicopsis* Sachse (subtribe Stilicopsina), but Blackwelder (1939) and Newton et al. (2001) keyed it as a member of Stilicina rather than Stilicopsina. The syntypes in the United States National Museum (Washington D.C.) have not been studied since Blackwelder (1939). Much later, *Philonthus testudo* Smetana (Fig. 8.3d) was described from specimens collected in the western part of the tortoises' distribution, collected from "dung 12 feet into the burrow" (Smetana 1995). All three staphylinid nidicoles are relatively pale in coloration consistent with the other invertebrates of this assemblage and are likely members of the specialized dung community as *Philonthus* Stephens and *Rugilus* Leach, a relative of *Acrostilicus* (Newton et al. 2001), includes many dung-attracted species. Like the mountain beaver assemblage, so few samples of staphylinids have been recovered from these burrows that many more species likely await discovery.

8.5 Debris Accumulators

8.5.1 Beaver and Muskrat

Although North American beavers are best known for their creation of "beaver ponds" through the action of damming, it is their lodge that is host to a distinct nidicolous fauna (Table 8.1, Fig. 8.4). Beavers and muskrats overwinter in a lodge and maintain its integrity (Banfield 1974), creating a stable environment within an accumulation of wet woody debris. Although muskrats will also use burrows within banks for their dens (Banfield 1974), it is unknown whether these burrows also support nidicoles. Unlike most other nidicolous beetles in this review, members of the beaver and muskrat staphylinid assemblage belong to groups that are associated with wet decaying plant matter rather than dung. Thus far, specialist nidicoles from the Aleocharinae, Micropeplinae, Oxytelinae, and Staphylininae have been collected from the walls of the nest chamber (Table 8.1, Fig. 8.4), often from recently abandoned lodges that are easier to access (Campbell 1968; Smetana 1971a, 1971b). Remarkably, the evolution of commensal nidicoles within beaver lodges is a Nearctic phenomenon only: no beetles are known to be specialized inhabitants of Eurasian beaver lodges, despite a long history of collecting in Europe. Unlike the commensal nidicoles, the highly modified leiodid beetle *Platypsyllus castoris* Ritsema lives entirely on its host on both species of beavers, is thought to be a skin-feeding ectoparasite, and may be Holarctic, Nearctic, or Palaearctic in origin (Peck 2006). The evolution of at least some species of beaver nidicoles may be very recent. A recently described species of *Hemiquedius* found in the lodges of beavers and muskrats in northeastern North America (Fig. 8.4d) differs very slightly in the form of the male genitalia from its sibling species and cannot be identified using the barcoding region of COI, suggesting recent speciation (Brunke et al. 2017). However, distinct and non-overlapping differences in ecology and external morphology demonstrate that it is a distinct lineage, sympatric with its sibling species (Brunke et al. 2017).

8.5.2 Woodrats

With the exception of the rare, rock crevice-dwelling northeastern species (Peles and Wright 2008), woodrats or "packrats" of the genus *Neotoma* Say & Ord construct a conspicuous den of sticks and other woody debris within which they place a nest composed of much finer material (Banfield 1974). This den is used by successive generations (Jaeger 1961) and thus can be a rather consistent accumulation of litter in the local area, which is often litter poor and rocky, or dry (Jaeger 1961; Frank and Thomas 1984). A regular "toilet area" is used, and this accumulation of dung appears to have created a community of dung-associated invertebrates, including the specialist nidicole scarabs *Ataenius brevicollis* (Wollaston) and *Onthophagus orpheus* (Panzer) (Frank and Thomas 1984). As far as known, woodrats host nidicoles from the greatest number of staphylinid subfamilies including Micropeplinae, Leptotyphlinae, Oxytelinae, Staphylininae, and Tachyporinae (Table 8.1, Fig. 8.5). Surprisingly, members of the Aleocharinae have yet to be reported as associates of *Neotoma*. *Linohesperus emarginatus* Smetana (Fig. 8.5a), the only species with detailed microhabitat data, was collected from dung inside the den (Smetana 1982). Based on the biology of related species, *Linohesperus similis* Smetana and *Anotylus neotomae* (Hatch) (Fig. 8.5c) may live in a similar way, while *Micropeplus neotomae* Campbell (Fig. 8.5d), *Bolitopunctus punctatissimus* Campbell, *Mycetoporus nidicola* Campbell (Fig. 8.5b), and *Cubanotyphlus largo* Frank and Thomas may be specifically associated with the debris of the midden or fungi growing on midden substrates. On the island of Key Largo (Florida), the dens created by the endangered Key Largo woodrat provide a moist, non-saline refuge for the minute *Cubanotyphlus largo* (Leptotyphlinae) in an otherwise dry, litter-poor forest on limestone bedrock (Frank and Thomas 1984). Litter dwelling is unusual for this normally soil-adapted subfamily, but *Cubanotyphlus* is thought to be only weakly adapted for soil dwelling (Saíz 1973), and at least one other species (Guatemala) was sifted from leaf litter (Gusarov 2003).

8.5.3 Bird Nests

In the Nearctic region, very few surveys have been conducted in the nests of birds, which are a rich source of nidicolous beetles in Europe (Hicks 1959). The majority of sampling has been conducted in Canada in the nests of owl species including barred, boreal, northern saw-whet, and great horned owls (Majka et al. 2006; Webster et al. 2009; 2016). In the cited Canadian studies, owl nest boxes were sampled after they were no longer in use, and thus, nidicoles could be expected to occur in abandoned nests as do those associated with beavers and muskrats. In other accounts, authors rarely report whether nests are occupied at the time of sampling or abandoned. Beetles are attracted to accumulations of organic matter including dropping-soaked bedding and regurgitated pellets of indigestible feathers, fur, and bones (Webster et al. 2016). Thus far, only one species of *Bisnius* (Staphylininae) (Fig. 8.6a) and four species of Aleocharinae are putatively specialized on bird nests in the Nearctic (Table 8.1, Fig. 8.6). *Bisnius subuliformis* (Gravenhorst), a Palaearctic relative of *B. palmi* (Smetana 1995), is also a specialist of bird nests (Hicks 1959). *Hylota ochracea* (Aleocharinae) (Fig. 8.6c) is rather commonly collected in Nearctic forests using flight traps but is consistently associated with bird nests in trees (Majka et al. 2006; Webster et al. 2009). Indeed, it was the most common native staphylinid in owl nests surveyed in Nova Scotia (Majka et al. 2006). Recently a second species of the genus, *Hylota cryptica* Klimaszewski and Webster, was described from Alberta and New Brunswick, Canada (Webster et al. 2016). It is only known from flight-based traps, but the genus *Hylota* Casey may prove to be entirely specialized on bird nest microhabitats. The morphologically similar and related genus *Haploglossa* is well-known in Central Europe as a nest specialist (Hicks 1959), and the two known North American species also appear to live this way (Klimaszewski and Ashe 1991). The Nearctic *Haploglossa barberi* (Fenyes) is known from the nest of a bank swallow, a Holarctic species which nests in riverbanks or quarry walls and hosts different *Haploglossa* species in the Nearctic and Palaearctic regions (Hicks 1959; Klimaszewski and Ashe 1991). Recent surveys in barred and great horned owl nests have revealed two new species of *Atheta* (*Dimetrota*

Mulsant & Rey) that are thus far only known from this microhabitat (Webster et al. 2016). An even greater number of Nearctic staphylinids found in bird nests are more generally associated with tree cavities or decaying organic matter in or on standing trees (e.g., *Atheta fanatica* Casey) (Webster et al. 2009).

8.6 Future Exploration of the Nearctic Nidicolous Staphylinid Fauna

Most vertebrate hosts of nidicolous staphylinids are widespread in North America, but these specialized beetles are often known from only a few sites due to limited sampling. *Heterothops marmotae* is still only known from a small suburban area of Ottawa (Ontario, Canada) (Smetana 1971a, 1973) that has since been developed—undoubtedly it is widespread in at least the northeastern portion of the groundhog's distribution. Based on published data, it would appear that species of the subgenus *Quedius* (*Megaquedius*) are specialized on pocket gopher burrows, but at least one specimen has been collected in a prairie dog burrow (unexamined by the authors) (R.M. Brattain, *pers. comm.*). Even more fragmentary is the coverage of burrow- and nest-forming vertebrates that have been sampled in the Nearctic. No records exist for the entire family Heteromyidae, the pocket mice and kangaroo rats, which create a complex network of burrows in the drier areas of the New World (Banfield 1974). The Arctic ground squirrel is a widespread potential host for nidicolous staphylinids wherever tundra permafrost cannot form (Banfield 1974), yet no records from their burrows are known to us. Raptor nests including ospreys and various hawks are known to host several nidicolous staphylinids in Europe (Hicks 1959; Owen and Taylor 1989), but sampling in Nearctic bird nests is mostly limited to Canadian owls (Majka et al. 2006; Webster et al. 2016). Very few records are available from burrows of the American badger, red fox, and the two western marmots: yellow-bellied from rocky habitats and hoary from alpine tundra (Banfield 1974). Despite the nearly North America-wide distribution of the North American beaver and muskrat, sampling has been concentrated in eastern Canada and the New England states; sampling beaver lodges in other ecoregions may be productive.

As more is discovered about the specialized relationship between staphylinid nidicoles and their mammal and bird hosts, it will become important for global and national conservation organizations to assess whether this diversity is under threat by human development and resource extraction. One positive aspect is that the overwhelming majority of nidicole hosts in North America are widespread and listed as "least concern" by the IUCN (2017). Groundhogs, some ground squirrels, and some species of pocket gophers have seen a population explosion with the rise of agriculture on the Canadian and American prairies and have become pests (Banfield 1974). Others, such as the gopher tortoise and Key Largo woodrat have experienced a dramatic decrease in their native habitat and are listed as threatened and endangered, respectively (IUCN 2017). Given its poor dispersal capabilities, *Cubanotyphlus largo* is probably limited to the island of Key Largo off the coast of Florida (Frank and Thomas 1984) and should be also considered as at high risk of extinction.

Unlike the mutualistic relationships of the mammal-riding staphylinids (Ashe and Timm 1988) and the often antagonistic relationships of "guest" staphylinids with their ant and termite hosts (reviewed in Parker 2016), commensal relationships between staphylinids and vertebrates have not been considered within an evolutionary framework. Several questions immediately arise about this phenomenon. Do specialized nidicoles always evolve from facultative nidicole ancestors? How old are these relationships? What types of microhabitat characteristics are involved in determining a nidicolous species' host range? Much of this research is impossible without adequate sampling as mentioned above, but a more developed understanding of phylogenetic relationships and taxonomy is also needed. It is still unknown which major paederine lineage *Acrostilicus*, with its single species, belongs to or whether it is a valid genus (Newton et al. 2001). It has not been re-collected. Cryptic nidicolous species remain overlooked within "distinctive" species such as *Quedius compransor*. It is difficult to identify a potential free-living sister group for highly derived nidicolous lineages such as *Quedius* (*Megaquedius*) without a broadly sampled phylogeny. The most immediate relative of *Quedius* (*Megaquedius*) may be the obscure Palaearctic genus *Velleiopsis* Fairmaire, which is also a suspected nidicole (Brunke, pers. obs.). The widespread taxonomic backlog in Nearctic collections notwithstanding, nidicolous species provide one of the last frontiers in the North American staphylinid fauna, and we hope that this review will inspire creative exploration of novel hosts and undersampled areas of their distribution.

Acknowledgments We would like to thank Aleš Smetana (Ottawa, Canada) for fruitful discussions and for sharing his knowledge of nidicolous Staphylinidae in both the Nearctic and Palaearctic. Jan Klimaszewski (Natural Resources Canada) is thanked for providing images of *Aleochara rubripennis*, *Atheta bubo*, *A. richardsoni*, and *Hylota ochracea*. Serge Laplante and Owen Lonsdale (Ottawa, Canada) provided access to imaging equipment.

References

Ashe JS, Timm RM (1988) *Chilamblyopinus piceus*, a new genus and species of amblyopinine (Coleoptera: Staphylinidae) from southern Chile, with a discussion of amblyopinine generic relationships. J Kansas Entomol Soc 61:46–57

Assing V, Schülke M (2012) Freude-Harde-Lohse-Klausnitzer – Die Käfer Mitteleuropas. Band 4. Staphylinidae I. Zweite neubearbeitete Auflage. Spektrum Akademischer Verlag, Heidelberg, I–XII, pp 560

Banfield AWF (1974) The mammals of Canada. University of Toronto Press, Toronto, p xv+438

Blackwelder RE (1939) A generic revision of the staphylinid beetles of the tribe Paederini. Proc US Natl Mus 87:93–125

Brunke A, Klimaszewski J, Dorval J-A et al (2012) New species and distributional records of Aleocharinae (Coleoptera, Staphylinidae) from Ontario, Canada, with a checklist of recorded species. ZooKeys 186:119–206

Brunke A, Smetana A, Carruthers-Lay D, Buffam J (2017) Revision of Hemiquedius Casey (Staphylinidae, Staphylininae) and a review of beetles dependent on beavers and muskrats in North America. ZooKeys 702:27–43

Campbell JM (1968) A revision of the New World Micropeplinae (Coleoptera: Staphylinidae) with a rearrangement of the world species. Can Entomol 100:225–267

Campbell JM (1973) A revision of the genus *Tachinus* (Coleoptera: Staphylinidae) from North America. Mem Entomol Soc Can 90:3–137

Campbell JM (1979) *Coprophilus castoris*, a new species of Staphylinidae (Coleoptera) from beaver lodges in eastern Canada. Coleopt Bull 33:223–228

Campbell JM (1988) New species and records of North American *Tachinus* Gravenhorst (Coleoptera: Staphylinidae). Can Entomol 120:231–295

Campbell JM (1991) A revision of the genera *Mycetoporus* Mannerheim and *Ischnosoma* Stephens (Coleoptera: Staphylinidae: Tachyporinae) of North and Central America. Mem Entomol Soc Can 156:1–169

Campbell JM (1993) A revision of the genera *Bryoporus* Kraatz and *Bryophacis* Reitter and two new related genera from America north of Mexico (Coleoptera: Staphylinidae: Tachyporinae). Mem Entomol Soc Can 166:1–85

Connoir MB, Kovarik PW, Chordas S et al (2014) Coleoptera (Histeridae, Leiodidae and Scarabaeidae) inhabiting the burrows of Baird's pocket gopher (Rodentia: Geomyidae: *Geomys breviceps*) in Arkansas. Insecta Mundi 389:1–27

Frank JH, Thomas MC (1984) *Cubanotyphlus largo*, a new species of Leptotyphlinae (Coleoptera: Staphylinidae) from Florida. Can Entomol 116:1411–1417

Griffiths G (1997) Anthomyiid flies (Diptera: Anthomyiidae) of the Yukon, pp 687–722. In: Danks HV, Downes JA (eds) Insects of the Yukon. Biological Survey of Canada (Terrestrial Arthropods), Ottawa, 1034

Gusarov V (2003) Cubanotyphlus guatemalae, a new species of leptotyphline staphylinid from Guatemala (Insecta: Coleoptera: Staphylinidae). Stud Neotrop Fauna Environ 38:125–128

Hatch MH (1957) The beetles of the Pacific Northwest. Part II: Staphyliniformia. University of Washington Press, Seattle, p ix+384

Hickman GC (1977) Burrow system structure of *Pappogeomys castanops* (Geomyidae) in Lubbock County, Texas. Am Midl Nat 97:50–58

Hicks EA (1959) Check-list and bibliography on the occurrence of insects in birds' nests. Iowa State College Press, Ames, 696

Hoebeke ER (1985) A revision of the rove beetle tribe Falagriini of America North of Mexico (Coleoptera: Staphylinidae: Aleocharinae). J N Y Entomol Soc 93 (2):913–1018

Hubbard HG (1896) Additional notes on the insect guests of the Florida land tortoise. Proc Entomol Soc Wash 3:299–303

IUCN (International Union for Conservation of Nature) (2017) The IUCN red list of threatened species. http://www.iucnredlist.org/. Accessed 17 May 2017

Jackson DR, Milstrey EG (1989) The fauna of gopher tortoise burrows, pp 86–98. In: Diemer JE, Jackson DR, Landers JL et al (eds) Proceedings of the gopher tortoise relocation symposium. Florida Game and Fresh Water Fish Commission, Gainesville, FL, Nongame Wildlife Program Technical Report No. 5, pp 109

Jaeger EC (1961) Desert wildlife. Stanford University Press, California, 320

Jenkins-Shaw J, Solodovnikov S (2017) The remarkable Australian rove beetle genus *Myotyphlus*: its cryptic diversity and significance for exploring mutualism among insects and mammals (Coleoptera: Staphylinidae). Aust Entomol 56:311–321. https://doi.org/10.1111/aen.12233

Jones JC, Dorr BS (2004) Habitat associations of gopher tortoise burrows on industrial timberlands. USDA National Wildlife Research Center – Staff Publications. Paper 502. http://digitalcommons.unl.edu/icwdm_usdanwrc/502

Jones CA, Franz R (1990) Use of gopher tortoise burrows by Florida mice (*Podomys floridanus*) in Putnam County, Florida. Fla Field Nat 18:45–68

Jones CG, Lawton JH, Shachak M (1997) Positive and negative effects of organisms as physical ecosystem engineers. Ecology 78:1946–1957

Klimaszewski J (1984) A revision of the genus *Aleochara* Gravenhorst of America North of Mexico (Coleoptera: Staphylinidae, Aleocharinae). Mem Entomol Soc Can 129:1–211

Klimaszewski J, Ashe JS (1991) The oxypodine genus *Haploglossa* Kraatz in North America (Coleoptera: Staphylinidae: Aleocharinae). Giornale Italiano di Entomologia 5:409–416

Klimaszewski J, Larson DJ, Labrecque M et al (2016) Twelve new species and fifty-three new provincial distribution records of Aleocharinae rove beetles of Saskatchewan, Canada (Coleoptera, Staphylinidae). ZooKeys 610:45–112

Kovarik P, Chordas S, Robison H et al (2008) Insects inhabiting the burrows of the Ozark pocket gopher in Arkansas. J Arkansas Acad Sci 62:75–78

Link R (2004) Living with wildlife – Species fact sheets. Washington Department of Fish and Wildlife, Olympia. Available: http://wdfw.wa.gov/living/species/. Accessed 31 May 2017

Majka CG, Klimaszewski J, Lauff RF (2006) New Coleoptera records from owl nests in Nova Scota, Canada. Zootaxa 1194:33–47

Marlowe MH, Murphy CA, Chatzimanolis S (2015) Sexual dimorphism and allometry in the sphecophilous rove beetle *Triacrus dilatus*. PeerJ 3:e1123. https://doi.org/10.7717/peerj.1123

Newton A, Thayer MK, Ashe JS et al (2001) Staphylinidae Latrielle, 1802. In: Arnett RH, Thomas MC (eds) American beetles: Archostemata, Myxophaga, Adephaga, Polyphaga: Staphyliniformia. CRC Press, New York, pp 272–418

Owen JA, Taylor S (1989) *Haploglossa picipennis* (Gyllenhal) (Col: Staphylinidae) in Ospreys' nests. Entomol Rec 101:53–55

Parker J (2016) Myrmecophily in beetles (Coleoptera): evolutionary patterns and biological mechanisms. Myrmecol News 22:65–108

Peck S (2006) Distribution and biology of the ectoparasitic beaver beetle *Platypsyllus castoris* Ritsema in North America (Coleoptera: Leiodidae: Platypsyllinae). Insecta Mundi 20:87–96

Peles JD, Wright J (2008) Den use behavior of Allegheny woodrats inhabiting rock outcrops in Pennsylvania. In: Peles JD, Wright J (eds) The Allegheny woodrat. Springer, New York, pp 75–91

Piaggio AJ, Coghlan BA, Miscampbell AE et al (2013) Molecular phylogeny of an ancient rodent family (Aplodontiidae). J Mammal 94:529–543

Powell GS, Brattain RM, Zaspel JM (2017) Beetles (Insecta: Coleoptera) associated with the plains pocket gopher, *Geomys bursarius* (Mammalia: Rodentia: Geomyidae), in Indiana. Ann Entomol Soc Am 110:269–275. https://doi.org/10.1093/aesa/saw094

Reichman OJ, Seabloom EW (2002) Ecosystem engineering: a trivialized concept? Response from Reichman and Seabloom. Trends Ecol Evol 17:308

Rosell F, Orsolya B, Collen P et al (2005) Ecological impact of beavers *Castor fiber* and *Castor canadensis* and their ability to modify ecosystems. Mammal Rev 35:248–276

Saíz F (1973) Biogeography of soil beetles in Mediterranean regions. In: di Castri F, Mooney HA (eds) Mediterranean type ecosystems: origin and structure. Springer, New York, pp 285–294

Schwartz CW, Schwartz ER (2016) In: Fantz DK, Jackson VL (eds) The wild mammals of Missouri, third revised edition. University of Missouri Press, Columbia, 496

Smetana A (1971a) Revision of the tribe Quediini of North America North of Mexico (Coleoptera: Staphylinidae). Mem Entomol Soc Can 79:1–303

Smetana A (1971b) Revision of the tribe Quediini of America North of Mexico (Coleoptera: Staphylinidae). Supplementum 1. Can Entomol 103:1833–1848

Smetana A (1973) Revision of the tribe Quediini of America North of Mexico (Coleoptera: Staphylinidae). Supplementum 2. Can Entomol 105:1421–1434

Smetana A (1981) Revision of the tribe Quediini of America North of Mexico (Coleoptera: Staphylinidae). Supplementum 5. Can Entomol 113:631–644

Smetana A (1982) Revision of the subfamily Xantholininae of America north of Mexico (Coleoptera: Staphylininae). Mem Entomol Soc Can 120:iv+389

Smetana A (1995) Rove beetles of the subtribe Philonthina of America North of Mexico (Coleoptera: Staphylinidae). Classification, phylogeny and taxonomic revision. Mem Entomol Int 3:1–945

Spieth HT (1979) The virilis group of *Drosophila* and the beaver castor. Am Nat 114:312–316

Webster R, Klimaszewski J, Bourdon C et al (2016) Further contributions to the Aleocharinae (Coleoptera, Staphylinidae) fauna of New Brunswick and Canada including descriptions of 27 new species. ZooKeys 573:85–216

Webster RP, Klimaszewski J, Pelletier G et al (2009) New Staphylinidae (Coleoptera) records with new collection data from New Brunswick, Canada. I. Aleocharinae. ZooKeys 22:171–248

Rove Beetles (Coleoptera, Staphylinidae) as Bioindicators of Change in Boreal Forests and Their Biological Control Services in Agroecosystems: Canadian Case Studies

9

Jan Klimaszewski, Adam J. Brunke, Timothy T. Work, and Lisa Venier

Abstract

The Staphylinidae, or rove beetles, is one of the largest and most biologically diverse of the beetle families. The world fauna consists of more than 63,495 described species, with more than 1774 species recorded in Canada. By virtue of their sheer abundance, number of species and trophic diversity, rove beetles play important roles in terrestrial ecosystems. These beetles also have great potential for use as indicators of forest change because of their affinity for specific habitats and microhabitats and their sensitivity to habitat change. With continued advances in identification tools such as taxonomic keys with high-resolution photos, rove beetles have been increasingly incorporated into biodiversity research projects, thus increasing both the scope and inferences drawn from these studies. In this chapter, we highlight recent biodiversity studies using rove beetles and discuss how these beetles have added to our perspectives on changing forest and agroecosystems. In forest ecosystems, rove beetle assemblages are affected by both large- and small-scale natural disturbances, forest succession and silviculture. Thus far in boreal ecosystems, responses of rove beetles have been evaluated primarily in terms of their responses to stand-replacing wildfire and clearcut harvesting. However, given the apparent affinity of rove beetles for microhabitats, increasing attention has been focused on smaller-scale disturbances and silvicultural interventions that may leave residual standing trees, create small gaps and/or modify soils or deadwood.

Rove beetles also are important biological control agents against pest populations. In Canadian agroecosystems, rove beetles have mostly been used in conservation biological control as a diverse assemblage of generalist predators or in more focused applications. We provide a review of the research conducted in Canadian agroecosystems and describe general patterns of the communities across a variety of crop types and compare these with other agricultural systems on a global scale. While the temporal and spatial dynamics of rove beetle assemblages

J. Klimaszewski (✉)
Natural Resources Canada, Canadian Forest Service, Laurentian Forestry Centre, Québec, QC, Canada
e-mail: jan.klimaszewski@canada.ca

A. J. Brunke
Canadian National Collection of Insects, Arachnids and Nematodes, Agriculture and Agri-Food Canada, Ottawa, ON, Canada

T. T. Work
Département des sciences biologiques, Université du Québec à Montréal, Montréal, QC, Canada

L. Venier
Natural Resources Canada, Canadian Forest Service, Great Lakes Forestry Centre, Sault Ste. Marie, ON, Canada

O. Betz et al. (eds.), *Biology of Rove Beetles (Staphylinidae)*,
https://doi.org/10.1007/978-3-319-70257-5_9

and the responses of their populations to different cultural practices are becoming better known, direct predator-prey interactions and subsequent impacts on pest populations remain poorly understood. Biological control of several dipteran pests by rove beetle parasitoids in the genus *Aleochara* remains one of the best-studied systems, likely due to the high specificity of *Aleochara* for dipteran hosts. Several general recommendations are provided for future research to better quantify and predict control by rove beetles.

9.1 Introduction

Rove beetles, along with weevils, leaf beetles and ground beetles, are among the largest and most biologically diverse beetle families in Canada and globally (Klimaszewski 2000; Bousquet et al. 2013). Newton et al. (2001) recorded over 46,200 known rove beetle species worldwide; however, there are now more than 63,495 described species in 3,762 genera (Newton, pers. com.). In Canada and Alaska, there are 1682 recorded rove beetle species (Bousquet et al. 2013) in 274 genera and 24 subfamilies (Klimaszewski 2000). Recent taxonomic revisions suggest that the number of described species in Canada and Alaska is now exceeding 1700 species (Webster et al. 2016). Yet, even with recent revisions, many species of the Canadian fauna still remain undescribed or have yet to be recorded from Canada. This is particularly true for the largest rove beetle subfamily, Aleocharinae, which may contain over 700 species in Canada and Alaska.

Rove beetles are readily distinguished morphologically from other Coleoptera. The family is best known for their shortened elytra and elongate flexible abdomen. Most species possess well-developed wings and good dispersal abilities (Halder 2011). Some (e.g. Aleocharinae) have developed defensive glands with chemicals to deter predators (Klimaszewski 2000). The majority of adults are nocturnal and generally avoid contact with light and prefer moist habitats like forest litter.

Many rove beetles (e.g. some Aleocharinae, Staphylininae, Paederinae, Scydmaeninae) are predators of other arthropods, but examples of other feeding relationships such as fungivory and herbivory are known. Gyrophaenina species are exclusively mycophagous, feeding on fungal spores and hyphae (Ashe 1984). Likewise, all species within the Scaphidiinae are obligate or facultative inhabitants and consumers of fungi (Newton 1984). A number of species are saprophagous and feed mainly on decomposing organic material (some Oxytelinae and Osoriinae) or are phytophagous on pollen, algae and rarely the leaves and stems of higher plants (some Omaliinae, Osoriinae, Oxytelinae) (Klimaszewski 2000; Frank and Thomas 1991). Larvae of *Aleochara* species are ectoparasitoids on pupae of cyclorrhaphous Diptera (Klimaszewski 1984). Some species occur under the bark of trees or logs (e.g. *Anomognathus*, *Dexiogyia*, *Homalota*, some *Leptusa*, *Placusa*) and may be predaceous or fungivorous (Klimaszewski et al. 2016a). Many other species are affiliated with ants (some members of Athetini, Oxypodini and Lomechusini). The primary feeding modes (trophic affiliations) of rove beetles are presented in Klimaszewski (2000) and Thayer (2005). However, information on feeding habits of many rove beetles is still expanding. For example, many species of the aleocharine tribe Athetini were historically considered predaceous, but this assumption was based on little empirical evidence. Recent molecular and microscopic analysis of the gut contents of abundant rove beetles (indicators of mature forests), including several athetine species in the boreal balsam fir forest of eastern Canada, revealed a lack of arthropod fragments or arthropod DNA in their posterior guts but presence of a diverse fungal community dominated by yeasts (Klimaszewski et al. 2013a; Stefani et al. 2016). Predominance of yeasts and fungal spores in the posterior gut of rove beetles suggests that fungi may play an important role in rove beetle dietary requirements and as endosymbionts (Klimaszewski et al. 2013b; Stefani et al. 2016).

High species richness, diversity of trophic relationships and sensitivity to changes in environmental conditions and microhabitats suggest that rove beetles may serve as useful ecological indicators and provide a large suite of potentially significant species in biological control efforts. In Sects. 9.2–9.8, we provide a short review of recent studies that have used rove beetles to evaluate the ecological impacts of different approaches to forest management. These examples come largely from boreal forests. In Sects. 9.9–9.13, we provide a review of the research conducted in Canadian agroecosystems and describe general patterns of the communities across a variety of crop types and compare these with other agricultural systems on a global scale.

9.2 Rove Beetles in the Forests of Canada

Rove beetles are useful as ecological indicators of changes in managed forests because of their high local abundance and diversity and ease of collection (Paquin and Dupérré 2001; Pohl et al. 2008). Often, rove beetles represent a large proportion of the total abundance and richness of organisms collected in pitfall traps, a method commonly used in biomonitoring studies. For example, Klimaszewski et al. (2005, 2007a, 2008a) reported that both abundance and species richness of rove beetles were much greater than those of ground beetles, another taxon commonly used in the evaluation of impacts of forest management, in yellow birch-balsam fir forests in Quebec. In the Quebec study (Klimaszewski et al. 2008a), rove beetles were ca. five times more abundant (9424 specimens) and were ca. three times more diverse (116 species) than ground beetles (Carabidae) (1875 specimens and 38 species). Unfortunately, rove beetle assemblages are often neglected in biomonitoring studies because of the difficulty in identifying species. However, this may change with advances in traditional tools such as taxonomic keys complemented by high-resolution imagery. Examples of recent modern generic and provincial treatments of Canadian Aleocharinae species are as follows: *Aleochara* (Klimaszewski 1984); *Placusa* (Klimaszewski et al. 2001); *Tinotus* (Klimaszewski et al. 2002); *Silusa* (Klimaszewski et al. 2003); *Leptusa* (Klimaszewski et al. 2004); *Oxypoda* (Klimaszewski et al. 2006); *Calodera* (Assing 2007, 2008); *Gnypeta* (Klimaszewski et al. 2008b); *Diglotta* and *Halobrecta* (Klimaszewski et al. 2008c); *Schistoglossa* (Klimaszewski et al. 2009a); *Gyrophaena* and *Eumicrota* (Klimaszewski et al. 2009b); *Alisalia* (Klimaszewski et al. 2009c); *Dinaraea* (Klimaszewski et al. 2013a); *Gnathusa*, *Mniusa* and *Ocyusa* (Klimaszewski et al. 2014); *Mocyta* (Klimaszewski et al. 2015a); *Clusiota* and *Atheta* (Klimaszewski and Majka 2007; Klimaszewski et al. 2015b); *Liogluta* (Klimaszewski et al. 2016b); Aleocharinae of Nova Scotia (Klimaszewski et al. 2007b; Majka and Klimaszewski 2008a, b, c, 2010; Majka et al. 2008); Aleocharinae of Yukon (Klimaszewski et al. 2008d, 2012); Aleocharinae of Ontario (Brunke et al. 2012); Aleocharinae of Newfoundland and Labrador (Klimaszewski et al. 2011, 2016c); Aleocharinae of New Brunswick (Klimaszewski et al. 2005; Webster et al. 2009, 2012, 2016a, b); and Aleocharinae of Saskatchewan and Alberta (Klimaszewski et al. 2015c, 2016b, d). As a result of the above publications, the number of species reported from Canada increased rapidly with modern diagnostic tools now available for species identification. Despite these advances, the aleocharine tribe Athetini, which includes the large genus *Atheta*, still remains poorly known, and additional studies across Canada are needed to address present taxonomic and environmental needs.

Advances in DNA barcoding may increase the use of rove beetles in biomonitoring, particularly given the extra time required to dissect and prepare genitalia from specimens to verify species identity. The BOLD (Barcode of Life Data Systems) and GenBank DNA databases of rove beetle species with registered barcodes are constantly expanding. The number of Staphylinidae species in BOLD is currently 2753, with 2104 of those having barcodes. However, clear protocols on genital dissection and preparation are widely available (Hanley and Ashe 2003) and can be followed by non-specialists including beginning students, volunteers or research technicians that may not possess a background in entomology. By following these protocols and newly available taxonomic keys, species-level identifications can often be made with little more effort than what is required for other taxa such as ground beetles (Coleoptera: Carabidae) and spiders (Araneae). With little training, these traditional morphological methods for species identification may still provide a cost-effective means of including rove beetles in biomonitoring.

Rove beetles may provide a higher resolution for forest biomonitoring studies than other commonly used arthropod groups (ground beetles and spiders) because of their strong affinities for specific stand types and the microhabitats that lie within (Pohl et al. 2007, 2008). They occur widely in the forest litter but are often associated with small-scale heterogeneity such as fruiting bodies of fungi and residual deadwood (Klimaszewski et al. 2005). Their seemingly tight relationship with microhabitats and forest structures such as downed wood make rove beetles useful tools for evaluating the effects of forest management. In many cases species can be characterized as forest specialists with strong affinity for intact forest stands (and the microhabitats that lie therein) that have not been recently disturbed by forest management. Species groups that are specific to undisturbed forests are key to assessing forest integrity as we apply a natural disturbance emulation paradigm in our forest management process. These are the species that best tell us how effective we are at preserving forest integrity in managed forests.

Rove beetle assemblages have been used to evaluate impacts on biodiversity of a variety of silvicultural approaches, including small-scale gap cuts that generate within-stand heterogeneity, site preparation of soils for replanting in gaps and larger cuts (Klimaszewski et al. 2008a), removal of residual biomass and overstorey trees in the context of forest biomass removal for bioenergy (Work et al. 2013) and the use of retention patches to emulate larger landscape patterns of fire skips in large harvest blocks (Pohl et al. 2007, 2008). These examples are covered in Sects. 9.4–9.7.

9.3 Diversity of Trophic Relationships and Sensitivity to Changes

Feeding preferences in rove beetles are highly diverse, but initial studies on rove beetle diet preferences are being revised with newer molecular tools. For example, the rove beetle gut microbiota play key roles in the metabolism and health of their hosts. The identification of the fungal taxa that inhabit the gut of rove beetles provides information on the diet of their host and on their potential endosymbionts and on the fungal communities that occupy microhabitats along with rove beetles. The first studies using DNA methods to investigate the gut contents of rove beetles were recently conducted using several abundant species in boreal balsam fir forests of Quebec (Klimaszewski et al. 2013b; Stefani et al. 2016). Nine rove beetle species representing three subfamilies were recently targeted to investigate the fungal diversity present in gut extracts: *Atheta strigosula*, *Atheta ventricosa*, *Liogluta terminalis*, *Lypoglossa franclemonti*, *Atrecus microcephalus*, *Gabrius brevipennis*, *Quedius labradorensis labradorensis*, *Ischnosoma longicorne* (=*fimbriatum*) and *Tachinus quebecensis* (Stefani et al. 2016). In total, we obtained 441 sequences of ITS rDNA representing 42 fungal operational taxonomic units (OTUs) from 9 species of rove beetles (Stefani et al. 2016). OTU richness per beetle species ranged from 4 in *Tachinus quebecensis* to 16 in *Atheta ventricosa*. The fungal mycobiota in the posterior gut was dominated by *Saccharomycetales* (12 OTUs) and *Sordariomycetes* (9 OTUs), and the core mycobiome of the posterior gut extracts was dominated by 3 OTUs related to yeast, i.e. *Candida mesenterica*, *Debaryomyces* spp. and *Ophiostoma pluriannulatum* (Fig. 9.1). Predominance of yeasts and podocarp fungal spores in the posterior gut of

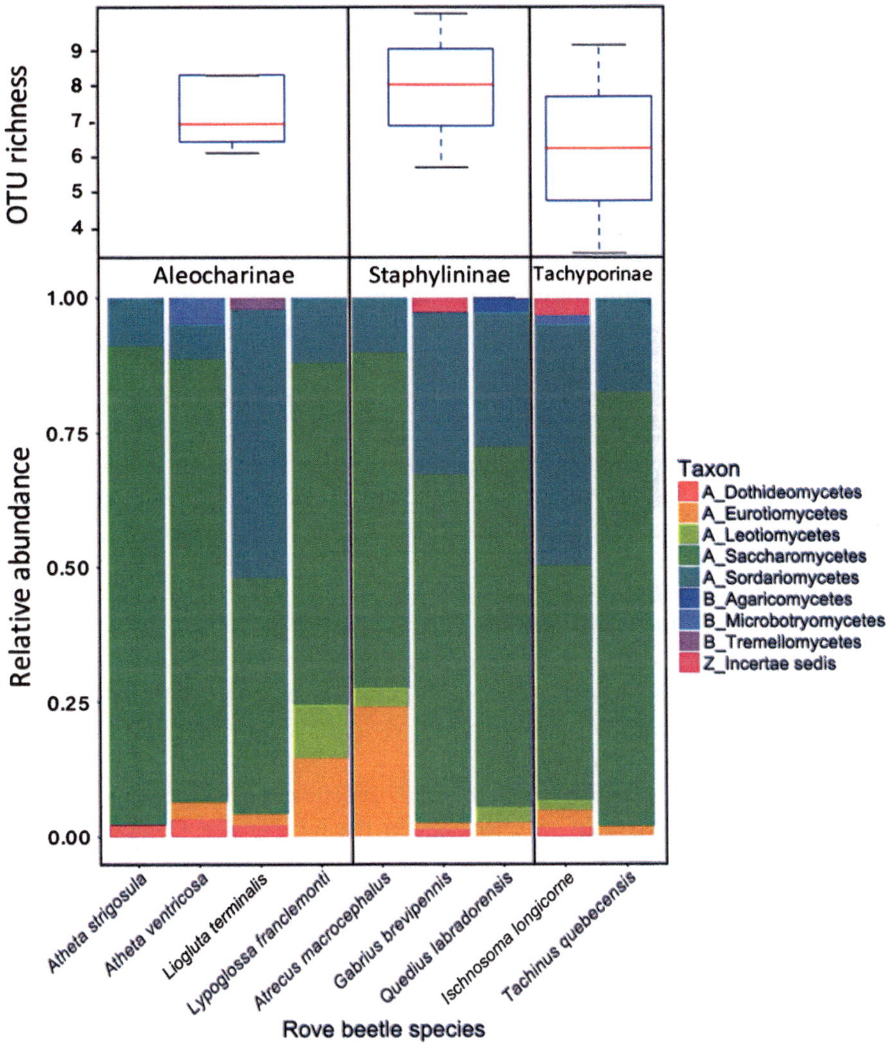

Fig. 9.1 Fungal OTU richness recorded within nine rove beetle species of three subfamilies of Staphylinidae (upper part of the figure) and relative abundance of the nine fungal classes recorded in the gut contents of the nine rove beetle species (bottom part of the figure). The A, B and Z upper case letters preceding the fungal class names designate the phyla *Ascomycota*, *Basidiomycota* and *Zygomycota*, respectively (after Stefani et al. 2016)

rove beetles suggests that these may play an important role in their dietary requirements and as endosymbionts (Klimaszewski et al. 2013b; Stefani et al. 2016). Particularly, *Sordariomycetes* and some podocarp fungi found in guts have an affiliation with decaying wood. This is an important link to understand interactions between rove beetles, fungi and woody debris and, consequently, how the depletion of woody debris in different forestry practices impacts rove beetles and other organisms involved in organic matter decomposition and subsequent nutrient cycling, as well as tree growth.

9.4 The Effects of Small-Scale Gap Cutting

Large-scale, even-aged silvicultural approaches such as clearcutting in Canada are often criticized as unsustainable by the public and by policymakers alike. Large clearcuts can reduce biodiversity through habitat loss if applied widely over forested landscapes primarily by homogenizing the stand structure and eliminating important microhabitats, including deadwood and other ecological legacies (Venier et al. 2014). This has led to wider implementation of alternative silvicultural approaches that create uneven-aged stands and maintain stand heterogeneity, such as partial cutting or gap cutting. Often, uneven-aged silvicultural approaches are inspired by natural disturbances caused by fires of variable intensity, insect outbreaks, windthrow events and even forest gap dynamics.

Small-scale disturbance events, such as gap dynamics, create heterogeneity in light and soil nutrient availability (Paré et al. 1993) and are important in the recruitment of structural features such as downed woody debris (Siitonen 2001). Ultimately, gaps influence future stand composition, typically by maintaining early-successional deciduous species that would otherwise be competitively replaced during succession (Lieffers et al. 1996; Kneeshaw and Bergeron 1998). Stand heterogeneity created by gaps is thought to have additional benefits for biodiversity and generally is more socially acceptable. Changes in the abundance, species richness and community composition of rove beetles were evaluated in response to three configurations of experimental gap cuts and clearcutting in an early-succession yellow birch-dominated boreal forest in Quebec, Canada (Klimaszewski et al. 2008a). In each gap cut, total forest removed was held constant (35% removal by partial cutting with a concomitant decrease in gap size), but the total number of gaps was increased (two gaps, 56 m diameter, 2513 m^2; four gaps, 40 m diameter, 1257 m^2; and eight gaps, 28 m diameter, 628 m^2) resulting in an experimental increase in the total amount of 'edge' within each stand. In the four- and eight-gap treatments, half of the gaps were scarified. In the two-gap treatment, one randomly selected half of each gap was scarified. The scarification was relatively light and affected only surface soil layers. Estimates of species richness using rarefaction were highest in the clearcut and two-gap treatments, decreased in smaller and more numerous gaps and were significantly higher in scarified areas than in unscarified areas. The total catch of all rove beetles was significantly higher in uncut forests than in the treated areas (Fig. 9.2). Composition of rove beetle assemblages among intermediate- and small-sized gap treatments (four- and eight-gap cuts) was more similar to that in uncut control forests than were those of larger gap treatments (two-gap) and clearcuts (Klimaszewski et al. 2008a). When the total area of forest removed was held constant, smaller, more numerous gaps were more similar to uncut control stands than to larger gaps and fell more closely within the natural forest heterogeneity. The smaller and more numerous gaps were a better emulation of natural stand dynamics than the larger gaps and clearcuts.

9.5 Community Structure and Indicator Species in Gap Cuts

Rove beetle composition responded significantly to increasing gap size. Composition among intermediate- and small-sized gap treatments (four- and eight-gap treatments) was more similar to that of uncut control forests than were those of larger gap treatments (two-gap) and clearcuts. Differences in rove beetle assemblage (defined primarily by a loss of closed-canopy species and an increase in species with an affinity for open habitats) relative to control plots were more pronounced in two-gap treatments and clearcuts than in four- and eight-gap treatments. These assemblage differences were made greater by scarification. Assemblages of rove beetles were less affected by numerous small gaps with increased edge relative to fewer larger gaps with less edge. It may be that the smaller gaps (28 m in diameter) were perceived by forest specialists as an extreme case of stand heterogeneity rather than perceptible gaps as rove beetles were much more abundant there than in larger gaps (Klimaszewski et al. 2008a). Smaller gaps may also be considered more permeable for dispersion than larger gaps.

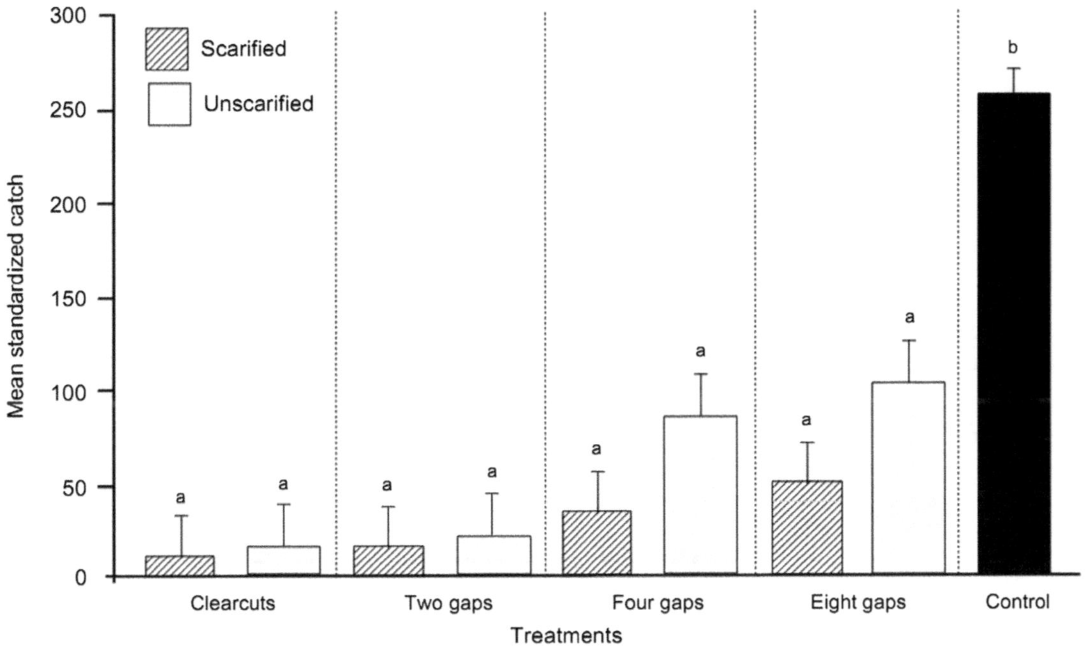

Fig. 9.2 Standardized catch (mean +/− standard error) of total rove beetles, forest specialists, forest generalists, hygrophilous species and open-habitat specialist in controls and each harvesting and scarification treatment (after Klimaszewski et al. 2008a)

Of the 116 species represented, 6 species dominated the total catch: *Tachinus fumipennis* (Say) (55.4% of the total catch), *Oxypoda convergens* Casey (10.6%), *Atheta pseudomodesta* Klimaszewski (6.6%), *Placusa tachyporoides* (Waltl) (3.7%), *Tachinus addendus* Horn (3.5%) and *Bisnius blandus* (Gravenhorst) (2.5%) (Klimaszewski et al. 2008a). Together these species represented over 82% of the overall catch, but there were 27 additional significant indicator species associated with different treatment as illustrated in Fig. 9.3.

9.6 Effects of Scarification in Gap Cuts

Scarification of forest soils is used to expose mineral soils and promote seed germination of particular species (Prévost et al. 2009). In the experimental gap cuts described above, an additional scarification treatment was applied, and responses of rove beetle assemblages were compared between scarified and non-scarified gap cuts (Klimaszewski et al. 2008a). In yellow birch-dominated forests, scarification is done to promote germination of yellow birch and red spruce trees. In scarified plots, rove beetle richness was greater than in non-scarified plots. Initially, this result seemed counter-intuitive, because we thought that homogenization of microsites by scarification would reduce species richness. However, when applied to small gaps, scarification may actually increase overall habitat heterogeneity at the stand level by creating recently disturbed soils in close proximity to intact closed-canopy forest. Scarification also altered overall composition of rove beetles through an interaction with gap size. Scarification effectively nullified any compositional differences attributed to larger gaps, thus making large, scarified gaps indistinguishable from clearcuts (Klimaszewski et al. 2008a). In smaller gaps, scarification acted as secondary, compounding

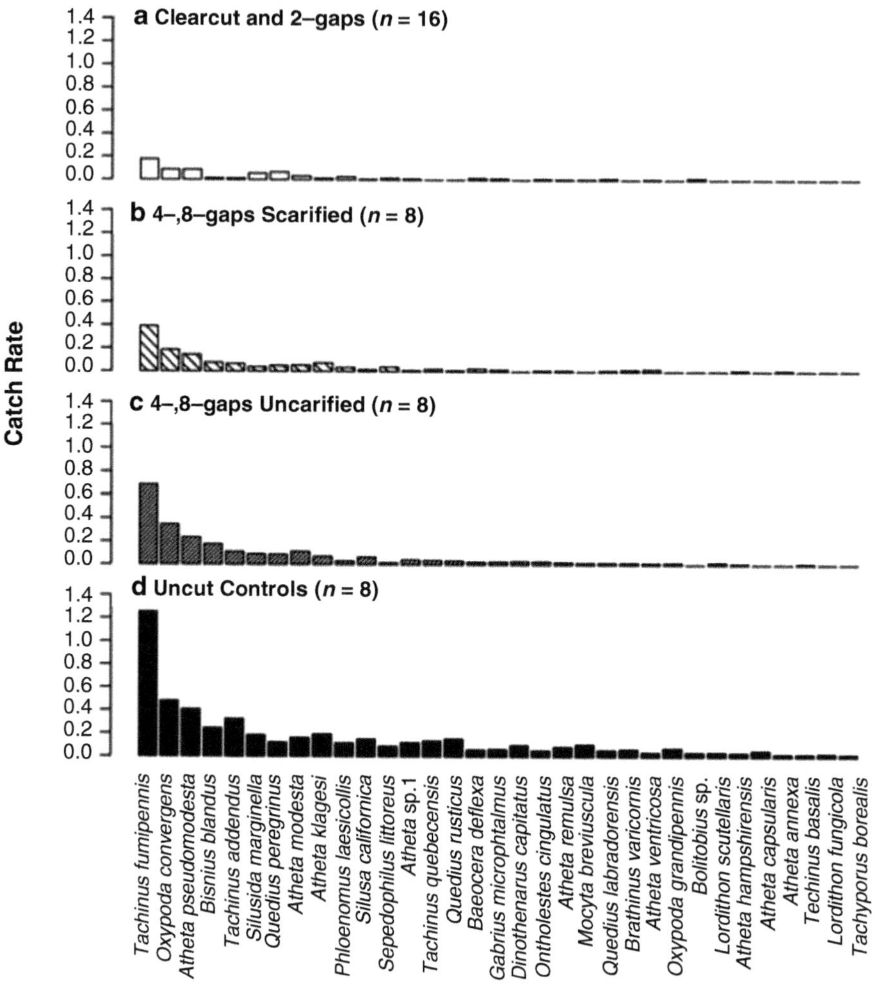

Fig. 9.3 Differences in mean catch rate of significant indicator species: (**a**) clearcut and two-gap treatments ($n = 16$), (**b**) four- and eight-gap scarified treatments ($n = 8$), (**c**) four- and eight-gap unscarified treatments ($n = 8$) and (**d**) uncut control stands ($n = 4$) (after Klimaszewski et al. 2008a)

disturbance event to alter composition between scarified and unscarified small gaps (Klimaszewski et al. 2008a).

9.7 Effects of Biomass Removal

Increasing interest in harvesting biomass for energy and non-timber forest products stems from the need to offset the decline in markets for traditional wood products and to mitigate climate change. While the development of bioenergy has potential for economic, social and environmental benefits, there remains concern about the impact of biomass harvesting on the long-term sustainability of the forest and the long-term sustainability of biodiversity. While wood residues from forest harvesting or salvaged wood from insect outbreaks or forest fires are considered as biomass ‘feedstocks’ for bioenergy production, they also provide reserves of nutrients and moisture, microsites and substrates for seedlings and habitat and resources for other organisms such as arthropods, bacteria and fungi. These attributes of woody debris play major roles in ecosystem functioning and in sustaining biodiversity (Stockland et al. 2012). Two studies using rove beetles were recently conducted to evaluate the impacts of different intensities of postharvest slash removal on rove beetle abundance, species richness and community composition, one in jack pine forest of the Island Lake Biomass Harvest Research and Demonstration Area, also known as Island Lake Site (ILS), in Ontario (Venier et al. 2017; Work et al. in preparation), and the other in a balsam fir-white birch forest at the Montmorency Teaching and Research Forest (MTRF), Quebec (Work et al. 2013; Klimaszewski et al. unpublished). The ILS evaluated changes in species composition of litter-dwelling arthropod taxa, including rove beetles, in response to a gradient of forest disturbance including the following treatments, from the least to the most disturbed: 85-year-old postfire forest (old control), 45-year-old postharvest plantation forest (young control), 3-year postfire treatment (fire control), 3-year post stem-only harvesting (SOH) with branches and tops left at the stump, 3-year post whole-tree biomass harvesting treatment (WTH) with branches and tops moved to roadside, 3-year post WTH with stumps removed and 3-year post WTH with forest floor removed by blading (Venier et al. 2017).

When rove beetle assemblages were compared between uncut stands and harvested stands with increasing removal of biomass, composition was markedly different between older uncut stands (85 years) and all other treatments, including the 45-year-old closed-canopy control. The vast majority of species and individuals were associated with the old-forest treatment, whereas very few individuals were captured in the other treatments. For example, of the 1231 individuals captured, 745 were captured in the five 85-year-old control plots, whereas only 486 were captured in the other 30 plots of all other treatments combined. Species strongly associated with older, uncut forest included *Atheta capsularis*, *Atheta strigosula*, *Lordithon fungicola*, *Oxypoda grandipennis*, *Atheta remulsa* and *Aleochara fumata* (occurred also in harvested sites) and were found only in very small numbers or not at all in harvested, fire disturbed and young control treatments. Only two species were associated with the harvested sites or younger stands, i.e. *Aleochara fumata* and *Mycetoporus consors* (Venier et al. 2017). *Atheta capsularis*, *A. strigosula*, *L. fungicola* and *O. grandipennis* are generally associated with forest litter, while *Aleochara fumata* is a more opportunistic and mobile species. Its life history depends on the availability of habitats with dipteran larvae and pupae, which are more randomly distributed in the forest. Moist litter is a very important habitat for the majority of rove beetles because their small bodies are prone to desiccation. Removal of the litter layer and exposure of mineral soil likely reduce soil moisture and habitat suitability for rove beetles. Among the harvested treatments, rove beetle composition did not differ between tree-length and full-tree harvesting. This could result from the presence of only a few generalist or accidental species and lack of true microhabitat specialists.

Lack of canopy and extensive biomass removal, including the removal of soil organic layers by blading, had strong effects on rove beetles. This treatment dramatically removed much of the habitat where a majority of rove beetles live. Removing the whole organic layer and deadwood from postharvest plots eliminates the buffer zone protecting subsequent layers of ground habitat from reduced humidity and increased temperature.

Interestingly, however, rove beetles responded similarly to other litter-dwelling taxa (ground beetles and spiders) to the gradient of disturbance across the other treatments. This response suggests that litter-dwelling communities respond strongly to forest floor disturbance and canopy cover and that harvesting does not effectively mimic fire disturbance (Venier et al. 2017), at least in this instance. These results also indicate that ground-dwelling arthropod communities are not differentially affected by full-tree harvesting and tree-length harvesting, but they are more sensitive to extreme levels of biomass removal such as stump removal and forest floor removal (Venier et al. 2017). Rove beetles in jack pine forests may serve as good indicators of mature forest conditions but are likely less effective as indicators of biomass removal due to their relatively low numbers in harvested plots (Venier et al. 2017). This illustrates that jack pine forest is different from broadleaved, mixed and other coniferous forests as habitat for rove beetles. It is a drier, more open forest with less ground floor litter, while the others, particularly broadleaved and mixed forests, provide a richer and moister ground litter, which is a more suitable habitat for these beetles.

In the 2011 MRTF study, changes in rove beetle abundance and assemblages were compared between two treatments and a control: clearcuts where branches and non-merchantable portions of the stem were left on site (stem-only harvesting or SOH); whole-tree harvesting (WTH), in which stems, tops and branches are removed; and uncut control plots (C) in about 42- to 60-year-old mature balsam fir stands in Quebec (Work et al. 2013). Overall catch rates were much greater in uncut plots (C) than in either SOH or WTH sites. The high overall abundance and differences in assemblage composition in uncut stands are attributable mainly to the presence of *Atheta capsularis*, *A. klagesi*, *A. strigosula* and *Tachinus fumipennis/frigidus* in controls. Among the clearcuts, catch rates in WTH were greater than in SOH sites, where three species (like *Pseudopsis subulata*, *Quedius labradorensis* and, to a lesser extent, *Gabrius brevipennis*) were more abundant when branches and finer woody debris had been removed. These differences between SOH and WTO in balsam fir-dominated stands contrast with the response of rove beetles in jack pine stands at the ILS experiment, where no differences in rove beetles were observed.

In 2013 (3-year post-clearcut), rove beetles were resampled at the MRTF experiment, with an inclusion of the additional treatment (ADR), in which all woody debris were removed from new clearcut plots and samples were also taken from different chronosequences of the surrounding forests (11-, 22-, 42- and 73-year-old stands) (Klimaszewski et al. in preparation). We collected 2069 specimens and 87 species of Staphylinidae in 10 subfamilies in harvesting treatments and 5291 specimens and 86 species in 12 subfamilies in chronosequence stands. The oldest chronosequence stand (73 years) had a distinct rove beetle community, and the communities in the younger chronosequence stands (11, 22, and 42 years) and in the control (C) stand were similar to each other and distinct from the harvesting treatments. Rove beetle abundance showed little relation to differences in fine or coarse woody material. However, staphylinid species richness was positively affected by higher fine woody debris volume. In this 3-year post-clearcut resampling, *Atheta capsularis*, *A. klagesi*, *A. ventricosa*, *Liogluta terminalis* (=*L. aloconotoides*), *Lypoglossa franclemonti*, *Placusa tachyporoides*, *Acidota quadrata*, *Proteinus* sp., *Quedius rusticus*, *Q. densiventris* and *Tachinus* species (mainly *T. frigidus* and *T. fumipennis*) showed strong affinities for uncut forests, whereas several other species were more or equally abundant in treatments: *Gabrius picipennis* (SOH, ADR, WTH), *G. brevipennis* (SOH, WTH, ADR), *Quedius labradorensis* (SOH, ADR, WTH), *Q. peregrinus* (WTH, SOH, ADR), *Ischnosoma longicorne* (SOH, ADR, WTH), *I. splendidum* (SOH, ADR, WTH), *Parascydmus corpusculus* (SOH, WTH, ADR) and *Pseudopsis subulata* (WTH, SOH, ADR). Harvesting and biomass removal significantly reduced the mean abundance of several species of rove beetles in the subfamilies Aleocharinae, Omaliinae and Tachyporinae and increased or maintained the mean abundance of predatory species of subfamilies Staphylininae and Scydmaeninae, compared with uncut control plots. The species from the latter two subfamilies may find more diverse prey species in harvested forests. However, overall catch of rove beetles was not significantly affected by harvesting treatment, likely because the abundance of tolerant and open-habitat species such as *Ischnosoma longicorne*, *Gabrius brevipennis*, *Pseudopsis subulata* and *Parascydmus corpusculus* increased and compensated for the decreased abundance of forest specialists such as *Atheta capsularis*, *A. klagesi* and *Tachinus frigidus* in harvested plots. Many species in the subfamilies

Aleocharinae, Omaliinae and Tachyporinae are dependent on wet litter, fungi and specific microhabitats that are more frequently found in mature closed-canopy forests than in open, recently harvested forests. These forest species require closed-canopy forests such as those found in control and chronosequence forest plots, less light, more humidity, moist litter including woody debris, specific species of fungi (associated with woody debris) and arthropod prey species that also prefer closed-canopy forests and associated microhabitats, such as under the bark of moist logs (e.g. *Placusa* species) or wet moss (e.g. *Acidota quadrata*) (Campbell 1982; Klimaszewski et al. 2001). Many rove beetles are associated with fungi (Newton 1984), and analyses of the gut contents of several forest specialists (e.g. *Atheta capsularis*, *A. klagesi*, *Tachinus frigidus*, *T. fumipennis*) revealed the presence of budding yeasts and podocarp spores of different fungi (Klimaszewski et al. 2013b). The reduced abundance of some forest specialists in harvested treatments may be due, in part, to the reduced abundance of fungi with which they are associated. *Ischnosoma longicorne* was captured in comparable numbers in SOH plots as in the uncut controls, but in lower numbers in WTH plots, suggesting that this species may be best suited to intermediate deadwood conditions. *Gabrius picipennis* is a common species occurring in a wide range of habitats and is the only species of *Gabrius* that exhibits a remarkable variability in almost all external morphological characters (Smetana 1995). *Placusa tachyporoides* is a subcortical species, and clearcut diminishes its preferred habitats, i.e. standing trees and logs (Klimaszewski et al. 2001). Again, details of habitat preferences of *Liogluta terminalis* remain elusive (Lohse et al. 1990, Work et al. 2013, as *L. aloconotoides*). Harvesting treatments significantly affected rove beetle species richness (Fig. 9.4), but not species evenness. This indicates that a majority of the rove beetle species studied were forest litter species and their occurrence in treated plots was significantly affected by environmental conditions changed by harvesting, including the

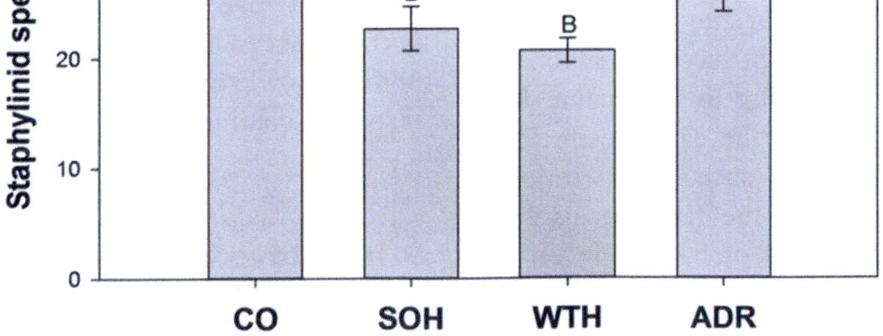

Fig. 9.4 Rove beetle species richness in the Montmorency Teaching and Research Forest, Quebec, that was subjected to different silvicultural treatments (2013): CO, untreated controls; SOH, stem-only harvesting; WTH, whole-tree harvesting; and ADR, all debris removed. The effect of harvesting intensity treatment on mean total season catch per plot of Staphylinidae (all species pooled), and individual species for which more than ten specimens were collected, was tested by one-way ANOVA and the Tukey post hoc test in SAS PROC GLM (SAS Institute 2002–2003). We report mean (±SE) catches per 100 trap days per plot of non-transformed data

amount of deadwoody debris and exposure to light. Light conditions for beetles improved within 3 years of harvesting in comparison with immediate postharvesting conditions in harvesting treatment plots because of fast-developing vegetation that increased the cover. Mean catches of several abundant forest specialist species were significantly greater in control stands than in the harvest treatments, and these species may be considered indicators of uncut forests and may thus be used to determine when regenerating sites have moved on to forested conditions.

9.8 Responses to Large-Scale Fire Events

Fire is the dominant disturbance in the Nearctic boreal forest (Pohl et al. 2008). It has a profound effect on rove beetles by destroying forest communities to an extent that depends on fire intensity and by resetting the successional trajectory to its earliest stages (Pohl et al. 2008). The burn pattern results in a patchwork of different communities at various stages in the successional cycle. In contrast to fire, forest harvesting does not directly destroy the rove beetle community, but to a large extent, it destroys the forest habitat (Pohl et al. 2008). This results in a unique rove beetle community characterized by a mix of forest species and open-ground specialists and overall high diversity in this period of flux (Pohl et al. 2008). In the years following harvesting, the rove beetle community goes through successional changes and becomes more similar to the mature forest community, but it skips the early postfire stage and proceeds along the successional trajectory more rapidly than after fire (Pohl et al. 2008). Other less direct effects of harvesting on rove beetles are a decrease in the proportion of landbase suitable for communities associated with older successional stages, alteration of forests by postharvest site preparations and planting of exotic tree species, edge and fragmentation effects that are detrimental to the remaining forest surrounding harvested areas and an influx of exotic arthropod species with affinities for disturbed sites (Pohl et al. 2008). It is recommended that future work explore the effects of postharvest forestry activities, fragmentation, edges and changes in riparian zones and wetlands on rove beetles in forested habitats (Pohl et al. 2008).

9.9 Rove Beetle Biological Control Services in Canada

In addition to being important components of nutrient cycling communities in the world's forests, rove beetles are abundant in anthropogenic habitats such as agroecosystems. Predatory or omnivorous species in these communities are known to provide biocontrol services to agriculture by suppressing pest populations, usually as generalists or occasionally as specialists, of diverse arthropods such as spider mites (Kishimoto and Adachi 2008), cereal aphids (Dennis and Wratten 1991), horn flies (Hu and Frank 1995), wireworms (Fox and MacLellan 1956) and cabbage maggots (Andreassen et al. 2010). Biocontrol services by rove beetles in Canadian agroecosystems are rarely documented and little understood, likely due to a long history of inadequate taxonomic knowledge that made identifications difficult or impossible, a situation that has only recently improved (see references in Sect. 9.1). This important first step forward has made detailed, species-level surveys more tractable, and rove beetle communities may soon become accessible to agroecologists as taxonomic knowledge and associated identification resources continue to improve.

9.10 Rove Beetle Assemblages in Canada's Diverse Agroecosystems

In Canada, rove beetle assemblages in soybean and blueberry fields and in dairy pastures have been surveyed in detail (D'Orsay 2012; Renkema et al. 2012; Brunke et al. 2014); nearby, those of vegetable crops have been sampled in

Pennsylvania, USA (Leslie et al. 2007). Conducted prior to the recent advances made in the taxonomy of the Aleocharinae, the most diverse and difficult subfamily of Staphylinidae in Canada, a detailed survey in Canadian raspberry fields necessarily omitted this group from their study. However, the results obtained by Levesque and Levesque (1996), who used multiple trap types, are noteworthy among North American studies for including detailed phenological information such as flight activity period and field colonization. Staphylinid assemblages in agroecosystems of northeastern North America contain high percentages of adventive species: 58.3% in dairy pastures, 43.1% in soybean fields, 39.6% in blueberry fields and 33% overall in diverse vegetable crops (Leslie et al. 2007; D'Orsay 2012; Renkema et al. 2012; Brunke et al. 2014). For most of these assemblages, more than half of the common species ($\geq$1% of total individuals) (Table 9.1) were adventive: 71.4% in dairy pastures, 57.1% in soybean fields, 58.8 in blueberry fields and 33% in diverse vegetable crops. However, native species may still contribute positively to biocontrol as the most abundant species captured in Canadian agroecosystems were *Strigota obscurata* Klimaszewski and Brunke, *Stenus erythropus* (Melsheimer) and *Stenus flavicornis* Erichson (Table 9.1), native rove beetles that favour open habitats. Several staphylinids were shared between northeastern assemblages, which indicate that although agroecosystems differ strongly in their physical structure and management practices, some species may be broadly tolerant to the challenges presented by frequent disturbances and unpredictable prey populations (Wiedenmann and Smith 1997). Generally, these species tend to be ubiquitous or prefer early-succession natural or disturbed habitats (Andersen 1991; Brunke et al. 2014). The most widespread common species across agricultural habitats were *Amischa analis* (Gravenhorst), *Dinaraea angustula* (Gyllenhal) and *Strigota ambigua* (Erichson), all Aleocharinae. Relatively few of the 34 staphylinid subfamilies are major elements in the fauna of Canadian agroecosystems, and common species are generally limited to seven subfamilies: Aleocharinae, Omaliinae, Oxytelinae, Paederinae, Staphylininae, Steninae and Tachyporinae.

9.11 Temporal and Spatial Dynamics

Staphylinids in Canadian agroecosystems are typically univoltine, with two peaks in activity each year: one after overwintering or summer aestivation and one after the eclosion of the next generation (Boiteau 1983; Levesque and Levesque 1996; Brunke et al. 2014). The majority of common species in those studies were most active between May and July and exhibited a decline in surface/flight activity as the growing season progressed. Late summer inactivity at the soil surface is typical of adult staphylinids in agroecosystems, though some Aleocharinae have been observed to maintain activity throughout the growing season (Levesque and Levesque 1996). Little is known about the activity of both adult and larval staphylinids below the soil surface. Exceptions to this pattern are known in the Tachyporinae and Omaliinae, which contain species such as *Tachinus corticinus* Gravenhorst and *Arpedium cribratum* Fauvel that prefer cooler temperatures and can be important members of an agroecosystem assemblage early or late in the season when many other staphylinids are less abundant (Levesque and Levesque 1996; Brunke et al. 2014). Staphylinid dispersal into Canadian agroecosystems by flight is known to occur predominantly in late spring/early summer (Boiteau 1983; Levesque and Levesque 1996). Dispersal of rove beetles into the interior of Canadian soybean fields does not appear to be challenging for common species of the assemblage, likely due to high dispersal power via flight (Brunke 2011). However, fewer staphylinids (mostly *Philonthus* spp., all flight-capable) were captured in the centre of potato fields in New Brunswick (Boiteau 1983) and may have preferred the more densely vegetated microhabitats of the field and woodland edges. A few species with low flight activity among individuals (e.g. *Neohypnus* spp., Xantholinini) or with small proportions of fully

Table 9.1 Common species of Staphylinidae (≥1% of total species captured in pitfall traps) in northeastern North American agroecosystems ranked from highest to lowest activity density. Rankings for vegetable crops were estimated from bar graphs in Leslie et al. (2007). Species occurring in three or more agroecosystems are in boldface, while other shared species are underlined. Adventive species are indicated by '+'. Modified from Brunke et al. (2014)

Soybean (Ontario)[a]	Vegetable field crops (Pennsylvania)[b]	Blueberries (Nova Scotia)[c]	Dairy pastures (Nova Scotia)[d]
Strigota obscurata Klimaszewski and Brunke	***Dinaraea angustula* (Gyllenhal)+**	*Stenus erythropus* Melsheimer	*Stenus flavicornis* Erichson
Apocellus sphaericollis (say)	*Hoplandria lateralis* (Melsheimer)	*Xantholinus linearis* (Olivier)+	*Stenus erythropus* Melsheimer
Drusilla canaliculata (Fabricius)+	***Strigota ambigua* (Erichson)**	*Mocyta fungi* (Gravenhorst)+	*Philonthus carbonarius* (Gravenhorst)+
***Dinaraea angustula* (Gyllenhal)+**	*Tachyporus nitidulus* (Fabricius)	*Drusilla canaliculata* (Fabricius)+	*Philonthus cognatus* Stephens+
Hoplandria lateralis (Melsheimer)	*Anotylus* spp.	*Philonthus carbonarius* (Gravenhorst)+	*Tachyporus dispar* (Paykull)+
Aleochara verna Say	*Aleochara verna* say	*Octhephilium fracticorne* (Paykull)+	*Philonthus varians* (Paykull)+
***Strigota ambigua* (Erichson)**	*Neohypnus* spp.	*Stenus semicolon* LeConte	*Tachinus rufipes* (DeGeer)+
Oxypoda brachyptera (Stephens)+	*Tachinus fimbriatus* Gravenhorst	*Anotylus rugosus* (Fabricius)+	*Rugilus angustatus* (Geoffrey)+
Anotylus tetracarinatus (block)+	*Aleochara curtula* (Gravenhorst)+	*Gabrius picipennis* (Mäklin)	***Amischa analis* (Gravenhorst)+**
Anotylus insecatus (Erichson)+	*Belonuchus rufipennis* (Fabricius)	*Oxypoda nigriceps* Casey	*Tachinus addendus* horn
***Amischa* spp.**[e]	*Gabrius nigritulus* (Gravenhorst)+	***Dinaraea angustula* (Gyllenhal)+**	*Gabrius picipennis* (Mäklin)
Anotylus rugosus (Fabricius)+		*Mycetoporus horni* Berhauer & Schubert	*Mocyta fungi* (Gravenhorst)+
Stethusa spuriella (Casey)		***Amischa analis* (Gravenhorst)+**	*Lathrobium* spp.
Scopaeus minutus Erichson+		*Ilyobates bennetti* Donisthorpe+	*Aleochara morion* Gravenhorst+
Tachinus corticinus Gravenhorst+		*Mycetoporus consors* LeConte	*Gyrohypnus angustatus* (Stephens)+
		***Strigota ambigua* (Erichson)**	
		Quedius curtipennis Bernhauer+	

[a]Brunke et al. (2014); [b]Leslie et al. (2007); [c]Renkema et al. (2012); [d]D'Orsay (2012)
[e]Included at least some individuals of *Amischa analis*

winged individuals (e.g. *Arpedium cribratum*) disperse from source populations primarily over the soil surface (Levesque and Levesque 1996) and may exhibit patchier local distributions. Overwintering may occur within fields in some species, but many are thought to disperse to nearby hedgerows or forest edges, which are known to support agriculturally significant staphylinids outside of the growing season (Holland et al. 2009; Brunke et al. 2014). Based on both passive (traps placed in crop vegetation) and active (sweeping, suction) sampling, some staphylinids are known to spend a considerable percentage or most of their active time in agroecosystems above the soil surface (Brunke 2011; D'Orsay 2012), potentially preying upon foliar pests such as aphids, caterpillars or the eggs of various taxa; Canadian species with high levels of foliar activity include *Amischa analis*, *Anotylus tetracarinatus*, *Mocyta fungi*, *Stenus flavicornis* and *Tachyporus dispar* Paykull.

9.12 Conservation Biological Control by the Generalist Predator Assemblage

Staphylinids are abundant and widespread members of the diverse predator assemblages that occur in Canada's agroecosystems, and as such they have been targeted under 'conservation bio-control' efforts, which aim to promote these assemblages through crop management techniques (Symondson et al. 2002). Such techniques, of which the impact on staphylinids has been investigated, include the provision of hedgerows or non-crop buffers, mulching and reduced weeding. In Europe, hedgerows provide staphylinids with overwintering habitat and refugia from disturbances during the growing season (Holland et al. 2009). The direct impact of non-crop buffers has not been investigated in Canada, but they are known to support important staphylinids of the soybean assemblage outside of the growing season (Brunke et al. 2014). Mulching in highbush blueberry, a management practice used to deter successful pupation in the blueberry maggot (Renkema et al. 2011), was observed to increase the abundance of rove beetles in Nova Scotia, Canada; species responded differently depending on whether compost or pine needles were applied (Renkema et al. 2012). Positive effects on the diversity and abundance of some species increased the following year, suggesting that the mulched areas were creating local, stable increases in available prey or preferred microclimates (Renkema et al. 2016). Non-weeded areas promoted the abundance of several staphylinids, including *Xantholinus linearis* (Olivier), *Anotylus rugosus* (F.), *Mocyta fungi* and *Gabrius picipennis* (Mäklin) (Renkema et al. 2012).

One of the greatest challenges to successful conservation biological control is the complex network of interactions between predator and potential prey and between individual predator species. For example, intraguild predation by the large carabid *Pterostichus melanarius* Illiger of smaller predator beetles, including staphylinids, suppressed the predation of pest *Delia* (Diptera) eggs by the small beetles (Prasad and Snyder 2004). This negative interaction may be improved by creating refugia for smaller predators such as those provided by weed cover (Prasad and Snyder 2004). Even more remarkable are intraguild interactions that have a positive impact on biological control services. In Mexican shaded coffee plantations, the aleocharine *Myrmedonota xipe* Mathis and Eldredge recognizes ant alarm pheromones to selectively prey upon phorid fly-parasitized *Azteca* ants, which are important predators of coffee-feeding insects (Mathis and Tsutsui 2016). It is unknown to which magnitude the staphylinids reduce parasitoid pressure on the ants, but this recent discovery emphasizes the fact that agroecology of conservation biological control is complex and that the poorly known natural history of most insects may be an impediment to its practical and predictable implementation.

We propose that the agroecological study of staphylinids faces an even greater issue: the basic diet of most rove beetle species is unknown, and much of what is stated in the literature is based on assumptions. Often, species are considered mycophagous or saprophagous based on observations of species of the same genus or subfamily. This is less problematic for the entirely predaceous and agriculturally important subfamilies Paederinae, Staphylininae and Steninae (Thayer 2005), but diet is complex and variable within the equally important subfamilies Aleocharinae, Omaliinae, Oxytelinae and Tachyporinae. For example, *Tachyporus dispar* (Tachyporinae), an important aphid predator in Europe, also feeds on the powdery mildew growing on the exudate produced by cereal aphids (Dennis et al. 1990). The diverse and widespread genus *Sepedophilus* (Tachyporinae) is often treated as uniformly mycophagous (Levesque and Levesque 1995; Clough et al. 2007;

Renkema et al. 2012), but a study based on gut contents and mouthpart morphology (Newton 1984) revealed that the species found in Nearctic and Palearctic agroecosystems (i.e. *S. marshami* and *S. testaceus*) belong to a predatory group with potential facultative mycophagy ('Group A'). Nearly every surveyed agroecosystem contains several species of *Anotylus* (Oxytelinae), which are normally considered to be entirely saprophagous in dung, compost or carrion (Hammond 1976; Clough et al. 2007; Renkema et al. 2016). However, observations by Horion (1967) and later in the field by Majka and Klimaszewski (2008b) confirm larval Diptera predation by *Anotylus insecatus*, a Palaearctic species that is adventive in North America (Campbell and Tomlin 1983). A closely related species of similar morphology, *Anotylus rugosus*, was observed to aggregate near oviposition sites of the cabbage root maggot (Guseva and Koval 2005). A recent study comparing conventional and organic agriculture in Europe found that organic fields contained more 'decomposers' (=saprophages, mostly *Anotylus* species), while conventional fields contained more predators (Clough et al. 2007). The higher organic content of organic fields was considered to support more decomposers, but it would also support higher prey populations for obligate or facultative predators. A better understanding of predation in common, agriculturally relevant oxyteline and tachyporine species would improve the foundation of studies seeking to reveal differences in responses to crop management among functional trophic groups.

9.13 Biocontrol Services in Simplified Systems

Far more is known about the biocontrol services of staphylinids in simplified systems such as those of the commercially reared *Dalotia coriaria* used in greenhouse integrated pest management and the Diptera predator/parasitoid genus *Aleochara*. *Dalotia coriaria* is a widespread Palaearctic species that has become adventive in the Nearctic and several other regions (Klimaszewski et al. 2013c); it is such a prolific predator of eggs and immatures that it can become a pest of laboratory insect colonies (Miller and Williams 1983). Easily reared on oatmeal but preferring living prey (Birken and Cloyd 2007), *D. coriaria* can be released in greenhouses as part of an integrated pest management strategy against fungus gnats, which are pests of potted plants (Jandricic et al. 2006). Mutual intraguild predation between released staphylinids and predatory mites is known to occur, but future research is needed to determine whether this impacts the overall suppression of greenhouse fungus gnat populations (Jandricic et al. 2006).

The staphylinid taxon that has been given the greatest amount of agroecological research attention is the diverse and widespread genus *Aleochara* due to the close association of its members with pest Diptera as predators and parasitoids. In the Northern Hemisphere, *A. bipustulata* (L.), *A. bilineata* Gyllenhal and *A. verna* Say (all subgenus *Coprochara*) are particularly well-studied as natural enemies of root maggots (*Delia* spp.), which are severe economic pests of canola and cause 20 to 100 million dollars of damage annually in Canada alone (Holliday et al. 2013). These species possess a specialized life cycle, with active, first-instar larvae that seek out cyclorrhapheous fly puparia (Klimaszewski 1984) and enter via a chewed hole, where they feed as ectoparasitoids of the pupae (Colhoun 1953). First-instar larvae overwinter in the puparia (Colhoun 1953). Second- and third-instar larvae are morphologically specialized for sedentary life as ectoparasitoids, and third-instar larvae pupate within their puparium (Colhoun 1953). Both *A. verna* and *A. bilineata* occur in the Canadian canola assemblage, but only the latter occurs consistently and emerges in synchronization with the oviposition period of its *Delia* hosts (Broatch et al. 2008). In Canadian canola, *A. bilineata* is the most prevalent parasitoid of *Delia radicum* (L.) due to competition with *Trybliographa rapae* (Westwood) (Hymenoptera, Figitidae), whose populations experience mortality if larvae of the former attack pupae that were already parasitized by the latter (Hemachandra et al. 2007). Although *A. bipustulata* does not occur in the Nearctic (Hemachandra et al. 2005), there is interest to introduce it intentionally into Canada as a natural enemy of root maggots in canola (Hemachandra et al. 2007). Further manipulation of *Aleochara* biological control services in canola may be made possible by gaining a better understanding of their host location mechanisms, which appear to be based on volatiles released by damaged canola plants (Broatch et al. 2010). An application of mustard seed meal can locally increase the number of adult *A. bilineata* (Holliday et al. 2011) and, potentially, parasitism and predation rates.

References

Andersen A (1991) Carabidae and Staphylinidae (Col.) frequently found in Norwegian agricultural fields. New data and a review. Fauna Norvegica Ser B 38:65–75

Andreassen LD, Kuhlmann U, Whistlecraft JW, Soroka JJ, Mason PG, Akinremi OO, Holliday NJ (2010) Spring emergence of Canadian *Delia radicum* and synchronization with its natural enemy, *Aleochara bilineata*. Can Entomol 142:234–249

Ashe JS (1984) Generic revision of the subtribe Gyrophaenina (Coleoptera: Staphylinidae: Aleocharinae) with review of described subgenera and major features of evolution. Quest Entomol 20:129–349

Assing V (2007) A revision of *Amarochara* of the Holarctic region III. A new species, a new synonym, and additional records (Coleoptera, Staphylinidae: Aleocharinae). Zootaxa 1411:25–32

Assing V (2008) The genus *Calodera* Mannerheim in Canada (Insecta, Coleoptera, Staphylinidae, Aleocharinae). In: Majka CG, Klimaszewski J (eds) Biodiversity, biosystematics, and ecology of Canadian Coleoptera. ZooKeys 2:203–208

Birken EM, Cloyd RA (2007) Food preference of the rove beetle, *Atheta coriaria* Kraatz (Coleoptera: Staphylinidae) under laboratory conditions. Insect Sci 14:53–56

Boiteau G (1983) Activity and distribution of Carabidae, Arachnida, and Staphylinidae in New Brunswick potato fields. Can Entomol 115:1023–1030

Bousquet Y, Bouchard P, Davies AE, Sikes D (2013) Checklist of beetles (Coleoptera) of Canada and Alaska. Pensoft, Sofia

Broatch JS, Dosdall LM, Yang R-C, Harker KN, Clayton GW (2008) Emergence and seasonal activity of the entomophagous rove beetle *Aleochara bilineata* (Coleoptera: Staphylinidae) in canola in Western Canada. Environ Entomol 37:1451–1460

Broatch JS, Dosdall LM, O'Donovan JTO, Harker KN, Clayton GW (2010) Responses of the specialist biological control agent, *Aleochara bilineata*, to vegetational diversity in canola agroecosystems. Biol Control 52:58–67

Brunke AJ (2011) Diversity, habitat use and potential biocontrol services of rove beetles (Coleoptera: Staphylinidae) in soybean agroecosystems and adjacent hedgerows. M. Sc. Thesis, School of Environmental Sciences, University of Guelph, Guelph, Ontario

Brunke AJ, Klimaszewski J, Dorval JA, Bourdon C, Paiero SM, Marshall SA (2012) New species and distribution records of Aleocharinae (Coleoptera, Staphylinidae) from Ontario, Canada, with a checklist of recorded species. ZooKeys 186:119–206

Brunke AJ, Bahlai CA, Klimaszewski J, Hallett RH (2014) Rove beetles (Coleoptera: Staphylinidae) in Ontario, Canada soybean agroecosystems: assemblage diversity, composition, seasonality, and habitat use. Can Entomol 146:652–670

Campbell JM (1982) A revision of the North American Omaliinae (Coleoptera: Staphylinidae). 3. The genus *Acidota* Stephens. Can Entomol 114:1003–1029

Campbell JM, Tomlin AD (1983) The first record of the Palearctic species *Anotylus insecatus* (Gravenhorst) (Coleoptera: Staphylinidae) from North America. Coleopt Bull 37:309–313

Clough Y, Kruess A, Tscharntke T (2007) Organic versus conventional arable farming systems: functional grouping helps understand staphylinid response. Agric Ecosyst Environ 118:285–290

Colhoun EH (1953) Notes on the stages and the biology of *Baryodma ontarionis* Casey (Coleoptera: Staphylinidae), a parasite of the cabbage maggot, *Hylemya brassicae* Bouché (Diptera: Anthomyiidae). Can Entomol 85:1–8

D'Orsay C (2012) The influence of pasture management intensity on species richness and abundance of beetles (Coleoptera: Carabidae, Staphylinidae and Curculionoidea). M. Sc. Thesis. Department of Biology, University of Prince Edward Island, Charlottetown, Prince Edward Island

Dennis P, Wratten SD (1991) Field manipulation of populations of individual staphylinid species in cereals and their impact on aphid populations. Ecol Entomol 16:17–24

Dennis P, Wratten SD, Sotherton W (1990) Feeding behaviour of the staphylinid beetle *Tachyporus hypnorum* in relation to its potential for reducing aphid numbers in wheat. Ann Appl Biol 117:267–276

Fox CJS, MacLellan CR (1956) Some Carabidae and Staphylinidae shown to feed on a wireworm, *Agriotes sputator* (L.), by the precipitin test. Can Entomol 88:228–231

Frank JH, Thomas MC (1991) The rove beetles of Florida (Coleoptera: Staphylinidae). Florida Department of Agriculture and Consumer Services, Division of Plant Industry. Entomology Circular No. 343

Guseva OG, Koval AG (2005) The functional response of polyphagous predators and their potential use in modern plant protection systems. In: The biological means of plant protection and technology of their use. Institute of Plant Protection, St. Petersburg, pp 18–27 [in Russian]

Halder AC (2011) Distribution and natural enemies in herbivores of oilseed rape fields in Sweden. Int J Innov Technol 1:29–36

Hammond PM (1976) A review of the genus *Anotylus* C.G. Thomson (Coleoptera: Staphylinidae). Bull Br Entomol Soc 33:139–187

Hanley RS, Ashe JS (2003) Techniques for dissecting adult aleocharine beetles (Coleoptera: Staphylinidae). Bull Entomol Res 93:11–18

Hemachandra KS, Holliday NJ, Klimaszewski J, Mason PG, Kuhlmann U (2005) Erroneous records of *Aleochara bipustulata* from North America: an assessment of the evidence. Can Entomol 137: 182–187

Hemachandra KS, Holliday NJ, Mason PG, Soroka JJ, Kuhlmann U (2007) Comparative assessment of the parasitoid community of *Delia radicum* in the Canadian prairies and Europe: a search for classical biological control agents. Biol Control 43:85–94

Holland JM, Birkett T, Southway S (2009) Contrasting the farm-scale spatio-temporal dynamics of boundary and field overwintering predatory beetles in arable crops. BioControl 54:19–33

Holliday AE, Pak DM, Holliday NJ (2011) The *Aleochara* (Coleoptera: Staphylinidae) attractant in mustard seed meal is not dimethyl disulphide. Proc Entomol Soc Manitoba 67:5–10

Holliday NJ, Andreassen LD, Dixon PL, Kuhlmann U (2013) *Delia radicum* (L.), cabbage maggot (Diptera: Anthomyiidae). In: Mason PG, Gillespie D (eds) Biological control programmes in Canada 2001–2012. CAB International, Wallingford, pp 142–151

Horion AD (1967) Faunistik der Mitteleuropaïschen Käfer. Volume II, Staphylinidae, Part 3 Habrocerinae bis Aleocharinae (Ohne Subtribus Athetae). P.C. W. Schmidt, Überlingen-Bodensee

Hu GY, Frank JH (1995) Biology of *Neohypnus pusillus* (Sachse) (Coleoptera: Staphylinidae) and its predation on immature horn flies in the laboratory. Coleopt Bull 49:43–52

Jandricic S, Scott-Dupree CD, Broadbent AB, Harris CR, Murphy G (2006) Compatibility of *Atheta coriaria* with other biological control agents and reduced-risk insecticides used in greenhouse floriculture integrated pest management programs for fungus gnats. Can Entomol 138:712–722

Kishimoto H, Adachi I (2008) Predation and oviposition by predatory *Stethorus japonicus*, *Oligota kashmirica benefica*, and *Scolothrips takahashii* in egg patches of various spider mite species. Entomol Exp Appl 128:294–302

Klimaszewski J (1984) A revision of the genus *Aleochara* Gravenhorst of America North of Mexico (Coleoptera: Staphylinidae, Aleocharinae). Mem Entomol Soc Canada 129:3–211

Klimaszewski J (2000) Diversity of the rove beetles in Canada and Alaska (Coleoptera, Staphylinidae). Mémoires de la Société Royale Belge d'Entomologie 39:3–126

Klimaszewski J, Majka CG (2007) Two new *Atheta* species (Coleoptera: Staphylinidae: Aleocaharinae) from eastern Canada: taxonomy, bionomics, and distribution. Can Entomol 139:45–53

Klimaszewski J, Pelletier G, Germain C, Humble LM, Winchester NN (2001) Diversity of *Placusa* (Coleoptera: Staphylinidae, Aleocharinae) in Canada, with descriptions of two new species. Can Entomol 133:1–47

Klimaszewski J, Pelletier G, Sweeney J (2002) Genus *Tinotus* (Coleoptera: Staphylinidae, Aleocharinae) from America North of Mexico: review of the types, distribution records, and a key to species. Can Entomol 134:281–298

Klimaszewski J, Pohl G, Pelletier G (2003) Revision of the Nearctic *Silusa* (Coleoptera, Staphylinidae, Aleocharinae). Can Entomol 135:159–186

Klimaszewski J, Pelletier G, Majka C (2004) A revision of Canadian *Leptusa* Kraatz (Col., Staphylinidae, Aleocharinae): new species, new distribution records, key and taxonomic considerations. Belg J Entomol 6:3–42

Klimaszewski J, Sweeney J, Price J, Pelletier G (2005) Rove beetles (Coleoptera: Staphylinidae) in red spruce stands, eastern Canada: diversity, abundance, and descriptions of new species. Can Entomol 137:1–48

Klimaszewski J, Pelletier G, Germain C, Work T, Hébert C (2006) Review of *Oxypoda* species in Canada and Alaska (Coleoptera, Staphylinidae, Aleocharinae): systematics, bionomics, and distribution. Can Entomol 138:737–852

Klimaszewski J, Langor D, Savard K, Pelletier G, Chandler DS, Sweeney J (2007a) Rove beetles (Coleoptera: Staphylinidae) in yellow birch dominated stands of southeastern Quebec, Canada: diversity, abundance, and description of a new species. Can Entomol 139:793–833

Klimaszewski J, Assing V, Majka CG, Pelletier G, Webster RP, Langor D (2007b) Records of adventive aleocharine beetles (Coleoptera: Staphylinidae: Aleocharinae) found in Canada. Can Entomol 139:54–79

Klimaszewski J, Langor D, Work TT, Hammond JHE, Savard K (2008a) Smaller and more numerous harvesting gaps emulate natural forest disturbances: a biological test case using rove beetles (Coleoptera, Staphylinidae). Divers Distrib 14:969–982

Klimaszewski J, Savard K, Pelletier G, Webster R (2008b) Species review of the genus *Gnypeta* Thomson from Canada, Alaska and Greenland (Coleoptera, Staphylinidae, Aleocharinae): systematics, bionomics and distribution. In: Majka CG, Klimaszewski J (eds) Biodiversity, biosystematics, and ecology of Canadian Coleoptera. ZooKeys 2:11–84

Klimaszewski J, Webster R, Assing V, Savard K (2008c) *Diglotta mersa* (Haliday) and *Halobrecta flavipes* Thomson, two new species for the Canadian fauna (Coleoptera, Staphylinidae, Aleocharinae). In: Majka

CG, Klimaszewski J (eds) Biodiversity, biosystematics, and ecology of Canadian Coleoptera. ZooKeys 2:115–188

Klimaszewski J, Godin B, Pelletier G, Savard K (2008d) Six new species and records of aleocharine beetles from the Yukon and Alaska (Coleoptera: Staphylinidae: Aleocharinae). Can Entomol 140:265–291

Klimaszewski J, Webster RP, Savard K (2009a) First record of the genus *Schistoglossa* Kraatz from Canada with descriptions of seven new species (Coleoptera, Staphylinidae, Aleocharinae). In: Majka CG, Klimaszewski J (eds) Biodiversity, biosystematics, and ecology of Canadian Coleoptera II. ZooKeys 22:45–79

Klimaszewski J, Webster RP, Savard K (2009b) Review of the rove beetle species of the subtribe Gyrophaenina Kraatz (Coleoptera, Staphylinidae) from New Brunswick, Canada: new species, provincial records and bionomic information. In: Majka CG, Klimaszewski J (eds) Biodiversity, biosystematics, and ecology of Canadian Coleoptera II. ZooKeys 22:81–170

Klimaszewski J, Webster RP, Savard K, Couture J (2009c) First record of the genus *Alisalia* Casey from Canada, description of two new species, and a key to all Nearctic species of the genus (Coleoptera, Staphylinidae, Aleocharinae). Zookeys 25:1–18

Klimaszewski J, Langor D, Pelletier G, Bourdon C, Perdereau L (2011) Aleocharinae beetles (Coleoptera, Staphylinidae) of the province of Newfoundland and Labrador. Pensoft, Sofia, 313 pp

Klimaszewski J, Godin B, Bourdon C (2012) Further contributions to the aleocharine fauna of the Yukon Territory, Canada (Coleoptera, Staphylinidae). ZooKeys 186:207–237

Klimaszewski J, Webster RP, Langor DW, Bourdon C, Jacobs J (2013a) Review of Canadian species of the genus *Dinaraea* Thomson, with descriptions of six new species (Coleoptera, Staphylinidae, Aleocharinae, Athetini). ZooKeys 327:65–101

Klimaszewski J, Morency J-M, Labrie P, Séguin A, Langor D, Work T, Bourdon C, Thiffault E, Paré D, Newton AF, Thayer MK (2013b) Molecular and microscopic analysis of the gut contents of abundant rove beetle species (Coleoptera, Staphylinidae) in the boreal balsam fir forest of Quebec, Canada. ZooKeys 353:1–24

Klimaszewski J, Brunke A, Assing V, Langor D, Newton AF, Bourdon C, Pelletier G, Webster R, Herman L, Perdereau L, Davies A, Smetana A, Chandler D, Majka C, Scudder G (2013c) Synopsis of adventive species of Coleoptera (Insecta) recorded from Canada. Part 2: Staphylinidae. Pensoft, Sofia

Klimaszewski J, Webster RP, Langor DW, Bourdon C, Hammond HEJ, Pohl GR, Godin B (2014) Review of Canadian species of the genera *Gnathusa* Fenyes, *Mniusa* Mulsant & Rey and *Ocyusa* Kraatz (Coleoptera, Staphylinidae, Aleocharinae). ZooKeys 412:9–40

Klimaszewski J, Webster RP, Bourdon C, Pelletier G, Godin B, Langor DW (2015a) Review of Canadian species of the genus *Mocyta* Mulsant & Rey (Coleoptera, Staphylinidae, Aleocharinae), with the description of a new species and a new synonymy. ZooKeys 487:111–139

Klimaszewski J, Webster RP, Sikes D, Bourdon C, Labrecque M (2015b) A review of Canadian and Alaskan species of the genera *Clusiota* Casey and *Atheta* Thomson, subgenus *Microdota* Mulsant & Rey (Coleoptera, Staphylinidae, Aleocharinae). ZooKeys 524:103–136

Klimaszewski J, Godin B, Langor D, Bourdon C, Lee S-I, Horwood D (2015c) New distribution records for Canadian Aleocharinae (Coleoptera, Staphylinidae), and new synonymies for *Trichiusa*. ZooKeys 498:51–91

Klimaszewski J, Langor DW, Bourdon C, Gilbert A, Labrecque M (2016a) Two new species and new provincial records of aleocharine rove beetles from Newfoundland and Labrador, Canada (Coleoptera, Staphylinidae, Aleocharinae). ZooKeys 593:49–89

Klimaszewski J, Langor DW, Hammond HEJ, Bourdon C (2016b) A new species of *Anomognathus* and new Canadian and provincial records of aleocharine rove beetles from Alberta, Canada (Coleoptera, Staphylinidae, Aleocharinae). ZooKeys 581: 141–164

Klimaszewski J, Larson D, Labrecque M, Bourdon C (2016c) Twelve new species and fifty three new provincial distribution records of Aleocharinae rove beetles of Saskatchewan, Canada (Coleoptera, Staphylinidae). ZooKeys 610:45–112

Klimaszewski J, Webster RP, Langor DW, Sikes D, Bourdon C, Godin B, Ernst C (2016d) A review of Canadian and Alaskan species of the genus *Liogluta* Thomson, and descriptions of three new species (Coleoptera, Staphylinidae, Aleocharinae). ZooKeys 573:217–256

Klimaszewski J, Langor D, Sweeney J, Sylvain ZA, Bourdon C, Paré D, Morrisette J, Dagnault S, Titus B (in preparation) Volume of fine woody debris influences composition of rove beetle assemblages (Coleoptera, Staphylinidae): a case study in the boreal forest of Quebec, Canada

Kneeshaw DD, Bergeron Y (1998) Canopy gap characteristics and tree replacement in the southeastern boreal forest. Ecology 79:783–794

Leslie TW, Hoheisel GA, Biddinger DJ, Rohr JR, Fleischer SJ (2007) Transgenes sustain epigeal insect biodiversity in diversified vegetable farm systems. Environ Entomol 36:234–244

Levesque C, Levesque G-Y (1995) Abundance, diversity and dispersal power of rove beetles (Coleoptera: Staphylinidae) in a raspberry plantation and adjacent sites in eastern Canada. J Kansas Entomol Soc 68:355–370

Levesque C, Levesque G-Y (1996) Seasonal dynamics of rove beetles (Coleoptera: Staphylinidae) in a raspberry plantation and adjacent sites in eastern Canada. J Kansas Entomol Soc 69:285–301

Lieffers VJ, Stewart JD, Macmillan RB, MacPherson D, Branter K (1996) Semi-natural and intensive silvicultural systems for the boreal mixedwood forest. For Chron 72:286–292

Lohse GA, Klimaszewski J, Smetana A (1990) Revision of Arctic Aleocharinae of North America (Coleoptera: Staphylinidae). Coleopt Bull 44:121–202

Majka CG, Klimaszewski J (2008a) New records of Canadian Aleocharinae (Coleoptera, Staphylinidae). ZooKeys 2:85–114

Majka CG, Klimaszewski J (2008b) Adventive Staphylinidae (Coleoptera) of the Maritime Provinces of Canada: further contributions. ZooKeys 2:151–174

Majka CG, Klimaszewski J (2008c) Introduced Staphylinidae (Coleoptera) in the Maritime Provinces of Canada. Can Entomol 140:48–72

Majka CG, Klimaszewski J (2010) Contributions to the knowledge of the Aleocharinae (Coleoptera, Staphylinidae) in the Maritime Provinces of Canada. ZooKeys 46:15–39

Majka CG, Klimaszewski J, Lauff RF (2008) The coastal rove beetles (Coleoptera, Staphylinidae) of Atlantic Canada: a survey and new records. In: Majka CG, Klimaszewski J (eds) Biodiversity, biosystematics, and ecology of Canadian Coleoptera. ZooKeys 2:115–150

Mathis KA, Tsutsui ND (2016) Dead ant walking: a myrmecophilous beetle predator uses parasitoid host location cues to selectively prey on parasitized ants. Proc R Soc B 283:1–7

Miller KV, Williams RN (1983) Biology and host preference of *Atheta coriaria* (Coleoptera: Staphylinidae), an egg predator of Nitidulidae and Muscidae. Ann Entomol Soc Am 76:158–161

Newton AF (1984) Mycophagy in Staphylinoidea (Coleoptera). Fungus-insect relationships: perspectives in ecology and evolution. Columbia University Press, New York, pp 302–353

Newton AF, Thayer MK, Ashe JS, Chandler DS (2001) 22. Staphylinidae Latreille, 1802. In: Arnett RH, Thomas MC (eds) American beetles. 1. Archostemata, Myxophaga, Adephaga, Polyphaga: Staphyliniformia. CRC Press LLC, Boca Raton, FL, pp 272–418

Paquin P, Dupérré N (2001) Beetles of the boreal forest: a faunistic survey carried out in western Quebec. Proc Entomol Soc Ontario 132:57–98

Paré D, Bergeron Y, Camiré C (1993) Changes in the forest floor of Canadian southern boreal forest after disturbance. J Veg Sci 4:811–818

Pohl GR, Langor D, Spence JR (2007) Rove beetles and ground beetles (Coleoptera: Staphylinidae, Carabidae) as indicators of harvest and regeneration practices in western Canadian foothills forests. Biol Conserv 137:294–307

Pohl G, Langor D, Klimaszewski J, Work T, Paquin P (2008) Rove beetles (Coleoptera: Staphylinidae) in northern forests. Can Entomol 140:415–436

Prasad RP, Snyder WE (2004) Predator interference limits fly egg biological control by a guild of ground-active beetles. Biol Control 2004:428–437

Prévost M, Raymond P, Lussier J-M (2009) Regeneration dynamics after patch cutting and scarification in yellow birch-conifer stands. Can J For Res 40:357–369

Renkema JM, Cutler GC, Lynch DH, MacKenzie K, Walde SJ (2011) Mulch type and moisture level affect pupation depth of *Rhagoletis mendax* Curran (Diptera: Tephritidae) in the laboratory. J Pest Sci 84:281–287

Renkema J, Lynch D, Cutler GC, Mackenzie K, Walde SJ (2012) Ground and rove beetles (Coleoptera: Carabidae and Staphylinidae) are affected by mulches and weeds in highbush blueberries. Environ Entomol 41:1097–1106

Renkema JM, Cutler GC, Lynch DH, MacKenzie K, Walde SJ (2016) Organic mulches in highbush blueberries alter beetle (Coleoptera) community composition and improve functional group abundance and diversity. Agric For Entomol 18:119–127

SAS Institute (2002–2003) Proprietary Software Release 9.1. SAS Institute Inc. Cary, NC, USA

Siitonen J (2001) Forest management, coarse woody debris and saproxylic organisms: Fennoscandian boreal forests as an example. Ecol Bull 49:11–41

Smetana A (1995) Rove beetles of the subtribe Philonthina of America north of Mexico (Coleoptera: Staphylinidae). Classification, phylogeny and taxonomic revision. Mem Entomol Int 3:x+946

Stefani FOP, Klimaszewski J, Morency M-J, Bourdon C, Labrie P, Blais M, Venier L, Séguin A (2016) Fungal community composition in the gut of rove beetles (Coleoptera: Staphylinidae) from the Canadian boreal forest reveals possible endosymbiotic interactions for dietary needs. Fungal Ecol 23:164–171

Stockland JN, Siitonene J, Jonsson BG (2012) Biodiversity of deadwood. Cambridge University Press, Cambridge

Symondson WOC, Sunderland KD, Greenstone MH (2002) Can generalist predators be effective biocontrol agents? Annu Rev Entomol 47:561–594

Thayer MK (2005) Staphylinidae Latreille, 1802. In: Beutel R, Leschen R (eds) Handbook of zoology: a natural history of the phyla of the animal kingdom. De Gruyter, New York, pp 296–344

Venier LA, Thompson ID, Fleming R, Malcolm J, Aubin I, Trofymow JA, Langor D, Sturrock R, Patry C, Outerbridge RO, Holmes SB, Haeussler S, De Grandpré L, Chen HYH, Bayne E, Arsenault A, Brandt JP (2014) Effects of natural resource development on the terrestrial biodiversity of Canadian boreal forests. Environ Rev 22:457–490

Venier LA, Work TT, Klimaszewski J, Morris D, Bowden J, Kwiaton M, Webster K, Hazlett P (2017) Ground-dwelling arthropod response to fire and clearcutting in jack pine: implications for ecosystem management. Can J For Res 47:1614–1631

Webster RP, Klimaszewski J, Pelletier G, Savard K (2009) New Staphylinidae (Coleoptera) records with new collection data from New Brunswick, Canada. 1. Aleocharinae. In: Majka CG, Klimaszewski J (eds) Biodiversity, biosystematics, and ecology of Canadian Coleoptera II. ZooKeys 22:171–248

Webster RP, Klimaszewski J, Sweeney JD, DeMerchant I (2012) New Staphylinidae (Coleoptera) records with new collection data from New Brunswick, and an addition to the fauna of Quebec, Canada: Aleocharinae. In: Klimaszewski J, Anderson R (eds) Biosystematics and ecology of Canadian Staphylinidae (Coleoptera) II. ZooKeys 186:83–118

Webster RP, Bouchard P, Klimaszewski J (eds) (2016a) The Coleoptera of New Brunswick and Canada: providing baseline biodiversity and natural history data. ZooKeys (Special Issue) 573:1–512

Webster RP, Klimaszewski J, Bourdon C, Sweeney JD, Hughes CC, Labrecque M (2016b) Further contribution to the Aleocharinae (Coleoptera, Staphylinidae) fauna of New Brunswick and Canada including descriptions of 27 new species. ZooKeys 573:85–216

Wiedenmann RN, Smith JW (1997) Attributes of natural enemies in ephemeral crop habitats. Biol Control 10:16–22

Work TT, Klimaszewski J, Thiffault E, Bourdon C, Paré D, Bousquet Y, Venier L, Titus B (2013) Initial responses of rove and ground beetles (Coleoptera, Staphylinidae, Carabidae) to removal of logging residues following clearcut harvesting in the boreal forest of Quebec, Canada. ZooKeys 258:31–52

A Worldwide Checklist of Parasites of Staphylinidae

10

J. Howard Frank

Abstract

This chapter attempts to document all the parasites of Staphylinidae worldwide. It is based on a 1982 work of the author and considers all changes in Staphylinidae taxonomy; in higher classifications of *Laboulbeniales* (*Ascomycetes*), the most species-rich order of parasites (*Laboulbeniales*); and in the family Proctotrupidae (Hymenoptera). The host(s) and geographical distribution of the parasites (*sensu latissimo*) are documented by genus/species in descending species richness.

10.1 Introduction

Based on a previous work (Frank 1982), this chapter attempts to document all the parasites of Staphylinidae worldwide. Since that work was sent to press, there have been many changes. The higher classification of Staphylinidae has changed to include subfamilies Pselaphinae, Scaphidiinae, Scydmaeninae, etc. The higher classification of *Laboulbeniales* (*Ascomycetes*), the most species-rich order of parasites (*Laboulbeniales*), has changed (Tavares 1985; Benjamin 2001) etc., and the family Proctotrupidae (Hymenoptera) has been revised (Townes and Townes 1982). There have been many additions of species and distributions of parasite/host records. All of these called for a new document, but yet the new document (here) is constrained by space. To conserve space, a parasite/host checklist is presented (there is no host/parasite checklist), and early references (before 1982) are not repeated if they were cited in the 1982 work. The objective is to report parasite/host associations to the species level for each country, as a basis for future investigation. This should help staphylinid specialists, who may be expected to know the classification of all staphylinids mentioned (but not that of their parasites), as well as mycologists, whose literature may fail to identify staphylinids below the level of genus (especially because their generic designation of staphylinids may be outdated). No new nomenclatural acts are made here. This chapter documents the host(s) and geographical distribution of the parasites (*sensu latissimo*) by genus/species in descending species richness as 70/510 (*Ascomycetes*, *Laboulbeniales*), 18/27 (Nematoda, Rhabditida), 11/21 (Insecta, Hymenoptera), and 7/23 (Nematoda, Tylenchida) and sundry smaller groups collectively with 33/34, for a grand total of 132/616.

10.2 Parasite/Host List

Phylum *Microsporidia*
Order *Dissociodihaplophasida*
Nosematidae

J. H. Frank (✉)
Department of Entomology and Nematology, University of Florida, Gainesville, FL, USA
e-mail: jhfrank@ufl.edu

O. Betz et al. (eds.), *Biology of Rove Beetles (Staphylinidae)*,
https://doi.org/10.1007/978-3-319-70257-5_10

Genus and species indet.: adult *Creophilus erythrocephalus* F., Australia (Meyer-Rochow 1972). This nosematid was reported from the eye of the host.

Phylum *Ascomycota*
Class *Laboulbeniomycetes*
Order *Laboulbeniales*

The available information suggests that *Laboulbeniales* are true external parasites, debilitating their insect hosts (Bro Larsen 1952; Scheloske 1969; Frank 1982; Tavares 1985). Currently divided into four families, of which only *Euceratomycetaceae* and *Laboulbeniaceae* contain parasites of Staphylinidae, and the latter contains all but one genus of them. The arrangement below lists genera alphabetically within the family *Laboulbeniaceae* and species alphabetically within genera, with their staphylinid hosts named alongside. Weir and Hammond (1997) calculated that about 26% of hosts in Asia and Europe are Staphylinidae, slightly exceeded only by Carabidae. Staphylinidae are hosts to 47 genera, whereas Carabidae are hosts to only 16 genera (Tavares 1985). There are now >2050 species of *Laboulbeniales* (Rossi and Santamaría 2012), and future knowledge is likely to expand the number enormously.

Euceratomycetaceae
Euceratomyces Thaxter, 1931
E. terrestris (Thaxter, 1894): adult ?*Lathrobium terminatum*, Gravenhorst, ME, USA (Thaxter 1894, 1931).

Laboulbeniaceae
Acallomyces Thaxter, 1902
A. gyrophaenae (Thaxter 1931): adult "*Gyrophaena parcipennis*," invalid species name, Jamaica (Thaxter 1931), *Gyrophaena* sp. indet., Japan (Sugiyama 1978a).
A. homalotae Thaxter, 1902: adult *Homalota* sp. (which may perhaps be referable to *Atheta*) NH, USA (Thaxter 1902); *Atheta transfuga* (Sharp) Japan (Sugiyama 1978b).
A. platyolae Thaxter, 1931: adult "*Platyola philippina*" invalid species name attributed to Bernhauer, Philippines (Thaxter 1931).
Acompsomyces Thaxter, 1901b
A. stenichni (Scheloske, 1969): adult *Stenichnus scutellaris* (Müller and Kunze) Germany, United Kingdom (Scheloske 1969, Weir 1994); *Stenichnus collaris* (Müller and Kunze) Poland, France (Majewski 1973b, 1994b, Santamaría and Rossi 1999).
= *Stigmatomyces stenichni* Scheloske, 1969, transferred to *Acompsomyces* by Tavares (1985).

Acrogynomyces Thaxter, 1931
A. arietinus Thaxter, 1931: adult *Scydmaenus* sp. (under the name of synonym *Eumicrus*) Cameroon (Thaxter 1931).
A. ellipsoideus Thaxter, 1931: adult *Scydmaenus* sp. (under the name of synonym *Eumicrus*) Cameroon (Thaxter 1931).
A. eumicralis Thaxter, 1931: adult *Scydmaenus* sp. (under the name of synonym *Eumicrus*) Cameroon (Thaxter 1931).
A. eumicri Thaxter, 1931: adult *Scydmaenus* sp. (under the name of synonym *Eumicrus*) Cameroon (Thaxter 1931).
A. eumicricola Thaxter, 1931: adult *Scydmaenus* sp. (under the name of synonym *Eumicrus*) Cameroon (Thaxter 1931).
A. hamatus Thaxter 1931: adult *Scydmaenus* sp. (under the name of synonym *Eumicrus*) Cameroon (Thaxter 1931).

Amorphomyces Thaxter, 1893
A. biformis Thaxter, 1931: adult *Falagria latemarginata* Bernhauer, Sumatra, Indonesia (Thaxter 1931).
A. falagriae Thaxter, 1893: adult *Falagria dissecta* Erichson, MA, USA (Thaxter 1893), and *Falagria* spp. in Argentina; OH, USA; Jamaica; and Spain (Spegazzini 1912, Thaxter 1931, Santamaría 2000).
A. hernandoi Santamaría, 2000: adult *Diglotta mersa* (Haliday) Spain (Santamaría 2000).

A. italicus Spegazzini, 1915b: adult ?*Atheta*, Italy (Spegazzini 1915b, Thaxter 1931 who considered that the host might have been a species of *Carpelimus*); *Carpelimus mundus* (Sharp) Ecuador (Thaxter 1931, Rossi 1978); *Carpelimus exiguus* (Erichson) Cameroon (Thaxter 1931, Rossi 1978); *Carpelimus halophilus* (Kiesenwetter) Algeria (Maire 1920, Rossi 1978); *Carpelimus bilineatus* Stephens, *C. alutaceus* (Fauvel), *C. pusillus* (Gravenhorst), and *Thinobius* sp., Spain (Santamaría 2000), Hebei Province, China (Shen and Ye 2006).

= *A. trogophloei* Spegazzini, 1917: adult *Carpelimus atramentarius* (Lynch) Argentina (Spegazzini 1917).

= *A. stipitatus* Thaxter, 1931: adult *Carpelimus exiguus* (Erichson) Sumatra, Indonesia (Thaxter 1931).

A. minusculus [original spelling was *minisculus*] Thaxter, 1931: adult *Ophioglossa bisulcata* (Erichson) as *Eudera bisulcata* (Erichson), Guatemala, and two "*Homalota*" spp., Sumatra, Indonesia (Thaxter 1931).

A. ophioglossae Thaxter 1912a: adult *Ophioglossa* sp., Argentina (Thaxter 1912a, 1931), *Ophioglossa cava* Sharp, Argentina (Spegazzini 1917, Thaxter 1931).

A. pronomaeae Thaxter, 1931: adult perhaps *Pronomaea thaxteri* Bernhauer although published as "*Pronomaea sumatrensis* Bernhauer," an invalid species name, Borneo (Thaxter 1931).

A. rubescens Thaxter 1912a: adult *Diestota* sp. indet., Argentina (Thaxter 1912a), *Homalota* sp. indet., Argentina (Thaxter 1912a), *Atheta conformis* (Erichson) and *A. lurida* (Erichson), Argentina (Spegazzini 1917, Thaxter 1931), *Atheta* sp. indet., Argentina (Thaxter 1931).

A. schistogeniae Thaxter, 1931: adult *Schistogenia crenicollis* Kraatz, Sarawak, Malaysia (Thaxter 1931).

A. stenusae Thaxter, 1931: adult "*Stenus borneensis* Bernhauer," an invalid name, Sarawak, Malaysia (Thaxter 1931).

NOTE that *A. floridanus* Thaxter (1893) and *A. obliqueseptatus* Thaxter (1900) have been transferred to *Dioicomyces*.

Apatelomyces Thaxter, 1931

A. ogmocerus Thaxter, 1931: adult *Ogmocerus* sp., Liberia (Thaxter, 1931).

Aporomyces Thaxter, 1931

A. perpusillus (Spegazzini, 1917): adult *Rhopalopherus gestroi* Bernhauer, Argentina (Spegazzini 1917).

Autophagomyces Thaxter 1901b

A. castellinii Rossi, 1982: adult *Trissemus punctipennis* (Raffray) Sierra Leone (Rossi 1982).

A. decarthricola (Spegazzini, 1917): adult pselaphine, probably *Decarthon rubripenne* Raffray, Argentina (Spegazzini 1917, Thaxter 1931).

A. hammondii Benjamin, 2001: adult scaphidiid indet., Sulawesi, Indonesia (Benjamin 2001).

A. tiwaiensis Rossi, 1990: adult *Trissemus assuetus* Castellini, Sierra Leone (Rossi 1990).

NOTE that *A. spegazzinii* was transferred to *Bordea* by Benjamin (2001).

Balazucia Benjamin, 1968b

B. bilateralis Benjamin, 1968b: adult *Phloeonomus* sp., Mexico (Benjamin 1968b).

B. japonica Terada, 1980: adult *Lordithon striatus* Olivier, Hokkaido, Japan (Terada 1980).

Blasticomyces Tavares, 1985

B. denigratus Majewski and Sugiyama, 1986: adult *Lispinus coarticollis* Kraatz, Sabah, Malaysia (Majewski and Sugiyama 1986).

B. lispini (Thaxter, 1915): adult *Nacaeus impressicollis* (Motschulsky) Java and Borneo, Indonesia, and Sri Lanka (Thaxter 1915, 1931) and Taiwan (Terada 1976); *Lispinus unipunctatus* Cameron and *Lispinus* sp., Sabah, Malaysia (Majewski and Sugiyama 1986); *Lispinus* sp. nr. *japonicus* Sharp, Iriomote Island, Japan (Majewski 1988b); *Eleusis* sp., Fujian and Suzhou provinces, China (Shen and Ye 2006).

Bordea Maire, 1916a (reinstated by Benjamin 2001). All reported hosts belong to Pselaphinae.

B. allenii Benjamin, 2001: adult *Cylindrarctus crinifer* Casey, LA, USA (Benjamin 2001).

B. bryaxalis (Thaxter, 1931): adult *Reichenbachia puncticollis* (LeConte) ME, USA (Thaxter 1931), *Reichenbachia borealis* Casey, IL and WI, USA (Benjamin 2001).

B. castellinii (Rossi, 1982): adult *Trissemus punctipennis* (Raffray) Sierra Leone (Rossi 1982, Benjamin 2001).

B. coronata Maire, 1916a: adult *Brachygluta aubei* Tournier, Algeria (Maire 1916a, Thaxter 1931), *Brachygluta perforata* (Aubé) Italy, *Brachygluta schuppeli* Aubé, Ibiza, Spain (Benjamin 2001).

B. denotata Haelewaters et al., 2014: adult *Bibloporus bicolor* (Denny), Netherlands (Haelewaters et al. 2014).

B. formosana (Sugiyama, 1982): adult of gen. et sp. indet. nr. *Lasinus*, Taiwan (Sugiyama 1982, Benjamin 2001, Shen and Ye 2006).

B. gigantea Benjamin, 2001: adult *Reichenbachia bicuspida* Park, Honduras (Benjamin 2001).

B. neocoronata Benjamin, 2001: adult *Decarthron defectum* Park, IL, USA (Benjamin 2001).

B. platensis Spegazzini, 1917 (formerly *Acallomyces* (*Bordea*) *platensis* Spegazzini 1917, so a change of status not of genus): adult pselaphine, Argentina (Spegazzini 1917), *Decarthron simplex* Raffray and *D. rubripenne* Raffray, Argentina (Benjamin 2001).

B. retroflexa Benjamin, 2001: *Melba* sp., MS, USA (Benjamin 2001).

B. spinigera Benjamin, 2001: adult *Reichenbachia spatulifer* Casey, MN, USA (Benjamin 2001).

B. strangulata (Thaxter, 1931): adult ?*Euconnus* (as *Eucomus*) Guatemala (Thaxter 1931).

B. thaxteri Benjamin, 2001: adult *Reichenbachia puncticollis* (LeConte) ME, USA (Benjamin 2001).

B. tiwaiensis (Rossi, 1990): adult *Trissemus assuetus* Castellini, Sierra Leone (Rossi 1990, Benjamin 2001).

B. weirii Benjamin, 2001: adult *Melba* sp. and *Melba thoracica* (Brendel) LA, USA and *Conoplectus canaliculatus* (LeConte) NY, USA (Benjamin 2001).

Camptomyces Thaxter, 1894. All reported hosts belong to *Astenus* (Paederinae).

C. brunneomarginatus Thaxter, 1926: adult *Astenus maculipennis* Kraatz, Philippines (Thaxter 1926).

C. europaeus Rossi and Cesari Rossi, 1980: adult *Astenus thoracicus* Baudi, Italy (Rossi and Cesari Rossi 1980).

C. falcatus Thaxter, 1926: adult *Astenus* sp., Philippines (Thaxter 1926).

C. guatemalensis Thaxter 1926: adult *Astenus* sp., Guatemala (Thaxter 1926).

C. melanopus Thaxter, 1894: adult *Astenus prolixus* (Erichson), MA, USA (Thaxter 1894); adult *Astenus paranensis* Lynch, Argentina (Spegazzini 1917).

C. recurvatus Thaxter, 1926: adult *Astenus bimaculatus* (Erichson), Philippines (Thaxter 1926).

C. subsigmoideus Thaxter, 1926: adult *Astenus* sp., Philippines (Thaxter 1926).

C. sumatrae Thaxter, 1926: adult *Astenus kraatzi* Bernhauer, Sumatra, Indonesia (Thaxter 1926).

Cantharomyces Thaxter, 1890

C. ancyrophori Picard, 1917: adult *Ochthephilus flexuosus* (Mulsant and Rey) France (Picard 1917).

C. andinus Thaxter, 1918: adult *Carpelimus puncticollis* Solier, Chile (Thaxter 1918).

C. aploderi Huldén, 1983: adult *Aploderus caesus* (Erichson) Karelia (Russia) (Huldén 1983).

C. bledii Thaxter, 1890 and 1896: adult *Bledius assimilis* Casey, IL, USA (Thaxter 1890).

C. chilensis Thaxter, 1918: adult *Thinodromus signatus* (Erichson) Chile (Thaxter 1918).

C. exiguus Thaxter, 1931: adult *Carpelimus exiguus* (Erichson) Cameroon (Thaxter 1931), *Carpelimus exiguus* (Erichson) Japan (Sugiyama 1973, 1974) and South Korea (Lee 1986), *Carpelimus* sp., Gansu Province, China (Shen et al. 2004, Shen and Ye 2006).

C. haytiensis Thaxter, 1931: adult *Carpelimus fulvipes* (Erichson) Haiti (Thaxter 1931).

C. japonicus Sugiyama, 1973: adult *Thinodromus sericatus* (Sharp) Japan (Sugiyama 1973); adult *Carpelimus* sp., China (Shen et al. 2004, Shen and Ye 2006).

C. magellanicus Thaxter, 1918: adult ?genus of Aleocharinae (this identification seems improbable given the restriction of other species to Oxytelinae) Argentina (Thaxter 1918).

C. numidicus Maire, 1920: adult *Thinodromus mannerheimi* (Kolenati) Algeria (Maire 1920); *Thinodromus arcuatus* (Stephens) and *Carpelimus anthracinus* (Mulsant and Rey) Poland (Majewski 1983); *Thinodromus arcuatus* (Stephens) Italy and Slovakia (Rossi 1978, Rossi et al. 2010); *Thinodromus hirticollis* (Mulsant and Rey) Greece (Castaldo et al. 2004).

C. occidentalis Thaxter, 1893: adult *Bledius bellicus* Blackwelder (as *B. armatus* Say) although possibly a misidentification of *B. strenuus* Casey, UT, USA (Thaxter 1893, 1896, Herman 1976).

C. orientalis Spegazzini, 1915b: adult ?*Quedius* sp. Italy (this host seems improbable given restriction of other species to Oxytelinae) (Spegazzini 1915b); *Carpelimus subtilicornis* (Roubal) Germany (Scheloske 1969); *Carpelimus corticinus* (Gravenhorst) Karelia (Russia) and *Carpelimus elongatulus* (Erichson) Finland (Huldén 1983); *Carpelimus bilineatus* Stephens and *C. corticinus* (Gravenhorst) England, Poland (Weir and Beakes 1993, Majewski 1987); *Carpelimus corticinus* (Gravenhorst) and *C. foveolatus* (Sahlberg) Belgium (De Kesel and Haghebaert 1991); *Carpelimus corticinus* (Gravenhorst) Greece (Castaldo et al. 2004); *Carpelimus obesus* (Kiesenwetter) Czech Republic and Slovakia (Rossi et al. 2010).

= *C. abbreviatus* Maire, 1920: adult *Carpelimus corticinus* (Gravenhorst) Algeria (Maire 1920).

C. pacei Rossi and Santamaría, 2000: adult *Spanioda andicola* Pace, Chile (Rossi and Santamaría 2000).

C. platystethi Thaxter, 1900: adult *Platystethus cornutus* (Gravenhorst) England (Thaxter 1900), Morocco (Maire 1916a); *Platystethus operosus* Sharp, Japan (Terada 1977); *Platystethus* sp., India (Kaur and Mukerji 1996) Tibet, China (Shen and Ye 2006).

C. robustus Majewski, 1987: adult *Carpelimus bilineatus* (Stephens) and *Carpelimus corticinus* (Gravenhorst) and *Carpelimus obesus* (Kiesenwetter) and *Carpelimus rivularis* (Motschulsky) Poland (Majewski 1987).

C. thaxteri Maire 1916b: adult *Thinodromus dilatatus* (Erichson) France (Maire 1916b); *Carpelimus bilineatus* (Stephens) and *Carpelimus rivularis* (Motschulsky) Belgium (De Kesel and Haghebaert 1991).

C. trogophloei Spegazzini, 1917: adult *Carpelimus atramentarius* Lynch, Argentina (Spegazzini 1917).

C. valdivianus Thaxter, 1918: adult *Thinodromus grandipennis* (Bernhauer) Chile (Thaxter 1918).

C. venetus Spegazzini 1915b: adult ?genus, Italy (Spegazzini 1915b); *Carpelimus rivularis* (Motschulsky) Poland and Belgium (Siemaszko and Siemaszko 1932, De Kesel and Haghebaert 1991).

Chaetomyces Thaxter, 1892

C. pinophili Thaxter, 1893: adult *Pinophilus latipes* (Gravenhorst) southeastern USA (Thaxter 1893); *Pinophilus* sp., Nicaragua (Thaxter 1908); *Pinophilus suffusus* Erichson, Argentina (Spegazzini 1917); *Pinophilus* sp., Brazil (Rossi and Bergonzo 2008).

= *C. borelli* (Colla, 1926): unidentified staphylinid, Bolivia (Colla 1926).

Clematomyces Thaxter, 1900

= *Schizomeromyces* Thaxter, 1931

C. argentinensis Spegazzini, 1917: adult *Pinophilus suffusus* Erichson, Argentina (Spegazzini 1917).

C. astenalis (Thaxter, 1931): adult *Astenus fimetarius* Fauvel, Cameroon (Thaxter 1931).

C. pinophili Thaxter, 1900: adult *Pinophilus* sp., Burma (Thaxter 1900).

Clonophoromyces Thaxter, 1931

C. grenadinus Thaxter, 1931: adult probably of *Bryoporus obscurus* (Erichson) (as "*Bryonomus serialis*" in error) Grenada (Thaxter 1931).

C. nipponicus Terada and Tavares, 1993: adult *Bryoporus gracilis* (Sharp) Honshu, Japan (Terada and Tavares 1993); *Bryoporus testaceus* LeConte, MA, USA (Haelewaters et al. 2015).

Compsomyces Thaxter, 1894

= *Moschomyces* Thaxter, 1894

C. insignis Thaxter, 1894: adult *Astenus prolixus* (Erichson) MA, USA (Thaxter 1894).

C. lestevae Thaxter, 1900: adult *Lesteva sicula* Gravenhorst ssp. *heeri* Fauvel, Scotland, England (Thaxter 1931) Belgium (De Kesel and Haghebaert 1991); *Lesteva hanseni* Lohse, France, Algeria, England (Maire 1916a, Balazuc 1873b); *Lesteva pubescens* Mannerheim, Scotland, Poland, Italy (Thaxter 1908, Majewski 1973b, Rossi 1978); *Lesteva villardi* Rey, France (Balazuc 1973b).

C. macropoda Thaxter, 1931 based on description by Spegazzini, 1917: adult *Astenus paranensis* Lynch, Argentina (Spegazzini 1917).

C. palamini Thaxter, 1931: adult *Palaminus* sp., Trinidad (Thaxter 1931).

C. stilicopsis Thaxter, 1931: on *Stilicopsis setigera* Sharp, Philippines (Thaxter 1931).

C. verticillatus (Thaxter 1890): adult *Astenus longiusculus* (Mannerheim) IL, ME, USA (Thaxter 1890); *Astenus paranensis* Lynch, Argentina (Spegazzini 1912); *Astenus signatus* Sahlberg, Argentina (Spegazzini 1917); *Astenus thoracicus* Baudi, Italy (Rossi and Cesari Rossi 1980); *Astenus martinezi* Uhagón, Spain (Santamaría 1995a).

Corethromyces Thaxter, 1892

C. acanthoglossae Thaxter, 1931: adult *Acanthoglossa intermixta* Eppelsheim, Cameroon (Thaxter 1931).

C. acuminatus Thaxter, 1931: adult *Diochus nanus* Erichson, Trinidad (Thaxter 1931).

C. aequatorialis Rossi and Weir, 2007: adult *Gnathymenus* sp., Ecuador (Rossi and Weir 2007).

C. argentinus Thaxter, 1912a: adult *Ochthephilum agile* Erichson, Argentina (Thaxter 1912a).

C. armatus Thaxter, 1912a: adult *Rugilus chilensis* (Solier) Argentina (Thaxter 1912a).

C. bolivianus Weir and Rossi, 2001: adult *Scopaeodes* sp., Bolivia (Weir and Rossi 2001).

C. brazilianus Thaxter, 1900: adult *Ochthephilum brasilianum* Lucas, Brazil and *Ochthephilum fasciatum* (Erichson) Venezuela and *Ochthephilum flohri* Sharp, Mexico and *Ochthephilum similipenne* Say, Mexico, and *Ochthephilum venustum* Sharp, Mexico (Thaxter 1900); *Ochthephilum paranense* Lynch, Argentina (Spegazzini 1912); *Ochthephilum basale* Blanchard, Argentina (Thaxter 1931).

C. brunneolus Thaxter, 1912a: adult *Rugilus elegans* Lynch, Argentina (Thaxter 1912a, 1931, Spegazzini 1917).

= *C. sternalis* Spegazzini, 1917.

C. buccalis Thaxter, 1931: adult *Rugilus ceylanensis* (Kraatz) Sumatra, Indonesia (Thaxter 1931).

C. calyculatus Thaxter, 1931: adult *Sepedophilus obesus* (Boheman) Cameroon (Thaxter 1931).

C. catalinae (Thaxter, 1912a): adult *Sepedophilus testaceus* (F.) Argentina (Thaxter 1912a, Tavares 1985).

C. chaetophilus Thaxter, 1931: adult *Coproporus bernhaueri* Scheerpeltz (as *Coproporus ventralis* Bernhauer) Sumatra, Indonesia (Thaxter 1931).

C. circinellus Thaxter, 1931: adult *Medon planatus* Bernhauer, Sumatra, Indonesia (Thaxter 1931).

C. cornutus Thaxter, 1931: adult *Sepedophilus obesus* (Boheman) Cameroon (Thaxter 1931).

C. cryptobii Thaxter, 1892: adult *Homaeotarsus pallipes* (Gravenhorst) VA, USA and *Homaeotarsus bicolor* (Gravenhorst) eastern USA (Thaxter 1892); *Cryptobium sharpi* Fauvel [perhaps misidentified] Burma (Spegazzini 1915a).

C. curtipes Thaxter, 1931: adult *Lithocharis ochracea* (Gravenhorst) Cameroon (Thaxter 1931) and Iriomote Island, Japan (Majewski 1988b); *Medon* sp., Taiwan (Terada 1976, 1978, Shen and Ye 2006).

C. dacnochili Thaxter, 1931: adult *Dacnochilus laetus* LeConte, Guatemala (Thaxter 1931).

C. decipiens Thaxter, 1915: adult *Medon birmanus* Fauvel, Borneo, Indonesia (Thaxter 1915); *Lithocharis curtus* (Kraatz) Java, Indonesia, and Sarawak, Malaysia (Thaxter 1915, 1931).

C. diochi Thaxter, 1931): adult *Diochus nanus* Erichson, Trinidad (Thaxter 1931).

C. divergens Thaxter, 1931: adult *Stilomedon triseriatus* Sharp, Guatemala (Thaxter 1931).

C. euaestheti Thaxter, 1931: adult *Euaesthetus americanus* Erichson, MA, USA (Thaxter 1931).

C. filifer Thaxter, 1931: adult *Stiliderus sculptipennis* (Kraatz) Philippines (Thaxter 1931).

C. flagellaris Thaxter, 1931: adult *Stiliderus sculptipennis* (Kraatz) Philippines (Thaxter 1931).

C. flectatus Thaxter, 1931: adult *Rugilus pruinosus* (Cameron) Sumatra, Indonesia (Thaxter 1931).

C. formicetorum Spegazzini, 1917: adult *Heterothops formicetorum* Bernhauer, Argentina (Spegazzini 1917).

C. fuscipes (Spegazzini, 1917): adult *Scopaeus lugubris* Lynch, Argentina (Spegazzini 1917, Thaxter 1931, Tavares 1985).

C. gracilicaulis Thaxter, 1931: adult *Rugilus sericeus* (Motschulsky) Sumatra, Indonesia (Thaxter 1931).

C. grenadinus Thaxter, 1931: adult *Diochus nanus* Erichson, Trinidad (Thaxter 1931).

C. guatemalensis Thaxter, 1931: adult ? *Stilomedon*, Guatemala (Thaxter 1931).

C. hernandoi Santamaría, 2006: adult *Medon* sp., Spain (Santamaría, 2006).

C. introversus Thaxter, 1931: adult *Medon angulipennis* Eppelsheim, Cameroon (Thaxter 1931).

C. jacobinus Thaxter, 1893: adult *Lobrathium jacobinum* (LeConte) CA, USA (Thaxter 1893); *Lobrathium collare* (Erichson) ME, USA (Thaxter 1896).

C. jamaicensis Thaxter, 1931: adult *Stilomedon insularum* (Cameron) Jamaica (Thaxter 1931).

C. kamerunensis Thaxter, 1931: adult "*Medon kamerunensis* Bernhauer," invalid name, Cameroon (Thaxter 1931). [A *Medon* species with the same name was described by Scheerpeltz in 1971, but was based on a collection made in 1949, and no reference was made by Scheerpeltz to a prior use of that name, even as a manuscript name, by Bernhauer.]

C. laminifer Thaxter, 1931: adult *Lithocharis ochracea* (Gravenhorst) Sarawak, Malaysia, and "*Medon sumatrensis* Bernhauer," invalid name, Sumatra, Indonesia (Thaxter 1931).

C. lepidus (Thaxter, 1931): adult *Sepedophilus kamerunensis* (Bernhauer), Cameroon (Thaxter 1931, Tavares 1985).

C. lingulatus Thaxter, 1931: adult *Coproporus* sp., Sumatra, Indonesia (Thaxter 1931).

C. longicaulis Thaxter, 1902: adult *Rugilus angulatus* (Erichson) MA, USA (Thaxter 1902); *Rugilus elegans* (Lynch) Argentina (Spegazzini 1917); *Rugilus cribratus* (Sharp) Guatemala (Thaxter 1931).

C. longicollis Thaxter, 1931: adult "*Medon vulneratus* Bernhauer," invalid name, Cameroon (Thaxter 1931).

C. luzonensis Thaxter, 1931: adult *Stiliderus sculptipennis* (Kraatz) Philippines (Thaxter 1931).

C. macropus Thaxter 1912a: adult *Heterothops thaxteri* Bernhauer, Argentina (Thaxter1912a).

C. medonis Thaxter, 1915: adult *Lithocharis curtus* (Kraatz) Sarawak, Malaysia and Java, Indonesia (Thaxter 1915), *Medon birmanus* Fauvel, Sarawak, Malaysia (Thaxter 1915, 1931).

C. minusculus [original spelling was *miniscululus*] Thaxter, 1931: adult *Scopaeus* sp., Argentina (Thaxter 1931).

C. nanus Thaxter, 1931: adult *Diochus nanus* Erichson, Trinidad (Thaxter 1931).

C. obscuristipes Santamaría, 2006: adult *Medon* sp., Spain (Santamaría 2006).

C. obtusus (Thaxter, 1900): adult *Leptobium illyricum* (Erichson), apparently from Algeria (Thaxter 1900), same host, Bosnia and Herzegovina (Bánhegyi 1960); *Leptobium melanocephalum* (Reiche and Saulcy), Israel (Siemaszko and Siemaszko 1932).

C. ophitis Thaxter, 1912a: adult *Opithes fauveli* Lynch, Argentina (Thaxter 1912a); *Opithes bergonzoi* Drugmand, Brazil (Rossi and Bergonzo 2008).

C. orientalis Thaxter, 1915: adult *Rugilus ceylanensis* (Kraatz) Java and Sumatra, Indonesia (Thaxter 1915).

C. otongaensis Rossi and Weir 2007: adult *Biocrypta* sp., Ecuador (Rossi and Weir 2007).

C. palumboi Rossi, 2010: adult *Sepedophilus* sp., Ecuador (Rossi 2010).

C. perelegans Thaxter, 1931: adult *Medon crassulum* Sharp, Trinidad (Thaxter 1931).

C. piesticola Thaxter, 1931: adult *Piestus bicornis* Laporte, Costa Rica, and *Piestus penicillatus* (Dalman) [as *P. striatus* (Gray)] Brazil, and *Piestus spinosus* (F.) [as *P. oxytelinus* (Perty)] Venezuela (Thaxter 1931); *Priochirus* sp., Peru (Sugiyama 1972).

C. platensis Thaxter, 1912a: adult *Lobrathium dimidiatum* (Say) Argentina and Uruguay and *Lobrathium nitidum* (Erichson) Argentina (Thaxter 1912a, 1931, Spegazzini 1912, 1917).

= *C. xantholini* Spegazzini, 1912: adult *Xantholinus subtilis* (Boheman), Argentina (Spegazzini 1912).

C. podophilus Spegazzini, 1917: adult *Rugilus elegans* (Lynch) Argentina (Spegazzini 1917).

C. propinquus (Thaxter, 1900): adult *Lathrobium* spp. (although all are probably *Leptobium* spp.) Europe (Thaxter 1900, 1908, 1931, Picard 1913b); *Leptobium illyricum* (Erichson) Algeria (Maire 1916a); *Leptobium gracile* (Gravenhorst) [as *L. guttulum* (Lacordaire)] Spain (Santamaría 1995a).

C. protrudens Thaxter, 1931: adult "*Medon sumatrensis* Bernhauer" invalid name, Sumatra, Indonesia (Thaxter 1931).

C. psilotracheli Thaxter, 1931: adult *Stiliderus crassus* (Kraatz) Sarawak, Malaysia (Thaxter 1931).

C. pulchellus Spegazzini, 1917: adult *Pinophilus bergi* Lynch, Argentina (Spegazzini 1917).

C. purpurascens Thaxter, 1900: adult *Biocrypta fulvipes* (Erichson) and *Ochthephilum* sp., Grenada (Thaxter 1900); *Ochthephilum densipenne* (Sharp) South Korea (Lee et al. 2005).

C. pygmaeus Thaxter, 1912a: adult *Rugilus chilensis* (Solier) Chile (Thaxter 1912a).

C. rectus Spegazzini, 1917: adult *Rugilus elegans* (Lynch) Argentina (Spegazzini 1917); *Rugilus ceylanensis* (Kraatz) and *Rugilus sericeus* (Motschulsky) Sumatra, Indonesia (Thaxter 1931).

C. recurvatus Thaxter, 1931: adult *Stiliderus crassus* (Kraatz) Sarawak, Malaysia (Thaxter 1931).

C. rostellatus Thaxter, 1931: adult *Sepedophilus obesus* (Boheman) Cameroon (Thaxter 1931).

C. rostratus Thaxter, 1912a: adult *Heterothops* sp., Guatemala (Thaxter 1912a); *Heterothops formicetorum* Bernhauer, Argentina (Spegazzini 1917).

C. sardous Rossi and Cesari Rossi, 1980: adult *Pseudobium labile* (Erichson) Sardinia, Italy (Rossi and Cesari Rossi 1980).

C. scimbalii W. Rossi and Cesari Rossi, 1978: adult *Scymbalium anale* (Nordmann) Italy (Rossi and Cesari Rossi 1978); adult *Lathrobium* sp., Guizhou Province, China (Shen and Ye 2006).

C. sciopori Thaxter, 1931: adult *Medon crassulum* Sharp, Trinidad (Thaxter 1931).

C. scopaei Thaxter, 1912a: adult *Scopaeus frater* Lynch, Argentina (Thaxter 1912a, Spegazzini 1917); Spain—but these Spanish specimens are later named as *S. striatus* (Santamaría 1992b, 1997).

C. scopaeicola Thaxter, 1931: adult *Scopaeus nitidulus* Motschulsky, Philippines (Thaxter 1931).

C. seticola Thaxter, 1931: adult "*Medon sumatrensis* Bernhauer," invalid name, Sumatra, Indonesia (Thaxter 1931).

C. setiger Thaxter, 1893: adult *Lathrobium nitidulum* LeConte, MA, USA (Thaxter 1893, 1896, Saccardo 1895, Benjamin 1971).

C. shazawae Majewski and Sugiyama, 1985: adult *Ochthephilum densipenne* (Sharp) Japan (Majewski and Sugiyama 1985b) and South Korea (Lee and Park 1991).

C. sigmoideus Thaxter 1912a: adult *Rugilus elegans* (Lynch) Argentina (Thaxter 1912a, 1931, Spegazzini 1917).

C. spectabilis Thaxter, 1931: adult *Sepedophilus obesus* (Boheman), Cameroon (Thaxter 1931).

C. speluncalis (Maire 1916a): adult *Heterothops praevius* Erichson ssp. *nigrus* Kraatz, Algeria (Maire 1916a).

C. stereocephalus Thaxter 1931: adult *Stereocephalus seriatipennis* Lynch, Argentina (Lynch 1917).

= *C. pallidus* (Spegazzini, 1917), homonym of *C. pallidus* Thaxter, 1893.

C. stilici Thaxter, 1901b: adult *Rugilus rufipes* Germar, Italy, Germany, Netherlands, Poland, Belgium (Thaxter 1901b, 1908, 1912a, Picard 1913b, Thaxter 1931, Middelhoek 1943b, Stadelmann and Poelt 1962, Scheloske 1969, Rossi 1975, Majewski 1980, De Kesel and Haghebaert 1991); *Rugilus similis* (Erichson) Poland and Belgium (Majewski 1980, De Kesel and Haghebaert 1991); *Rugilus angularis* (Erichson) MA, USA (Thaxter 1908); *Rugilus chilensis* (Solier) Argentina, Chile (Spegazzini 1912a); *Rugilus elegans* (Lynch) Argentina and Uruguay (Spegazzini 1917).

C. stilicicola (Thaxter, 1902): adult *Rugilus angularis* (Erichson) MA, USA (Thaxter 1902, 1912a, 1931); *Rugilus elegans* (Lynch) Argentina and Uruguay (Spegazzini 1917).

C. strangulatus Thaxter, 1931: adult *Stilomedon triseriatum* Sharp, Guatemala (Thaxter 1931).

C. striatus Santamaría, 1997: adult *Scopaeus* spp., Spain (Santamaría 1997).

C. thayerae Rossi and Weir, 2007: adult *Medon obscuriventer* Fairmaire and Germain, Chile (Rossi and Weir 2007).

C. thinocharinus Thaxter, 1915: adult *Thinocharis pygmaea* Kraatz and *Thinocharis curticeps* (invalid name attributed to Bernhauer), Java and Sumatra, Indonesia (Thaxter 1915).

C. unciger Thaxter, 1912a emended by Trotter 1926: adult *Rugilus elegans* (Lynch) Argentina (Thaxter 1912a).

C. uncinulus Thaxter, 1931: adult "*Stilicus thaxteri* Bernhauer," invalid name, likely a species of *Rugilus*, Cameroon (Thaxter 1931).

C. urophilus (Spegazzini, 1917): adult *Scopaeus* sp., Argentina (Spegazzini 1917, Thaxter 1931).

C. venezolanus Thaxter, 1931: adult "*Stilicus venezolanus* Bernhauer," invalid name, likely a species of *Rugilus*, Venezuela (Thaxter 1931).

C. verrucifer Spegazzini, 1917: adult *Rugilus elegans* (Lynch) Argentina (Spegazzini 1917, Thaxter 1931).

C. vesiculifer (Thaxter, 1931): adult *Sepedophilus obesus* (Boheman) Cameroon (Thaxter 1931, Tavares 1985).

C. vilis Thaxter, 1931: adult *Lithocharis vilis* (Kraatz) Sumatra, Indonesia (Thaxter 1931).

NOTE: Tavares (1985) transferred four species to *Sphaleromyces*: *S. indicus* Thaxter 1901b; *S. lathrobii* Thaxter, 1894; *S. occidentalis* Thaxter, 1895; and *S. rhinoceralis* (Thaxter, 1912a). She also resurrected the generic name *Rhadinomyces* Thaxter and transferred into it *C. pallidus*, *C. cristatus*, and *C. gracilis*.

Cryptandromyces Thaxter, 1912
= *Peyerimhoffiella* Maire, 1916

C. batrisi (Thaxter, 1931): adult *Batrisus* sp., Cameroon (Thaxter 1931, Tavares 1985).

C. batrisocenus (Thaxter, 1931): adult *Batrisocenus difformipes* Raffray, and *Batrisocenus elevatus* Raffray, Sarawak and Borneo, Indonesia (Thaxter 1931, Tavares 1985).

C. biblioplecti Majewski, 1987: adult *Biblioplectus ambiguus* (Reichenbach) and *Plectophloeus fischeri* (Aubé) Poland (Majewski 1987, 1994a,b); "*Plectophloeus tuberculatus* Reichenbach" an invalid name, Spain (Santamaría 2001).

C. bryaxidis Majewski, 1999: adult *Bryaxis bulbifer* (Reichenbach) Poland (Majewski 1999); *Biblioporus bicolor* (Denny) and *Bryaxis baudueri* (Reitter), Spain (Santamaría 2001).

C. cauliculatus (Thaxter, 1931): adult *Scydmaenus* sp. (as *Eumicrus* sp.) Cameroon (Thaxter 1931, Tavares 1985).

C. euplecti Santamaría, 2001: adult *Euplectus duponti* Aubé and *E. infirmus* Raffray, Spain; *E. sanguineus* Denny, Belgium; *E. karsteni* Reichenbach, France; *Plectophloeus zoufali* Machulka, Italy; and *Amauronyx kraatzi* Saulcy, Spain (Santamaría 2001).

C. geniculatus Thaxter, 1912: adult *Connophron* sp., Argentina (Thaxter 1912, 1931, Tavares 1985), *Euconnus* sp., Hainan Island, China (Shen and Ye 2006).

C. incurvatus (Thaxter 1931): adult *Scydmaenus* sp. (as *Eumicrus* sp.), Java and Sumatra, Indonesia (Thaxter 1931, Tavares 1985).

C. isabellae Rossi, 1990: adult *Scydmaenus* sp., Sierra Leone (Rossi 1990).

C. javanus Thaxter, 1915: adult of undetermined scydmaenine, Java, Indonesia and Japan (Thaxter 1915, Sugiyama 1973).

C. nigromarginatus Thaxter, 1931): adult *Scydmaenus* sp. (as *Eumicrus* sp.) Java, Indonesia and Cameroon (Thaxter 1931), and Congo Kinshasa (formerly Zaire) (Rossi and Santamaría 1992); *Scydmaenus suspicionis* Castellini, Sierra Leone (Rossi 1982).

C. sarawakensis (Thaxter, 1931): adult *Scydmaenus* sp. (as *Eumicrus* sp.) Sarawak, Malaysia (Thaxter 1931, Tavares 1985).

C. scydmaenarius (Thaxter, 1931): adult *Scydmaenus* sp. (as *Eumicrus* sp.) Sarawak, Malaysia (Thaxter 1931, Tavares 1985).

C. scydmaenicola (Spegazzini 1917): adult *Scydmaenus* sp. (as *Eumicrus* sp.) Argentina and Trinidad (Spegazzini 1917, Thaxter 1931, Tavares 1985).

C. subsigmoideus (Spegazzini, 1917): adult *Eurhexius putzeyssi* Schaufuss (as "*E. putzussi*"), Argentina (Spegazzini 1917, Thaxter 1931, Tavares 1985).

C. zethopsi (Thaxter, 1931): adult *?Zethopsinus* sp., Cameroon (Thaxter 1931, Tavares 1985).

NOTE: Tavares (1985) separated the monotypic genus *Peyerimhoffiella* based on *P. elegans* Maire.

Cucujomyces Spegazzini, 1917

C. phycophilus Weir and Rossi, 1997: adult "*Macralymma brevipenne* Cameron" [invalid name, is perhaps *Macralymma punctiventris* Cameron] and *Omaliomimus conicus* (Fauvel) New Zealand (Weir and Rossi 1997); *Crymus kronei* (Kiesenwetter), Antipodes Islands, New Zealand (Hughes et al. 2004); *Macralymma punctiventris* Cameron, Auckland Islands, Antipodes Islands, New Zealand (Hughes et al. 2004); *Nesomalium campbellensis* Steel, Campbell Island, New Zealand (Hughes et al. 2004); *Nesomalium pacificum* (Kiesenwetter) Auckland Islands, New Zealand (Hughes et al. 2004); *Omaliomimus venator* (Broun) Auckland Islands, Antipodes Islands, Campbell Island, New Zealand (Hughes et al. 2004).

Diandromyces Thaxter, 1918

D. chilenus Thaxter 1918: adult *Leptoglossula sculpticollis* (Fauvel) Chile (Thaxter 1918).

D. onorei Rossi, 2010: adult *Hypotelus* sp., Ecuador (Rossi 2010).

Diaphoromyces Thaxter, 1926

D. lispini (Thaxter 1912a): adult *Nacaeus tenellus* (Erichson) Argentina, Guatemala (Thaxter 1912a, Spegazzini 1917, Thaxter 1926).

D. zirophori (Thaxter, 1916): *Piestus* sp. indet., Trinidad (Thaxter 1916, 1926, Tavares 1985).

Diclonomyces Thaxter, 1931

D. eumicrophilus (Thaxter, 1931): adult *Scydmaenus* sp. (as *Eumicrus* sp.) Cameroon (Thaxter 1915, 1931).

D. stilomedonis Thaxter, 1931: adult *Stilomedon strigicollis* Sharp, Guatemala (Thaxter 1931).

D. subgaleatus (Thaxter, 1915): adult scydmaenine, Java, Indonesia (Thaxter 1915, 1931).

Dimeromyces Thaxter, 1896

= *Eudimeromyces* Thaxter, 1918

= *Jeanneliomyces* Lepesme, 1945 (nomen nudum).

D. coprопori Thaxter, 1920a: adult *Coproporus* sp. Cameroon (Thaxter 1920, 1924).

D. gracilis Thaxter, 1920a: adult aleocharine, Cameroon (Thaxter 1920a, 1924).

D. gyrophaenae Thaxter, 1920a: adult *Gyrophaena* sp., Cameroon (Thaxter 1920, 1924).

D. osellae Rossi, 2010: adult *Plesiomalota cotopaxiensis* Pace, Ecuador (Rossi 2010).

D. roreri Thaxter, 1920a: adult *Gyrophaena* sp., Trinidad (Thaxter 1920a, 1924).

D. versicolor Majewski and Sugiyama, 1985: adult *Pseudoplandria spiniventris* (Bernhauer) Japan (Majewski and Sugiyama 1985b).

NOTE: Lepesme (1945) reported "*Jeanneliomyces tachyoryctidii*" (nomen nudum) on larvae of a scydmaenine on Mount Elgon in Uganda.

Dimorphomyces Thaxter, 1893

D. acutus Sugiyama and Majewski, 1985: adult *Spatulonthus longicornis* (Stephens) Bali, Indonesia (Sugiyama and Majewski 1985a).

D. argentinensis Spegazzini, 1912: adult *Zyras argentinus* Lynch (as *Myrmedonia argentina*), Argentina (Spegazzini 1912, Thaxter 1924).

D. baliensis Sugiyama and Majewski, 1985: adult aleocharine (gen. et sp. indet.) Bali, Indonesia (Sugiyama and Majewski 1985a).

D. bledii Thaxter, 1920a: adult *Bledius emarginatus* (Say), KS, USA (Thaxter 1920a, 1924).

D. brevirostris Thaxter, 1920a: adult *Coproporus* sp., Guatemala, Brazil, Jamaica (Thaxter 1920a).

D. carolinae Rossi, 2010: adult *Parosus* sp., Ecuador (Rossi 2010).

D. decipiens (Thaxter, 1920a): adult *Eleusis* sp., Cameroon (Thaxter 1920a, 1924).

D. denticulatus Thaxter, 1893: adult *Falagria dissecta* Erichson, MA, USA (Thaxter 1893).

D. eleusinus Thaxter, 1920b: adult of *Inopeplus* (Salpingidae), **not** *Eleusis* (Staphylinidae), **so do not count this parasite as having a staphylinid host.**

D. grenadinus Thaxter, 1920a: adult ?genus of Aleocharinae, Grenada (Thaxter 1920a, 1924).

D. muticus Thaxter, 1894: adult *Falagria dissecta* Erichson, MA, ME, USA (Thaxter 1894, 1896, 1924).

D. myrmedoniae Thaxter, 1900: adult "*Myrmedonia flavicornis*," invalid name, perhaps a species of *Zyras*, Guatemala (Thaxter 1900, 1908, 1924); *Tachyusa constricta* Erichson and *Gnypeta rubrior* Tottenham and *Mocyta fungi* (Gravenhorst) Poland (Majewski 1983); *Tachyusa balteata* Erichson, Greece and Italy (Castaldo et al. 2004).

D. philippinensis Rossi and Santamaría, 2000: adult *Pseudoplandria drugmandi* Pace, Philippines (Rossi and Santamaría 2000).

D. platensis Spegazzini, 1917: adult *Apocellus parvipennis* Bernhauer, Argentina (Spegazzini 1917); *Apocellus* sp., Brazil (Rossi and Bergonzo 2008).

=*D. vulgatissimus* Spegazzini, 1917: adult *Apocellus argentinus* Bernhauer and

Apocellus mendozanus Steinheil and *Apocellus opacus* Bernhauer, Argentina and Uruguay (Spegazzini 1917, Thaxter 1924).

=*D. furcatus* Thaxter, 1920a: adult *Apocellus* sp., Guatemala (Thaxter 1920a).

D. pygosteni Thaxter, 1926: adult *Typhlonemys pubescens* (Wasmann) as "*Pygostenus kamerunensis*," invalid name, but probably *Pygostenus thaxteri* Bernhauer, a synonym, Cameroon (Thaxter 1926).

D. phloeoporae Thaxter, 1900 (originally as *D. thleoporae* but emended): adult *Phloeoporus corticalis* (Gravenhorst) Madeira (Thaxter 1900, 1908, 1924, Hincks 1960).

D. rotundatus Sugiyama and Majewski, 1985: adult *Spatulonthus longicornis* (Stephens) Bali, Indonesia (Sugiyama and Majewski 1985a).

D. trogophloei Spegazzini, 1917: adult *Carpelimus tenuipunctus* Bernhauer, Argentina (Spegazzini 1917).

D. verticalis Thaxter, 1912a: adult *Atheta lurida* (Erichson) Argentina and *Atheta* spp., Argentina and Guatemala and *Oxypoda* spp., Argentina (Thaxter 1912a, 1924, Spegazzini 1917).

Dioicomyces Thaxter, 1901

D. floridanus (Thaxter, 1893): adult *Bledius basalis* LeConte, FL, USA (Thaxter 1893, 1901b, 1908); and *Bledius* sp., Ipswich, England, from Thaxter's type series (Santamaría 2002).

D. obliqueseptatus (Thaxter, 1900): adult of genus nr. *Myrmedonia*, Brazil (Thaxter 1900, 1901b, 1908, Santamaría 2002).

Diphymyces Spegazzini, 1917

D. penicillifer Weir and Rossi, 1997: adult *Stenomalium helmsi* (Cameron) New Zealand (Weir and Rossi 1997), *Nesomalium pacificum* (Kiesenwetter) New Zealand; *Allodrepa decipiens* Steel, Antipodes Islands, New Zealand (Hughes et al. 2004).

Diplomyces Thaxter, 1895

D. actobianus Thaxter, 1895: adult *Erichsonius nanus* (Horn) MA, USA (Thaxter 1895, 1896, 1931).

D. atanygnathi Thaxter, 1931: adult *Atanygnathus pictus* Motschulsky [cited as *A. ruficollis* (Kraatz), a synonym], Cameroon [identification is unlikely because this is an Asian species], and Sarawak, Malaysia (Thaxter 1931).

D. clavifer Rossi and Cesari Rossi, 1978: adult *Erichsonius signaticornis* (Mulsant and Rey) Italy (Rossi and Cesari Rossi 1978); *Erichsonius cinerascens* (Gravenhorst) Finland, Sweden (Huldén 1983) Poland (Majewski 1983); *Erichsonius* sp., Spain (Santamaría 1999).

Dipodomyces Thaxter, 1931

D. phloeocharidis Majewski, 1982: adult *Phloeocharis subtilissima* Mannerheim, Poland (Majewski 1982, Tavares 1985).

Ecteinomyces Thaxter, 1902

E. bonariensis Spegazzini, 1917: adult *Pseudodinusa ?richteri* Bernhauer, Argentina (Spegazzini 1917). Tavares (1985) declared that the species does not belong to *Ecteinomyces*, but did not transfer it elsewhere.

Euceratomyces Thaxter, 1931

E. terrestris (Thaxter, 1894): adult *Lathrobium terminatum* Gravenhorst, ME, USA (Thaxter 1894, 1931); *Lathrobium angustatum* Lacordaire and *Lathrobium* sp., Spain (Santamaría 1995).

Euhaplomyces Thaxter, 1901b

E. ancyrophori Thaxter, 1901b: adult *Ochthephilus aureus* (Fauvel) Scotland, France (Thaxter 1901b, 1908, Picard 1913b, Hake 1923, Balazuc 1974).

Eumonoicomyces Thaxter, 1901b

E. papuanus Thaxter, 1901b: adult *Oxytelus* sp., New Britain, and Sumatra, Indonesia (Thaxter

1901b, 1908, 1931); *Anotylus nitidulus* (Gravenhorst) Poland Poland (Majewski 1987); *Anotylus rugosus* (F.) Denmark (Rostrup 1916); *Platystethus* sp., Tibet, China (Shen and Ye 2006)

= *E. argentinensis* Spegazzini, 1912: adult *Anotylus insignitus* (Gravenhorst), Uruguay, and *Platystethus fallax* Lynch, Argentina (Spegazzini 1912, 1917, Thaxter 1931). There exists some doubt about this synonymy and even about generic assignment (Tavares 1985).

E. platystethi Thaxter, 1931: adult *Platystethus spectabilis* Kraatz, Sumatra, Indonesia (Thaxter 1931).

Euphoriomyces Thaxter, 1931

E. chaekyui Lee and Na, 2011: adult *Scaphisoma rufum* Achard, South Korea (Lee and Na 2011).

E. huggertii Majewski, 1983: adult *Proteinus brachypterus* (F.) and *Acrulia inflata* (Gyllenhal) Poland (Majewski 1983).

E. sugiyamae Majewski, 1988a: adult *Scaphisoma rufum* Achard, Japan and South Korea (Majewski 1988a, Lee et al. 2007).

Euzodiomyces Thaxter, 1900

E. capillarius Cépède and Picard, 1908a: adult *Lobrathium multipunctum* (Gravenhorst) France, Belgium (Cépède and Picard presented in a talk in 1907 but published formally 1908a, 1908b, Thaxter 1931, Collart 1945, Balazuc 1973b); *Lathrobium geminum* Kraatz, Belgium (De Kesel and Rammeloo 1991); *Lobrathium anale* (Lucas) Algeria (Maire 1916a, Balazuc 1973b).

E. lathrobii Thaxter, 1900: adult *Lathrobium angusticolle* Boisduval and Lacordaire, France (Picard 1917, Lepesme 1941, Balazuc 1973b); *L. armatum* Say, IL and MO, USA (Benjamin and Shanor 1951, Balazuc 1973b); *L. brunnipes* (F.) Belgium, England, Germany (Thaxter 1908, Picard 1913b, Collart 1945, Scheloske 1969, De Kesel and Rammeloo 1991); *L. elongatum* (L.) Netherlands, Poland (Middelhoek 1943a, b, Stadelmann and Poelt 1962, Balazuc 1973b, Majewski 1994a), Belgium (De Kesel and Haghebaert 1991); *Lathrobium fovulum* Stephens, Germany (Scheloske 1969, Balazuc 1973b); *L. geminum* Kraatz, Netherlands, Germany (Middelhoek 1943b, Stadelmann and Poelt 1962, Scheloske 1969, Balazuc 1973b); *L. impressum* Heer (= *L. filiforme* Gravenhorst) England, Poland (Thaxter 1900, 1908, Picard 1913b, Majewski 1973b, Balazuc 1973b); *L. laevipenne* Heer, Switzerland, Netherlands (Baumgartner 1923, Middelhoek 1943b, Stadelmann and Poelt 1962, Balazuc 1973b); *L. longulum* Gravenhorst, Germany (Scheloske 1969, Balazuc 1973b) Finland (Huldén 1983); *L. simile* LeConte, IL, USA (Benjamin and Shanor 1951, Balazuc 1973b); *L. zetterstedti* Rye (= *L. punctatum* Zetterstedt) England (Thaxter 1900, 1908, Cépède and Picard 1908b, Picard 1913b, Hake 1923); *Lathrobium* spp. indet. Germany, Japan, and MI, USA (Poelt 1952a, Stadelmann and Poelt 1962, Benjamin and Shanor 1951, Balazuc 1974, Sugiyama 1973, 1974); *Lobrathium anale* (Lucas) Algeria (Maire 1916a, Balazuc 1973b); *L. manueli* (Fauvel), Italy (Rossi 1975); *Lobrathium multipunctum* (Gravenhorst) Belgium (De Kesel and Rammeloo 1991); *Lobrathium yoshidai* Adachi, Japan (Majewski 1988a); *Lobrathium* sp., Morocco (Santamaría and Rossi 1999); *Homaeotarsus bicolor* (Gravenhorst) IL, USA (Benjamin and Shanor 1951, Balazuc 1973b); *Achenium humile* (Nicolai) Czech Republic (Rossi et al. 2010); *Xantholinus* sp. indet., Germany (Scheloske 1969, Balazuc 1973b); *Hemiquedius ferox* (LeConte), IL, USA (Benjamin and Shanor 1951, Balazuc 1973b).

Haplomyces Thaxter, 1893, reported only on *Bledius* spp. (Oxytelinae).

H. californicus Thaxter 1893: adult *Bledius ornatus* LeConte, CA, USA (Thaxter 1893, 1896).

H. texanus Thaxter, 1893: adult *Bledius bicornis* (Germar) Germany (Thaxter 1908, Picard

1913b, Balazuc 1974); *B. cribricollis* Heer, France (Maire 1916b, Thaxter 1931, Scheloske 1969); *B. gallicus* (Gravenhorst) [= *B. fracticornis* (Paykull)] Netherlands (Middelhoek 1943b, 1947a, Stadelmann and Poelt 1962, Balazuc 1974); *B. opacus* (Block) England (Thaxter 1908, Picard 1913, Hake 1923, Balazuc 1974); *B. spectabilis* Kraatz, France (Balazuc 1974); *B. subterraneus* Erichson, Germany (Thaxter 1906, Balazuc 1974); *B. unicornis* Germar (= *B. juvencus* Erichson) France (Thaxter 1908, Balazuc 1973); *Bledius cribricollis* Heer and *Bledius graellsi* Fauvel, Italy (Rossi and Cesari Rossi 1980); *Bledius atricapillus* (Germar) and *Bledius longulus* Erichson, Poland (Majewski 1984); *Bledius arcticus* Sahlberg, *Bledius diota* Schiødte, *Bledius filipes* Sharp, *Bledius gallicus* (Gravenhorst), *Bledius kutsae* Kangas, *Bledius longulus* Erichson, *Bledius opacus* (Block), *Bledius pallipes* (Gravenhorst), *Bledius poppiusi* Bernhauer, *Bledius vilis* Mäklin, Finland (Huldén 1983); *Bledius* sp., Heilongjiang, China (Shen and Ye 2006).

H. virginianus Thaxter, 1893: adult *Bledius emarginatus* (Say) VA, USA (Thaxter 1893, 1896). Balazuc (1974) denied the presence of this species in France, citing an earlier published claim to the contrary as a misidentification of *H. texanus*).

Idiomyces Thaxter, 1893, reported only on *Deleaster* spp. (Oxytelinae).

I. peyritschii Thaxter, 1893: adult *Deleaster dichrous* (Gravenhorst) (= *D. adustus* Küster) Austria, Germany, France, Hungary, Italy, Netherlands, Poland, Caucasus (Russia), Switzerland, United Kingdom, former Yugoslavia, Greece (Thaxter 1893, 1896, 1908, 1931, Picard 1913b, Baumgartner 1923, Hake 1923, Siemaszko and Siemaszko, 1932, Bánhegyi 1940, 1960, Middelhoek 1943, 1960, Stadelmann and Poelt 1963, Balazuc 1974, Rossi 1978, Weir 1994, De Kesel and Rammeloo 1991, Castaldo et al. 2004); *Deleaster yokoyamai* Adachi, Japan (Terada 1977).

Ilyomyces Picard, 1917, reported only on Steninae.

I. dianoi Weir: adult *Dianous* sp., Sulawesi, Indonesia (Weir 1995).

I. lavagnei (Picard 1913a): adult *Stenus aceris* Stephens (= *S. elegans* Fairmaire) France (Picard 1913a, 1917, Thaxter 1931, Lepesme 1941, Balazuc 1971c, 1974).

I. mairei Picard, 1917: adult *Stenus elegans* Rosenhauer, France (Picard 1917); *Stenus aceris* Stephens (= *S. elegans* Fairmaire) France (Thaxter 1931, Lepesme 1941, Balazuc 1971c, 1974); *Stenus clavicornis* (Scopoli) MA, USA (Haelewaters 2013, Haelewaters et al. 2015).

I. victoriae Weir: adult *Stenus* sp. Sulawesi, Indonesia (Weir 1995).

Kainomyces Thaxter, 1901, reported only on *Eleusis* (Osoriinae).

K. alutellae Thaxter, 1931: adult *Eleusis alutella* Fauvel, Cameroon (Thaxter 1931); *Eleusis coarctata* Sharp, Japan (Sugiyama 1973).

K. hyalinus Terada, 1978: adult *Eleusis humilis* (Erichson) Taiwan (Terada 1978, Shen and Ye 2006).

K. isomali Thaxter, 1901b: adult *Eleusis conradti* Fauvel, Tanzania (Thaxter 1901b); "*Eleusis schwabi* Bernhauer," invalid name, and *Eleusis* sp. Cameroon (Thaxter 1931); adult *Eleusis kraatzi* Fauvel, Philippines and Sumatra, Indonesia (Thaxter 1931) and Taiwan (Sugiyama 1978a, Shen and Ye 2006); *Eleusis lunigera* Fauvel, Philippines and Sumatra, Indonesia (Thaxter 1931); *Eleusis* sp., Java, Indonesia and Sarawak, Malaysia (Thaxter 1931); "*Maseochara sumatrensis* Bernhauer," invalid name (Thaxter 1931).

Kleidiomyces Thaxter, 1908, reported on Aleocharinae and Oxytelinae, perhaps by error.

K. ambiguus Rossi, 2010: adult *Anotylus* sp., Ecuador (Rossi 2010).

K. furcillatus (Thaxter, 1908): adult *Aleochara repetita* Sharp, Panama (Thaxter 1908, 1912a, 1931).

K. hoplandriae Thaxter 1931: adult "*Hoplandria carinicollis* Bernhauer," invalid name, Trinidad, and "*Hoplandria quadridentata* Bernhauer," invalid name, Grenada (Thaxter 1931).

K. venezolanus Thaxter 1931: adult "*Atheta venezolana* Bernhauer," invalid name, Venezuela (Thaxter 1931).

Kyphomyces Tavares, 1985

K. ansatus (Thaxter, 1931): adult *Lithocharis ochracea* (Gravenhorst) and *Scopaeus apicipennis* Sharp, Trinidad (Thaxter, 1931).

K. appendiculata (Thaxter, 1931): adult *Coproporus sumatrensis* Bernhauer, Sumatra, Indonesia (Thaxter 1931).

K. argentinensis (Spegazzini, 1917): adult *Coproporus* sp. (as "*Coproporus argentinus*," an invalid name) Argentina (Spegazzini 1917, Thaxter 1931, Tavares 1985).

K. bicornis (Thaxter, 1931): adult *Coproporus rutilus* (Erichson) Trinidad (Thaxter 1931).

K. carinatus (Thaxter, 1931): adult *Coproporus* sp., Sumatra, Indonesia (Thaxter 1931).

K. coprorori (Thaxter, 1912a): *Coproporus rutilus* (Erichson) Argentina and Guatemala (Thaxter 1912a, 1931).

K. devexirostris (Spegazzini, 1917): adult *Coproporus* sp. (as "*Coproporus argentinus*") (Spegazzini 1917); *Coproporus rutilus* (Erichson) as [*C. terminalis* (Erichson)] Trinidad (Thaxter 1931).

K. filarius (Thaxter, 1912a): adult *Copropr orus rutilus* (Erichson) Argentina (Thaxter 1912a, 1931).

K. grenadinus (Thaxter, 1931): adult *Coproporus pulchellus* (Erichson) Grenada (Thaxter 1931).

K. philippinus (Thaxter, 1931): adult *Coproporus latus* Motschulsky, Philippines (Thaxter 1931).

K. platensis (Spegazzini, 1917): adult *Coproporus platensis* Bernhauer, Argentina (Spegazzini 1917); *Achenomorphus latro* (Sharp) Venezuela (Thaxter 1931).

K. rhizophorus (Thaxter, 1931): adult *Rugilus ceylanensis* (Kraatz), Sumatra, Indonesia (Thaxter 1931).

K. stilici (Thaxter, 1931): adult *Rugilus ceylanensis* (Kraatz), Sumatra, Indonesia (Thaxter 1931).

K. thinocharinus (Thaxter 1912a): adult *Thinocharis* exilis (Erichson) Argentina (Thaxter 1912a, 1931, Spegazzini 1917).

Laboulbenia Montagne and Robin, 1853

L. achenii Maire, 1920: adult *Achenium tenellum* Erichson, Algeria (Maire 1920).

L. atlantica Thaxter, 1908: adult *Lobrathium multipunctum* (Gravenhorst) Madeira, France, Belgium, Canary Islands (Thaxter 1908, Picard 1913b, Collart 1945, De Kesel and Rammeloo 1991, Arndt and Santamaría 2004) and on *Zargus schaumi* Wollaston (as "*Gargus schaumii* Woll.") (Carabidae), Madeira (Thaxter 1908)

L. atrosepta Majewski, 1989: adult *Rugilus erichsoni* Fauvel, Poland (Majewski 1989).

L. barbara Middelhoek and Boelens, 1943: adult *Philonthus punctus* (Gravenhorst) Netherlands (Middelhoek 1943); *Philonthus fumarius* (Gravenhorst) Italy (Rossi 1975); *Spatulonthus longicornis* (Stephens) South Korea (Lee et al. 2005).

L. bergiana Spegazzini, 1917: adult *Pinophilus bergi* Lynch, Argentina (Spegazzini 1917).

L. cafii Thaxter, 1899: adult *Cafius seminitens* Horn and *C. canescens* Mäklin, CA, USA, and *C. bisulcatus* Solier, Chile (Thaxter 1899); *Cafius sericeus* (Holme) Great Britain, Italy, Spain (Thaxter 1899, 1908, Picard 1913b, Hake 1923, Colla 1926, 1934, Rossi 1978, Santamaría 1998); *Cafius xantholoma* (Gravenhorst) Belgium, France, Canary Islands (Collart 1945, Balazuc 1974, De Kesel and Rammeloo 1991, Arndt and Santamaría 2004); *Phucobius simulator* Sharp, Japan (Sugiyama 1973); *Cafius* sp., Hong Kong, China (Shen and Ye 2006).

L. cristata Thaxter, 1893: adult *Paederus littorarius* Gravenhorst and *P. obliteratus* LeConte, ME, USA (Thaxter 1893); *Paederus australis* Guérin-Méneville, Australia; *Paederus coarctatus* Erichson, Brazil; *Paederus duplex* Eppelsheim, Ethiopia; *Paederus erythoderus* Erichson, Mexico;

Paederus luridiventris Sharp, Panama; *Paederus rutilicornis* Erichson, Colombia (Thaxter 1908); *Paederus fuscipes* Curtis (= *P. longipennis* Erichson), India, Italy, Switzerland, France, Hungary, Japan, South Korea, Taiwan and Hunan, China, and Malaya, Malaysia, and Greece (Thaxter 1908, Picard 1913b, Spegazzini 1914, Baumgartner 1923, 1927, Colla 1925, 1926, 1934, Siemaszko and Siemaszko 1932, Bánhegyi 1940, Kurosa 1958, Stadelmann and Poelt 1962, Sugiyama 1973, Balazuc 1974, Sugiyama and Shazawa 1977, Sugiyama and Majewski 1985b, Lee 1986, Majewski 1988a, Castaldo et al. 2004, Shen and Ye 2006); *Paederus bruchi* Bernhauer, Argentina (Spegazzini 1912, 1917); *Paederus littoralis* Gravenhorst, Italy, Belgium, France, Hungary, former Yugoslavia, Switzerland, USSR (Picard 1913b, Spegazzini 1914, 1915a, Siemaszko and Siemaszko 1932, Colla 1926, 1934, Baumgartner 1934, Bánhegyi 1940, Lepesme 1941, Collart 1945, Stadelmann and Poelt 1962, Balazuc 1974, De Kesel and Rammeloo 1991, De Kesel and Haghebaert 1991); *Paederus riparius* (L.) Italy, France, Netherlands, Belgium, Hungary, Austria, Czech Republic, Poland (Picard 1913b, Spegazzini 1914, Siemaszko and Siemaszko 1932, Colla 1934, Kossen 1936, Bánhegyi 1940, Middelhoek 1943b, Collart 1945, Stadelmann and Poelt 1962, Balazuc 1974, De Kesel and Rammeloo 1991, Majewski 1994a, Rossi and Máca 2006); *Paederus madagascariensis* Erichson, Madagascar (Spegazzini 1915a, Balazuc 1982); *Paederus lusitanicus* Aubé, Portugal (Siemaszko and Siemaszko 1932); *Paederus limnophilus* Erichson, Poland (Siemaszko and Siemaszko 1932); *Paederus brevipennis* Lacordaire, Hungary, Germany (Banhegyi 1940, Poelt 1952a); *Paederus parallelus* Weise, Japan and South Korea (Sugiyama 1973, Lee 1986, Lee et al. 2002); *Paederus alternans* Walker, Malaya, Malaysia (Sugiyama and Majewski 1985b); *Paederus tamulus* Erichson, Bangladesh (Lepesme 1941) and Hainan Island, China (Shen and Ye 2006); *Paederidus rubrothoracicus* (Goeze (= *P. longicornis* Aubé) Italy, Poland, France, former Yugoslavia, Greece (Spegazzini 1915, Siemaszko and Siemaszko 1932, Colla 1934, Stadelmann and Poelt 1962, Balazuc 1974, Castaldo et al., 2004); *Paederidus ruficollis* (F.) (= *P. gemellus* Kraatz) Austria, France, Italy, Switzerland, Poland, Hungary, Germany (Thaxter 1893, 1896, Picard 1913b, Spegazzini 1914, Picard 1917, Baumgartner 1923, Siemaszko and Siemaszko 1932, Colla 1934, Bánhegyi 1940, Stadelmann and Poelt 1962, Scheloske 1969, Balazuc 1974); *Megalopaederus poweri* (Sharp) Japan (Sugiyama 1973); *Paederus* sp., Spain (Santamaria 1992a).

L. dolicaontis Maire, 1920: adult *Leptobium densiventre* (Fauvel) Algeria (Maire 1920); *Leptobium gracile* (Gravenhorst), Spain (Santamaría 1993).

L. dubia Thaxter, 1902: adult *Philonthus politus* (L.) England (Thaxter 1902), Poland and Germany (Siemaszko and Siemaszko 1932, Scheloske 1969); *Philonthus fuscipennis* (Mannerheim) France, Belgium, Poland, Latvia (Picard 1917, Siemaszko and Siemaszko 1928, Briedis 1932, Collart 1945); *Philonthus mannerheimi* Fauvel, Poland, Germany (Siemaszko and Siemaszko 1928, Scheloske 1969); *Philonthus cognatus* (Stephens) Belgium (De Kesel and Rammeloo 1991); *Philonthus* spp. Turkey and Spain (Santamaría 1993, 1996a).

= *L. philonthicola* Spegazzini, 1915a: adult *Philonthus fuscipennis* (Mannerheim) France (Lepesme 1941); *Philonthus decorus* (Gravenhorst) Germany (Scheloske 1969); *Philonthus politus* (L.) France (Balazuc 1974).

L. ecitonis Blum, 1924: adult *Ecitophya gracillima* Mann, and the host ant *Eciton hamatum* F. of this inquiline, Ecuador (Rossi 1991).

L. geodromici Baumgartner, 1923: adult *Geodromicus plagiatus* (F.), Switzerland (Baumgartner 1923).

L. gracilis Spegazzini, 1915b: adult *Sunius* sp., Italy (Spegazzini 1915b).

L. gregaria Rossi, 2011: adult *Philonthus* sp., Ecuador (Rossi 2011).

L. gridellii Colla, 1926: adult *Platystethus spinosus* Erichson, Italy (Colla 1926).

L. jarrigei Balazuc, 1975): adult *Lispinus propinquus* Cameron, Reunion (Balazuc 1975, 1982).

L. kenyensis Rossi and Santamaría, 2000: adult *Mimogonus rossii* Bordoni, Kenya (Rossi and Santamaría 2000).

L. lathropini Thaxter, 1912a: adult *Lathropinus fulvipes* Erichson, Argentina (Thaxter 1912a). = *L. oedodactyli* Spegazzini, 1912: adult "*Latrobium*," this misspelling corrected and name of parasite synonymized by Spegazzini (1917). *L. lathropinicola* Spegazzini, 1917: adult *Lathropinus major* Blanchard, Argentina (Spegazzini 1917).

L. latonae Thaxter, 1902: adult *Pseudocryptobium spinolae* (Guérin-Méneville) Colombia (Thaxter 1902, 1908); *Pseudocryptobium* sp., Ecuador (Proaño Castro and Rossi 2008).

L. littoralis De Kesel and Haelewaters, 2014: adult *Cafius xantholoma* (Gravenhorst) Belgium, Netherlands, France, Italy (De Kesel and Haelewaters 2014). [Until 2014, this species was frequently misidentified as *L. slackensis* Cépède and Picard, 1908a, a parasite of Carabidae].

L. micrandra Rossi, 2011: adult *Lobrathium* sp., Ecuador (Rossi 2011).

L. moiwae Terada, 1980: adult *Domene crassicornis* Sharp, Japan (Terada 1980).

L. oedichiri Thaxter, 1902: adult *Oedichirus* sp., Brazil (Thaxter 1903, 1908).

L. oedodactyli Thaxter, 1899: adult *Oedodactylus fuscobrunneus* Fairmaire and Germain, Chile, Argentina (Thaxter 1899, 1908, Spegazzini 1912).

L. outambensis Rossi, 1986: adult *Lobrathium* sp., Sierra Leone (Rossi 1986).

L. parriaudi Balazuc, 1974: adult *Bledius fergussoni* Joy (= *B. arenarius* (Paykull) France (Balazuc 1974); *Bledius* spp., Turkey and Spain (Santamaría 1989, 1996a).

L. pedicellata Thaxter, 1892: adult *Bembidion* sp. (Carabidae) ME, USA, but reported from *Aleochara tenuicornis* Kraatz, Algeria (Maire 1920).

L. philonthi Thaxter, 1893: adult *Philonthus aequalis* Horn, "Lake Superior," USA, and *P. cunctans* Horn and *P. debilis* (Gravenhorst) "New England," USA (Thaxter 1893); *Philonthus micans* (Gravenhorst) "New England," USA, Poland, France, Italy (Thaxter 1893, 1896); *Philonthus furvus* Nordmann, Mexico and California, USA (Thaxter 1896); *Philonthus accedens* Sharp and *Philonthus occultus* Sharp, Guatemala, and *Philonthus incertus* Solsky and *Philonthus ochromerus* Sharp, Mexico (Thaxter 1908); *Philonthus convexicollis* Lynch, and *Philonthus hepaticus* Erichson, and *Philonthus parvimanus* Sharp) Argentina (Spegazzini 1912); *Philonthus quadraticeps* Boheman, Argentina (Spegazzini 1917); *Philonthus fulvipes* (F.) Germany, Poland, Turkey, Italy (Scheloske 1969, Majewski 1973a, Balazuc 1974, Rossi 1975); *Philonthus punctus* (Gravenhorst) France (Balazuc 1974); *Philonthus quisquiliarius* (Gyllenhal) (= *P. dimidiatus* Boisduval and Lacordaire), Germany, France (Scheloske 1969, Balazuc 1974); *Philonthus concinnus* (Gravenhorst) and *Philonthus nigritus* (Gravenhorst) and *Philonthus oblitus* Jarrige, Italy (Rossi 1975); *Philonthus rubripennis* (Stephens) and *Philonthus rufimanus* Erichson, Greece (Castaldo et al. 2004); *Philonthus wuesthoffi* Bernhauer, South Korea (Lee and Na 1998, Lee et al. 2002, 2011b), *Philonthus micans* (Gravenhorst) Netherlands (Haelewaters et al. 2014); *Philonthus fumarius* (Gravenhorst) Czech Republic (Rossi and Máca 2006); *Philonthus carbonarius* (Gravenhorst) and *P. lomatus* Erichson, MA, USA and *P. aurulentus* Horn, Quebec, Canada, and *Oligotergus fasciatus* (Nordmann) Venezuela (Haelewaters et al. 2015), *Philonthus*,

Spatulonthus, *Paragabrius*, Turkey and Spain (Santamaría et al. 1991, Santamaría 1996a).

L. platyprosopi Thaxter, 1902: adult *Platyprosopus beduinus* Nordmann, Sudan (Thaxter 1902, 1908).

L. quedii Thaxter, 1893: adult *Anaquedius vernix* (LeConte) as "*Quedius vernilis* LeC," IL, USA (Thaxter 1893, 1896).

L. richardii Rossi and Santamaría, 2000: adult *Phlaeopterus castaneus* Casey, OR, USA and *Phlaeopterus* sp., ID, USA and *Vellica longipennis* Casey, CA, USA, and *Unamis* sp., CA and UT, USA (Rossi and Santamaría 2000).

L. stenolophi Spegazzini, 1914: adult *Spatulonthus longicornis* (Stephens) South Korea (Lee and Na 1998, Lee et al. 2002).

L. subterranea Thaxter, 1896: adult *Pseudanophthalmus* sp. (Carabidae) (Thaxter 1896); *Rugilus geniculatus* (Erichson), *Rugilus similis* (Erichson) and *R. orbiculatus* (Paykull), Europe (Thaxter 1908); *Rugilus similis* (Erichson) Finland (Huldén 1983); *Rugilus rufipes* Germar, Netherlands (Middelhoek 1943b); *Rugilus orbiculatus* (Paykull) Belgium (De Kesel and Haghebaert 1991).

= *? L. stilicicola* Spegazzini, 1915: adult *Rugilus fragilis* (Gravenhorst) [as *R. angustatus* (Fourcroy)] Italy (Spegazzini 1915), synonymized by Colla (1934); *Rugilus rufipes* Germar, Greece (Castaldo et al. 2004).

L. taenodemae Thaxter, 1899: adult *Taenodema cinerea* Sharp, Brazil (Thaxter 1899, 1908); *Taenodema* sp., Ecuador (Bernardi et al. 2014).

L. trogacti Rossi, 2011: adult *Trogactus* sp., Ecuador (Rossi 2011).

L. vulgaris Peyritsch, 1873: adult *Bembidion* spp. (Carabidae) and *Deleaster dichrous* (Gravenhorst) (Peyritsch 1873); *Lesteva villardi* Rey, France, some doubt of fungal identity (Balazuc 1974).

NOTE: At least three species, *L. pedicicillata*, *L. subterranea*, and *L. vulgaris* seem to have Carabidae as primary hosts, with the issue divided for *L. atlantica*.

Meionomyces Thaxter, 1931

M. astenalis Thaxter 1931, adult *Astenus pulchellus* Kraatz, Sumatra, Indonesia (Thaxter 1931).

M. asteni Thaxter, 1931: adult *Astenus cognatus* Sharp, Guatemala (Thaxter 1931).

M. astenicola Thaxter, 1931: adult *Astenus pulchellus* Kraatz, Sumatra, Indonesia (Thaxter 1931).

M. thaxterelli Thaxter, 1931: adult *Octavius sulcicollis* (Bernhauer), Cameroon (Thaxter 1931).

NOTE: *M. dibelonetis* Thaxter, 1931 was transferred to *Phaulomyces* by Tavares (1985).

Mimeomyces Thaxter, 1912

M. andinus (Spegazzini 1917): adult *Cheilocolpus impressifrons* (Solier), Argentina, Chile (Spegazzini 1917, Thaxter 1918, 1931).

M. atropurpureus (Thaxter, 1900): adult *Quedius basiventris* Sharp and *Q. graciliventris* Sharp, Panama (Thaxter 1900, 1931).

M. brachydiri (Thaxter, 1900): adult *Nordus antennatus* (Sharp), Peru (Thaxter 1900, 1908, 1931).

M. chiriquensis (Thaxter, 1901b): adult *Quedius flavicaudus* Sharp, Panama (Thaxter 1901b, 1912a, 1931).

M. decipiens Thaxter, 1912a: adult "*Quedius sorecocephalus* Bernhauer," invalid name, Argentina (Thaxter 1912a, 1931).

M. deplanatus Tavares, 1985: adult "*Quedius sorecocephalus* Bernhauer," invalid name, Argentina (Tavares 1985).

M. formicetorum (Spegazzini, 1917): adult *Heterothops formicetorum* Bernhauer, Argentina (Spegazzini 1917, Thaxter 1931, Tavares 1985).

M. gregarius Rossi, 2010: adult *Philonthus* sp., Ecuador (Rossi 2010).

M. latonae (Thaxter, 1901b): adult *Pseudocryptobium spinolae* (Guérin-Méneville), Colombia (Thaxter 1901b, 1908, 1931).

M. macropus (Thaxter, 1912a): adult *Heterothops thaxteri* Bernhauer, Argentina (Thaxter 1912a, 1931).

M. quedionuchi (Thaxter, 1901b): adult *Quedius impunctus* Solsky, Mexico and "*Quedius sorecocephalus* Bernhauer" invalid name, Argentina (Thaxter 1901b, 1912a, 1931).

M. trogacttii Rossi, 2010: adult *Trogactus* sp., Ecuador (Rossi 2010).

M. valdivianus (Thaxter 1918): adult *Cheilocolpus impressifrons* (Solier), Chile (Thaxter 1918, 1931).

M. zeelandicus Middelhoek and Boelens, 1943b: *Heterothops binotatus* (Gravenhorst), Netherlands and *H. quadripunctulus* (Gravenhorst) Poland, Spain (Middelhoek and Boelens 1943b, Majewski 1989, Santamaría 1995a).

Misgomyces Thaxter, 1900

M. dyschirii Thaxter, 1931: on adults of the carabid genus *Dyschirius*, which often is associated with *Bledius* as a predator within its tunnels; *Bledius graellsi* Fauvel, Spain (Santamaría 1995b).

= *M. lavagnei* Picard 1913b: adult *Bledius spectabilis* Kraatz, France, Italy (Picard 1913b, Maire 1916a, Scheloske 1969, Balazuc 1973b, Rossi 1975, Tavares 1985).

NOTE: Tavares (1985) transferred 14 species, all of which have staphylinid hosts, to other genera.

Monoicomyces Thaxter, 1900

M. aleocharae Thaxter, 1901b: adult *Aleochara bohemani* Bernhauer and Scheerpeltz (= *A. rufipes* Boheman) Tanzania (Thaxter 1901b, 1908).

M. amauroderae Thaxter, 1915: adult *Amaurodera kraepelini* Fauvel, Java, Indonesia (Thaxter 1915, 1931).

M. asymmetricus Thaxter, 1931: adult *Atheta opaca* Fauvel (surely misidentified because host range is East Africa) and *Atheta platygaster* Kraatz and "*Homalota pseudocribrum* Bernhauer" (invalid name), all in Sumatra, Indonesia (Thaxter 1931).

M. athetae Thaxter, 1900: adult *Aloconota insecta* (Thomson) England, Poland (Thaxter 1900, Siemaszko and Siemaszko 1932); *Atheta cinnamoptera* (Thomson) England (Thaxter 1931); *Atheta tibialis* (Heer) Czech Republic (Rossi and Máca 2006); *Atheta aterrima* (Gravenhorst) and *Atheta gagatina* (Baudi) Poland (Majewski 1994a); *Mocyta fungi* (Gravenhorst) Poland (Majewski 1984); *Evanystes circellaris* (Gravenhorst) Germany (Scheloske 1969).

M. barberi Thaxter, 1931: adult *Oxytelus nimius* Casey, MD, USA (Thaxter 1931).

M. benjaminii Santamaría, 1996b: adult *Atheta* sp., OR, USA (Santamaría 1996b).

M. bolitocharae Majewski, 1994: adult *Bolitochara obliqua* (Erichson) Poland (Majewski 1994b).

M. britannicus Thaxter, 1900: adult *Aloconota insecta* (Thomson) England (Thaxter 1900); "*Homalota* sp.," France, Germany (Picard 1917, Lepesme 1941, Scheloske 1969); *Atheta longicornis* (Gravenhorst) Finland and western Russia (Huldén 1983); *Atheta fungi* (Gravenhorst) Belgium (De Kesel and Haghebaert 1991); *Atheta* sp., Spain (Santamaría 1992a); *Acrotona pseudotenera* (Cameron) Netherlands (Haelewaters et al. 2014); *Drusilla canaliculata* (F.) Greece (Castaldo et al. 2004).

M. californicus (Thaxter, 1901b): adult *Oxytelus* sp., CA, USA (Thaxter 1901b, 1931); *Anotylus sculpturatus* (Gravenhorst) Belgium (De Kesel and Haghebaert 1991). According to Thaxter (1931) a record from *Anotylus inustus* (Gravenhorst) in Algeria by Maire (1920) is of some other fungal species, not *M. californicus*.

M. caloderae Thaxter, 1912b: adult *Calodera* spp., and *Atheta* sp., Argentina, and "*Atheta chilensis* Bernhauer," invalid name, Chile (Thaxter 1912b, 1931).

= *M. ocaleae* Spegazzini, 1917: adult *Ocalea funebris* Lynch, Argentina (Spegazzini 1917, Thaxter 1931).

M. ceylonensis Santamaría, 1996b: adult *Atheta inornata* Kraatz, Sri Lanka (Santamaría 1996b).

M. chosunensis Lee, 1986: adult *Atheta* sp., South Korea (Lee 1986)

M. denticulatus Thaxter, 1915: adult *Stenomastax nigrescens* (Fauvel), Java, Indonesia, and *Atheta platygaster* Kraatz, Sumatra, Indonesia (Thaxter 1915, 1931); *Homalota* sp., Bali, Indonesia (Sugiyama and Majewski 1985a).

M. diestotae Thaxter, 1931: adult *Diestota testacea* (Kraatz), Sumatra, Indonesia (Thaxter 1931).

M. doryloniae Thaxter, 1931: adult *Ocyplanus amaneensis* (Eichelbaum) Cameroon (Thaxter 1931).

M. dorylonillae Thaxter, 1931: adult perhaps of *Dorylonilla spinipennis* Wasmann, Cameroon (Thaxter 1931).

M. echidnoglossae Thaxter 1901b: adult *Blepharymenus* sp. (as "*Echidnoglossa americana* Fauvel," invalid name) Colorado, USA (Thaxter 1901b, 1908).

M. eleusinus Thaxter, 1931: adult "*Eleusis reynoldsii* Bernhauer," invalid name, Venezuela (Thaxter 1931).

M. falagriae Thaxter, 1931: adult *Falagria coarticollis* Fauvel, Cameroon, and *Falagria* spp. indet., Jamaica, and OH, USA (Thaxter 1931).

M. focarilei Rossi, 1981: adult *Leptusa piceata* Mulsant and Rey, Switzerland, and *L. montisgrappae* Pace, Italy (Rossi 1981).

M. fragilis Scheloske, 1969: adult *Ocalea picata* (Stephens) Germany (Scheloske 1969), France (Santamaría 1996b), Belgium (De Kesel and Haghebaert 1991); *Ocalea concolor* Kiesenwetter, Spain (Santamaría 1996b); *Oxypoda opaca* (Gravenhorst) Poland (Majewski 1994a).

M. furcatus (Thaxter, 1931): adult *Anotylus insignitus* (Gravenhorst) Jamaica, Haiti, Guatemala (Thaxter 1931); *Oxytelus laqueatus* (Marsham) Finland and western Russia (Huldén 1983).

M. gibbosus Thaxter, 1931: adult *Stenomastax cribrum* (Fauvel) Sumatra, Indonesia (Thaxter 1931).

M. gnypetae Thaxter, 1931: adult *Gnypeta modesta* Bernhauer, Sumatra, Indonesia (Thaxter 1931, Balazuc 1971b).

M. homalotae Thaxter, 1900: adult *Atheta dilutipennis* (Motschulsky) (as *Homalota putrescens* Wollaston, a synonym), Azores (Thaxter 1900, 1908, 1931, Siemaszko and Siemaszko 1932, Petch 1944, Hincks 1960, Balazuc 1974); *Atheta* sp., Argentina (Thaxter 1912a); *Atheta lurida* (Erichson) Argentina (Spegazzini 1917); *Atheta amicula* (Stephens) and *A. parens* Mulsant and Rey, and *A. pertyi* (Heer), Algeria (Maire 1920); *Atheta gagatina* (Baudi) and *A. triangulum* (Kraatz) Netherlands (Middelhoek 1943a, b); *Atheta xanthopus* (Thomson) Netherlands (Middelhoek 1947); *Atheta nigrifrons* (Erichson) Germany (Scheloske 1969); *Atheta transfuga* Sharp, Japan (Sugiyama 1978b); *Atheta paracrassicornis* Brundin, Finland (Huldén 1983); *Atheta lewisiana* Cameron and *Atheta reitteriana* Bernhauer, Japan (Majewski 1988a); *Atheta luridipennis* (Mannerheim) Czech Republic (Rossi et al. 2010); *Atheta cinnamoptera* (Thomson) and *Aloconota mihoki* (Bernhauer) Slovakia (Rossi et al. 2010); *Atheta aeneipennis* (Thomson) and *Atheta aterrima* (Gravenhorst) and *Tachyusa coarctata* (Erichson) Poland (Majewski 1994a); *Evanystes circellaris* (Gravenhorst) Germany (Scheloske 1969); "*Homalota* sp." and *Carpelimus* sp. ME, NH, USA (Thaxter 1908); "*Homalota* sp.," Italy (Colla 1934); *Acrotona* sp., Congo-Kinshasa (Rossi and Santamaría 1992); *Drusilla canaliculata* (F.) England (Weir and Beakes 1993); *Geostiba oertzeni* (Eppelsheim); and *Atheta* sp., Greece (Castaldo et al. 2004).

= *M. ternatus* Spegazzini, 1915b: adult ?genus, ? subfamily (Spegazzini 1915b, Thaxter 1931, Colla 1934, Stadelmann and Poelt 1962). Synonymy due to Santamaría (1994).

M. infuscatus Spegazzini, 1912: adult *Xantholinus subtilis* Boheman, Uruguay, and *X. bonariensis* Gemminger and Harold (= *X. gracilis* Boheman) and *X. andinus* Fauvel, Argentina (Spegazzini 1912, 1917, Thaxter 1912a, 1931); *Xantholinus longiventris* Heer, Poland (Majewski 1989); *Xantholinus* sp., Spain (Santamaría 1996b).

M. invisibilis Thaxter, 1900: adult *Homalota putrescens* Wollaston, a synonym of *Atheta dilutipennis* (Motschulsky), Azores (Thaxter 1900), but this host association was in the same publication attributed to *Monoicomyces homalotae* (Thaxter 1900) presumably by confusion; this confusion was resolved by Thaxter (1931) in stating that the host of *M. invisibilis* in reality is *Oxytelus* sp.; *Anotylus hostilis* (Bernhauer) Sumatra, Indonesia, and *Anotylus sulcifer* (Fauvel) Venezuela (Thaxter 1931); *Oxytelus laqueatus* (Marsham) Poland (Siemaszko and Siemaszko 1932); *Anotylus insecatus* (Gravenhorst) and *Oxytelus piceus* (L.) Poland (Majewski 1994a); *Platystethus arenarius* (Fourcroy) Finland, Belgium (Huldén 1983, De Kesel and Haghebaert 1991); *Oxytelus* sp., Hainan Island, China (Shen and Ye 2006).

= *M. affinis* Spegazzini, 1915b: adult ?genus, ? subfamily, Italy (Spegazzini 1915b, Thaxter 1931).

= *M. argentinensis* (Spegazzini, 1912): adult *Platystethus fallax* Lynch, Argentina (Spegazzini 1912, 1917, Thaxter 1931).

M. javanus Thaxter, 1931: adult *Oxytelus* sp., Java, Indonesia (Thaxter 1931).

M. labiatus Majewski, 1984: adult *Mocyta fungi* (Gravenhorst) and *Gnypeta rubrior* Tottenham and *Thinonoma atra* (Gravenhorst) Poland (Majewski 1984); *Acrotona parvula* (Mannerheim) Poland (Majewski 1994a); *Atheta* sp. Spain and OR, USA (Santamaría (1996b).

M. kamerunensis Thaxter, 1931: adult ?genus nr. *Oxytelus*, Cameroon (Thaxter 1931).

M. leptochiri Thaxter, 1905: adult *Borolinus javanicus* (Laporte) and *B. minutus* (Laporte) Java, Indonesia (Thaxter 1905, 1908); *Borolinus bicornis* Nakane and Sawada, Japan (Terada 1977, Sugiyama 1978a); *Leptochirus* sp., Java and Sumatra, Indonesia (Thaxter 1905, 1908); *Priochirus unicolor* (Laporte) Java, Indonesia (Thaxter 1905, 1908); *Priochirus* sp., Taiwan (Sugiyama 1972, 1978a, Shen and Ye 2006); and *Priochirus silvestris* Bernhauer, Taiwan (Terada 1981, Shen and Ye 2006).

M. leptotracheli Thaxter, 1931: adult "*Leptotrachela javana* Bernhauer," invalid name, Java, Indonesia (Thaxter 1931).

M. luxurians Rossi and Santamaría, 1992: adult *Drusilla* sp., Congo-Kinshasa (formerly Zaire) (Rossi and Santamaría 1992).

M. matthiatis (Majewski, 1989): adult *Platystethus arenarius* (Fourcroy) Poland (Majewski 1989).

M. myllaenae Santamaría, 2006: adult *Myllaena elongata* (Matthews) Spain (Santamaría 2006), Netherlands (Haelewaters et al. 2014).

M. nigrescens Thaxter, 1902: adult *Calodera* sp., ME, NH, USA (Thaxter 1902, 1908, Scheloske 1969, Rossi 1978); *Tachyusa* sp., ME, NH, USA (Thaxter 1902, 1908, Scheloske 1969, Rossi 1978); *Meronera sharpi* Lynch, Argentina (Thaxter 1912a, Spegazzini 1917, Thaxter 1931, Scheloske 1969, Rossi 1978); *Ocalea funebris* Lynch, Argentina (Spegazzini 1912, Scheloske 1969); *Atheta inornata* Kraatz, Sri Lanka (Thaxter 1931, Scheloske 1969); *Falagria splendens* Kraatz, Italy (Rossi 1978); *Falagria nigra* (Gravenhorst) Poland (Majewski 1994a); *Atheta grata* Cameron and *Taxicera* sp., Japan (Majewski 1988a); *Atheta nigra* (Kraatz) England, Poland (Weir and Beakes 1993, Majewski 1994a); *Atheta intermedia* (Thomson) Czech Republic (Rossi et al. 2010); unidentified Aleocharinae, Poland, Spain (Majewski 1984, Santamaría 1992a); *Atheta pertyi* (Heer) and

A. *sordidula* (Erichson) Spain (Santamaría 1996b).

M. osorianus Thaxter, 1931: adult *Neosorius opaculus* (Eppelsheim) Cameroon (Thaxter 1931); *Neosorius arebianus* Cameron, Congo-Kinshasa (formerly Zaire) (Rossi and Santamaría 1992).

M. oxypodae Thaxter, 1902: adult *Oxypoda* sp., NH, USA (Thaxter 1902, 1908, 1931); *Atheta orbata* (Erichson) Balearic Is., Spain, and *Atheta nigritula* (Gravenhorst) Spain (Santamaría 1996b). *Brachygluta fossulata* (Reichenbach) Finland (Huldén 1983).

M. oxytelis Huldén, 1983: adult *Oxtelus fulvipes* Erichson, Finland (Huldén 1983).

M. plagiusae Thaxter, 1931: adult *Neosilusa* sp., as "*Plagiusa gracilicornis* Bernhauer" invalid genus, Sumatra, Indonesia (Thaxter 1931); *Plagiusa* sp. [invalid genus in Staphylinidae, perhaps *Neosilusa* sp.] Japan (Majewski 1988a).

M. ramosus Thaxter, 1931: adult *Falagria latemarginata* Bernhauer, Sumatra, Indonesia (Thaxter 1931).

M. sanctaehelenae Thaxter, 1900: adult *Oxytelus alutaceifrons* Wollaston, St. Helena (Thaxter 1900, 1908); *O. laqueatus* (Marsham) (= *O. luteipennis* Erichson) Germany (Picard 1913b, Maire 1916a, Siemaszko and Siemaszko 1932); *Oxytelus piceus* (L.) Algeria (Picard 1913b, Maire 1916a); *Oxytelus piceus* (L.) Finland (Huldén 1983); *Oxytelus ferrugineus* Kraatz, Sarawak, Malaysia, and "*Oxytelus semisulcatus* Bernhauer," invalid name, Cameroon (Thaxter 1931); *Oxytelus lucens* Bernhauer, Taiwan (Sugiyama 1981, Shen and Ye 2006); *Oxytelus javanus* Cameron, Bali, Indonesia (Sugiyama and Majewski 1985a); *Oxytelus* sp., Sierra Leone (Rossi 1994).

= *M. roccae* Colla, 1925: *Platystethus* sp. (as *Platistetus*) Italy (Colla 1925, Thaxter 1931).

M. similis Thaxter, 1905: adult ?*Homalota* sp., ME, USA (Thaxter 1905, 1908).

M. singularis Thaxter, 1931: adult *Oxytelus grandis* Eppelsheim, Cameroon (Thaxter 1931).

M. stenusae Thaxter, 1915: adult *Neosilusa ceylonica* (Kraatz) Java, Indonesia (Thaxter 1915, 1931).

M. trogacti Rossi, 2010: adult *Trogactus* sp., Ecuador (Rossi 2010).

M. unilateralis Spegazzini, 1915b: adult ?genus, ?subfamily, Italy (Spegazzini 1915b, Thaxter 1931, Stadelmann and Poelt 1962).

M. venetus Spegazzini, 1915: adult ?*Atheta*, Italy (Spegazzini 1915, Thaxter 1931).

M. yamamotoi Majewki and Sugiyama, 1985: adult *Thamiaraea diffinis* Sharp, Japan (Majewki and Sugiyama 1985b).

M. zealandicus Thaxter, 1918: adult *Atheta fungi* (Gravenhorst) New Zealand; *A. obtusidens* Eppelsheim, South Africa; *A. sumatrensis* Bernhauer, Sumatra, Indonesia, *Atheta* sp., Chile (Thaxter 1918, 1931); *Atheta pasadenae* Bernhauer (given as *A. pseudolaticollis* Erber and Hinterseher) Spain and CA, USA (Santamaría 1996b).

Neohaplomyces Benjamin, 1955

N. cubensis Benjamin, 1955: adult *Medon schwarzi* Blackwelder, Cuba (Benjamin 1955).

N. medonalis Benjamin, 1955: adult *Medon* sp., AZ, CA, USA (Benjamin 1955); *Medon brunneus* (Erichson) Turkey (Santamaría and Rossi 1999); *Medon* sp., Spain, Greece (Santamaría and Girbal 1987, Santamaría and Rossi 1999); *Thinocharis* sp., Madagascar (Santamaría and Rossi 1999).

N. neomedonalis Benjamin, 1955: adult *Medon schwarzi* Blackwelder, Cuba (Benjamin 1955);

Osoriomyces Terada, 1981

O. rhizophorus Terada, 1981: adult *Osorius formosae* Bernhauer, Taiwan (Terada 1981, Shen and Ye 2006).

Peyerimhoffiella Maire, 1916a

P. elegans Maire 1916a: adult *Brachygluta* spp., Algeria (Maire 1916a); *Brachygluta reichei* Motschulsky, *B. dichroa* (Saulcy), *B. lefebvrei* (Aubé), *B. fossulata* (Reichenbach), *B. xanthoptera* (Reichenbach), *Rybaxis longicornis* (Leach), *Tychus niger* (Paykull), *Bryaxis bulbifer* (Reichenbach), and *B. puncticollis* (Denny) Germany (Scheloske 1969); *Brachygluta fossulata* (Reichenbach) Finland (Huldén 1983) and Netherlands (Haelewaters et al. 2014); *Brachygluta xanthoptera* (Reichenbach) Belgium (De Kesel and Rammeloo 1991); *Brachygluta tibialis* (Aubé) Greece and Italy (Castaldo et al. 2004); *Brachygluta pandellei* (Saulcy), *Faronus besucheti* Castellini, *Rybaxis laminata* (Motchulsky), *Rybaxis longicornis* (Leach), and *Trissemus antennatus* (Aubé) Italy (Castaldo et al. 2004).

= R. *brachyglutae* (Siemaszko and Siemaszko, 1926): adult *Brachygluta xanthoptera* (Reichenbach) Poland (Siemaszko and Siemaszko 1926, Scheloske 1969, Tavares 1985); *Brachygluta fossulata* (Reichenbach) Poland (Majewski 1994a,b).

Peyritschiella Thaxter, 1890

= *Dichomyces* Thaxter, 1893

= *Rheophila* Cépède and Picard, 1908

P. amazonica Thaxter, 1900: adult of unidentified staphylinid, ?Brazil (Thaxter 1900, 1908).

P. angolensis (Thaxter, 1900): adult *Philonthus* sp., Angola (Thaxter 1900).

P. anisopleura (Spegazzini, 1915b): adult *Philonthus hepaticus* Erichson, Argentina (Spegazzini 1915b, 1917).

P. argentinensis (Spegazzini, 1912): adult *Xantholinus subtilis* Boheman, Uruguay and Argentina (Spegazzini 1912, Thaxter 1931).

P. arimensis (Thaxter, 1931): adult *Belonuchus cognatus* Sharp, Trinidad (Thaxter 1931).

P. australiensis (Thaxter, 1901b): adult *Quedius ruficollis* (Gravenhorst) Australia (Thaxter 1901b, 1908).

P. belonuchi (Thaxter, 1901b): adult *Belonuchus fuscipes* Fauvel, New Guinea (Thaxter, 1901b, 1908)

P. bicolor (Thaxter, 1931): adult *Philonthus borneensis* Bernhauer, Sarawak, Malaysia and Philippines (Thaxter 1931).

P. bifida (Thaxter, 1901b): adult *Philonthus* sp., New Britain (Thaxter 1901b); *Philonthus micanticollis* Sharp, Japan (Sugiyama 1978b); *Bisnius* sp., Jiangsu Province, China (Shen and Ye 2006).

P. biformis (Thaxter, 1900): adult *Philonthus* sp., NY, USA, and *Philonthus umbratilis* (Gravenhorst) England and Scotland and Madeira and St. Pierre and Miquelon, Poland, Finland (Thaxter 1900, 1908, Hake 1923, Siemaszko and Siemaszko 1932, Stadelmann and Poelt 1962, Huldén 1983); *Philonthus micanticollis* Sharp, Japan and South Korea (Sugiyama 1973, Lee and Lee 1981, Lee et al. 2002), *Philonthus peliomerus* Kraatz, South Korea (Lee 1986).

P. cafiana (Thaxter, 1900): adult *Cafius puncticeps* White, South Africa (Thaxter 1900, 1908).

P. chilensis (Thaxter, 1918): adult *Cheilocolpus impressifrons* (Solier) Chile (Thaxter 1918, 1931).

P. dubia (Thaxter, 1900): adult *Philonthus* sp., NY, USA (Thaxter 1908); corrected to *Philonthus politus* (L.) (as *Philonthus aeneus* Rossi), NY and MA, USA and later Belgium and Germany (Thaxter 1900, 1908, Collart 1945, Scheloske 1969, De Kesel and Rammeloo 1991).

P. eulissi (Thaxter, 1931): adult *Eulissus chloropterus* Erichson, Australia (Spegazzini 1915b, Scheloske 1960); *Xantholinus illucens* Erichson, as "*Eulyssus illucens*," Trinidad and Grenada (Thaxter 1931).

P. exilis (Thaxter, 1900): adult *Bisnius fimetarius* (Gravenhorst) (as *Philonthus xanthomerus* Kraatz) Mexico (Thaxter 1900); *Belonuchus rufipennis* (F.) and *Philonthus oxyporinus* Sharp (Thaxter 1908).

P. fumosa (Thaxter, 1931): adult *Hesperus* sp., Sumatra, Indonesia (Thaxter 1931).

P. furcifera (Thaxter, 1893): adult *Philonthus debilis* (Gravenhorst) MA, USA (Thaxter 1893, Saccardo 1895, Benjamin 1971); *Philonthus centralis* Sharp, Mexico (Thaxter 1908); *Philonthus cinctipennis* Fauvel origin unstated (Thaxter 1908, Scheloske 1969); *Philonthus discoideus* (Gravenhorst) MA, USA, Scotland, Finland, Canary Islands, Algeria (Thaxter 1908, Picard 1913b, Maire 1916a, Thaxter 1931, Scheloske 1969, Huldén 1983 Arndt and Santamaría 2004); *Philonthus rectangulus* Sharp, Germany, Netherlands, Japan, South Korea, Poland (Thaxter 1908, Middelhoek 1943b, Stadelmann and Poelt 1962, Scheloske 1969, Sugiyama 1973, Lee 1986, Majewski 1989); *Philonthus oreophilus* Fauvel, Australia (Spegazzini 1915a, Scheloske 1969); *Philonthus convexicollis* Lynch and *Spatulonthus longicornis* (Stephens), Argentina (Spegazzini 1917); *Philonthus hepaticus* Erichson, Argentina and Uruguay (Spegazzini 1912, 1917); *Philonthus quisquiliarius* (Gyllenhal) Poland (Majewski 1989); *Philonthus amicus* Sharp, Japan (Sugiyama 1973) and South Korea (Lee 1986, Lee and Choi 1992); *Philonthus albipes* (Gravenhorst) and *Philonthus puella* Nordmann, Finland (Huldén 1983); *Philonthus rectangulus* Sharp, Japan (Sugiyama 1973) and South Korea (Lee 1986); *Philonthus flavocinctus* Motschulsky, Taiwan (Sugiyama and Shazawa 1977, Shen and Ye 2006); *Philonthus azabuensis* Dvořák, South Korea (Lee 1986); *Philonthus sericans* Sharp, Japan (Majewski 1988a); *Philonthus* sp., Hubei Province, China (Shen and Ye 2006); *Gabrius osseticus* (Kolenati) Austria (Stadelmann and Poelt 1962).

P. gracilis (Thaxter, 1915): adult *Hesperus* sp., Java, Indonesia (Thaxter 1915, 1931).

P. heinemanniana De Kesel, 1998: adult *Xantholinus longiventris* Heer, Belgium (De Kesel 1998); *Leptolinus nothus* (Erichson) Italy, Spain (Santamaría and Rossi 1999).

P. homalotae (Thaxter, 1901b): adult *Nehemitropia sordida* (Marsham), MA, USA (Thaxter 1901b, 1908); *Atheta* sp., South Korea (Lee and Choi 1992a); "aleocharine," South Korea (Lee 1986).

P. hybrida (Thaxter, 1900): adult *Philonthus ventralis* (Gravenhorst) England, Madeira, North America (Thaxter 1900, 1908, Picard 1913b, Hake 1923): *Philonthus aeneipennis* Boheman, Gulf of Oman, Japan, Taiwan (Thaxter 1900, 1908, Terada 1976, 1978); *Philonthus gemellus* Kraatz, Sri Lanka (Thaxter 1908); *Philonthus ventralis* (Gravenhorst) Finland (Huldén 1983); *Philonthus tardus* Kraatz, South Korea (Lee 1986); *Philonthus* sp., Taiwan (Terada 1976, Shen and Ye 2006).

P. infecta (Thaxter, 1894): adult *Gyrohypnus fracticornis* (Müller) (as *Xantholinus obsidianus* Melsheimer) MA, USA, and *Xantholinus* sp., Argentina (Thaxter 1894, 1906, 1908, 1931); *Leptacinus* sp., China (Shen and Ye 2006).

P. insignis (Thaxter, 1900): adult of unknown subfamily and genus, Sarawak, Malaysia (Thaxter 1900).

P. japonica Terada, 1980: adult *Philonthus spadiceus* Sharp, Japan (Terada 1980); *Philonthus japonicus* Sharp, South Korea (Lee et al. 2011b).

P. javanus (Thaxter, 1900): adult *Philonthus* sp., ?Java, Indonesia (Thaxter 1900).

P. lampropygi Thaxter, 1931: adult *Philothalpus analis* (Erichson) Panama (Thaxter 1931).

P. lepida (Thaxter, 1931): adult *Philonthus circumductus* Fauvel, Philippines (Thaxter 1931).

P. madagascariensis (Thaxter, 1900): adult *Philonthus pollux* Fauvel (as "*Philonthus sikorae* Fauvel," invalid name) Madagascar (Thaxter 1900); *Philonthus turbidus* Erichson, Egypt (Maire 1916b, Thaxter 1931).

P. mexicana (Thaxter, 1901b): adult *Philonthus atriceps* Sharp, Mexico (Thaxter 1901b, 1908).

P. nigrescens (Thaxter, 1893): adult *Philonthus debilis* (Gravenhorst) MA, USA (Thaxter 1893), Scotland, Germany (Scheloske 1969), and Finland (Huldén 1983); adult *Philonthus flavocinctus* Motschulsky, Japan (Sugiyama 1978b) and Taiwan (Shen and Ye 2006); *Philonthus* sp., Jiangxi Province, China (Shen and Ye 2006).

= *P. inaequalis* (Thaxter, 1894): adult *Philonthus debilis* (Gravenhorst) MA and ME, USA (Thaxter 1894).

P. pallida (Thaxter, 1931): adult *Philonthus aeneipennis* Boheman, Philippines (Thaxter 1931).

P. peruviana (Thaxter, 1900): adult *Nordus simplex* (Sharp) Peru (Thaxter 1900); *Plociopterus laetus* Sharp, Brazil (Thaxter 1908).

P. pretiosa (Thaxter, 1931): adult "*Staphylinus sumatrensis* Bernhauer," invalid name, Sumatra, Indonesia (Thaxter 1931).

P. princeps (Thaxter, 1895): adult *Philonthus sordidus* (Gravenhorst) Algeria, Germany, Italy, Netherlands, and MA and CA, USA (Thaxter 1895, 1896, 1908, Picard 1913b, Maire 1916a, Colla 1925, 1934, Middelhoek 1943a, Poelt 1952a, Stadelmann and Poelt 1962); *Philonthus cephalotes* (Gravenhorst) Belgium, Finland, Poland, Czech Republic (Thaxter 1908, Picard 1913b, Collart 1945, Middelhoek 1947a,b, Majewski 1973b, Huldén 1983, De Kesel and Rammeloo 1991, Rossi and Máca 2006); *Philonthus quadraticeps* Boheman, Argentina (Spegazzini 1912, 1917, Thaxter 1931); *Philonthus politus* (L.) Belgium (Collart 1945, De Kesel and Kammeloo 1991); *Philonthus bonariensis* Bernhauer, Argentina (Spegazzini 1917); *Philonthus tardus* Kraatz, South Korea (Lee 1986); *Philonthus* sp., Peru (Sugiyama 1972); *Spatulonthus longicornis* (Stephens) Madeira, Italy (Thaxter 1908, Colla 1934, Rossi 1975); *Quediomacrus puniceipennis* Solsky, Mexico (Thaxter 1908).

P. protea Thaxter, 1900: adult *Anotylus rugosus* (F.), England, France, Poland, Romania, Hungary, Belgium, German, Italy, Netherlands, Czech Republic, Spain (Thaxter 1900, 1908, 1931, Cépède and Picard 1908b, Picard 1913b, 1917, Siemaszko and Siemaszko 1931, Lepesme 1941, Middelhoek 1943b, 1947b, Bánhegyi 1944, 1949, Collart 1945, Stadelmann and Poelt 1962, Scheloske 1969, Balazuc 1973a, 1974, Rossi 1975, De Kesel and Rammeloo 1991, De Kesel and Haghebaert 1991, Weir 1994, Rossi and Máca 2006, and Santamaría 1992a as "*Styloxys rugosus*"); *Anotylus insecatus* (Gravenhorst) France, Poland, North Africa, and MA, USA (Picard 1913b, Thaxter 1931, Siemaszko and Siemaszko 1932, Stadelmann and Poelt 1962, Scheloske 1969, Balazuc 1974; Haelewaters et al. 2015); *Bledius bicornis* (Germar) Europe, Germany (Thaxter 1900, 1908, Siemaszko and Siemaszko 1932, Stadelmann and Poelt 1962, Scheloske 1969); *Bledius spectabilis* Kraatz, North Africa, France, Austria (Picard 1913b, 1917, Thaxter 1931, Siemaszko and Siemaszko 1932, Lepesme 1941, Stadelmann and Poelt 1962, Scheloske 1969, Balazuc 1974); *Manda mandibularis* (Gyllenhal) Europe (Thaxter 1900, 1908, Picard 1913b, Scheloske 1969); *Oxytelus fulvipes* Erichson, Poland (Siemaszko and Siemaszko 1932, Scheloske 1969); *Planeustomus miles* Scriba, Tunisia (Maire 1916b, Thaxter 1931, Siemaszko and Siemaszko 1932, Scheloske 1969); *Philonthus wuesthoffi* Bernhauer, Tibet (Lee et al. 2006).

= *P. oxyteli* (Cépède and Picard, 1908a): adult *Anotylus rugosus* (F.) France (Cépède and Picard, 1908a).

P. quedii (Scheloske, 1969): adult *Quedius vexans* Eppelsheim, Germany (Scheloske 1969).

P. staphylini (Thaxter, 1931): adult "*Staphylinus sumatrensis* Bernhauer," invalid name, Sumatra, Indonesia (Thaxter 1931).

P. subinaequilatera (Spegazzini, 1915a): adult *Philonthus abyssinus* Fauvel (= *P. holomelas* Eppelsheim) as "*Philonthus holomallus*," Somalia (Spegazzini 1915a); *Philonthus convexicollis* Lynch, Argentina (Spegazzini 1917); *Philonthus* sp., Spain (Santamaría 1999).

P. thyreocephali (Thaxter, 1931): adult "*Thyreocephalus venezolanus*," invalid name, Venezuela (Thaxter 1931).

P. trichodoniae (Thaxter, 1931): adult *Trichodonia schwabi* Wasmann, Cameroon (Thaxter 1931).

P. vulgata (Thaxter, 1900): adult *Philonthus varians* (Paykull) England (Thaxter 1900); *Philonthus flavolimbatus* Motschulsky, Mexico and Panama (Thaxter 1900, 1908); *Philonthus cruentatus* (Gmelin) Europe (Thaxter 1900, 1908, Picard 1913b); *Philonthus quisquiliarius* (Gyllenhal) England, France, Czech Republic (Thaxter 1900, 1908, Picard 1913b, Hake 1923, Balazuc 1974, Rossi et al. 2010); *Philonthus ebeninus* (Gravenhorst) Siberia, Russia, Europe (Thaxter 1908); *Philonthus hepaticus* Erichson, Argentina and Nicaragua (Thaxter 1908, Spegazzini 1912, 1917); *Philonthus agilis* (Gravenhorst) Austria and France (Thaxter 1908, Picard 1913b, Balazuc 1974); *Philonthus aeruginosus* Nordmann and *P. convexicollis* Lynch, Argentina (Spegazzini 1912, 1917); *Philonthus sanguinolentus* (Gravenhorst) France (Picard 1917, Lepesme 1941, Balazuc 1974); *Philonthus moquerysi* Fauvel, Cameroon (Thaxter 1931); *Philonthus albipes* (Gravenhorst) Netherlands (Middelhoek 1943a); *Philonthus cephalotes* (Gravenhorst) Netherlands (Middelhoek 1943b) and Finland (Huldén 1983); *Philonthus spinipes* Sharp, Japan (Sugiyama 1973) and South Korea (Lee et al. 1982, Lee 1986); *Philonthus minutus* Boheman, Algeria, Taiwan (Maire 1916a, Colla 1925, Sugiyama and Shazawa 1977); *Philonthus sordidus* (Gravenhorst) Germany, Uruguay, Netherlands (Spegazzini 1917, Middelhoek 1943a,b, Stadelmann and Poelt 1962, Scheloske 1969); *Philonthus politus* (Linnaeus), England (Hincks 1960); *Philonthus corruscus* (Gravenhorst) Slovakia (Rossi et al. 2010); *Philonthus solidus* Sharp, Japan (Majewski 1988a); *Philonthus subuliformis* (Gravenhorst) Finland (Huldén 1983); *Spatulonthus longicornis* (Stephens) St. Helena, Sri Lanka, Italy, Madeira, Scotland, Australia, Finland, Taiwan, Missouri (USA), and Bali (Indonesia) (Thaxter 1900, 1908, 1931, Picard 1913b, Colla 1925, 1926, Hincks 1960, Sugiyama and Shazawa 1977, Huldén 1983, Majewski and Sugiyama 1985a, Shen and Ye 2006); *Leptacinus* sp., Hainan, China (Shen and Ye 2006).

P. xanthopygi Thaxter, 1901b: adult *Xanthopygus calidus* (Erichson) (as synonym *X. solskyi* Sharp), perhaps Brazil (Thaxter 1901b).

P. zyricola Thaxter 1931: adult *Zyras sulcicollis* Fauvel, Cameroon (Thaxter 1931); *Zyras corniger* Bernhauer, Sierra Leone (Rossi 1994).

Phaulomyces Tavares, 1985

P. dibelonetis (Thaxter, 1931): adult "*Dibelonetes piceus* Bernhauer," invalid name, Grenada (Thaxter 1931). Transferred from genus *Meionomyces* by Tavares (1985).

P. euaestheti Thaxter, 1931: adult *Euaesthetus americanus* Erichson, MA, USA (Thaxter 1931); *Euaesthetus ruficapillus* Boisduval and Lacordaire, Germany, Poland (Scheloske 1969, Balazuc 1971c, Majewski 1994a), transferred from genus *Corethromyces* by Tavares (1985).

P. leonensis Rossi, 1994: adult *Pseudespeson rossii* Lecoq, Sierra Leone (Rossi 1994).

Polyascomyces Thaxter, 1900

P. trichophyae Thaxter, 1900: adult *Trichophya pilicornis* (Gyllenhal) England (Thaxter 1900).

Porophoromyces Thaxter, 1926

P. tmesiphori (Thaxter, 1926): adult *Tmesiphorus* sp., Cameroon (Thaxter 1926); *Centrophthalmus grandis* Pic, Sierra Leone (Rossi 1994).

NOTE: *P. formosanus* Sugiyama (1982) was transferred to *Bordea* by Benjamin (2001).

Pselaphidomyces Spegazzini, 1917

P. pselapti Spegazzini, 1917: adult *Pselaptus tuberculifer* Raffray, in nests of an ant, *Acromyrmex lundi* Guérin-Méneville, Argentina (Spegazzini 1917); *Reichenbachia puncticollis* (LeConte) (as *Bryaxis*) northeastern USA (Thaxter 1931).

Rhachomyces Thaxter, 1895

= *Acanthomyces* Thaxter, 1893 [preoccupied]

R. arbusculus Thaxter, 1896: adult of gen. and sp. nr. *Lathrobium*, Liberia (Thaxter 1896, 1931).

R. bordonii Rossi and Santamaría, 2000: adult *Medhiama schawalleri* Bordoni, Nepal, and *Yunna micophora* Bordoni, Yunnan, China, and *Nepalinus* sp. Burma, and *Atopolinus insulanus* Bordoni, Taiwan (Rossi and Santamaría 2000).

R. cayennensis Thaxter, 1900: adult *Ochthephilum* sp. indet., French Guiana (Thaxter 1900, 1908, 1931).

R. cryptobianus Thaxter, 1900: adult "*Cryptobium capitatum*," invalid name, Brazil (Thaxter 1900, 1908, 1931).

R. dolicaontis Thaxter, 1896: adult *Dolicaon lathrobioides* Laporte, South Africa (Thaxter 1896, 1908, 1931).

R. falcatus Rossi and Proaño Castro, 2009: adult *Palaminus* sp., Ecuador (Rossi and Proaño Castro 2009).

R. feloi Rossi, 2006: adult *Domene benahoarensis* Oromí and Martín Canary Islands (Rossi 2006).

R. furcatus (Thaxter, 1893): adult *Othius punctulatus* (Goeze) [= *O. fulgidus* (Paykull) = *O. fulvipennis* (F.)] France, Germany, Italy, Poland, Algeria, western Russia, Slovakia (Thaxter 1908, 1931, Picard 1913b, Maire 1920, Siemaszko and Siemaszko 1928, 1932, Lepesme 1942, Stadelmann and Poelt 1962, Scheloske 1969, Balazuc 1973b, Rossi 1975, Huldén 1983, De Kesel and Haghebaert 1991, Rossi et al. 2010); *O. angustus* Stephens [= *O. melanocephalus* (Gravenhorst)] England, Italy (Thaxter 1908, 1931, Rossi 1975); *O. myrmecophilus* Kiesenwetter, England, Germany, France, Belgium (Thaxter 1908, Scheloske 1969, Balazuc 1973b, De Kesel and Rammeloo 1991, De Kesel and Haghebaert 1991); *O. lapidicola* Markel and Kiesenwetter, France, Finland, Czech Republic (Balazuc 1973b, Huldén 1983, Rossi and Máca 2006); *O. pilifer* Quedenfeldt, Maire 1916a, Algeria; *O. transsilvanicus* Ganglbauer, Romania (Bánhegyi 1949).

R. glyptomeri Thaxter, 1901b: adult *Glyptomerus cavicola* Müller, former Yugoslavia (Thaxter 1901b, 1908, 1931).

R. lasiophorus (Thaxter, 1892): adult of a carabid beetle *Atranus pubescens* (Dejean) CT, USA (Thaxter 1892, 1895, 1931), perhaps misidentified on *Othius punctulatus* (Goeze) in Germany (Stadelmann and Poelt 1962) and should be compared with *R. lathrobii* (Tavares 1985).

R. lathrobii Thaxter, 1895: adult *Lathrobium* sp., NH, USA, and *Lobrathium longiusculum* (Gravenhorst) NH, USA (Thaxter 1894, 1895, 1931).

R. medonalis Thaxter, 1931: adult "*Medon tristis* Bernhauer" invalid name, Grenada (Thaxter 1931).

R. ocypi Sugiyama, 1973: adult *Ocypus scutiger* Sharp, Japan (Sugiyama 1973).

R. oedochiri Thaxter, 1901b: *Oedochirus* sp., Brazil (Thaxter 1901b, 1908, 1931).

R. philonthinus Thaxter, 1900: adult *Spatulonthus longicornis* (Stephens) and *Philonthus* sp., UK (Thaxter 1900, 1908); *Philonthus albipes* (Gravenhorst) Sweden, Finland (Thaxter 1908, Huldén, 1983); *Philonthus cruentatus* Gmelin [= *P. bipustulatus* (Panzer)], Netherlands, France, Italy, Finland (Middelhoek 1943b, Balazuc 1973b, Rossi 1975, Huldén 1983); *Philonthus fulvipes* (F.) Finland, western Russia (Huldén 1983); *Philonthus fumarius* (Gravenhorst) and *Philonthus rectangulus* Sharp, Belgium (De Kesel and Haghebart 1991); *Philonthus marginatus* (Strøm) Netherlands, France, Belgium, Poland (Middelhoek 1943b, Collart 1945, Balazuc 1973b, De Kesel and Rammeloo 1991, Majewski 1994a); *Philonthus rubripennis* (Stephens) and *Philonthus rigidicornis*

(Gravenhorst) and *Spatulonthus longicornis* (Stephens) Finland and *Philonthus micans* (Gravenhorst) Finland, western Russia (Huldén 1983); *Philonthus minutus* Boheman (= *P. mutans* Sharp) China (Thaxter 1908); *Philonthus varians* (Paykull) England, France, Poland, Belgium, Germany, Netherlands (Siemaszko and Siemaszko 1932, Middelhoek 1943a, 1943b, 1947, Petch 1944, 1945, Shaw 1952, Scheloske 1969, Stadelmann and Poelt 1962, Balazuc 1973b, De Kesel and Rammeloo 1991, De Kesel and Haghebart 1991, Majewski 1994a); *Philonthus varius* (Gyllenhal) France (Balazuc 1973b); *Philonthus umbratilis* (Gravenhorst) and *Philonthus quisquiliarius* (Gyllenhal) Poland (Majewski 1994a); *Philonthus wuesthoffi* Bernhauer, Japan (Sugiyama 1973, 1974); *Philonthus micanticollis* Sharp, South Korea (Lee and Park 1991); *Philonthus* sp. indet., Romania (Bánhegyi 1949, Lee 1986); *Bisnius fimetarius* (Gravenhorst) western Russia (Huldén 1983) Poland, Czech Republic, Slovakia, and Belgium (Majewski 1994a, Rossi et al. 2010, De Kesel and Haghebaert 1991); *Gabrius exiguus* (Nordmann) Europe (Thaxter 1908), and *Gabrius fauveli* (Coquerel), Réunion (Balazuc 1982); *Gabrius* sp., Congo-Kinshasa (formerly Zaire) (Rossi and Santamaría 1992); *Amichrotus apicipennis* Sharp and *Amichrotus* sp., Japan (Thaxter 1908, Ishikawa 1952); *Diatrechus griveaudi* Jarrige and *Platydracus peyrierasi* Jarrige, Madagascar (Balazuc 1982); *Sepedophilus nitidicollis* (Jarrige), Réunion (Balazuc 1982); *Philonthus* sp., Tibet, China (Shen and Ye 2006).

R. pilosellus (Robin, 1871): adult *Lathrobium* sp., origin not stated but probably France (Robin 1871); transferred to *Rhachomyces* by Thaxter (1895); *Lathrobium castaneipenne* Kolenati and *L. elongatum* (L.) Poland (Siemaszko and Siemaszko 1932); *L. fulvipenne* (Gravenhorst) France, Belgium, Germany, Italy (Thaxter 1908, Picard 1913b, Collart 1947, Scheloske 1969, Rossi 1975; De Kesel and Rammeloo 1991); *Lathrobium geminum* Kraatz, Belgium (De Kesel and Rammeloo 1991).

= *R. brevipes* (Thaxter 1893): adult *Lathrobium fulvipenne* (Gravenhorst) Germany (Thaxter 1893, 1895).

R. venustus Rossi and Proaño Castro, 2009: adult *Megalopinus robustus* (Motschulsky), Ecuador (Rossi and Proaño Castro 2009).

Rhadinomyces Thaxter, 1893. Generic name resurrected by Tavares (1985).

R. pallidus Thaxter, 1893: adult *Lathrobium punctulatum* LeConte and *L. angulare* LeConte, MA, USA, as well as *L. fulvipenne* (Gravenhorst) Germany (Thaxter 1893); *Lathrobium angustatum* Boisduval and Lacordaire, England (Hake 1923), *Lathrobium brunnipes* (F.) France, England, Germany, Belgium (Picard 1913b, Thaxter 1931, Hincks 1960, Scheloske 1969, Balazuc 1973b, De Kesel and Rammeloo 1991); *Lathrobium magistrettorum* Koch, Switzerland (Rossi 1975); *Lathrobium fovulum* Stephens, Poland (Majewski 1984), *Lathrobium terminatum* Gravenhorst, Greece and western Russia (Castaldo et al. 2004) and Czech Republic (Rossi and Máca 2006); *Lathrobium pallipes* Sharp, Japan (Majewski and Sugiyama 1985b).

= *R. cristatus* Thaxter, 1893: adult *Lathrobium nitidulum* LeConte, MA and ME, USA; *Lathrobium castaneipenne* Kolenati, Poland, Belgium, Germany (Siemaszko and Siemaszko 1932, Collart 1945, Stadelmann and Poelt 1962, De Kesel and Rammeloo 1991); *Lathrobium elongatum* (L.) Poland, Germany (Siemaszko and Siemaszko 1932, Stadelman and Poelt 1962); *Lathrobium elongatum* (L.) and *Lathrobium ripicola* Czwalina, Poland (Siemaszko and Siemaszko 1932); *Lathrobium terminatum* Gravenhorst, England (Hincks 1960); *Lathrobium fulvipenne* (Gravenhorst) and *L. geminum* Kraatz, Belgium (De Kesel and Rammeloo 1991); placed as subspecies of *R. pallidus* by Balazuc (1973b). The name *Lathrobium punctulatum* LeConte was recognized as that of a valid species by Watrous (1980); *Lathrobium* sp., Japan (Majewski and Sugiyama 1985b).

R. gracilis (Thaxter, 1912a) as a variety of *Corethromyces platensis* Thaxter, raised to a specific rank by Spegazzini (1917), transferred to *Rhadinomyces* by Tavares (1985): adult *Lobrathium nitidum* (Erichson) Argentina (Thaxter 1912a, Spegazzini 1917); *Lobrathium dimidiatum* (Say) and *Stereocephalus seriatipennis* Lynch, Argentina (Spegazzini 1917).

Rickia Cavara, 1899
= *Distichomyces* Thaxter, 1905

R. carpanetoi Rossi and Santamaría, 1992: adult *Euconnus* sp., Congo-Kinshasa (formerly Zaire) (Rossi and Santamaría 1992)

R. compressa Thaxter, 1915: adult *Leptochirus* sp., Java, Indonesia (Thaxter 1915, 1926).

R. fastigiata Thaxter, 1926: adult *Coproporus secretus* Bernhauer, Philippines (Thaxter 1926).

R. huggerti Balazuc, 1980: adult *Omalium riparium* (Thomson), Sweden, France, and *Omalium caesum* Gravenhorst, Italy, and *Omalium ferrugineum* Kraatz, Italy (Balazuc 1980).

R. hyperborea Balazuc, 1980: adult *Micralymma marinum* (Strøm) Norway, England (Balazuc 1980, Weir 1994); *Micralymma brevilingue* Schiødte, eastern Siberia (Russia) (Balazuc 1980, Tavares 1985).

R. inclusa Thaxter, 1916: adult *Coproporus latus* Motschulsky, Philippines (Thaxter 1916, 1926).

R. introversa Thaxter, 1916: adult *Coproporus latus* Motschulsky, Philippines (Thaxter 1916, 1926).

R. kistneri Rossi, 1991: adult *Mimaenictus wilsoni* Kistner and Jacobson, Malaysia (Rossi 1991).

R. leptochiri (Thaxter, 1908): adult *Leptochirus* spp., Java, Indonesia (Thaxter 1908, 1912, 1926); *Priochirus unicolor* (Laporte) Malaya, Malaysia (Sugiyama and Majewski 1985b).

NOTE: *R. lispini* Thaxter (1912a) was transferred to *Diaphoromyces* by Thaxter (1926).

R. lordithonis Majewski and Sugiyama, 1985): adult *Lordithon* sp., Japan (Majewski and Sugiyama 1985b).

R. minutissima Sugiyama and Yamamoto, 1982: adult *Lispinus* sp., Sabah, Malaysia (Sugiyama and Yamamoto 1982).

R. nigrescens Thaxter, 1916: adult *Coproporus hypocyptoides* Bernhauer, Sumatra, Indonesia (Thaxter 1915, 1926).

R. nigrofimbriata Thaxter, 1916: adult *Coproporus latus* Motschulsky, Philippines, and *Coproporus* sp., Sumatra, Indonesia (Thaxter 1916, 1926).

R. pallescens Thaxter, 1916: adult *Mimocyptus globulus* Cameron, host genus as "*Mymocyphus*," Philippines (Thaxter 1916, 1926).

R. peyerimhoffii Maire, 1916c: adult *Scaphisoma agaricinum* (L.) and *Scaphisoma flavonotatum* (Pic), Algeria (Maire 1916c, Thaxter 1926); *Scaphisoma agaricinum* (L.) Finland and western Russia and *Rickia inopinatum* Löbl, Finland (Huldén 1983); *Scaphisoma agaricinum* (L.), *Scaphisoma assimile* Erichson, *Scaphisoma boreale* (Lundblad), *Scaphisoma boleti* (Panzer), and *Scaphisoma subalpinum* Reitter, Poland (Majewski 1994a, b); *Scaphisoma* sp., Czech Republic (Rossi and Máca 2006); *Scaphisoma agaricinum* (L.) Greece, *Scaphisoma boleti* (Panzer) Italy, and *Scaphisoma loebli* Tamanini, Italy (Castaldo et al. 2004); *Scaphidium femorale* Lewis, South Korea (Lee et al., 2011a); *Scaphidium* sp., South Korea (Lee 1986).

R. phloeonomi Thaxter, 1926: adult *Phloeonomus singularis* Kraatz, Sumatra, Indonesia (Thaxter 1926); *Paraphloeostiba sonani* Bernhauer, Japan and Taiwan (Sugiyama 1978b, Shen and Ye 2006); *Paraphloeostiba* sp., Hainan Island, China (Shen and Ye 2006).

R. proteini Majewski, 1983: adult *Proteinus brachypterus* (F.) and *Proteinus macropterus* (Gyllenhal) Poland (Majewski 1983, 1994a); *Proteinus crassicornis* Sharp, Japan (Majewski 1988a).

R. rostellata Thaxter, 1926: adult *Holosus* sp., Sumatra, Indonesia (Thaxter 1926); *Holosus olisthaeriformis* Motschulsky, Malaya, Malaysia (Sugiyama and Mochizuka 1979).

R. rostrata Thaxter, 1915: adult *Atanygnathus ruficollis* (Kraatz), generic name as *Tanygnathus*, Sarawak, Borneo, Java, Indonesia (Thaxter 1915, 1926).

R. sakkae Sugiyama and Majewski, 1985: adult *Lordithon affinis* (Cameron) and *Lordithon* sp., Malaya, Malaysia (Sugiyama and Majewski 1985b).

R. scydmaeni Thaxter, 1916: adult: "*Scydmaenus bicolor*" (perhaps *Euconnus bicolor* LeConte) ME, USA (Thaxter 1916, 1926).

= *R. rhachomycoides* Spegazzini, 1917: indet. scydmaenine, Argentina (Spegazzini 1917, Thaxter 1926).

R. seticola Thaxter, 1926: adult *Coproporus* sp., Sumatra, Indonesia (Thaxter 1926).

R. sugiyamae Tavares, 1985, new name *for Rickia lispinae* Sugiyama 1973, not *Rickia lispini* Thaxter, 1912a: adult *Lispinus asper* Sharp, as "*Lispinus aper*," Japan (Sugiyama 1973, Tavares 1985).

R. tachini Terada, 1980: adult *Tachinus impunctatus* Sharp, Japan (Terada 1980).

R. wulaiensis Sugiyama, 1978a: adult *Priochirus tonkinensis* Bernhauer, Taiwan (Sugiyama 1978a, Shen and Ye 2006).

R. zanettii Rossi and Cesari Rossi, 1978: adult *Omalium excavatum* Stephens Italy (Rossi and Cesari Rossi, 1978); *Omalium laeviusculum* Gyllenhal, Scotland (Weir and Beakes 1993).

Sandersoniomyces Benjamin, 1968a

S. divaricatus Benjamin, 1968a: adult *Quedius* sp., CA, USA (Benjamin 1968a).

Scaphidiomyces Thaxter, 1912a. All known hosts belong to Scaphidiinae.

S. baeocerae Thaxter, 1912a: adult *Baeocera* sp., Argentina (Thaxter 1912a, 1931, Spegazzini 1917); *Scaphisoma unicolor* Achard, South Korea (Lee et al. 2011b).

S. baeoceridii (Thaxter, 1931): adult *Baeoceridium depressipes* Reitter, Cameroon (Thaxter 1931).

S. platensis (Spegazzini, 1917): adult scaphidiine, Argentina (Spegazzini 1917, Thaxter 1931).

S. pusillus Rossi, 2010: adult *Baeocera* sp., Ecuador (Rossi 2010).

S. scaphicomae Thaxter, 1931: adult *Scaphicoma* sp., Cameroon (Thaxter 1931).

Scelophoromyces Thaxter, 1912a

= *Skelophoromyces* Thaxter, 1931

S. osorianus Thaxter 1912a: adult *Osorius sexpunctatus* Bernhauer, Argentina, *Osorius intermedius* Erichson, Trinidad, and *Osorius* sp., Brazil? (Thaxter 1912a, 1931).

Smeringomyces Thaxter, 1908

S. anomalus (Thaxter, 1902): adult "*Conosoma pubescens* Paykull," a name that was at that time erroneously applied to various species of *Sepedophilus* in MA, USA (Thaxter 1902, 1908); *Sepedophilus testaceus* (as *Conosoma testaceum* (F.), a synonym) Italy (Rossi and Cesari Rossi 1980) and Greece (Castaldo et al. 2004); *Sepedophilus* sp., Spain (Santamaría 1992a).

S. chaetophilus (Thaxter, 1931): adult *Coproporus bernhaueri* Scheerpeltz (*as* "*C. ventralis* Bernhauer," an invalid name), Sumatra, Indonesia (Thaxter 1931).

S. lingulatus (Thaxter, 1931): adult *Coproporus* sp., Sumatra, Indonesia (Thaxter 1931).

S. trinitatis Thaxter, 1931: adult "*Conosoma trinidadense* Bernhauer," an invalid name, likely a species of *Sepedophilus*, Trinidad (Thaxter 1931); *Sepedophilus tibialis* (Sharp) Japan (Terada 1977).

Sphaleromyces Thaxter, 1894

Synonymized into *Corethromyces* by Thaxter (1912a p. 180) but resurrected by Tavares (1985) for the four species named below.

S. indicus Thaxter 1901b: adult *Pinophilus* sp., India and Burma (Thaxter 1901b), transferred to *Corethromyces* by Thaxter (1912a), transferred to *Sphaleromyces* by Tavares (1985).

S. lathrobii Thaxter, 1894: adult *Lathrobium nitidulum* LeConte and *L. punctulatum*

LeConte, ME, USA (Thaxter 1894), transferred to *Corethromyces* by Thaxter (1912a), transferred to *Sphaleromyces* by Tavares (1985); *Lathrobium quadratum* (Paykull) Europe (Thaxter 1908, 1912a, 1931, Picard 1913b, Petch 1944, Hincks 1960; Majewski 1982); *Lathrobium* sp., Spain (Santamaría 1995).

S. occidentalis Thaxter, 1895: adult *Pinophilus densus* LeConte, UT, USA (Thaxter 1895), transferred to *Corethromyces* by Thaxter (1912a), transferred to *Sphaleromyces* by Tavares (1985).

S. rhinoceralis (Thaxter, 1912a): adult *Pinophilus suffusus* Erichson, Argentina (Thaxter 1912a), transferred to *Sphaleromyces* by Tavares (1985).

Stemmatomyces Thaxter, 1931

S. euconni (Picard, 1917): adult *Euconnus wetterhali* (Gyllenhal), France (Picard 1917); unidentified scydmaenine, Philippines (Thaxter 1931), transferred to *Stemmatomyces* (Tavares 1985).

Stichomyces Thaxter, 1901

S. capensis Thaxter, 1931: adult *Sepedophilus angustus* (Bernhauer) South Africa (Thaxter 1931).

S. conosomatis Thaxter, 1901b: adult *Sepedophilus testaceus* (F.) MA, ME, USA (Thaxter 1901b, Maire 1920, Thaxter 1931, Benjamin 1971) Greece (Castaldo et al. 2004); *Sepedophilus tibialis* Sharp, Japan (Terada 1977); *Sepedophilus littoreus* (L.) England and MA, USA (Weir and Beakes 1993, Haelewaters et al. 2015); *Sepedophilus bipustulatus* (Gravenhorst) Greece (Castaldo et al. 2004); *Sepedophilus pedicularius* (Gravenhorst) Belgium (De Kesel and Rammeloo 1991) and Italy (Castaldo et al. 2004); *Sepedophilus nigripennis* (Stephens) Belgium, Netherlands (De Kesel and Haelewaters 1991, Haelewaters et al. 2012).

S. europaeus Majewski, 1973a: adult *Sepedophilus testaceus* (F.) Poland (Majewski 1973a).

S. sumatrae Thaxter, 1931: adult *Sepedophilus tenue* (Bernhauer) Sumatra, Indonesia (Thaxter 1931).

NOTE: *Stichomyces catalinae*, *S. lepidus*, and *S. vesiculifer* were transferred to *Corethromyces* by Tavares (1985).

Stigmatomyces Karsten, 1869

S. caudicola Spegazzini, 1917: adult *Stereocephalus seriatipennis* Lynch, Argentina (Lynch 1917). Thaxter (1917) and Tavares (1985) suggested that the fungus belongs to *Corethromyces*, but made no transfer.

S. montevidensis Spegazzini, 1917: adult *Lobrathium dimidiatum* (Say), Argentina (Spegazzini 1917). Thaxter (1931) suggested close affinity to *Corethromyces*, whereas Tavares (1985) suggested it belongs to *Sphaleromyces*, but no transfer was made.

NOTE: All the species described from Coleoptera were removed from *Stigmatomyces* by Tavares (1985), yet she did not assign the two above to another genus, so they are listed here lest they be forgotten.

Sugiyamaemyces Tavares and Balazuc, 1989

S. oroussetii Tavares and Baluzuc, 1989: adult *Clidicus formicarius* Pascoe, Borneo (Tavares and Baluzuc 1989).

Symplectromyces Thaxter, 1908

S. lapponicus Huldén, 1983: adult *Quedius boops* (Gravenhorst) Finland (Huldén, 1983).

S. rarus Huldén, 1983: adult *Quedius fuliginosus* (Gravenhorst) Finland (Huldén, 1983) and Poland (Majewski 1987).

S. vulgaris (Thaxter, 1908): adult *Quedius cinctus* (Paykull) [= *Q. impressus* (Panzer)] Portugal, *Q. cruentus* (Olivier), Europe, *Q. dubius* Heer, France, *Q. fulgidus* (F.) Germany, Spain, *Q. fuliginosus* (Gravenhorst) Germany, Q. *peregrinus* (Gravenhorst) Canada, *Q. ventralis* (Aragona) (= Q. *truncicola* Fairmaire and Laboulbène)

Great Britain, *Quedius* sp., India and Canada, and ?*Philonthus* sp., Hungary (Thaxter 1908); *Quedius collaris* (Erichson) Poland (Majewski 1974, 1990a), *Quedius longicornis* Kraatz, France and Germany (Scheloske 1969, Balazuc 1971b), *Q. maurorufus* (Gravenhorst) Germany (Scheloske 1969), *Q. mesomelinus* (Marsham) North America, former Yugoslavia, Netherlands, Poland, Germany, Italy, Belgium, Finland, Czech Republic (Middelhoek 1943a,b, Collart 1945, Stadelmann and Poelt 1962, Scheloske 1969, Balazuc 1974, Majewski 1974, 1990a, Huldén, 1983, De Kesel and Rammeloo 1991, Rossi et al. 2010); *Quedius fulgidus* (F.) and *Quedius xanthopus* Erichson, Poland (Majewski 1974, 1990a); *Quedius nitipennis* (Stephens) Belgium (De Kesel and Rammeloo 1991); *Q. ochropterus* Erichson and *Q. scitus* (Gravenhorst) Italy (Rossi 1978); *Quedius maurus* Sahlberg, Czech Republic (Rossi et al. 2010); *Quedius curtipennis* Bernhauer and *Quedius tristis* (Gravenhorst) Belgium (De Kesel and Haghebaert 1991).

Teratomyces Thaxter, 1893

T. actobii Thaxter, 1894: adult *Erichsonius nanus* (Horn) MA, ME, USA (Thaxter 1894, 1896); *Erichsonius cinerascens* (Gravenhorst) England, Germany, Italy (Thaxter 1908, 1931, Hake 1923, Scheloske 1969, 1969, Balazuc 1971c, Rossi and Cesari Rossi 1980); *Erichsonius signaticornis* (Mulsant and Rey) England (Hake 1923); *Erichsonius brachycephalus* Frank, MA, USA (Haelewaters et al. 2015).

= *T. atropurpureus* Maire, 1920): *Erichsonius signaticornis* (Mulsant and Rey) Mauretania (Maire 1920, Thaxter 1931).

=*T. brevicaulis* Thaxter, 1894: adult *Erichsonius nanus* (Horn) ME, USA, and *Erichsonius basalis* (Motschulky) Sri Lanka (Thaxter 1894, 1931); *Erichsonius cinerascens* (Gravenhorst) Finland and Karelia (Russia) (Huldén 1983) and Poland (Majewski 1994a).

T. atanygnathi Thaxter, 1931: adult *Atanygnathus ruficollis* (Kraatz) Philippines and Sarawak, Malaysia (Thaxter 1931).

Fig. 10.1 *Teratomyces mirificus* Thaxter (Laboulbeniales) (marked by an asterisk) on the margin of the abdomen of *Acylophorus zdenae* Smetana (Staphylininae) collected in Florida

T. heterothopsis Rossi, 2010: adult *Heterothops* sp., Ecuador (Rossi 2010).

T. insignis Thaxter, 1901b: adult *Quedius edwardsi* Sharp and *Quedius* sp., New Zealand (Thaxter 1901b, 1908, 1931).

T. mirificus Thaxter, 1893: adult *Acylophorus pronus* Erichson, MA, ME, USA (Thaxter 1893, 1896); *Acylophorus glaberrimus* (Herbst), Spain (Santamaría and Rossi 1999) Greece (Castaldo et al. 2004); *Acylophorus bumbunae* Bordoni and *Acylophorus rossii* Bordoni, Sierra Leone (Rossi 1994); *Acylophorus* sp., Yunan Province, China (Shen and Ye 2006) (Fig. 10.1).

T. petiolatus Thaxter, 1901b: adult *Quedius* sp., New Zealand (Thaxter 1901b, 1908).

T. philonthi Thaxter, 1900: adult *Philonthus* sp., Hungary (conceivably this was a *Gabrius* sp. like the other recorded hosts because *Gabrius* at that time was considered a junior synonym of *Philonthus*) (Thaxter 1900, 1908, Picard 1917, Thaxter 1931); *Gabrius trossulus* (Nordmann), Netherlands, Germany, Finland (Middelhoek 1943a, b, Stadelmann and Poelt 1962, Scheloske 1969 Huldén 1983); *Gabrius pennatus* Sharp, Germany (Scheloske 1969); *Gabrius* sp., Congo-Kinshasa (formerly Zaire) (Rossi and Santamaria 1992); *Gabrius nigritulus* (Gravenhorst) Poland (Majewski 1984); *Gabrius nitidulus* (Gravenhorst) Belgium,

England (De Kesel and Rammeloo 1991, Weir 1994).

T. quedianus Thaxter, 1896: adult *Hemiquedius ferox* (LeConte), MA, USA (Thaxter 1896, 1931); *Quedius fuligidus* (F.) Hungary (Moesz 1931, Bánhegyi 1940).

T. zealandicus Thaxter, 1894: adult *Quedius insolitus* Sharp, New Zealand (Thaxter 1894, 1906, 1931).

Tetrandromyces Thaxter, 1912
= *Dicrandromyces* Thaxter, 1931
= *Triandromyces* Thaxter, 1931 NOTE: These synonymies are disputed by Rossi and Santamaría (2000).

T. brachidae Thaxter, 1912a: adult *Brachida reyi* Sharp, Argentina (Thaxter 1912a, 1931).

T. weirianus Rossi and Santamaría, 2000: adult *Pseudoplandria drugmandi* Pace, Philippines (Rossi and Santamaría 2000).

Zeugandromyces Thaxter, 1912a

Z. australis Thaxter, 1912a: adult *Scopaeus laevis* Sharp, Argentina (Thaxter 1912a, 1915, 1931).

Z. orientalis (Thaxter, 1931): adult *Scopaeus nitidulus* Motschulsky (as *S. subfasciatus* Kraatz) Philippines (Thaxter 1931); *Scopaeus currax* Sharp, Japan (Majewski and Sugiyama 1985b); *Scopaeus dilutus* Motschulsky, Malaya, Malaysia (Sugiyama and Majewski 1985b).

Z. pseudomedalis (Thaxter, 1931): adult *Lithocharis thoracica* (Casey) ME, USA (Thaxter 1931, Tavares 1985).

Z. stilici (Thaxter, 1915): adult *Rugilus ceylanensis* (Kraatz) (as *Stilicus*) Sarawak, Malaysia; Sumatra, Indonesia; and Sri Lanka (Thaxter 1915, 1931, Tavares 1985); "*Stilicus*" sp., Yunan, China (Shen and Ye 2006).

NOTE: These four species were transferred here from *Stigmatomyces* by Tavares (1985).

Order Pyxidiophorales
Pyxidiophoraceae

Entomocosma Spegazzini, 1918

?*Entomocosma* sp.: adult "*Tachinus pallipes*" (but see Campbell 1973), MA, USA (Thaxter 1920).

Thaxteriola Spegazzini, 1918

T. infuscata Spegazzini, 1918: adult *Belonuchus haemorrhoidalis* (F.) and *Philonthus convexicollis* Lynch, Argentina (Spegazzini 1918).

ORDER AND FAMILY NOT ASSIGNED

Chantransiopsis Thaxter, 1914. In *Ascomycota* but order and family not assigned as of 2013.

C. bonaerensis Spegazzini, 1918: adult *Xantholinus subtilis* Boheman and *Nacaeus tenellus* (Erichson), although the latter is possibly *Lispinus spegazzinii* Bernhauer, Argentina (Spegazzini 1918).

C. decumbens Thaxter, 1914: adult of unidentified genus, Java, Indonesia (Thaxter 1914).

C. platensis Spegazzini, 1918: adult *Belonuchus haemorrhoidalis* (F.) Argentina (Spegazzini 1918).

C. stipatus Thaxter, 1914: adult of genus near *Tachinus*, Java, Indonesia (Thaxter 1914).

C. xantholini Thaxter, 1914: adult *Xantholinus* sp., MA, USA (Thaxter 1914).

Class *Sordariomycetes*
Order *Hypocreales*
Clavicipitaceae
Beauveria Vuillemin, 1912

B. *bassiana* (Balsamo) Vuillemin, 1912: adult *Tachyporus hypnorum* (F.) and *Anotylus rugosus* F. and *Gyrohypnus angustatus* (Stephens) Denmark (Steenberg et al. 1996).

Metarhizium Sorokin, 1883

M. *anisopliae* (Metchnikoff) Sorokin, 1883: adult *Anotylus rugosus* (F.) and *Anotylus insecatus* (Gravenhorst) Denmark (Steenberg et al. 1996).

Cordycipitaceae
Cordyceps Link, 1833

C. confragosa (Mains) (synonymn *Lecanillium lecanii* (Zimmerman) Zare and Gams, 1861): adult *Tachyporus chrysomelinus* (L.) Denmark (Steenberg et al. (1996).

Ophiocordycipitaceae
Hirsutella Patouillard, 1892
H. eleutheratorum (Nees) Petch, 1932 [= *H. guignardi* (Maheu, 1906)]: adult *Quedius mesomelinus* (Marsham) France (Maheu 1906).

Subphylum Entomophthoromycotina [is not assigned to a phylum]
Order Entomophthorales
Entomophthoraceae
Pandora Humber, 1989
P. philonthi (Balazy, 1993): adult *Anotylus rugosus* (F.) Belgium (Steenberg et al. 1996).

Zoophthora Battko, 1964
Z. anglica (Petch, 1944): adult of unidentified staphylinid genus, England (Leatherdale 1958).

Basidiobolaceae

This genus has been characterized as a toxic black mold, perhaps not specific to its apparent hosts.

Basidiobolus Eidam, 1886
= *Amphoromorpha* Thaxter, 1914
B. entomophila (Thaxter, 1914): adult *Diochus conicollis* Motschulsky, Philippines (Thaxter 1914, 1920b).
B. mirabilis (Siemaszko and Siemaszko, 1928): *Philonthus politus* L. [an erroneous citation by Stadelmann and Poelt 1962—not recorded from this host].
B. subminuta (Spegazzini, 1918): adult *Echiaster depressus* Solier, Argentina (Spegazzini 1918).

Phylum Apicomplexa
Class Conoidasida
Order Eugregarinorida
Actinocephalidae

Members of Actinocephalidae are parasites in the digestive system of the host. When they are present in large numbers, they are likely to have a debilitating effect.

Actinocephalus Stein, 1848
A. crassus (Ellis, 1912): adult *Leptochirus edax* Sharp, Guatemala (Ellis 1913).
A. stelliformis (Schneider, 1875): larva and adult *Ocypus olens* (Müller) France (Schneider 1875) and adult *Philonthus tenuicornis* Mulsant and Rey, Poland (Foerster 1938), and adult *Staphylinus caesareus* Cederhjelm and *Staphylinus erythropterus* L., Poland (Wellmer 1911).

Stylocystis Léger, 1899
S. ensifera (Ellis, 1912): adult *Leptochirum edax* Sharp, Guatemala (Ellis 1913), and *Anotylus tetracarinatus* (Block) Poland (Foerster 1938).

Gregarinidae

Members of Gregarinidae are parasites in the digestive system of the host. When they are present in large numbers, they are likely to have a debilitating effect.

Gregarina Dufour, 1828
G. omalina Foerster, 1938: adult *Omalium rivulare* (Paykull) and *Heterothops dissimilis* (Gravenhorst) Poland (Foerster 1938).

Phylum Platyhelminthes
Class Cestoda
Choanotaeniidae
Choanotaenia Railliet, 1896
C. infundibulum (Bloch, 1779): adult *Apocellus sphaericollis* (Say) is an intermediate host of this cestode (which is a parasite of domestic chickens) USA (Horsfall and Jones 1937).

Hymenolepididae
Soricinia Spassky and Spasskaya, 1954
S. diaphana (Cholodkowsky, 1906): adult *Tachinus pallipes* (Erichson) is an intermediate host of this cestode (which is a parasite of *Sorex* shrews) Czechoslovakia (Prokopič 1967).

Variolepis Spassky and Spasskaya, 1954
S. variabilis (Mayhew, 1925): adult *Oxytelus* sp. is an intermediate host of this cestode (which is a parasite of *Corvus* crows) USA (Jones 1934).

Phylum Nematoda
Class Secernentea
Order Tylenchida

Most members of Tylenchida are parasites of plants. Those listed are parasites of insects, apparently attacking the larval or pupal stage and exiting from the adult.

Allantonematidae

Members of Allantonematidae are endoparasites. They may have severe consequences on host fitness, including female sterility, reduced male fertility and mating success, and reduced survival.

Allantonema Leuckart, 1887
A. philonthi Wachek, 1955: adult *Gabrius osseticus* Kolenati and *Philonthus debilis* (Gravenhorst) and *Bisnius fimetarius* (Gravenhorst) Germany (Wachek 1955).

Metaparasitylenchus Wachek, 1955
M. boopini Wachek, 1955: adult *Carpelimus fuliginosus* (Gravenhorst) Germany (Wachek 1955).

Palaeoallantonema Poinar, 2011. A fossil genus. The species named below was detected next to an adult staphylinid beetle, from which it was thought to have emerged, in Dominican amber dated as 20–30 MY old.
P. dominicana Poinar, 2011: near to an adult *Neoxantholinus* in amber, Dominican Republic (Poinar and Brodzinsky 1986, Poinar 2011).

Parasitylenchoides Wachek, 1955
P. koerneri Wachek, 1955: adult *Anotylus tetracarinatus* (Block) Germany (Wacheck 1955).
P. paederi Wachek, 1955: adult *Paederus littoralis* Gravenhorst, Germany (Wachek 1955).
P. rheocharae Wachek, 1955: adult *Aleochara spadicea* (Erichson) Germany (Wachek 1955)
P. steni Wachek, 1955: adult *Stenus biguttatus* (L.) and *Stenus bimaculatus* and *Stenus boops* (Ljungh) and *Stenus pallitarsis* Stephens, Germany (Wachek 1955).
P. sp. indet.: adult *Tachyporus hypnorum* (F.) Germany (Lipkow 1968).

Proparasitylenchus Wachek, 1955. These attack pupae of their hosts and emerge from adults.
P. athetae Wachek, 1955: adult *Nehemitropia sordida* (Marsham) Germany (Wachek 1955); *Aloconota* sp., France, and *Plataraea brunnea* (F.) Germany (Puthz 1982).
P. medonis Wachek, 1955: adult *Medon ripicola* (Kraatz), Germany (Wachek 1955).
P. mymedoniae Wachek, 1955: adult *Zyras lugens* (Gravenhorst) Germany (Wachek 1955).
P. oxyteli Wachek, 1955: adult *Anotylus cornutus* (Gravenhorst) Germany (Wachek 1955).
P. platystethi Wachek, 1955: adult *Platystethus cornutus* (Gravenhorst) Germany (Wachek 1955).
P. trogophloei Wachek, 1955: adult *Carpelimus bilineatus* Stephens, Germany (Wachek 1955).
P. californicus Poinar et al. 2015: adult *Tarphiota geniculata* (Mäklin) CA, USA, but likely infected the prepupa or pupa (Poinar et al. 2015)

Stictylus Thorne, 1941
= *Sphaerulariopsis* Wachek, 1955
S. philonthi auct.: typographical error by some authors, not Wachek, no such species.

Entaphelenchidae

The female parasite of *Entaphelenchus* is free-living until it invades the pupa of the host; *Entaphelenchus* has not been found in host larvae. Details of the development of *Roveaphelechus* are unknown.

Entaphelenchus Wachek, 1955
E. aliantae Wachek, 1955: adult *Alianta incana* (Erichson) Germany (Wachek 1955).
E. bledii Wachek, 1955: adult *Bledius opacus* (Block) Germany (Wachek 1955).
E. oxyteli Wachek, 1955: adult *Oxytelus piceus* (L.) Germany (Wachek 1955).

E. philonthi Wachek, 1955: adult *Philonthus concinnus* (Gravenhorst) and *Bisnius fimetarius* (Gravenhorst) and *Philonthus fulvipes* (F.) Germany (Wachek 1955).
E. platystethi Wachek, 1955: adult *Platystethus cornutus* (Gravenhorst), Germany (Wachek 1955).
E. xantholini Wachek, 1955: adult *Gyrohypnus punctulatus* (Paykull) and *Xantholinus tricolor* (F.) Germany (Wachek 1955).

Roveaphelenchus Nickle, 1970
R. jonesi Nickle, 1970: adult of *Aleochara tristis* Gravenhorst NE, USA (Nickle 1970a).

Order Rhabditida

Most members of Rhabditida are parasites of plants. Those listed are parasites of insects, apparently attacking the larval or pupal stage and exiting from the adult.

Diplogastridae
Acrostichus Rahm, 1928
A. monhysteroides (Bütschli, 1874): adult *Oxytelus* sp. and *Aleochara* sp., Germany (Sachs 1950) and *Atheta* sp. Germany (Sachs 1950).

Allodiplogaster Paramonov and Sobolev, 1954
A. henrichae (Sachs, 1950): adult *Oxytelus* sp. and *Atheta* sp. Germany (Sachs 1950).

Diplogaster Schultze, 1857
D. coprophages de Man, 1876: adult *Oxytelus* sp. and *Aleochara* sp. and *Atheta* sp. Germany and *Emus hirtus* (L.) Germany (Sachs 1950).

Diplogasteritus Paramonov, 1952
D. zurstrasseni (Sachs, 1950): adult *Emus hirtus* (L) Germany (Sachs 1950).

Diplogastrellus Paramonov, 1952
D. gracilis (Bütschli, 1874): adult *Quedius cinctus* (Paykull) Germany (Hirschmann 1952).

Goffartia Hirschman, 1952
G. heteroceri Hirschmann, 1952: adult *Paederus riparius* (L.) Germany (Hirschmann 1952).

Mesodiplogaster Weingartner, 1955
M. maupasi (Potts, 1910) [= *M. entomophaga* (Steiner, 1929)]: adult *Drusilla canaliculata* (F.) Netherlands (Schuurmans Stekhoven 1929).

Rhabditoides Rahm, 1928
R. stigmatus (Steiner, 1930): adult *Lordithon thoracicus* (F.) Germany (Körner 1954).

Neodiplogastridae
Fictor Paramonov, 1952
F. stercorarius (Bovien, 1937): adult *Aleochara* sp. and *Emus hirtus* (L.) Germany (Sachs 1950).

Glauxinema Allgén, 1947
G. schwemmlei (Sachs, 1950): adult *Oxytelus* sp., Germany (Sachs 1950).

Koerneria Meyl, 1960
K. erlangensis (Sachs, 1950): adult *Aleochara* sp. and *Atheta* sp. Germany (Sachs 1950).

Panagrolaimidae
Halicephalobus Timm, 1956
H. similigaster (Andrássy, 1952): *Euryusa sinuata* Erichson and *Batrisodes* sp., Germany (Köhler 2012).

Peloderidae
Coarctadera Dougherty, 1953
C. coarctata (Leuckart, 1891): adult *Anotylus sculpturatus* (Gravenhorst) England (Triffitt and Oldham 1927); *Oxytelus* sp. and *Atheta* sp. and *Emus hirtus* (L.) Germany (Sachs 1950).
C. tretzeli (Sachs, 1950): *Emus hirtus* (L.) Germany (Sachs 1950).
C. voelki (Sachs, 1950): *Emus hirtus* (L.) Germany (Sachs 1950).

Rhabditidae
Caenorhabditis Osche, 1952
C. dolichura (Schneider, 1866): adult *Aleochara* sp., Germany (Sachs 1950).

Diploscapter Cobb, 1913
D. lycostoma Völk, 1950: adult *Philonthus quisquiliarius* (Gyllenhal) Germany (Völk 1950).

Oscheius Andrássy, 1976
O. wohlgemuthi (Völk, 1950): adult *Philonthus quisquiliarius* (Gyllenhal) Germany (Völk 1950).

Rhabditis Dujardin, 1845
R. aspera Bütschli, 1873: adult *Philonthus quisquiliarius* (Gyllenhal) Germany (Völk 1950).
R. cylindrica Cobb, 1898 [= *R. neuhausi* (Sachs, 1950)]: adult *Oxytelus* sp. and *Aleochara* sp. and *Atheta* sp. Germany (Sachs 1950).
R. gracilicauda de Man, 1876: adult *Oxytelus* sp. and *Atheta* sp., Germany (Sachs 1950).
R. hartmanni Sachs, 1950: adult *Aleochara* sp., Germany (Sachs 1950).
R. labiata Völk, 1950: adult *Philonthus quisquiliarius* (Gyllenhal) Germany (Völk 1950).
R. longispina Reiter, 1928: adult *Philonthus fulvipes* (F.) Germany (Hirschmann 1952).
R. spiculigera Steiner, 1936: adult *Oxytelus* sp. and *Aleochara* sp. and *Atheta* sp., Germany (Sachs 1950).
R. sp. indet.: adult *Spatulonthus longicornis* (Stephens) HI, USA (Zwaluwenburg 1928).

Rhabditophanes Fuchs, 1930
= *Cheilobus* Cobb, 1924, not Rafinesque, 1817

R. aphodii (Sachs, 1950): adult *Emus hirtus* (L.) Germany (Sachs 1950).

Class Adenophorea
Order Mermithida
Mermithidae

Members of Mermithidae attack the immature stages of insects, develop within the hosts, and kill the host from which they emerge. Thus, they are parasitoids.

Oesophagomermis Artyukhovsky, 1969
Oesophagomermis sp.: larva of *Philonthus splendens* (F.) Germany (Lipkow 1982).

Orthomermis Poinar, 1965
O. oedobranchus Poinar, 1965: larva of *Omalium caesum* Gravenhorst, England (Poinar 1965, 1975). This mermithid also attacks dipterous hosts and has a wider distribution including Australia.

Nemata Incertae Sedis
Agamonema Diesing, 1851 [a collective group of larval nematodes]
A. migrans Christie, 1930: larva, perhaps of *Tachinus luridus* Erichson (although as "*T. flavipes*") VA, USA (Christie 1930). This nematode also attacks Silphidae and Scarabaeidae.

Phylum Nematomorpha
Class Gordioidea
Gordiidae
Gordius Linnaeus, 1758
G. aquaticus Linnaeus, 1766: adult *Creophilus maxillosus* (L.) Austria (Assmuss 1858, Zwaluwenburg 1928).

Phylum Arthropoda
Class Arachnida
Order Sarcoptiformes
Histiostomatidae
Histiostoma Kramer, 1876
H. sapromyzarum (Dufour, 1839): adult *Quedius mesomelinus* (Marsham) France (Théodoridès 1954, 1955).
H. sp. indet.: adult *Paederus ruficollis* (Fabricius) France (Théodoridès 1955).

Order Prostigmata
Pyemotidae
Pygmephorus Kramer, 1877

Pygmephorus sp.: adult of *Platystethus cornutus* (Gravenhorst) Iraq (Shamshuddin et al. 1967). The authors suggest that these mites may be predators.

Scutacaridae

Imparipes Berlese, 1903

Imparipes armatus Karafiat, 1959: adult of *Platyprosopus mexicanus* Sharp, Mexico (Smetana 1982).

Order Mesostigmata

Eviphididae

Thinoseius Halbert, 1920

T. fucicola (Halbert, 1920): adult of *Cafius xantholoma* (Gravenhorst), England (Egglishaw 1966).

Family indet. of Gamasina

Genus and species indet.: adult of *Staphylinus dimidiaticornis* Gemminger (= *S. parumtomentosus* Stein) Wales (Green and Wilkinson 1950).

Class Insecta

Order Coleoptera

Colydiidae

Aglenus Erichson, 1845

A. brunneus (Gyllenhal, 1813): this eyeless, wingless beetle is phoretic on a cucujid beetle and on *Xylodromus concinnus* (Marsham) in England (Woodroffe 1967).

Order Diptera

There seem to be no valid records of Diptera as parasitoids of Staphylinidae.

Order Hymenoptera

Ichneumonidae

The large family Ichneumonidae has thus far been reported to attack only two species of Staphylinidae. Its larvae are parasitoids of host larvae.

Barycnemis Förster, 1869

B. blediator (Aubert, 1970): larva of *Bledius* sp., France (Aubert 1970, Horstmann 1981); *Bledius spectabilis* Kraatz, England (Wyatt 1982).

Phygadeuon Gravenhorst, 1829

P. subspinosus (Gravenhorst, 1829) [= *P. grandiceps* Thomson, 1884]: a record from a larva of *Quedius ochripennis* Ménétriès, France (Falcoz 1924), is suspect because Muesebeck et al. (1951) state that members of *Phygadeuon* are parasitoids of muscoid Diptera.

Braconidae

Braconidae are a large family of parasitoid wasps. Their specialization to host is at least somewhat phylogenetic because some subfamilies attack larvae of Coleoptera, whereas others attack eggs or larvae of Hymenoptera, Diptera, or Lepidoptera. Here are noted some that attack adult staphylinids.

Blacus Nees, 1818

B. ruficornis (Nees, 1811): adult *Tachyporus obtusus* (L.) Germany (Lipkow 1965, 1968).

Centistes Haliday, 1835

C. cuspidatus (Haliday, 1833) [= *C. lucidator* (Nees, 1834)]: adult of *Tachyporus chrysomelinus* (L.) and *T. hypnorum* (F.) and *T. obscurus* (L.) and *T. solutus* Erichson, Germany (Lipkow1965, 1968).

Perilitus Nees, 1818

P. sp. indet.: adult *Paederus alfierii* Koch, Egypt (Kamal 1951).

Eulophidae

Eulophidae are a large family attacking eggs or larvae of Coleoptera and of many other orders.

Necremnus Thomson, 1878

N. cosconius (Walker, 1839) [= N. *amempsimus* (Walker, 1839)]: larva of *Quedius brevis* Erichson, England (Donisthorpe 1938). The wasp emerged in the laboratory from the pupal stage of a specimen collected in the field as a larva.

Myrmaridae

All known species of Mymaridae are minute parasitoids of insect eggs of orders Heteroptera,

Homoptera, Psocoptera, Coleoptera, Diptera, and Orthoptera.

Litus Haliday, 1833

L. krygeri Kieffer, 1913: egg of *Ocypus olens* (Müller) ?France (Paulian 1941).

Diapriidae

These very small wasps are parasitoids of larvae and pupae of many insects.

Spilomicrus Westwood, 1832

S. stigmaticalis Westwood, 1832 [= *S. tripartitus* Kieffer, 1911]: larva of *Quedius puncticollis* (Thomson 1867) France (Paulian 1941). Two wasps emerged in the laboratory from the pupal stage of larvae collected in the field.

Proctotrupidae

Townes and Townes (1982) revised the family Proctotrupidae (= Serphidae) and found that virtually all the species associated with Staphylinidae had been misidentified or mis-assigned to genus by earlier authors. For that reason, the list of species and their hosts provided by Townes and Townes (1982) is accepted, and records published after 1981 are added. Eggs are oviposited into host larvae, and wasp larvae develop as endoparasitoids. Hosts of the vast majority of species (> 300) of Proctotrupidae are unknown.

Cryptoserphus Kieffer, 1907

C. aculeator (Haliday, 1839): early records from *Creophilus maxillosus* are due to nomenclatural confusion.

Exallonyx Kieffer, 1904

E. ater (Gravenhorst, 1807): larva of *Creophilus maxillosus* (L.) England, Mongolia; *Ocypus olens* (Müller) Mongolia.

E. brevicornis (Haliday 1839): larva of *Quedius vexans* Eppelsheim, Czechoslovakia.

E. confusus Nixon, 1938: the species has been much confused in the literature, so literature records of hosts before 1982 should be discounted.

E. crenicornis (Nees, 1834): larva of *Staphylinus* sp., Sweden.

E. ligatus (Nees 1834): the species has been much confused with *E. trifoveatus* and *E. quadriceps*, so literature records of hosts before 1982 should not be accepted without reexamination of parasitoid specimens.

E. longicornis (Nees, 1834): larva of "Staphylinidae," Germany.

E. microcerus Kieffer 1908: larva of "Tachyporini," Sweden; *Xantholinus* ? *linearis* (Olivier) Sweden.

E. obsoletus (Say, 1836): larva of "*Staphylinus* sp." USA, *Platydracus violaceus* (Gravenhorst), MI and NY, USA; *Quedius laevigatus* (Gyllenhal), Ohio, USA (Hoebeke and Kovarik 1988).

E. pallidistigma Morley, 1911: larva of *Ocypus ater* (Gravenhorst) Germany.

E. quadriceps (Ashmead, 1893): larva of *Quedius* sp. England.

E. trifoveatus Kieffer, 1904: larva of *Quedius* sp., Denmark, *Quedius* sp., Germany; *Quedius vexans* Eppelsheim, Czechoslovakia; *Philonthus turbidus* Erichson, Hawaii.

Phaneroserphus Pschorn-Walcher, 1958

P. calcar (Haliday, 1839): larva of *Bolitochara obliqua* Erichson, England; *Quedius simplicifrons* Fairmaire, Germany.

Platygastridae

Most of the Platygastridae are parasitoids of larvae of Cecidomyiidae (Diptera). This record is unusual.

Trimorus Förster, 1856

T. fulvimanus (Kieffer, 1908): egg of *Acylophorus wagenscheiberi* Kiesenwetter, Poland (Staniec 2005).

10.3 Conclusion

The methods of detection of insect and nematode parasitoids differ sharply from the methods for detection of fungal parasites. Anyone who rears staphylinid larvae may encounter adult Hymenoptera or nematodes emerging from the resultant pupal or adult beetles. The easiest way to encounter *Laboulbeniales* is to examine a pinned and labeled insect collection. This may account for a relatively larger number of host/parasitoid records of *Laboulbeniales*. It is obvious that a team approach, involving specialists in disparate taxa, will be the most productive and reliable in terms of accurate identification. The best way to advance the study of parasites of Staphylinidae in general may be for a young staphylinid specialist to establish a *database* of all the parasites, perhaps using this text as a basis. Then, that person should collaborate with specialists in the various parasite groups and urge staphylinid collectors to send parasites to those specialists. Construction of a useful computerized database was not possible before the first version (Frank 1982), and now this writer is retired. Construction of the database should be supported by the constructor's institution to ensure that it is not abandoned at change of personnel or with retirement. In the coming years, the available information will expand enormously.

Acknowledgments I thank T. Majewski, W. Rossi, S. Santamaría, K. Sugiyama, K. Terada, and A. Weir for reprints of many of their publications. I thank L. Cheng for translating appropriate parts of the work by Shen and Ye (2006) from Chinese to English. I am very grateful to U. Irmler and O. Betz for editorial help.

References[1]

Arndt E, Santamaría S (2004) Laboulbeniales (Ascomycota) of the Canary Islands. Vieraea 32:107–115

Aubert J-F (1970) Ichneumonides pétiolées inédites. Bull Soc Entomol Mulhouse 1970:65–73

Balazuc J (1980) Laboulbéniales nouvelles (Ascomycètes), parasites de coléoptères et de diptères. Bull Mus Hist Nat, Paris (ser. 4, Botanique) 2:209–219

Balazuc J (1982) Laboulbeniales (Ascomycetes) de Madagascar, des Comores et des Mascareignes. Bull Mens Soc Linnéenne Lyon 51:6–27

Balazy S (1993) Entomophthorales. Flora of Poland (Flora Polska). Fungi (Mycota) 24:1–356

Benjamin RK (2001) *Autophagomyces*, *Bordea* and a new genus, *Rossiomyces* (Laboulbeniales). El Aliso 19:99–136

Bernardi M, Barragán A, Rossi W (2014) New records of Laboulbeniales (Fungi, Ascomycetes) from Ecuador and other countries. Webbia: J Pl Taxon Geog 69:281–290

Blum G (1924) Zwei neue Laboulbenien aus Brasilien. Centralbl Bakteriol Parasitenk und Infektionskrankheiten, 2-Abt. 62:300–302

Castaldo D, Rossi W, Sabatini F (2004) Contribution to the knowledge of the Laboulbeniales from Greece. Pl Biosyst 138:261–269

Cavara F (1899) Di una nuova Laboulbeniacea: *Rickia Wasmannii*, nov. gen. e nov. spec. Malpighia 13:173–188

De Kesel A (1998) *Peyritschiella heinemanniana* sp. nov. (Ascomycetes, *Laboulbeniales*), a new ectoparasitic species from *Xantholinus longiventris* (Coleoptera, Staphylinidae). In: Fraiture A. (ed) Paul Heinemann Memorial Symposium. Belg J Bot 131(2):176–180

De Kesel A, Haelewaters D (2014) *Laboulbenia slackensis* and *L. littoralis* sp. nov. (Ascomycota, Laboulbeniales), two sibling species as a result of ecological speciation. Mycologia 106:407–414

De Kesel A, Haghebaert G (1991) Laboulbeniales (Ascomycetes) of Belgian Staphylinidae (Coleoptera). Bull Ann Soc Roy Entomol Belg 127:253–270

De Kesel A, Rammeloo K (1991) Checklist of the Laboulbeniales (Ascomycetes) of Belgium. Belg J Bot 124:204–214

Falcoz L (1924) Observations biologiques sur divers insectes des environs de Vienne en Dauphiné. Bull Soc Entomol Fr 1924:221–224

Frank JH (1982) The parasites of the Staphylinidae (Coleoptera). A contribution towards an encyclopedia of the Staphylinidae. Univ Fla Agr Exp Stn, Tech Bull 824: i–vii, 1–118

Haelewaters D (2013) First record of the genus *Ilyomyces* for North America parasitizing *Stenus clavicornis*. Bull Insectol 66:269–272

Haelewaters D, van Wielink P, van Zuijlen VW et al (2012) New records of Laboulbeniales for the Netherlands. Entomol Ber 72:175–183

Haelewaters D, Vorst O, De Kesel A (2014) New and interesting Laboulbeniales (Fungi, Ascomycota) from the Netherlands. Nova Hedwigia 98:113–125

Haelewaters D, Zhao SY, De Kesel A et al (2015) Laboulbeniales (Ascomycetes) of the Boston Harbor Islands I: Species parasitizing Coccinellidae and Staphylinidae, with comments on typification. Northeast Nat 22:459–477

Hoebeke ER, Kovarik PW (1988) *Exallonyx obsoletus* (Hymenoptera: Serphidae), a larval parasite of the rove beetle *Quedius laevigatus* (Coleoptera: Staphylinidae). Entomol News 99:217–220

[1] Note: Almost all pre-1981 references may be found cited in Frank (1982). The parasites of the Staphylinidae (free online at: http://ufdc.ufl.edu/UF00027462). Those earlier references are not repeated below to save space.

Horsfall MW, Jones MF (1937) The life history of *Choanotaenia infundibulum*, a cestode parasitic in chickens. J Parasitol 23:435–450

Horstmann K (1981) Revision der europäischen Tersilochinae II (Hymenoptera: Ichneumonidae). Spixiana, Suppl 4:1–76

Hughes M, Weir A, Leschen R et al (2004) New species and records of Laboulbeniales from the subantarctic islands of New Zealand. Mycologia 96:1355–1359

Huldén L (1983) Laboulbeniales (Ascomycetes) of Finland and adjacent parts of the U.S.S.R. Karstenia 23:31–136

Jones MF (1934) Cysticercoids of the crow cestode, *Hymenolepis variabilis* (Mayhew, 1925) Fuhrmann, 1932. (Hymenolepidae). Proc Helminthol Soc Wash 1:62–63

Kaur S, Mukerji KG (1996) Studies on Indian Laboulbeniales. 2. Three unrecorded species. Nova Hedwigia 62:151–156

Köhler A (2012) Nematodes in the heads of ants associated with sap flux. Nematology 14:191–198

Lee Y-B (1986) Taxonomy and geographical distribution of the Laboulbeniales in Asia. Korean J Pl Taxon 16:89–185

Lee Y-B, Choi D-S (1992) Studies on taxonomy and distribution of the Laboulbeniales collected in Korea (species from Cheonnam Province). Korean J Mycol 20:183–194

Lee Y-B, Lee JY (1981) Studies on the Laboulbeniales in Korea (1). Korean J Mycol 9:177–192

Lee Y-B, Na Y-H (1998) Notes on some species including unrecorded species of the Laboulbeniales (Ascomycetes) collected in Korea. Korean J Mycol 26:108–118

Lee Y-B, Na Y-H (2011) A new species of the genus *Euphoriomyces* (Laboulbeniales Ascomycotina) collected in Korea. Mycobiology 39:59–60

Lee Y-B, Park S-H (1991) Three species of the Laboulbeniales collected in Korea. Korean J Mycol 19:18–21

Lee Y-B, Lee C-I, Lee J-Y (1982) Studies on the Laboulbeniomycetes in Korea (II). Korean J Mycol 10:1–6

Lee Y-B, Kim KT, Lim CK (2002) Interesting species of the Laboulbeniales from the Upo Swamp. Mycobiology 30:128–132

Lee Y-B, Na Y-H, Lim C-K (2005) Notes on three species of the Laboulbeniales newly collected in Korea. Mycobiology 33:61–64

Lee Y-B, Na Y-H, Lim C-K (2006) Notes on two species of the Laboulbeniales from Tibet. Mycobiology 34:41–43

Lee Y-B, Na Y-H, Jang I-H et al (2007) Notes on three species of the genus *Euphoriomyces* (Laboulbeniales) from Korea. Mycobiology 35:1–5

Lee Y-B, Lim C-K, Na Y-H (2011a) Notes on some species of the Laboulbeniales collected from Korea (2011). Korean J Mycol 39:136–140

Lee Y-B, Cha S-T, Park S-H et al (2011b) Studies on interesting species of the Laboulbeniales collected from Korea. Mycobiology 39:219–225

Lepesme P (1945) Laboulbéniales parasites de coléoptères. V. p 149–150. In: Jeannel R et al (eds) Mission scientifique de l'Omo, vol VI, fasc 57. Faune des terriers des rats-taupes. Mém Mus Hist Nat, Paris (n.s.) 19:1–214

Lipkow E (1982) Lebensweise von *Philonthus*-Arten und anderen Staphylinidae (Coleoptera) des Dungs. Drosera 1982:47–54

Maheu J (1906) Contribution à l'étude de la flore souterraine de France. Ann Sci Nat, Botanique (ser 9) 3:1–192

Maire R (1916) Sur une nouvelle Laboulbéniale parasite des Scaphidiidae. Bull Sci Fr Belg (ser 7) 49:290–296

Majewski T (1973a) Rare and new Laboulbeniales from Poland. III. Acta Mycologica 9:111–124

Majewski T (1973b) Rare and new Laboulbeniales from Poland. IV. Acta Mycologica 9:229–238

Majewski T (1974) Rare and new Laboulbeniales from Poland. V. Acta Mycologica 10:267–282

Majewski T (1980) Rare and new Laboulbeniales from Poland. VI. Acta Mycologica 10:141–153

Majewski T (1982) Rare and new Laboulbeniales from Poland. VII. Acta Mycologica 17:53–62

Majewski T (1983) Rare and new Laboulbeniales from Poland. VIII. Acta Mycologica 19:183–191

Majewski T (1984) Rare and new Laboulbeniales from Poland. IX. Acta Mycologica 20:231–241

Majewski T (1987) Rare and new Laboulbeniales from Poland. X. Acta Mycologica 23:97–108

Majewski T (1988a) Some Laboulbeniales (Ascomycotina) collected in Japan. I. Species from Shizuoka Prefecture. Trans Mycol Soc Japan 29:33–54

Majewski T (1988b) Some Laboulbeniales (Ascomycotina) collected in Japan III. Species on coleopterous insects from Iriomote Island. Trans Mycol Soc Japan 29:249–264

Majewski T (1989) Rare and new Laboulbeniales from Poland. XI. Acta Mycologica 25:43–55

Majewski T (1994a) The occurrence of the Laboulbeniales (Fungi, Ascomycetes) in natural forest communities in the Białowieża National Park, NE Poland. Fragm Flor Geobot 39:479–501

Majewski T (1994b) The Laboulbeniales of Poland. Polish Bot Stud 7:1–466

Majewski T (1999) New and rare Laboulbeniales (Ascomycetes) from the Bialowieza Forest (NE Poland). Acta Mycol 34:7–39

Majewski M, Sugiyama K (1985a) Notes on the Laboulbeniomycetes of Bali Island (Indonesia). III. Trans Mycol Soc Japan 26:169–178

Majewski T, Sugiyama K (1985b) The Laboulbeniomycetes of eastern Asia. IV. On ten species from Japan including four new species. Trans Mycol Soc Japan 26:295–313

Majewski T, Sugiyama K (1986) Notes on the Laboulbeniomycetes of Borneo IV. Trans Mycol Soc Japan 27:425–439

Peyritsch J (1873) Beiträge zur Kenntnis der Laboulbenien. Sitzungsber Kaiserl Akad Wissensch, Mathem-Naturwiss Classe, Abt 1. (Vienna) 68:227–254

Poinar GO Jr (2011) The evolutionary history of nematodes. Brill, Leiden

Poinar GO, Brodzinsky J (1986) Fossil evidence of nematode (Tylenchida) parasitism in Staphylinidae (Coleoptera). Nematologica 31:353–355

Poinar G Jr, Datlen N, Espinoza M et al (2015) *Proparasitylenchus californicus* n. sp. (Tylenchida: Allantonematidae) parasitic in the intertidal rove beetle *Tarphiota geniculata* (Mäklin) (Coleoptera: Staphylinidae) in California, USA. Syst Parasitol 92:57–64

Proaño Castro AC, Rossi W (2008) New records of Laboulbeniales (Fungi, Ascomycetes) from Ecuador. Biodiversity of South America, I. Mem Biodivers 1:11–18

Prokopič J (1967) *Tachinus pallipes*, a new intermediate host of the cestode *Soricinia diaphana* (Cholodkowsky, 1906). Folia Parasitol 14:234

Puthz V (1982) Entoparasitische Nematoden bei einigen Aleocharinen (Staph.). Entomol Bl 78:31

Robin CP (1871) Traité du Microscope. Baillière et fils, Paris. (A 2nd edition (1877) with different pagination is available free online)

Rossi W (1981) Una nuova specie di *Monoicomyces* (Ascomycetes, Laboulbeniales). Candollea 36:375–378

Rossi W (1982) Laboulbeniales (Ascomycetes) della Sierra Leone. Quad Accad Nazion Lincei 255:9–22, pl. 1–4

Rossi W (1986) Su alcune specie di Laboulbeniales nuove e interessanti della Sierra Leone. Quad Accad Nazion Lincei 260:77–96, pl. 1–3

Rossi W (1990) New or interesting Laboulbeniales (Ascomycetes) from Sierra Leone. Quad Accad Nazion Lincei 265:5–13, pl. 1–2

Rossi W (1991) A new species and a new record of Laboulbeniales (Ascomycetes) parasitic on myrmecophilous Staphylinidae. Sociobiology 18:197–202

Rossi W (1994) A new contribution to the knowledge of the Laboulbeniales (Ascomycetes) from Sierra Leone. Quad Accad Nazion Lincei 267:4–17

Rossi W (2006) New cavernicolous Laboulbeniales (Fungi, Ascomycetes). Nova Hedwigia 83:128–136

Rossi W (2010) New Laboulbeniales (Ascomycota) parasitic on Staphylinidae from Ecuador. Mycol Progr 9:407–415

Rossi W (2011) New species of *Laboulbenia* from Ecuador, with evidence for host switch in the Laboulbeniales. Mycologia 103:184–194

Rossi W, Bergonzo E (2008) New or interesting Laboulbeniales from Brazil. El Aliso 26:1–8

Rossi W, Cesari Rossi MG (1980) Nuovo contributo alla conoscenza delle laboulbeniali (Ascomycetes) parassite di stafilinidi italiani (Insecta, Coleoptera). Giorn Bot Ital 114:187–192

Rossi W, Máca J (2006) Notes on the Laboulbeniales (Ascomycetes) from the Czech Republic. Sydowia 58:110–124

Rossi W, Proaño Castro AC (2009) New species of *Rhachomyces* from Ecuador, one of which is dimorphic. Mycologia 101:674–680

Rossi W, Santamaría S (1992) New or interesting Laboulbeniales (Ascomycetes) from Zaire. Nova Hedwigia 55:519–525

Rossi W, Santamaría S (2000) New Laboulbeniales parasitic on Staphylinidae. Mycologia 92:786–791

Rossi W, Santamaría S (2012) *Rodaucea*, a new genus of the Laboulbeniales. Mycologia 104:785–788

Rossi W, Weir A (2007) New species of *Corethromyces* from South America. Mycologia 99:131–134

Rossi W, Máca J, Vavra J (2010) New records of Laboulbeniales (Ascomycota) from the Czech Republic and Slovakia. Polish Bot J 55(2):343–351

Rostrup O (1916) Bitrag til Danmarks svampeflora. I. Dansk Bot Arkiv 2:1–56

Santamaría S (1989) El orden Laboulbeniales (Fungi, Ascomycotina) en la Península Ibérica e Islas Baleares. Ed Espec Soc Catalana Micol 3:1–396

Santamaría S (1992a) Laboulbeniales (Ascomycotina) ibéricos. Nuevos datos corológicos. Orsis 7:139–144

Santamaría S (1992b) New and interesting Laboulbeniales (Fungi, Ascomycotina) from Spain. Nova Hedwigia 54:479–492

Santamaría S (1993) Contribución al conocimiento de los Laboulbeniales (Fungi, Ascomycotina) ibéricos. III. Orsis 8:21–31

Santamaría S (1994) Studies on the genus *Monoicomyces* (Laboulbeniales, Ascomycotina): on the species described by Spegazzini. Mycotaxon 50:89–92

Santamaría S (1995a) New and interesting Laboulbeniales (Fungi, Ascomycotina) from Spain, III. Nova Hedwigia 61:65–83

Santamaría S (1995b) A new generic arrangement for two species formerly described in *Laboulbenia*. Mycologia 87:697–701

Santamaría S (1996a) Bases corológicas de Flora Micológica Iberica, Números 895-1113. Cuad Trab Flora Micol Ibérica 10:1–126

Santamaría S (1996b) Taxonomic notes on *Monoicomyces* with descriptions of two new species. Mycol Res 100:1179–1193

Santamaría S (1997) Species of *Corethromyces* parasitic on rove beetles of the genus *Scopaeus*. Mycologia 89:325–331

Santamaría S (1998) Laboulbeniales I. Laboulbenia. Flora Micológica Ibérica, vol 4. J. Cramer, Madrid

Santamaría S (1999) New and interesting Iberian Laboulbeniales (Fungi, Ascomycota). Nova Hedwigia 68:351–364

Santamaría S (2000) Taxonomy and morphology of *Amorphomyces* (Laboulbeniales). Mycol Res 104:1389–1398

Santamaría S (2001) New and interesting Laboulbeniales (Fungi, Ascomycota) from Spain, IV. Nova Hedwigia 72:375–389

Santamaría S (2002) A taxonomic revision of the genus *Dioicomyces* (Laboulbeniales). Mycol Res 106:615–638

Santamaría S (2006) New or interesting Laboulbeniales (Fungi, Ascomycota) from Spain. V. Nova Hedwigia 82:349–363

Santamaría S, Girbal J (1987) Contribución al conocimiento de los Laboulbeniales (Ascomycotina) ibéricos, II. An Jardín Bot Madrid 44:11–22

Santamaría S, Rossi W (1999) New or interesting Laboulbeniales (Ascomycota) from the Mediterranean region. Pl Biosyst 133:163–171

Santamaría S, Balazuc J, Tavares II (1991) Distribution of the European Laboulbeniales (Fungi, Ascomycotina). An annotated list of species. Treb Inst Bot Barcelona 14:1–123

Shamshuddin M, Al-Hafidh R, Al-Adhami MA (1967) Biology of *Craetopycrus cornutus* (Gravenhorst) (Coleoptera: Staphylinidae) and its associated species of mite of the genus *Pygmephorus* (Acarina: Pyemotidae). Appl Entomol Zool 2:218–222

Shen Y-H, Ye D-H (eds) (2006) Flora fungorum sinicorum, Laboulbeniales, vol 28. Science Press, Beijing

Shen Y-H, Li T-H, Song B (2004) Two new records of *Cantharomyces* to China. Mycosystema 23:303–305

Siemaszko J, Siemaszko W (1926) Laboulbeniales polonici et palaeartici. Polskie Pism Entomol 6:188–211

Smetana A (1982) Revision of the subfamily Xantholininae of America north of Mexico (Coleoptera: Staphylinidae). Mem Entomol Soc Canada 120: i–iv, 1–389

Staniec B (2005) A description of the developmental stages of *Acylophorus wagenschieberi* Kiesenwetter, 1850 (Coleoptera: Staphylinidae), with comments on its biology, egg parasite and distribution in Poland. Dt Entomol Ztschr 52:97–113

Steenberg T, Langor V, Esbjerg P (1996) Entomopathogenic fungi in predatory beetles (Col.: Carabidae and Staphylinidae) from agricultural fields. Entomophaga 40:77–85

Sugiyama K (1981) Notes on Laboulbeniomycetes of Formosa III. Trans Mycol Soc Japan 22:311–319

Sugiyama K (1982) The second species of the genus *Porophoromyces* (Laboulbeniomycetes). Trans Mycol Soc Japan 23:231–244

Sugiyama K, Majewski M (1985a) Notes on the Laboulbeniomycetes of Bali Island (Indonesia). II. Trans Mycol Soc Japan 26:125–144

Sugiyama K, Majewski M (1985b) The Laboulbeniomycetes of Peninsular Malaysia II. Trans Mycol Soc Japan 26:449–462

Sugiyama K, Mochizuka H (1979) The Laboulbeniomycetes (Ascomycotina) of peninsular Malaysia. Trans Mycol Soc Japan 20:339–355

Sugiyama K, Yamamoto H (1982) Notes on the Laboulbeniomycetes of Borneo II. The genus *Rickia*. Trans Mycol Soc Japan 23:301–311

Tavares II (1985) Laboulbeniales (Fungi, Ascomycetes). Mycol Memoir 9:1–627

Tavares II, Balazuc J (1989) *Sugiyamaemyces*, a new genus of Laboulbeniales (Ascomycetes) on *Clidicus* (Scydmaenidae). Mycotaxon 34:565–576

Terada K (1980) New or interesting species of the Laboulbeniales found on some coleopterous insects of Japan. Trans Mycol Soc Japan 21:193–203

Terada K (1981) *Osoriomyces*, a new genus of Laboulbeniales from Taiwan. Mycotaxon 13:412–418

Terada K, Tavares II (1993) A new species of *Clonophoromyces* (Ascomycetes, Laboulbeniales) discovered in Japan. Trans Mycol Soc Japan 34:357–362

Townes H, Townes M (1982) A revision of the Serphidae (Hymenoptera). Mem Am Entomol Inst 32:i–iv, 1–541

Watrous LE (1980) *Lathrobium* (*Tetartopeus*): natural history, phylogeny and revision of the Nearctic species (Coleoptera, Staphylinidae). Syst Entomol 8:303–338

Weir A (1994) Further records of Laboulbeniales from collections of British Coleoptera. Mycol Res 98:433–444

Weir A (1995) Two new species of *Ilyomyces*. Mycol Res 99:789–792

Weir A, Beakes GW (1993) New British Laboulbeniales. Mycol Res 97:1045–1055

Weir A, Hammond P (1997) A preliminary assessment of species-richness patterns of tropical beetle-associated Laboulbeniales (Ascomycetes). In: Hyde KD (ed) Biodiversity of tropical microfungi. Hong Kong University Press, Hong Kong, pp 121–139

Weir A, Rossi W (1997) New and interesting Laboulbeniales from New Zealand. Can J Bot 75:791–798

Weir A, Rossi W (2001) New and interesting Bolivian Laboulbeniales. Mycologia 93:171–180

Wyatt TD (1982) The salt marsh beetle. Antenna 6:303–304

Part III

Biology, Reproduction and Development

The Biology of Steninae

11

Oliver Betz, Lars Koerner, and Konrad Dettner

Abstract

Current knowledge of the biology of the megadiverse beetle subfamily Steninae is reviewed here with regard to its systematics, general morphology, life history, behaviour, (chemical) ecology and evolution into various ecomorphs. Comprising >3000 species worldwide, the staphylinid genus *Stenus* is one of the most speciose animal genera on Earth. Steninae are well characterized by a number of adult and larval autapomorphies. Adult *Stenus* beetles are diurnal, optically oriented, epigeic predators of springtails and other small arthropods. The most obvious autapomorphic character defining *Stenus* is its protrusible elongated labium with the paraglossae being modified into adhesive pads. This prey-capture apparatus can be rapidly ejected towards potential prey by increased haemolymph pressure. The paired anal glands of Steninae are described morphologically and with respect to their secretion chemistry. The alkaloid and terpenoid secretions significantly act as defensive compounds against both bacteria and various predators. The unique skimming behaviour of selected species on water surfaces is described in detail, and the chemotaxonomic value of all gland constituents is discussed.

We describe the developmental stages of these beetles from egg to adult and provide details of the functional morphology of the prey-capture apparatus including its adhesion performance via viscous forces.

Steninae beetles prefer moist habitats and inhabit waterside environments such as reeds or sparsely vegetated sites on river or lake margins or the litter and humus layer in tropical forests. More than 70 distinct behavioural patterns can be assigned to the functional categories of feeding, reproduction, grooming, resting and protection. The ecomorphology of Steninae is reviewed in terms of their compound eyes, the labial prey-capture apparatus including the sticky pads (paraglossae) at its tip, the legs including the tarsi and the abdominal tergites.

11.1 Introduction

The taxon Steninae comprises 3300 species worldwide. Its megadiverse genus *Stenus* (about

O. Betz (✉) • L. Koerner (✉)
Evolutionsbiologie der Invertebraten, Institut für Evolution und Ökologie, Universität Tübingen, Tübingen, Germany
e-mail: oliver.betz@uni-tuebingen.de; Lars.Koerner@springer.com

K. Dettner
Lehrstuhl Tierökologie II, Universität Bayreuth, Bayreuth, Germany
e-mail: k.dettner@uni-bayreuth.de

O. Betz et al. (eds.), *Biology of Rove Beetles (Staphylinidae)*,
https://doi.org/10.1007/978-3-319-70257-5_11

3000 species including 24 fossil ones) is one of the most species-rich animal genera on earth. It has a worldwide distribution, with the exception of Antarctica and New Zealand (Puthz 1971, 2010). The body sizes of Steninae span 1.5 mm (*Stenus pernanus*) to 10 mm (*Dianous robustus*) (Fig. 11.1). Adult *Stenus* beetles are diurnal, epigeic and large-eyed, optically oriented predators of springtails and other small arthropods. In adaptation to their predatory behaviour, these beetles possess a fascinating feature (or structure), a rod-like elongated labium that can be rapidly protruded out of the head (Figs. 11.2 and 11.8). It functions in capturing prey by the means of two adhesive pads at its tip. This key innovation may have largely contributed to the evolutionary success of these beetles, since it has opened up a feeding niche to them and makes possible the effective exploitation of elusive prey such as springtails (e.g. Weinreich 1968). Special pygidial glands exhibited by both *Stenus* and "*Dianous*" beetles are used not only for predator defence but, at least in some species, for skimming across the water surface (e.g. Billard and Bruyant 1905).

Research on this intriguing beetle taxon started in 1797 with the description of the genus *Stenus* by Latreille (1797), after Carl von Linné had described the first *Stenus* species as *Staphylinus biguttatus*. Early studies of the unusual mouthparts found in these beetles were conducted by Meinert (1884, 1887) and others. They found their way into early textbooks of entomology (Fig. 11.2).

In the nineteenth century, the genus was subdivided into six subgenera (e.g. Motschulsky 1857, 1860; Rey 1884); they do not represent monophyla and thus have nowadays only some practical relevance for determination purposes (e.g. Lott and Anderson 2011). In the first half of the twentieth century, studies on the taxonomy, morphology and ecology of Steninae are mainly associated with the taxonomist Ludwig Benick (e.g. Benick 1929). His work was continued by the staphylinidologist Volker Puthz, who has described 1722 Steninae species so far and has continuously refined the classification system of this taxon in more than 20 revisions (e.g. Puthz 1971, 1980, 2000a, 2001, 2008, 2013; Naomi et al. 2017). In addition to this continuous taxonomic and systematic work, our biological knowledge of the Steninae has been diversified in many fields such as phylogenetic systematics (e.g. Clarke and Grebennikov 2009; Koerner et al. 2013; Lang et al. 2015; Żyła et al. 2017), population genetics (Serri et al. 2016), autecology (e.g. Renkonen 1934, 1950; Horion 1963), applied ecology (e.g. Schatz et al. 2003; Qui and

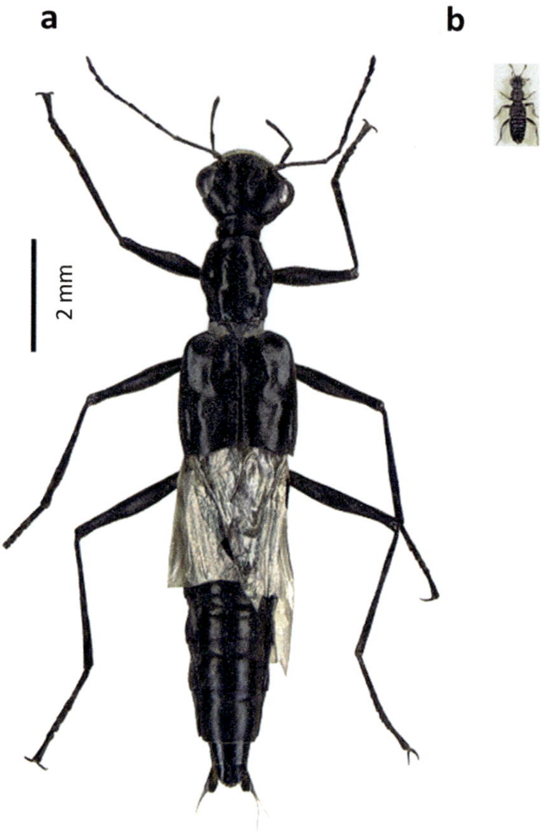

Fig. 11.1 Habitus of the largest and the smallest representatives of Steninae. (**a**) *Dianous robustus* (body length: 10 mm) and (**b**) *Stenus pernanus* (body length 1.5 mm). Both the beetles are shown in their natural size relationship (Image of *D. robustus* courtesy of Karin Wolf-Schwenninger)

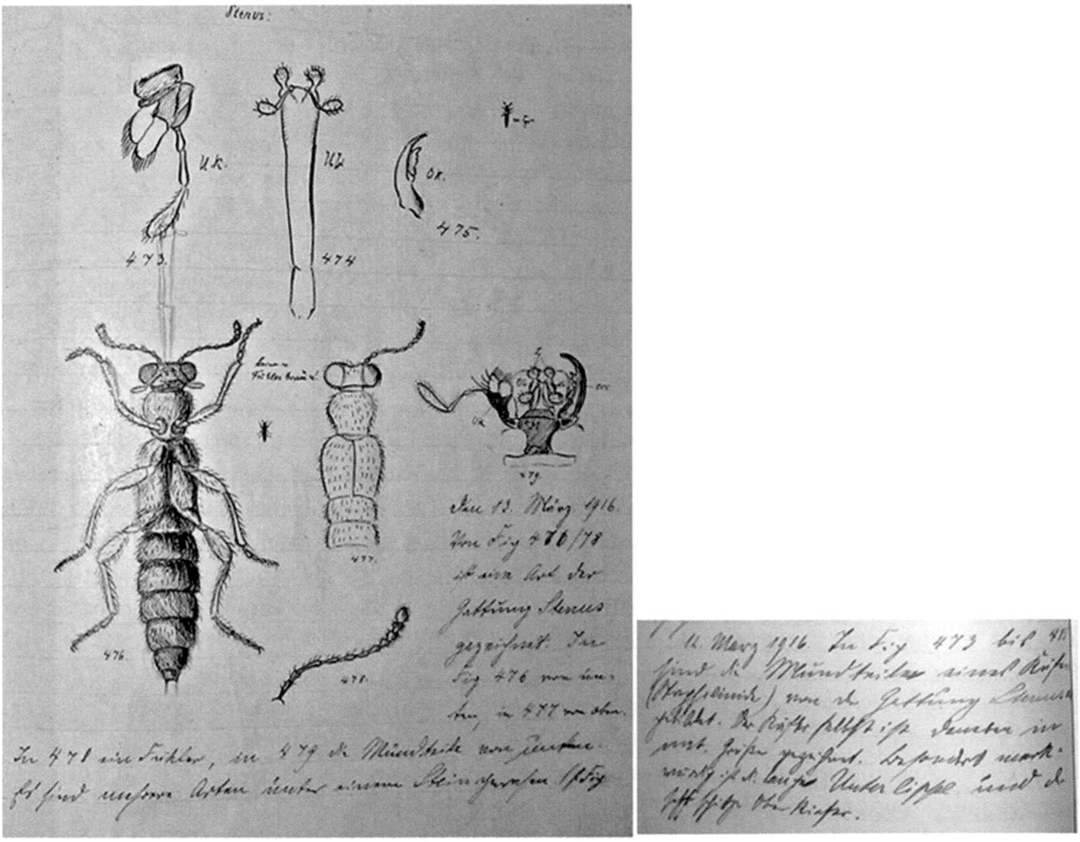

Fig. 11.2 Excerpt of the unpublished "Biological diary" of the German insect morphologist Hermann Weber. His drawings and preliminary notes on the external morphology of *Stenus* beetles were later introduced into his entomology textbook (Weber 1933)

Zheng 2006), life history (Betz and Fuhrmann 2001), behaviour (e.g. Bauer and Pfeiffer 1991; Betz 1999), ecomorphology (e.g. Betz 2002, 2006), chemical ecology (e.g. Lusebrink et al. 2007, 2008a, b), biophysics (Dietz et al. 2016) and biomimetics (Betz et al. 2009). In particular, the functional morphology and biological role of the unusual labial prey-capture apparatus has attracted much attention (e.g. Schmitz 1943; Weinreich 1968; Bauer and Pfeiffer 1991; Betz 1996; Kölsch and Betz 1998; Koerner et al. 2012a, b, 2017).

In the present contribution, our aim has been to correlate the literature concerning the biology of Steninae and hence to reveal the current interim status of research on these beetles.

11.2 Biodiversity, Phylogenetics and Zoogeography

11.2.1 Current Status of Systematics and Phylogenetic Position

Steninae MacLeay 1825 is a subfamily of the "Staphylinine group" (i.e. subfamily groups of Lawrence and Newton 1982) within the Staphylinidae (rove beetles), containing the extant genera *Stenus* Latreille, 1797 and *Dianous*[1]

[1] Throughout this text, the genus name *Dianous* is presented in quotation marks according to the presumed paraphyly of *Stenus* with respect to *Dianous* (cf. Koerner et al. 2013; Lang et al. 2015).

Leach, 1819 and additional, probably a new, but as yet undescribed genus from Australia (Clarke and Grebennikov 2009; McKenna et al. 2015; Żyła et al. 2017), and two fossil genera, i.e. †*Eocenostenus* (Cai et al. 2014) and †*Festenus* (Żyła et al. 2017). Steninae currently comprises about 3300 described species (about 3000 *Stenus* spp. and 300 "*Dianous*" spp.).

With regard to their phylogenetic position within the Staphylinine group, the monophyletic Steninae are most closely related to monophyletic Euaesthetinae (Fig. 11.3). Both these groups may represent sister taxa (Clarke and Grebennikov 2009; Grebennikov and Newton 2009; McKenna et al. 2015; Żyła et al. 2017).

Various fossils demonstrate that Steninae and Euaesthetinae, and other subfamilies of the Staphylinine group (e.g. Scydmaeninae, Solieriinae, Megalopsidiinae), were well established as early as the Late Cretaceous (Clarke and Chatzimanolis 2009; Thayer et al. 2012; Jałoszyński et al. 2016; Jałoszyński and Peris 2016; Yamamoto and Solodovnikov 2016; Żyła et al. 2017). Accordingly, all subfamilies of the Staphylinine group are hypothesized to have originated by the Late Jurassic to the Lower Cretaceous and as being well established in the mid-Cretaceous (Clarke and Chatzimanolis 2009; Grebennikov and Newton 2009; Solodovnikov and Schomann 2009; Solodovnikov et al. 2013; Thayer et al. 2012).

Whereas †*Eocenostenus fossilis* has been described from the late Eocene of France (37–34 mya) (Cai et al. 2014), Żyła et al. (2017) have described the extinct genus †*Festenus* from Late Cretaceous Burmese amber (earliest Cenomanian, ca. 99 mya); the latter genus possibly possessed the *Stenus*-like prey-capture apparatus. The phylogenetic analysis of morphological data of extinct and extant taxa of Steninae and allied subfamilies of Staphylinidae has resolved the extinct genus †*Festenus* as a member of Steninae, and has been suggested to form a stem lineage of Steninae. The presence of †*Festenus* in the Later Cretaceous implies an early origin of the prey-capture apparatus in Steninae, with this apparatus possibly driving evolution towards the crown *Stenus* (Żyła et al. 2017).

Whereas no "*Dianous*" fossils are known so far, the oldest fossil records of *Stenus* beetles date back 70–100 million years to the Cretaceous *Stenus inexpectatus* Schlüter, 1978 from Bezonnais, France, and *Stenus imputribilis* Ryvkin, 1988 from Obeshchayushchiy Creek, Russia. Both these fossil taxa are characterized by large eyes, a three-segmented antennal club and antennal insertions positioned on the frons behind the anterior margin of eyes, a character combination similar to those of *Stenus* (Żyła et al. 2017). However, the poor degree of preservation of these fossils makes their more confident identification and characterization impossible (Żyła et al. 2017). Definite *Stenus* with a prey-capture apparatus are known from about 50 mya old Baltic amber (Schlüter 1978; Ryvkin 1988; Puthz 2010; Chatzimanolis 2018). Younger (sub-) fossil material is known from Pleistocene deposits (e.g. Puthz 1971).

The monophyly of Steninae is supported by many larval and adult autapomorphies (Kasule 1966; Puthz 1981; Hansen 1997; Leschen and Newton 2003; Thayer 2005; Clarke and Grebennikov 2009; Grebennikov and Newton 2009) and is also suggested by molecular analyses (Grebennikov and Newton 2009; Koerner et al. 2013; Zhang and Zhou 2013; Lang et al. 2015). According to the phylogenetic analysis of Clarke and Grebennikov (2009) and Żyła et al. (2017), the autapomorphic characters of adult Steninae are (1) the dorsal insertion of the antennae (between the anterior margins of the compound eyes), (2) the presence of apodemes arising from interantennal pits, (3) the absence of the tentorial bridge (Fig. 11.4), (4) the presence of a tentorial loop (Fig. 11.4), (5) the insertion of the labial palps closely together near the anterior margin of the labium, (6) the pronotal marginal carina meeting the pronotosternal suture anterolaterally, (7) the apex of the mesoventral intercoxal process abutting the apex of the metaventral process and (8) the opening of the

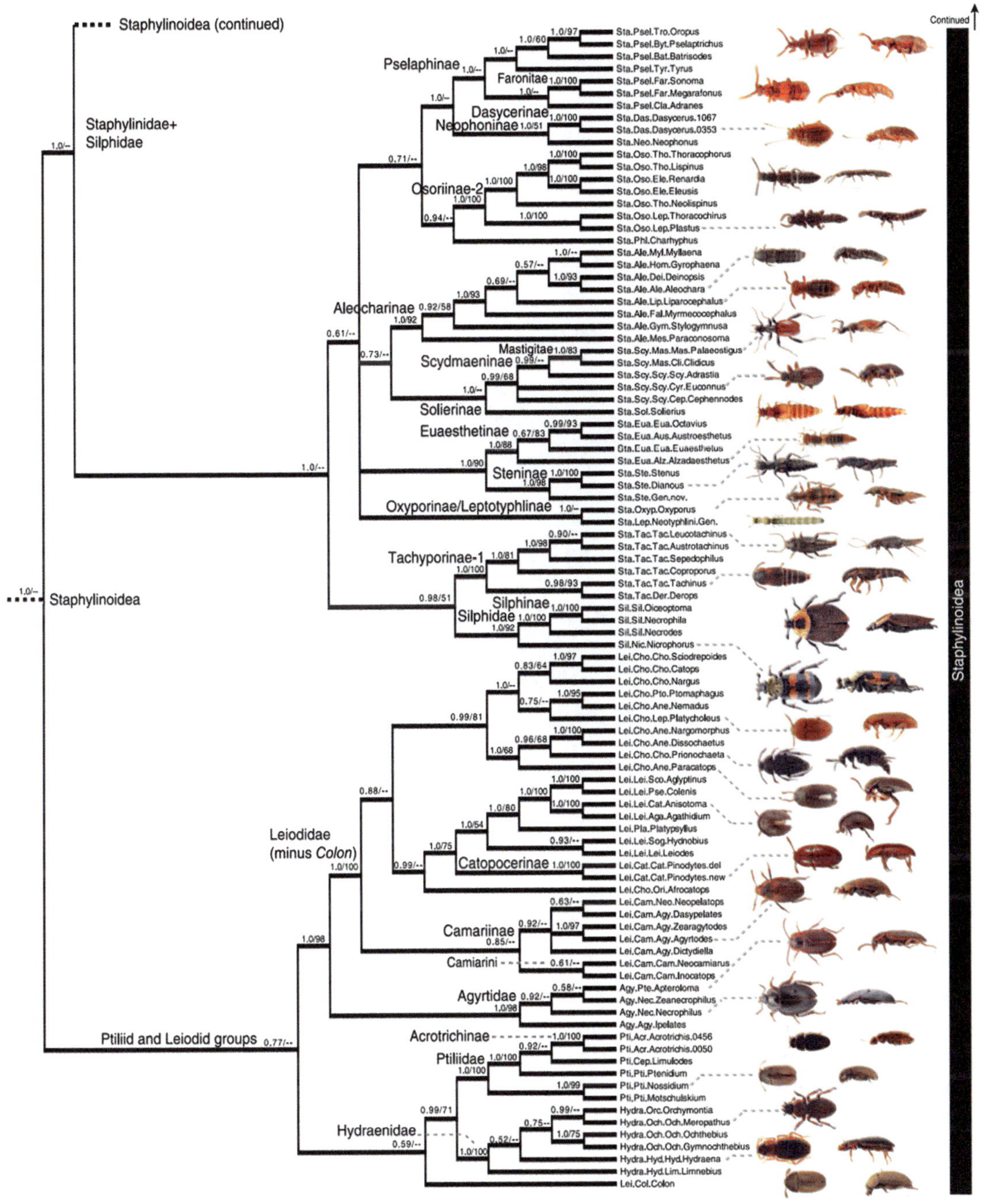

Fig. 11.3 Bayesian 50% MR consensus tree showing the relationships among major lineages of beetles in the infraordinal group Staphyliniformia. Bayesian posterior probabilities (first/left number) and partitioned maximum likelihood bootstrap support (second/right number) are shown above branches. This tree is based on analyses of DNA sequence data from 28S rDNA and the nuclear protein-encoding gene CAD [Excerpt from McKenna et al. (2015). With kind permission from John Wiley and Sons]

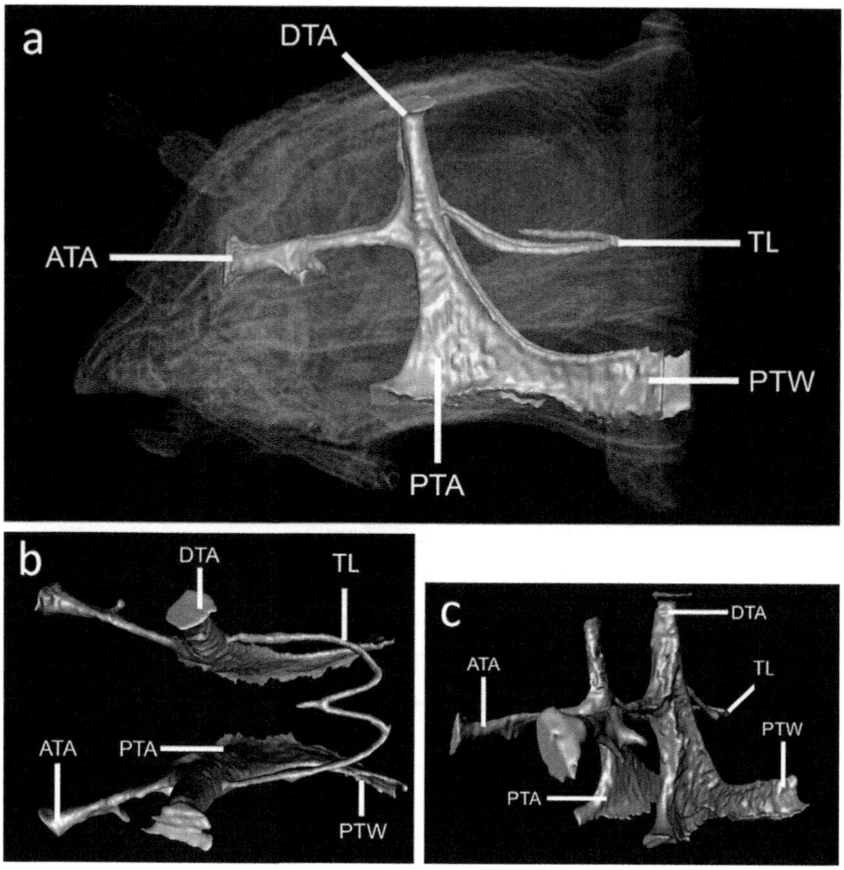

Fig. 11.4 Synchrotron µCT images of the tentorium of *Stenus cicindeloides*. (**a**) Lateral, (**b**) dorsal, (**c**) fronto-lateral view. Note that a tentorial bridge connecting both the posterior tentorial walls is lacking, whereas a more dorsal tentorial loop that connects both the dorsal tentorial arms is present. Identical conditions were found in *Dianous coerulescens*. Abbreviations: *ATA* anterior tentorial arms, *DTA* dorsal tentorial arms, *PTA* posterior tentorial arms, *PTW* posterior tentorial walls, *TL* tentorial loop

paired pygidial defence glands into the rectum. Several larval characters add to this autapomorphic complex, i.e. (9) the setae on the cranium and the tergites being differentiated in length into long and thick versus short and thin; (10) the first antennomere being constricted, so that the antennae appear four-segmented; (11) the antennomeres 1 and 2 being especially elongate; (12) the second maxillary palpomeres being bent and much longer than wide; (13) the tibia having apical setae being longer than the claws; and (14) the abdominal segment IX showing a latero-ventral projection on each side (Clarke and Grebennikov 2009).

Whereas the genus *Stenus* has been considered monophyletic on the basis of its possession of the labial adhesive prey-capture apparatus and several other adult (mostly related to the prey-capture apparatus) and larval characters, the genus "*Dianous*" is not defined by any autapomorphies (Puthz 1981; Clarke and Grebennikov 2009; Żyła et al. 2017). Molecular analyses based on three genes (COI, 16S rRNA, Histone H3) indicate the evolutionary origin of "*Dianous*" within *Stenus*, suggesting a secondary loss of the specialized prey-capture apparatus and rendering *Stenus* paraphyletic (Koerner et al. 2013; Lang et al. 2015). This view is consistent with chemotaxonomic analyses (Lusebrink 2007), the morphology of the paired pygidial glands (Jenkins 1957) and the tentorial characters shown in Fig. 11.4 (see also Żyła et al. 2017) as established in both genera. If the hypothesis of paraphyly of *Stenus* with respect to "*Dianous*" gains further support, taxonomists will be forced to abandon the genus "*Dianous*," which, in this case, would become a subgenus of the genus *Stenus*.

Three as yet undescribed Australian species with protrusible labia bearing adhesive cushions probably form an additional monophyletic genus of Steninae. This hypothesis is confirmed by phylogenetic analyses based on DNA data from 28S rDNA and the nuclear protein-encoding gene CAD (Fig. 11.3), because the undescribed Australian species have a basal position within the Steninae (indicated as "Sta.Ste.Gen.nov." in Fig. 11.3). Their partial resemblance to the Euaesthetinae (e.g. reduced eye size in at least one species) may have resulted from convergent lifestyles in the litter layer (Clarke and Grebennikov 2009). The basal arrangement of this undescribed genus, which possesses a protrusible prey-capture apparatus similar to that of *Stenus*, further supports the assumption of a secondary loss of the specialized prey-capture apparatus of "*Dianous*."

With respect to the intrageneric classification, *Stenus* was traditionally subdivided into morphologically defined subgenera (e.g. Schülke and Smetana 2015). The genus *Stenus* was grouped into subgenera based on various morphological features by staphylinid specialists (Rey 1884; Motschulsky in Hermann 2001; Lusebrink 2007; Puthz 2008). Originally, the genus was grouped into six subgenera, i.e. *Stenus, Nestus, Tesnus, Hemistenus, Hypostenus* and *Parastenus* (see also the determination key of Lohse 1964, which uses an outdated subgenus concept), mainly based on morphological features. These characteristics, for example, are the appearance of the fourth segment of the metatarsi (simple or bilobed), the relative length of the first and fifth tarsomere of the metatarsi, the relative length of the metatarsi and the metatibiae and the presence or absence of abdominal paratergites (Cameron 1930; Lohse 1964; Zhao and Zhou 2004; Koerner et al. 2013). Later, subgenera were taxonomically revised resulting in five valid subgenera: *Stenus* s. str., *Hemistenus* Motschulsky 1860, *Hypostenus* Rey 1884, *Metatesnus* Adam 1987 and *Tesnus* Rey 1884 (Puthz 2001, 2008). Here, *Nestus* belongs to *Stenus* s. str., *Hemistenus* is redefined to *Metatesnus* and *Parastenus* is renamed to *Hemistenus*. However, recent findings indicate that this classification is artificial and probably does not reflect authentic phylogenetic relationships. Currently, the genus is grouped into 157 monophyletic species groups that better reflect the phylogenetic relationships within this group (Puthz 2008), a view that is supported by a few available intrageneric molecular studies (Koerner et al. 2013; Lang et al. 2015; Serri et al. 2016).

In "*Dianous*," according to the morphology of the frons (and some other characters), the species can be classified into species groups I and II (Puthz 1981, 2000b; Shi and Zhou 2011; Tang et al. 2011; Puthz 2016). In contrast to "*Dianous*" group II (about 70% of all "*Dianous*" species), the members of "*Dianous*" group I (about 30% of all "*Dianous*" species) have large *Stenus*-like eyes and were traditionally considered to belong to the genus *Stenus* until the recognition of their lack of the typical prey-capture apparatus of this genus (Puthz 1981). "*Dianous*" group II has been morphologically subdivided into nine "species complexes" that probably form phylogenetic entities (Puthz 2016).

11.2.2 Species Numbers and Zoogeography

The subfamily Steninae has a worldwide distribution, occurring (from the lowlands up to alpine elevations) in the northern temperate, the southern temperate and the tropical zones (Thayer 2005). With respect to geographical regions, Steninae occur in the Nearctic (188/2),

Neotropical (601/0), Palearctic (including China) (1337/151), Afrotropical (358/0), Oriental (including China and Pacific Islands) (629/59) and Australian regions (29/0) [numbers behind each region refer to species numbers (*Stenus*/ "*Dianous*") according to Puthz (in litt.)]. The total number of Holarctic species amounts to 33 (*Stenus*) and 0 ("*Dianous*"), respectively.

Stenus currently comprises about 3000 species. Its worldwide distribution (with the exception of Antarctica and New Zealand) is indicative of a Gondwanan origin, whereby Puthz (2010) assumes its centre of origin in Asia. The majority of the species have bilobed tarsi in combination with an abdomen that does not show distinct raised margins (Puthz 2017). Most species inhabit the southern hemisphere (Puthz 2017).

"*Dianous*" currently comprises >300 species and is distributed in the Oriental, the Palearctic and the Nearctic regions (not in the Ethiopian region) with its main distribution in Asia (India, China and Southeast Asia). The centre of the distribution of the genus "*Dianous*" is in the mountain areas between the Palearctic and the Oriental regions south of 31° northern latitude (Indochina Peninsula and southern slopes of the Himalaya) (Puthz 1981, 2000b; Shi and Zhou 2011) (Fig. 11.5). From here, it has probably dispersed in (1) a south-eastern direction, (2) a north-western direction and (3) via the Bering Straits to North America (Puthz 1981).

The overall range of the distribution of *Stenus* species that occur in Central Europe is provided in Horion (1963) and Puthz (2012). The world-wide distribution of the currently known "*Dianous*" species is provided by Puthz (2000b, 2015, 2016) and Shi and Zhou (2011).

11.3 Morphological Characterization of Eggs, Larvae, Pupae and Adults of Extant Steninae

11.3.1 Eggs

Depending on the species, the eggs of *Stenus* beetles (Fig. 11.11a, b, f–g) come in various colours such as whitish, brownish or yellow–orange (Weinreich 1968). The submillimetre-sized eggs have an ellipsoid shape and may be laid singly or in clusters of up to 20 eggs (Betz 1999), conglutinated to each other and to the substrate by an adhesive that forms a small appendage at one pole of the egg (Weinreich 1968, Fig. 21) (Fig. 11.11f).

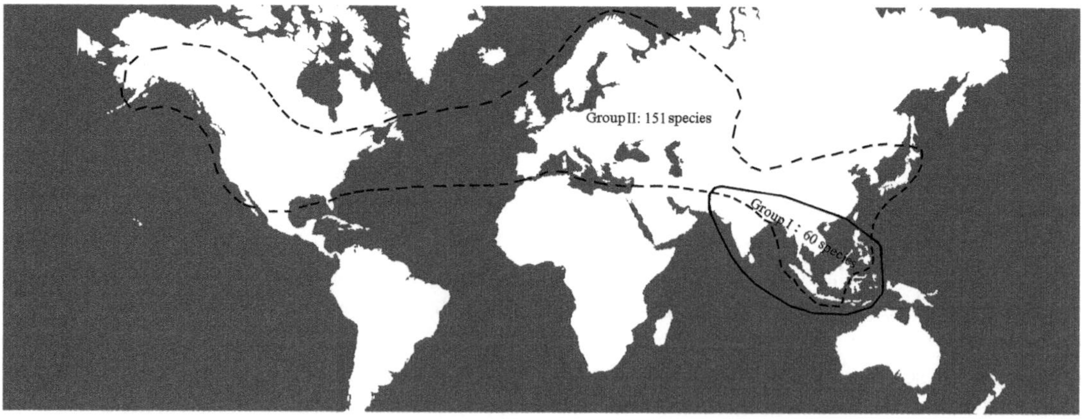

Fig. 11.5 The distribution of the genus "*Dianous*." Solid line, distribution of "*Dianous*" species group I; dashed line, distribution of "*Dianous*" species group II [From Shi and Zhou (2011, Fig. 25). With kind permission from John Wiley and Sons]

11.3.2 Larvae

Steninae larvae pass through three instars (L_1–L_3) before they hatch into the pupa. Only a very few larval descriptions are available at the species level (e.g. Welch 1966; Weinreich 1968). The larval morphology of selected *Stenus* species has been described in detail by Weinreich (1968) (Fig. 11.6). Specific characteristics of larval Steninae are also given in Kasule (1966), Newton (1990), Frank (1991), Beutel and Molenda (1997), Leschen and Newton (2003) and Thayer (2005). Their basic colour is whitish, whereby the more strongly sclerotized parts are brown–black. The entire body surface including the setae appears glossy and sticky.

The head capsule (Fig. 11.6b), including the fused and somewhat advanced labrum, is rounded and bears a characteristic chaetotaxy. It has no neck. Laterally, on each side, the head bears six stemmata that are arranged in an oval shape. The long and slender mandibles are falciform; along the inner side of their apical third, they are denticulate but do not bear any mola or prostheca. In the maxillae, the three-membered palpus rests on a compact palpifer. In the labium, the palpifer bears a two-palpomered palpus. The ligula is broad and bilobed. The antennae are with three antennomeres but appear as with four antennomeres because of a basal constriction of the first antennomere. Their first and second antennomeres are markedly elongated, whereas the third one is very short. The pre-apical antennomere laterally bears a sensory appendage. The frontal suture is Y-shaped. Together with the elongated coronal suture, it divides the

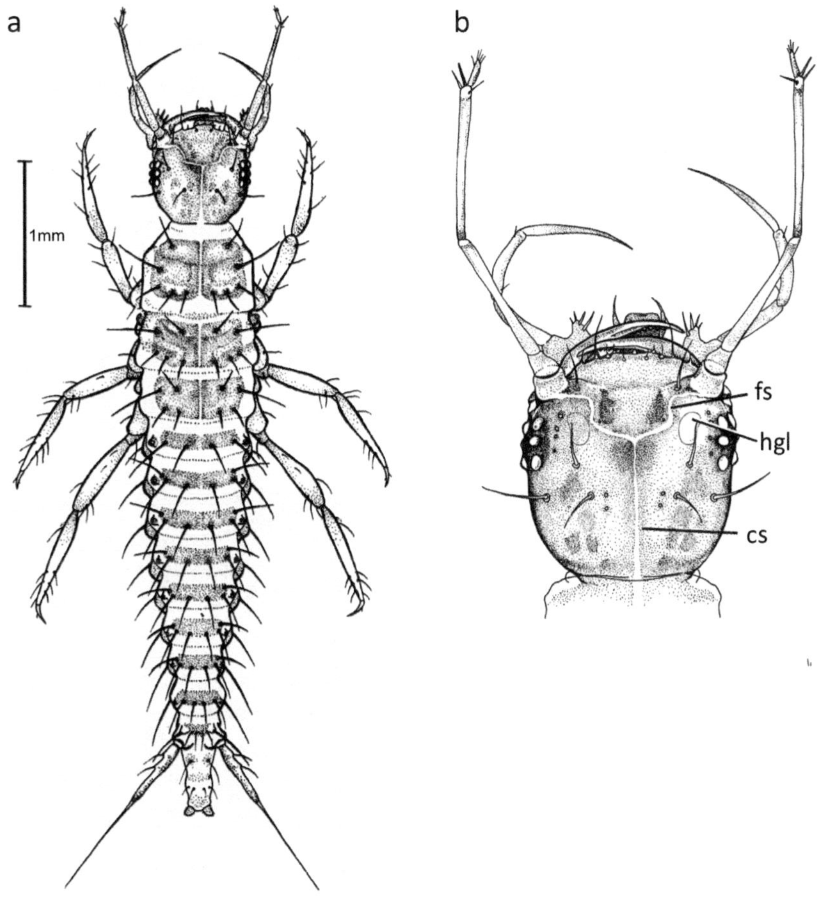

Fig. 11.6 Dorsal habitus of (**a**) entire larva and (**b**) larval head of *Stenus comma*. From Weinreich (1968). In (**b**), the internal position of the head gland according to Beutel and Molenda (1997) is added to the original drawing. Abbreviations: *cs* coronal suture, *fs* frontal suture, *hgl* internal position of head gland

dorsal side of the head capsule into three areas. A pair of glands occurs lateral of the frontal suture. Medially, the ocelli are accompanied by especially arranged pores. Such pores (single or paired) are also found on the vertex.

The dark-coloured thoracic tergites bear a longitudinal median ecdysial line. The anterolateral corners of all three thoracomeres exhibit a pair of gland openings (not shown in Fig. 11.6). Between the pro- and mesocoxae, one pair of spiracles opens at the ends of spiracular tubes. The slender legs bear a uniarticulate tarsus.

The Steninae abdomen is relatively short, making up less than 1.2 times the length of the thorax (Thayer 2005). Abdominal spiracles are equally developed on segments I–VIII. Apart from their characteristic chaetotaxy, the abdominal tergites bear a pair of anterolateral gland openings. The urogomphi of the ninth abdominal segment are single-membered and originate from a basal dome-like structure.

No "egg bursters" are known from the head or the thoracic tergites. Kasule (1966) provides a key with characters for distinguishing between *Stenus* and "*Dianous*" larvae.

11.3.3 Pupae

The pupal instar of *Stenus* was described by Welch (1966) and Weinreich (1968). These pupae belong to the exarate non-sclerotized type (Fig. 11.7). The head and the mouthparts

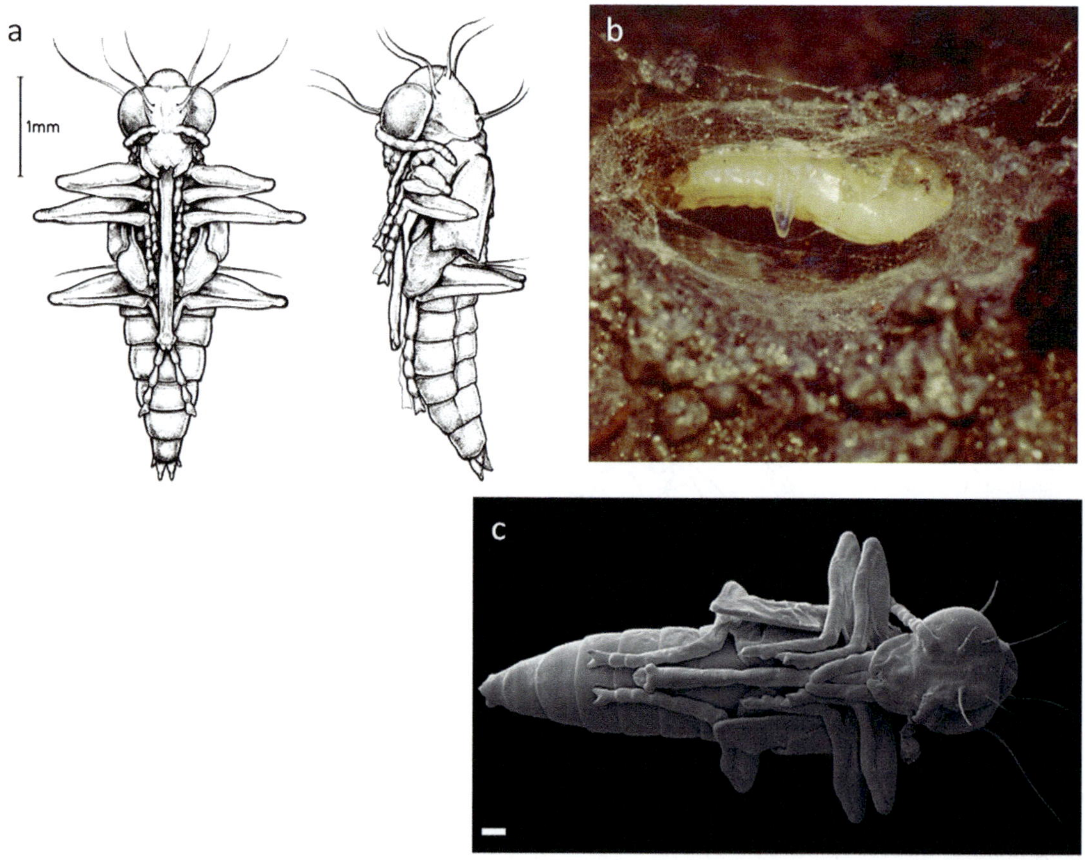

Fig. 11.7 Pupae of *Stenus* species. (**a**) Ventral and lateral view of *S. comma*. From Weinreich (1968). (**b**) Lateral view of cocooned pupa of *S. comma* with cocoon carefully opened. Photo courtesy of Martin Pfeiffer. (**c**) Scanning electron microscopic (SEM) image of ventral aspect of pupa of *S. guttula*. Scale bar = 200 μm

show a hypognathous orientation. Freshly hatched pupae have a whitish appearance but become darkly pigmented later on. The *Stenus* pupa is instantly identifiable by its laterally protruding eyes and the elongate labium that it bears on its ventral side (the pupa of *Dianous coerulescens* does not show such an elongate labium). On the dorsal head capsule (between the eyes), the pronotum and the first (and potentially the second) abdominal tergite, the pupae possess one to two pairs of elongate spiny projections that darken in 1–2 days after hatching.

11.3.4 Adults

Adult Steninae are characterized by their laterally protruding large eyes, antennae with 11 antennomeres (with two thicker basal antennomeres and a weak club of 3 antennomeres) originating on the vertex between the eyes, a cylindrical pronotum without lateral margins [the pronotum of †*Eocenostenus* is strongly transverse (Cai et al. 2014) and the pronotum of †*Festenus* bears marginal carina reaching the anterolateral prothoracic margin (Żyła et al. 2017)], the small cone-shaped procoxae that are sitting in entirely closed procoxal cavities, cylindrical short elytra not projecting to the metathorax and without an epipleural keel [such epipleural keel is present in †*Festenus* (Żyła et al. 2017)], the markedly separated metacoxae, the pentamerous tarsi, a slender cylindrical abdomen with six visible sterna and paired internal defensive glands that empty into the anus and a usually symmetric copulatory apparatus (i.e. the aedeagus) (e.g. Puthz 1971, 2012; Newton et al. 2001; Lott and Anderson 2011; Thayer 2016). The abdominal musculature of *S. clavicornis* does not show the diagonal crossing type (as characteristic for many other staphylinids with a freely movable abdomen) but the regular longitudinal muscle pattern (Blum 1979). A more comprehensive list of autapomorphic adult characters of Steninae is provided by Clarke and Grebennikov (2009).

The aedeagus is composed of the median lobe (i.e. the penis) and the paired lateral lobes (i.e. the parameres), but, unlike those from other staphylinids, the parameres insert at the dorsal side of the median lobe (Puthz 1971; Naomi 2018). The median lobe consists of a large basal bulb with musculature; a contractor muscle is considered to heighten the pressure of body fluid within the median lobe, by which the internal sac (i.e. the endophallus) is evaginated during copulation. The internal sac is basically furnished with paired expulsion hooks, longitudinal bands, additional teeth, hooks or tufts, or it is modified into a protrusible tube-like flagellum (Puthz 1971; Naomi 2018). Female Steninae possess unsclerotized or well-sclerotized spermatheca (Puthz 1971). Compared with *Stenus*, the aedeagi of "*Dianous*" are more uniform both in outline and in internal structures (Puthz 2000b).

Stenus and "*Dianous*" are mainly distinguished by the presence or absence of the elongated protrusible labium bearing a pair of adhesive cushions at its tip (Figs. 11.2, 11.8, 11.23 and 11.24). In addition, in *Stenus* only, the mentum is divided by a medial longitudinal carina. Other differences such as the relative size of the compound eyes, the presence or absence of long terminal styles at the ninth abdominal segment and the morphology of the genital organs are not consistent throughout the species of the two genera (Puthz 1981) supporting the view that *Stenus* may be paraphyletic with respect to "*Dianous*."

11.3.5 Functional Morphology of the Labial Prey-Capture Apparatus of *Stenus* Beetles

The labial prey-capture apparatus of adult *Stenus* beetles is composed of a distal elongated rod-like prementum that is connected to the head by an equally elongated membranous connective membrane (Fig. 11.8a). This membrane can be invaginated like the inside-out finger of a glove. At its tip, the prementum bears two adhesive pads (i.e. the modified paraglossae) that attain their sticky function via the viscosity of a

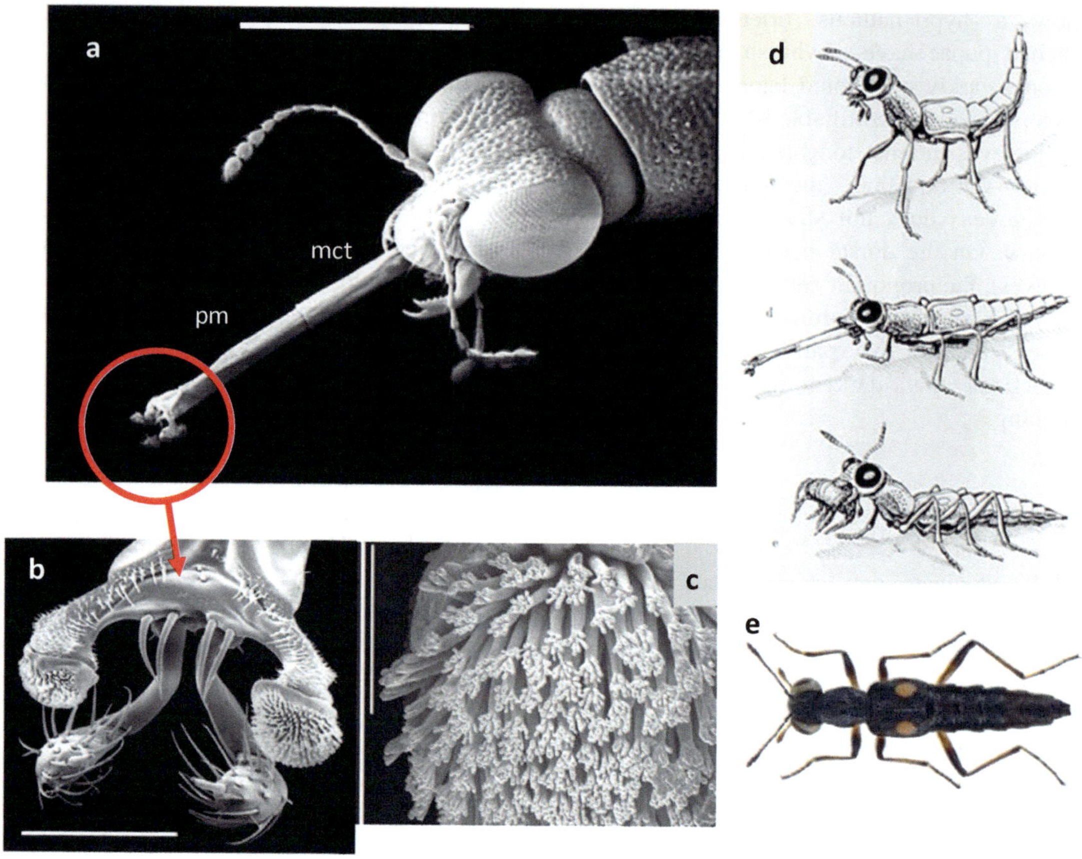

Fig. 11.8 Adhesive prey-capture apparatus of *Stenus* spp. (**a–c**) SEM images. (**a, b**) *S. comma*. From Bauer and Pfeiffer (1991); (**c**) *S. nitidiusculus*. (**a**) Head with protruded labium. Scale bar = 1 mm. (**b**) Dorso-frontal view of the apex of the prementum with the paraglossae modified into sticky pads. Scale bar = 100 µm. (**c**) Ventrolateral detailed aspect of a sticky pad. Note the terminally branched adhesive trichomes. Scale bar = 20 µm. (**d**) Prey-capture sequence in *S. comma*. After approaching the prey to a critical distance (top), the prey-capture apparatus is rapidly protruded (middle), and the glued prey is withdrawn to the mandibles (bottom). From Weinreich (1968). (**e**) Habitus image of *S. guttula*. Length: 4.3 mm. Courtesy of Udo Schmidt (www.kaefer-der-welt.de). Abbreviations: *mct* membranous connecting tube, *pm* prementum

gel-like adhesive secretion that is released through lateral pores and that distributes itself over the whole surface of the paraglossa, which is differentiated into terminally branched trichomes (Fig. 11.8c). The connecting membrane and the prementum contain bundles of ductules that transport the adhesive secretion produced by prominent secretory glands in the head (described in Weinreich 1968; Kölsch 2000; reviewed in Betz 2010) to the sticky cushions. Internally, the sticky cushions are made up of a complex reticulum of endocuticular fibres (Betz 1996; Kölsch and Betz 1998). Together with the mesocuticular nature of their outer wall and their resilin content (Koerner et al. 2012a), this makes these structures highly flexible and elastic, so that they can closely adapt themselves to the shape and surface irregularities of the prey. Moreover, according to their loose arrangement, the sticky cushions can be inflated immediately prior to a strike by increased haemolymph pressure. On their external surface, the sticky cushions are differentiated into a large number of adhesive trichomes (Fig. 11.8c). Each trichome branches out terminally, thereby dramatically increasing the total number of adhesive contacts. The number of both trichomes and terminal branches is species-specific and may amount to many thousand adhesive contacts per sticky cushion (Bauer and Pfeiffer 1991; Betz 1996; Koerner et al. 2017).

Upon prey-capture, the rod-like labium is hurled out of the body and has to be retracted after prey-capture by dorsal and ventral retractor muscles in order to bring the prey into the range of the mandibles (Fig. 11.8d).

On the basis of several morphological and behavioural studies (e.g. Schmitz 1943; Weinreich 1968; Bauer and Pfeiffer 1991; Betz 1996, 1998a; Kölsch and Betz 1998; Kölsch 2000; Betz and Kölsch 2004; Betz et al. 2009; Koerner et al. 2012a, b, 2017), a functional model can be suggested for the working principles of the prey-capture apparatus (Fig. 11.9) (Kölsch and Betz 1998). When not in use, the labium is withdrawn back into the head, where it is wrapped by the connecting membranous tube. In order to capture prey, the beetles rapidly protrude their prementum from this resting position towards the prey within only 3–5 ms. The prey adheres to the sticky cushions and is seized by the mandibles after immediate retraction of the prementum. The rapid protrusion of the labium is made possible by a catapult mechanism that involves the antagonistic action of increased haemolymph pressure on the one hand and the contraction of large retractor muscles on the other (Fig. 11.9).

During protrusion, the labium is not simply pressed out of the body by the inflowing haemolymph, because the lumen in the inner tube in the transition zone between the prementum and the membranous connecting tube is filled with tissues (retractor muscles, nerve tissue, gland ductules, tracheae) as indicated by the yellow area in Fig. 11.9c. This largely hinders any influx of haemolymph. However, this changes when the labium is in its

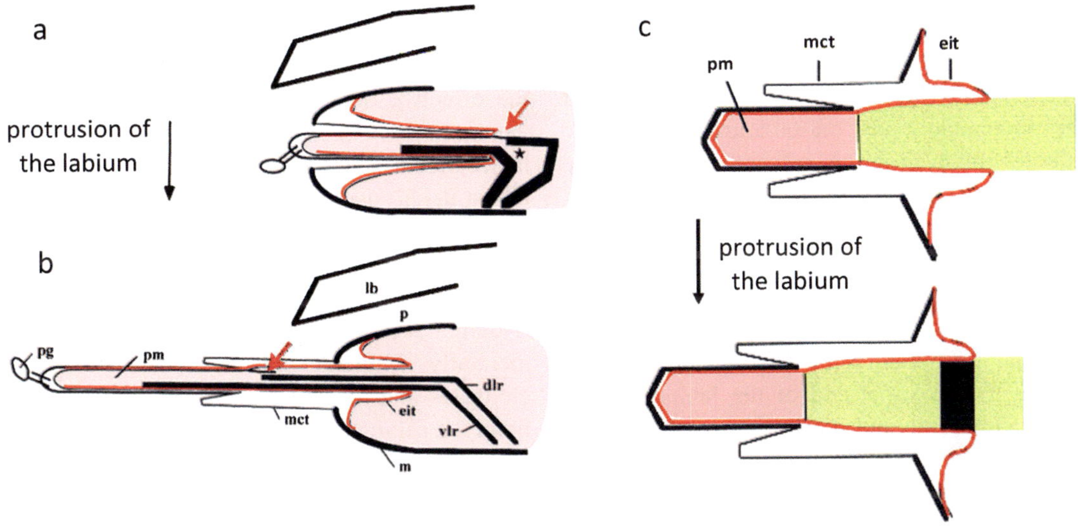

Fig. 11.9 Schematic sagittal section through the labial prey-capture apparatus in *Stenus* spp. Functional model at the resting position (**a**) and the active capture position (**b**). The pink area indicates the region filled with haemolymph. The arrow points to the dorsal apodeme of the prementum onto which the dorsal labial retractors are inserted. For further explanations, see text. Details of the way that the internal lumens of the protruding labium is filled are shown in (**c**). The extra inner tube (red line) turns inside out proximally. The yellow area represents the complex of tissues that fill the inner part of the proximal half in a fully protruded labium. The black rectangle in the lower scheme shows the position at which the "increasing" volume within the extra inner tube is created during protrusion. Note that, during protrusion, the membranous connecting tube (black line) unfolds distally, whereas the extra inner tube (red line) does so proximally. For further explanations, see text. Abbreviations: *dlr* dorsal labial retractors, *eit* extra inner epidermal tube, which directly wraps the inner complex of tissue, *lb* labrum, *m* mentum, *mct* membranous connecting tube, *pg* paraglossa, *pm* prementum, *vlr* ventral labial retractors. Modified from Kölsch and Betz (1998)

resting position (Fig. 11.9a). In this position, an open cavity forms between the dorsal and the ventral retractor muscles[2] (asterisk in Fig. 11.9a); through this cavity, the prementum can be filled with haemolymph, while it is still in its resting position. The increase of the haemolymph pressure within the prementum probably also causes the observed erection of the adhesive pads, so that their adhesive surfaces are frontally directed towards the prey (pg in Fig. 11.9b). Despite the increased haemolymph pressure, in this phase, the labium is not yet pressed out of the head, as long as the contracted labial retractor muscles antagonize this pressure. Only once these muscles relax, is the labium (preloaded by haemolymph pressure like a bent bow) catapulted out of the body.

Hence, we are dealing with a catapult mechanism in which the elastic elements of the labium (including the retractor muscles and the resilin-containing membranous connecting tube) are set under an initial stress by an increase of the haemolymph pressure. This haemolymph pressure is probably caused by the contraction of the abdominal body musculature, so that the muscles involved in this antagonism are located at very different parts of the body.

The mechanism of the protrusion of the labium is detailed in Fig. 11.9c. The increasing inner space within the extra inner tube (i.e. the epidermis that has become separate from the cuticular membranous connecting tube) is created at the proximal end of the labium within the head capsule and not in the transition zone between the membranous connecting tube and the prementum. This is a consequence of the extra inner tube (i.e. the epidermis) and the membranous connecting tube (i.e. the cuticle) not lying in parallel (Fig. 11.9). Rather, the complex of the inner tissues and the narrow extra inner tube form an integrated whole, so that the inner tissue complex does not have to glide within the extra inner tube but moves together with it during both protrusion and retraction.

The trichomes at the surface of the sticky cushions are usually deeply immersed in the adhesive secretion, and hence only the tips of their branches protrude (Fig. 11.10a). The adhesive secretion is released to the exterior via pores that are located in the lateral flank of the sticky cushions (arrow in Fig. 11.10b). From here, it spreads all over their surface. Because of the large amount of the involved secretion, the adhesive working mechanism can be assigned to the viscosity of the adhesive that comes into action during the rapid withdrawal of the labium towards the head after the strike (Fig. 11.10c). Ultrastructural images show that droplets of a lipid-like substance emulsify in a larger aqueous protein fraction (Fig. 11.10b) (Kölsch 2000). Additional histochemical tests have identified water-soluble sugars, proteins and lipids indicating that the secretion is a complex mix of more than one chemical phase (Betz et al. 2009). An emulsion of this kind might be beneficial for the effective distribution of the secretion over various types of surface (hydrophilic *versus* hydrophobic) but might also help precisely to adjust its viscosity. Indeed, the high viscosity of the adhesive secretion has been demonstrated in high-speed video recordings that have shown the sticky pad being retracted from a surface (Koerner et al. 2012a). As is the case with commercially available adhesives, the secretion elongates and forms long parallel fibres, before finally breaking away from the contact area of the substrate. This observation supports the importance of the cohesive forces that give the secretion a high level of internal strength. Additionally, the drag of the fluid is transferred to the pad cuticle whose upper layer is considered to be very soft and compliable, deforming viscoelastically in the direction of the pull (as can be seen in the high-speed video footage of Koerner et al. (2012a). The elongation of the entire pad upon pull-off is further supported by its resilin content and additional structural peculiarities discussed in Kölsch and Betz (1998).

[2] The dorsal retractor muscles insert at an apodeme that arises from the dorsal wall of the proximal base of the prementum (see arrow in Fig. 11.9a, b), whereas the ventral muscles insert further distally at the inner side of the ventral wall of the prementum (Fig. 11.9a, b).

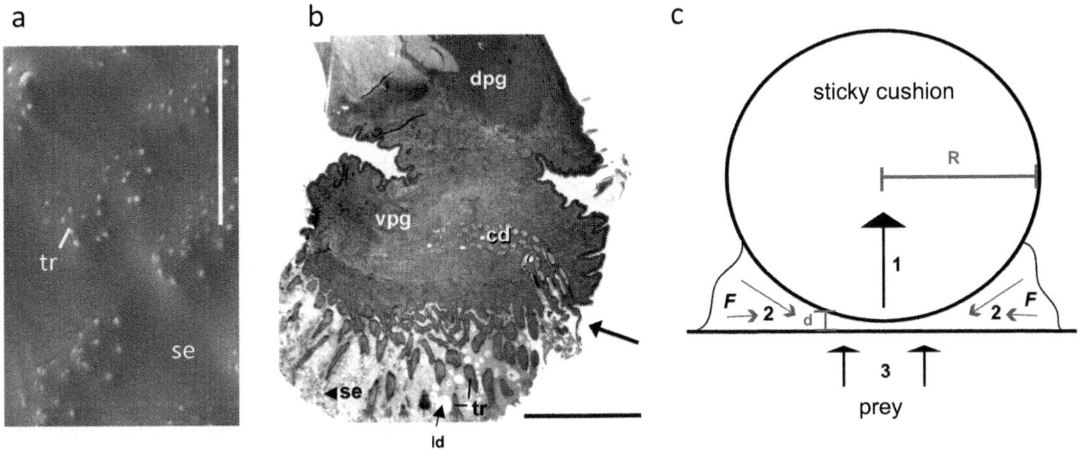

Fig. 11.10 Structure and function of the paraglossae modified into sticky cushions in *Stenus* spp. (**a**) Cryo-scanning electron microscopic view of a part of the surface of a sticky cushion showing the tips of the adhesive trichomes emerging from the adhesive secretion. Scale bar = 3 μm. (**b**) Transmission electron microscopic image of a transverse section through a sticky cushion in *S. juno*. The arrow points to the location on the paraglossa at which the secretion is discharged via the lateral pores. Scale bar = 20 μm. For further explanations, see text. (**c**) Representation of the proposed adhesive mechanism based on the viscosity of the secretion. At withdrawal of the sticky cushion (arrow 1), the emerging gap between both the sticky cushion and the substrate has to be filled by the secretion (arrow 2) provided that it is sufficiently fluid. If the secretion is too viscous to fill the gap immediately, this gap cannot increase, and the prey is drawn instead towards the adhesive pad (arrow 3). Abbreviations: *cd* chitinous ductules, *dpg* dorsal part of paraglossa, *F* adhesive fluid, *R* radius of the contact area, *d* distance between both the surfaces, *ld* presumed lipid droplet, *se* adhesive secretion, *tr* adhesive trichomes, *vpg* ventral part of paraglossa (from Kölsch 2000; Betz and Kölsch 2004)

11.4 Life History, Development and Larval Biology

11.4.1 Life History and Development

In the temperate zone, adult *Stenus* beetles are long-lived and iteroparous (with overlapping generations), reproducing throughout the season (starting in April) (Larsen 1959, 1963; Weinreich 1968). Betz and Fuhrmann (2001) compared three *Stenus* species with respect to their life history. Their dissections of field-collected females over the entire season revealed that, in Northern Germany, egg production was highest in June and had ceased by the end of summer. The average egg number per female and season/week amounted to 29/1.5 (*S. pubescens*), 63/5 (*S. comma*) and 189/14 (*S. juno*), whereby the egg number was negatively correlated to egg size. Whereas *S. comma* and *S. juno* beetles deposited their clutches on the underside of moist bark and reed leaves, *S. pubescens* beetles did so into convoluted leaves or hollow stalks (Betz and Fuhrmann 2001). In all these species, one to two egg clusters were laid per week.

Larval development time is short; the whole development from egg to imago can be accomplished within 1 month (Weinreich 1968). Taking the development times (determined at a temperature regime of 18/10 °C) of eggs (8–11 days), larvae (L_1, 4–5 days; L_2, 3–4 days; L_3, 6–7 days) and pupae (7–8 days) and the observed preoviposition time of about 1 month as a basis, we can assume that the generation time is 2 months. This means that an overwintered adult can produce two to three further generations per season (Betz and Fuhrmann 2001). At a constant temperature of 20 °C, the developmental time from egg to imago is accomplished within 3 weeks (Lang 2014). Whereas most *Stenus* species reproduce and develop from spring to summer, a few Central European species develop in the winter months (e.g. *S. impressus*) (Larsen 1959).

Based on their different locomotory activities, metabolic/consumption rates and egg sizes/numbers, *S. juno* beetles show a reproductive effort that is twice as high as that of *S. comma* and *S. pubescens* (Betz and Fuhrmann 2001). This indicates that different lifestyles are established in *Stenus* entailing different reproductive costs. For inhabitants of open sites, such as *S. comma*, the demand for increased mobility has probably called for spending high maintenance energy at the expense of reproductive effort. For plant-mounting species such as *S. pubescens*, reproductive constraints may arise from an "energy-minimizing" strategy (i.e. low metabolic rates in concurrence with moderate reproductive efforts) that has evolved in adaptation to a diminished food supply.

For "*Dianous*" species, no solid life history data (including egg numbers and developmental times) are available as for *Stenus*.

11.4.2 Larval Biology

Steninae larvae are very rarely found in the field (e.g. Blair 1917), so that our knowledge on their biology is based on laboratory observations. Within an egg cluster, the larvae (as observed in *S. comma*, *S. juno* and *S. pubescens*) mostly hatch synchronously within 1 h. Upon hatching (the entire process takes about 4 min), the larvae penetrate the chorion with their heads first, followed by the extraction of their legs, which are then used to erect the flexed antennae and maxillary palps. Finally, the hind body is drawn out of the chorion (Fig. 11.11c).

Only a few minutes after hatching, the larvae begin to search for prey such as springtails. While searching for prey, the head performs bidirectional pendulum movements to increase its range of perception. While doing so, the antennae and the maxillary palps are directed forward and do not make contact with the ground. Prey detection does not seem to involve visual clues, since a predatory strike is only released upon a direct mechanical contact. Usually, the first contact is accomplished with the antennae, whereby this contact seems to be below the threshold for the prey. Upon prey contact, the larva rears up its forebody and tries to overwhelm the prey from above. In this phase, the pygopodium attaches to the ground, thus supporting the body. During the strike, the mandibles are first widely opened and subsequently rapidly closed to seize the prey. Larger prey animals are kept away from the ground by the beetle raising its head. Sometimes, the first pair of legs is used to hold the prey. Once the prey is captured, it is held in place by the labium (from below) and by both the maxillae (from the sides). The mandibles penetrate the prey cuticle, so that the prey can be extra-orally digested. In this phase, the prey is further squeezed by the mouthparts and the forelegs and is finally sucked out.

Similar to the imagoes, *Stenus* larvae involve adhesive mechanisms in prey-capture (Larsen 1959, 1963). Springtails often adhere to the mouthparts, the antennae, the legs or other parts of the integument before they are taken off with the legs and/or captured with the mandibles (Betz and Kölsch 2004; Lang 2014). Accordingly, the entire body surface of *Stenus* larvae often has a glossy appearance, and sometimes secretion droplets can be found at the apex of the body setae. Although fast-fleeing prey such as springtails can be captured without any involvement of adhesive mechanisms, adhesion certainly contributes to the high capture success of these larvae towards elusive prey such as springtails. Under laboratory conditions, the capture success of *S. pubescens* and *S. comma* attains 70–90% (Betz and Kölsch 2004). The origin of the adhesive secretion in *Stenus* larvae has not as yet been investigated. Potential candidates are glands associated with the paired openings found dorsally on the head, the three thoracic tergites and the abdominal tergites I–IX. When reared in the laboratory, the larvae of all three stages show an aggressive cannibalistic behaviour.

Before hatching into the next stage, the larvae search a retreat to spin themselves a cocoon. At least 1 day prior to pupation, the L_3 larvae cease to feed and search for an appropriate pupation

Fig. 11.11 Selected developmental stages of *Stenus juno*. (**a**) Egg cluster, (**b**) mature egg with L_1 larva showing ocelli and antennae shining through, (**c**) hatching L_1 larva disposing of the egg schell, (**d**) L_2 larva with pigmented and fully sclerotized cuticle, (**e**) full-grown L3 larva. Note, at the abdominal tip, the pygopodium that is used to attach the larva to the substratum. Scale bars = 1 mm. (**f–g**) SEM images of *S. bimaculatus* eggs. (**f**) Egg cluster. Scale bar = 200 μm. (**g**) Single egg. Scale bar = 100 μm [Images (**a–e**) from Lang (2014)]

site, where they can spin themselves a cocoon. A *S. comma* pupa sitting in its cocoon is shown in Fig. 11.7b.

During spinning, the larvae perform bidirectional dabbing movements with their pygopodium. The silk is released from the posterior segmental glands of the ninth abdominal segment (Frank and Thomas 1984). In *S. comma*, the silk can be additionally camouflaged by plaiting tiny substrate particles into it. Cocoon building in *Dianous coerulescens* (including the involved gland structures) has thoroughly been described by Jenkins (1958). As also observed in some *Stenus* beetles (e.g. Blair 1917), the cocoon of *D. coerulescens* is a double structure consisting of an outer and an inner envelope. The silk is produced by paired glands opening in the ninth tergite anterior to the insertion of the urogomphi. Spinning is performed under the aid of the urogomphi and the movable abdomen. While spinning the outer envelope, the larva moves in several directions encircling the chosen pupation site. Once the outer envelope is finished (which takes a couple of hours), the larva fabricates the inner envelope that is situated more closely to its own body. The building of the entire cocoon is finished within 12 h (Jenkins 1958). The larva finally turns into the immobile prepupal stage, with pupal hatching occurring about 1 day after cocoon building.

11.5 Ecology and Behaviour

11.5.1 Ecology

General Lifestyle Adult Steninae beetles are diurnal, free-living and optically oriented predators of springtails and other small invertebrates (mainly soft-bodied arthropods such as aphids and dipteran larvae) (e.g. Betz 1998a; Yang 2003; Ryvkin 2012). As long as the prey species do not exceed a certain size limit, the beetles seem to be rather non-specific, although they reject chemically protected prey immediately after seizure (cf. Larsen 1959; Weinreich 1968; Heethoff et al. 2011). Consumption rates of adult *Stenus* species amount to six medium-sized springtails d^{-1} mg^{-1} (Betz and Fuhrmann 2001) and seven aphids d^{-1} and individual *Stenus* (Yang 2003). Because they potentially feed on aphids (in the laboratory, these beetles show a Holling Type II functional response), *Stenus* beetles have been considered as natural enemies for the control of pests in the cultivation of vegetables and rice in China (Li and Zheng 2005).

Parasites In terms of parasites, only infestations by the endoparasitic nematode *Parasitylenchoides* Wachek and members of the fungus genus *Ilyomyces* Picard (Ascomycetes: Laboulbeniaceae) have been reported for Steninae, whereby *Ilyomyces* seems to be host-specific on Steninae (Frank 2018). Investigating two *Stenus* species, Koerner et al. (2016) identified 13 different bacteria species in the adhesive secretion of the labial sticky cushions, although whether these are merely commensals or influence the performance of these pads by other means is unknown.

Habitat Preferences Steninae usually prefer moist habitats and inhabit waterside environments such as reeds or sparsely vegetated sites on river or lake margins or the litter and humus layer in tropical forests. Thayer (2005) specifies the habitat types known to be settled as (1) canopy, (2) foliage of the lower storey, (3) debris (e.g. leaf) layer and grass clumps on ground, (4) logs, (5) debris (and/or moss) and the ground surface close to the water edges, (6) dung and fermenting plant material and (7) the nests of ants (e.g. the Palearctic species *Stenus aterrimus*), birds and mammals (including middens and bivouacs). Rarely, Steninae have been found nearby fungi (Thayer 2005). Few are adventive, probably being spread by humans (e.g. *S. melanopus*) (Puthz 1971). "*Dianous*" species seem to be strongly hygrobiont (and often bryo- and petrimadicolous including logs and rocks immersed in water) and are particularly associated with the spray zone of running water and cascades (e.g. Champion 1919; Puthz 1971, 1981, 2000b; Rougemont 1983, 1985). In Central Europe, *D. coerulescens* beetles can also be found associated with the logs of beaver dams. Other (sporadically reported) "*Dianous*" habitats are dead leaves and wet mosses in forests and tussocks, provided that these habitats are sufficiently moist.

In the Palearctic region, most *Stenus* species occur in littoral freshwater environments, where they inhabit waterside environments such as reeds or sparsely vegetated sites on river and lake margins (e.g. Horion 1963). Several *Stenus* species have also been recorded in very dry habitats (e.g. *S. atratulus*) (Puthz 1971). Other surveys provide more detailed information on the habitat choice of mostly European species (Benick 1929; Renkonen 1934, 1950; Horion 1963; Hammond 1975; Anderson 1984; Koch 1989; Betz 1994; Lott and Anderson 2011; Puthz 2012).

11.5.2 Behaviour

By direct observations of the adult focal animals of eight Central European *Stenus* species inhabiting different biotopes in terraria, a genus-specific behavioural inventory was acquired that included more than 70 distinct behavioural patterns (Betz 1999). These could be assigned to the higher functional categories of feeding, reproduction, grooming, resting and protection. Before this general survey, only single observed behaviours in Steninae had been reported (e.g. Billard and Bruyant 1905; Benick 1922; Delahon 1927; Urban 1928; Renkonen 1934; Voris 1934; Jenkins 1960; Linsenmair 1963; Weinreich 1968; Lecoq 1991). Focal animal sampling revealed that, during the day, the beetles spend their greatest amount of time grooming (mostly 5–30% of the observation time) and searching (for prey) (50–90% of the observation time) (Betz 1999). While searching, the beetles move in a stop and go manner ("saltatory search" according to O'Brien et al. 1989), whereby the movements of potential prey animals are visually detected in the locomotion pauses. Alternatively, these beetles seem to set an ambush, as they remain at a given place and await prey movements in the surroundings. The predatory behaviour of *Stenus* beetles has been investigated in greater detail by using (high-speed) video techniques (Bauer and Pfeiffer 1991; Betz 1998a; Koerner et al. 2012a). Once the beetles have detected a moving prey, they approach it in a series of runs and stops, orient their longitudinal body axis until it is in alignment with the prey (Fig. 11.22) and, as soon as the critical attack distance is reached, lunge forward to seize the prey. Prey seizure can be performed either by abruptly ejecting the elongated labium or by quickly grasping the prey with the mandibles (Fig. 11.12a) (Bauer and Pfeiffer 1991; Betz 1998a). While hunting in

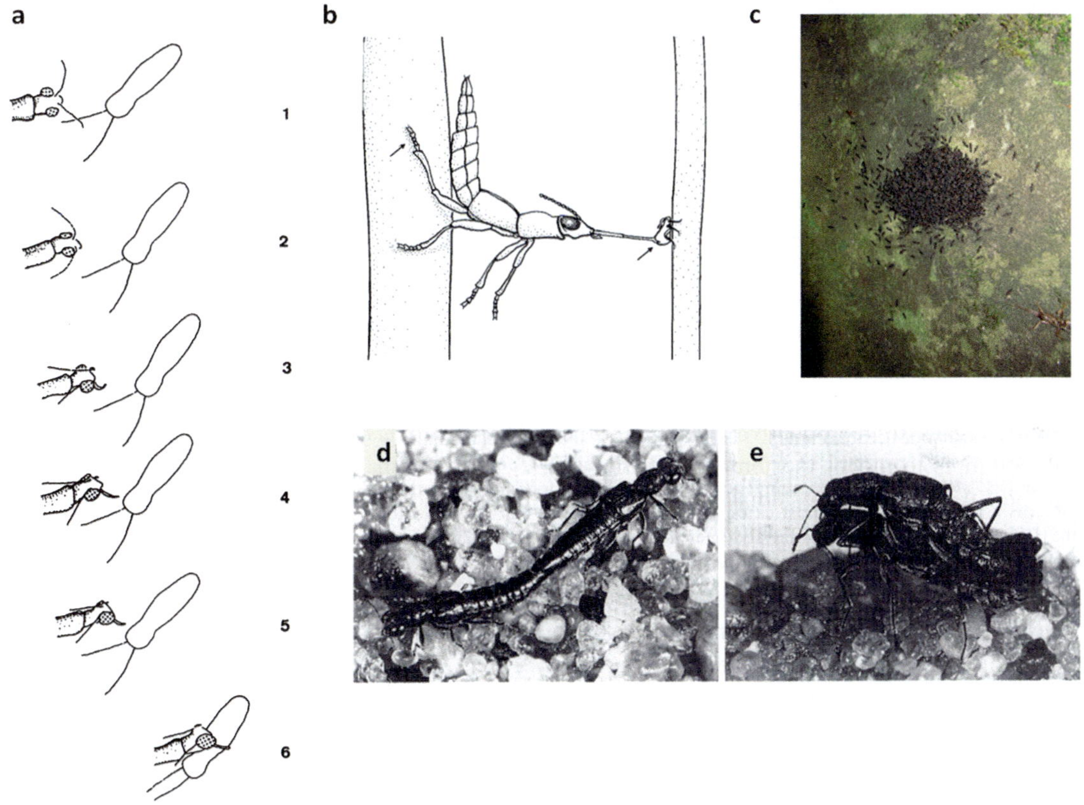

Fig. 11.12 Selected behavioural patterns observed in adult *Stenus* beetles. (**a**) Schematic drawing of the head of a *S. canaliculatus* beetle grasping a large springtail directly with its mandibles. Sequence drawn from high-speed video footage. (**b**) *S. cicindeloides*: characteristic prey-capture situation occurring while hunting an aphid in vegetation. The visible length of the labium amounts to 1.3 mm. Arrows indicate structures (tarsi and paraglossae) in which adhesive forces are required. (**c**) Gregariousness shown by *Dianous fellowesi* in Guangxi, China (courtesy of Jian-Qing Zhu). (**d**) *S. bifoveolatus* pair maintaining a copulatory end-to-end position. (**e**) *S. comma* pair showing a copulatory parallel position. Figures (**d**)–(**e**) from Betz (1999). For further explanations see text

the vegetation or in plant debris, the prey-capture instances may be different, forcing the beetles to adapt to the specific situation by, for example, detaching the foretarsi from the substratum and, additionally, bending the thorax against the abdomen in order to attain otherwise unreachable prey (Fig. 11.12b). Attacks with the labium allow larger attack distances than attacks performed with the mandibles (Bauer and Pfeiffer 1991; Betz 1998a). This reduces the chance of premature detection by the prey and makes it possible to catch prey that would otherwise be out of reach (cf. Fig. 11.12b).

Once a prey item is captured, it is kneaded by the mandibles and the maxillae entailing rotatory movements ("rotary-mill method" according to Leschen and Newton 2003). Digestion is extraoral, followed by the gradually sucking in of the predigested prey material.

In addition to searching (for prey), self-grooming accounts for a substantial part of the time budget of a beetle and consists of quite a variety of different movement patterns (Betz 1999). This behaviour seems to be mainly aimed at spreading a pygidial secretion over the entire body surface. It makes use of the considerable bendability of the abdomen, whose tip may even reach the dorsal surface of the elytra. This will be dealt with in more detail in Sect. 11.6.

With regard to reproductive behaviour, the male attempts to mount the female and tries to copulate. A non-receptive female may forestall this by several antagonistic patterns. Depending on the species, the final copulation position may take the form of an end-to-end position (with the heads facing away from each other) (Fig. 11.12d) or a parallel position (Fig. 11.12e). Copulations in the parallel position do not last longer than several minutes, whereas mating in the end-to-end position may take up to an hour (even with the female continuing its locomotion and dragging the male behind her) (Betz 1999).

One special feature of Steninae beetles is their ability to move on the water surface (cf. Billard and Bruyant 1905; Urban 1928; Jenkins 1960; Linsenmair 1963; Linsenmair and Jander 1963; Betz 1999; Lang et al. 2012; Dietz et al. 2016). This involves three different methods, i.e. walking (with only the non-wettable tarsi and, additionally, the abdominal tip touching the surface), swimming (performing characteristic swimming movements involving both the legs and the abdomen) and skimming. Skimming involves the release of a spreading-active abdominal secretion (cf. Jenkins 1960) and is dealt with in more detail in Sect. 11.6. We have observed in *Dianous fornicifrons* and *D. srivichaii* that the long terminal styles at the ninth abdominal segment are laterally spread out during skimming, whereas otherwise they are kept parallel.

Another behaviour for both *Stenus* and "*Dianous*" is gregariousness (e.g. Lecoq 1991; Puthz 2000b, Figs. 29–30; Cuccodoro 2017). In this case, numerous individuals flock closely together (touching each other) and may, in this way, form incrustation-like aggregates covering rocks or leaves (Fig. 11.12c). This might be triggered by adverse conditions such as drought or coldness and might represent a kind of protective behaviour. According to Cuccodoro (2017), the aggregations of Steninae (which might be formed of up to more than 100,000 specimens) resemble a composite of several kinds of behavioural patterns, such as hibernation, aestivation, reproductive swarming and possibly hilltopping, with some most likely intermixed.

Recent focal animal observations of nine Thai "*Dianous*" species (including both groups I and II) in terraria revealed large overall correspondences with the behavioural inventory of *Stenus* beetles that was published by Betz (1999).

11.6 Chemical Ecology

Like other members of Staphylinidae, representatives of Steninae are characterized by short elytra resulting in a relatively unprotected and partly flexible abdomen. These characters, which represent true evolutionary success, allow the beetles to colonize habitats with small interstices (Steninae: banks of streams, wet grasslands, swamps, bogs) but make them susceptible to predation and parasitism because of the missing protective elytra. In order to avoid these hazards, adults of the free-living carnivorous Steninae and many other rove beetle taxa have evolved abdominal defensive glands that synthesize and store a variety of highly effective defensive compounds (Dettner 1987, 1993; Thayer 2016). In addition, several exocrine glands have recently been described in other rove beetle taxa (see Thayer 2016), such as Staphylininae (Quennedey et al. 2002), Paederinae (Schomann and Solodovnikov 2016) or Piestinae (Caron et al. 2008).

11.6.1 Morphology of the Anal Glands

Adults of all at least extant Steninae, including the genera *Stenus* and "*Dianous*" Leach, possess defensive glands at the tip of their abdomen. The complex glandular system consists of two large translucent sac-like reservoirs (r1) expanding through the last four abdominal segments (Figs. 11.13 and 11.14) and a second smaller reservoir pair (r2), one each at the base of the big reservoirs and the corresponding secretory tissues (g1, g2; Jenkins 1957; Schildknecht 1970; Schierling and Dettner 2013). Basically the reservoirs are characterized by partly eversible membranes, which enable the beetles to deplete volatile secretions, when haemolymph pressure has been previously increased. It is interesting that the large reservoirs contain deterrent and antimicrobial alkaloids (Fig. 11.13), whereas the small reservoirs contain various terpenes (Fig. 11.13).

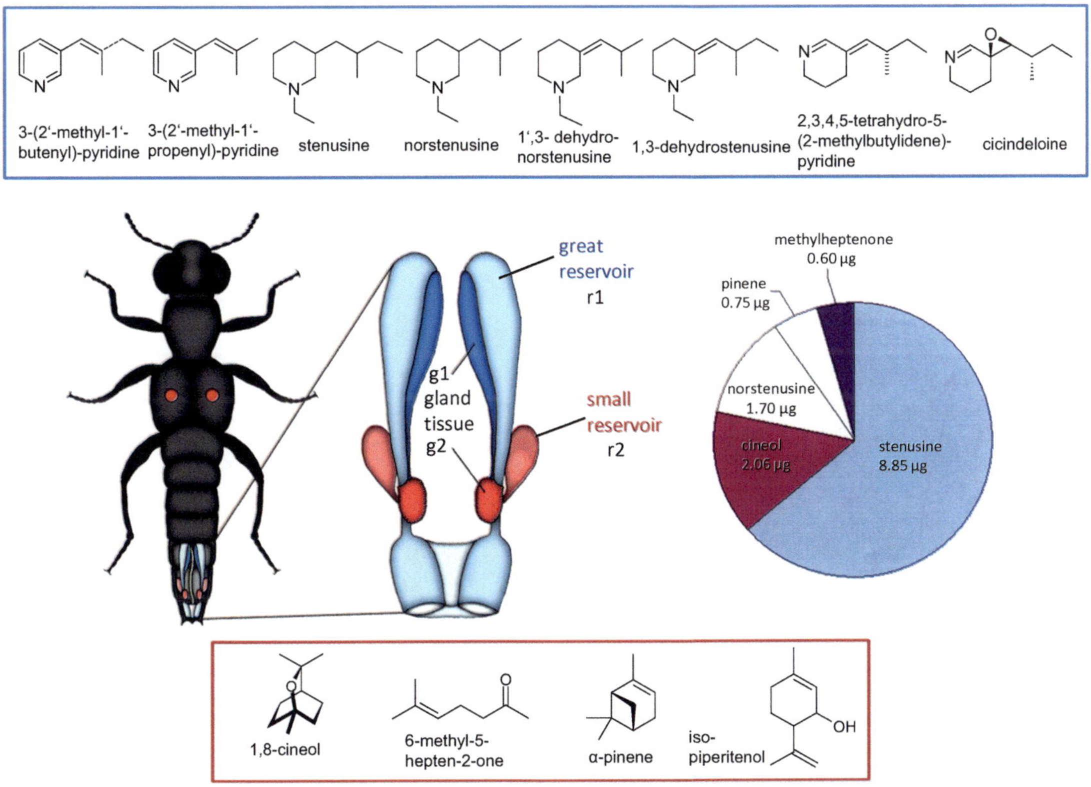

Fig. 11.13 Anal glands and secretions of *Stenus* rove beetles (Schierling 2013). Paired anal glands are characterized by great reservoirs (r1) associated with longitudinal gland tissues (g1, blue). In addition, small paired reservoirs (r2) exist with associated oval gland tissues (g2, centre; red). The identified alkaloids from the large glands are presented above, depending on species. Typical terpenoid compounds from the small reservoirs are illustrated below. As an example, the circle right symbolizes quantitative amounts of two alkaloids and three terpenoid compounds from the anal gland secretion of *S. comma*

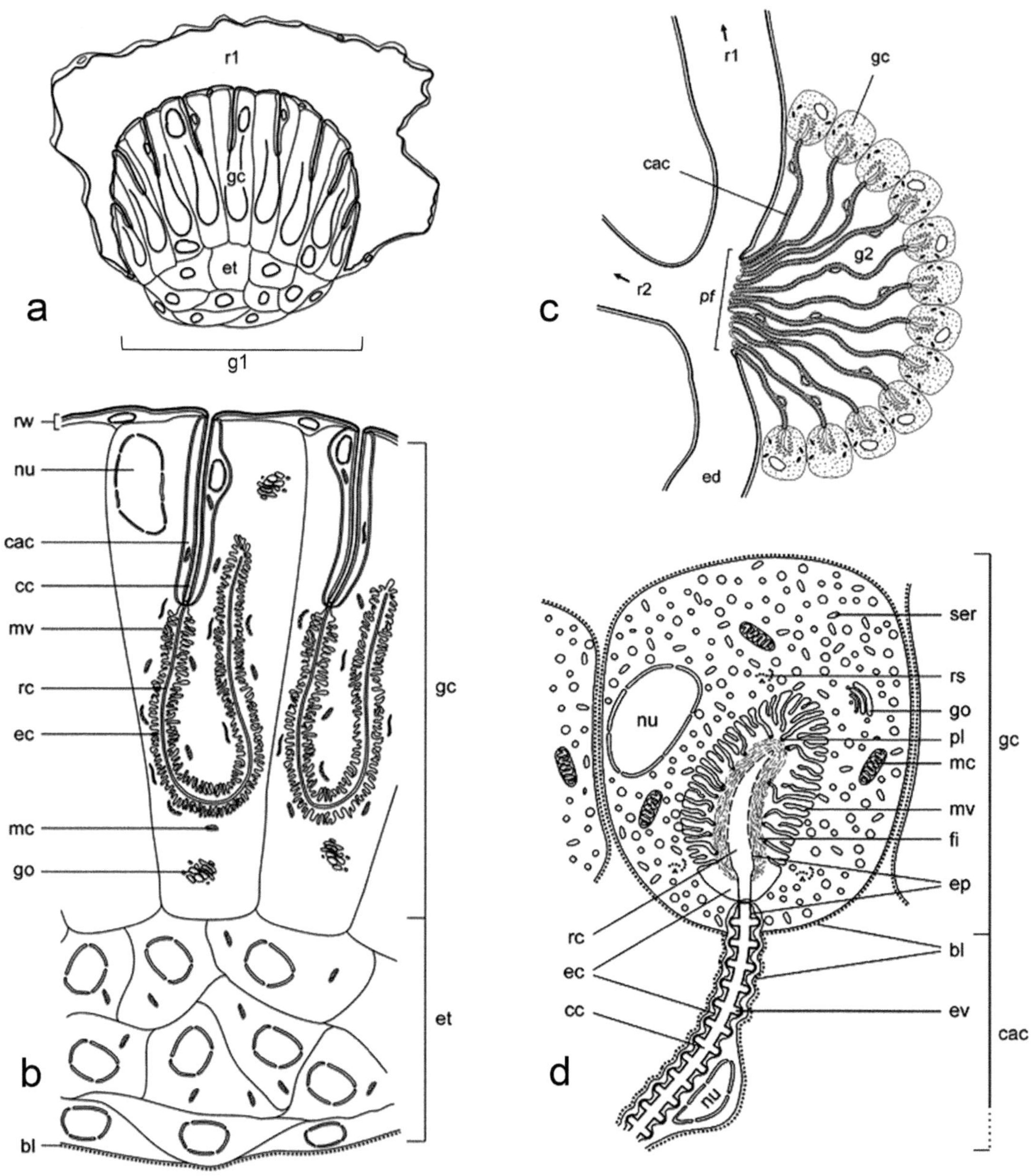

Fig. 11.14 Representation of the large gland system (r1/g1) of *Stenus comma* LeConte and *S. biguttatus* Linnaeus according to Schierling and Dettner (2013). (**a**) Overview of cross section through r1/g1. The band-shaped gland tissue (gc) and the protective epithelium (et) are located in an invagination of the r1 wall (modified according to Jenkins 1957). (**b**) Detail of gland cells of the large reservoir and protective epithelium with reservoir wall (rw), nucleus (nu), canal cell (cac), secretion-contacting canal (cc), microvilli of gland cell membrane (mv), secretion-receiving canal (rc), extracellular cavity (ec), mitochondria (mc), Golgi apparatuses (go) and basal intima (bl). (**c**) Representation of the small gland system r2/g2 of *S. comma* and *S. biguttatus* with secretion-contacting canals gathering on a pore field on the efferent duct membrane of r1 opposite to r2. Pore field (pf), efferent duct (ed) of r1, canal cells (cac), secretory tissue filling r2 (g2), gland cells (gc). (**d**) Diagram of the secretion production, receiving and conducting structures of g2. Smooth endoplasmic reticulum (ser), ribosomes (rs), Golgi apparatus (go), plaques at the microvilli apices (pl), mitochondria (mc), microvilli (mv), filament layer (fi), epicuticle (ep), basal lamina (bl), evagination of the conducting canal (ev), receiving canal (rc), extracellular cavity (ec), conducting canal (cc) and nuclei (nu)

The band-shaped glandular tissue (g1) is longitudinally associated with the large reservoir and situated within an invagination of the reservoir membrane (Fig. 11.14a). Each secretory active cell of g1 bears an extracellular cavity, in which the products are secreted by exocytosis and drained by an epicuticular duct. The distal secretion-receiving structure is located in the middle of the extracellular cavity of the gland cell. It is lined with a porous granular epicuticular material of about 50 nm thickness, surrounded by a 120–300-nm-wide filamentous mass that does not resist maceration. In order to achieve a high exocytosis rate, the surface of the gland cell membrane surrounding the extracellular cavity is increased by numerous microvilli (Fig. 11.14b, d). Inside and around the microvilli, the gland cells exhibit large, partially elongated mitochondria with cristae of variable length and frequency. The stability and flexibility of the reservoir wall are attributable to a presumably single layer of epithelium cells containing numerous myofibrils arranged in a right-angled manner. The epithelium cells are covered by a basal lamina. Retraction of the glands is accomplished by retractor muscles (Jenkins 1957; Whitman et al. 1990). Although the large gland system r1/g1 is predominant in every *Stenus* species (Figs. 11.13 and 11.14), the smaller system r2/g2 (Figs. 11.13 and 11.14) is often reduced and difficult to localize (Fig. 11.15). Probably, for this reason, it has been reported only for *S. comma* and *S. biguttatus* (Schildknecht 1970; Schildknecht et al. 1975, 1976; Whitman et al. 1990; Lusebrink 2007), but it is present in all investigated *Stenus* species and might serve as a valuable character for phylogenetic studies of the Steninae.

In contrast to the gland system r1/g1, great interspecific differences are found concerning the morphology and ultrastructure of the smaller gland system r2/g2, which have been especially investigated in *S. comma* (Fig. 11.15a) and *S. biguttatus* (Fig. 11.15b). The small gland system r2/g2 consists of a clear secretion-filled reservoir r2 (up to 330 μm long and 130 μm wide), which opens into the efferent duct of the large reservoir r1. The associated secretory tissue g2 encloses the efferent duct of r1 at the opening of r2, with the main part being located opposite to r2. G2 consists of numerous globular gland cells (16 × 22 μm in diameter) that make up a "cauliflower-like" structure visible by light microscopy. Each gland cell is equipped with an epicuticular duct that collects the secretion and conducts it into the reservoir through at least one

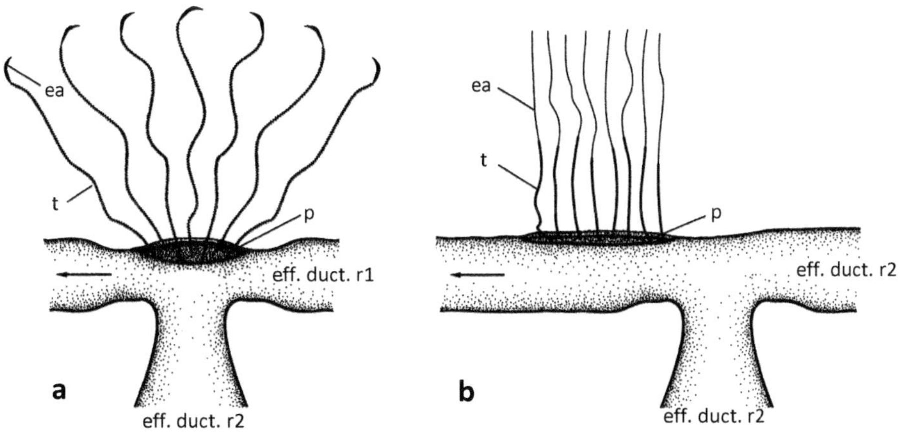

Fig. 11.15 KOH-macerated parts of anal glands of selected representatives of Steninae (**a** *Stenus comma*, **b** *S. biguttatus*) showing the arrangements and reductions of the small reservoirs r2 (eff. duct. r2) and associated tubules (t) and end apparatus (ea). Tubule pores are symbolized by p and are arranged opposite to the efferent duct of r2 (**a**), respectively, shifted away from efferent duct r2 as in other Steninae (**b**). Both efferent ducts of the large reservoirs (eff. duct r1) and the direction of secretion emissions (arrows) are indicated (according to Schierling 2013)

canal cell. The canals measure up to 60 μm in total length and can be divided into a distal secretion-receiving, a proximal secretion-conducting and a transition section. The distal secretion-receiving structure of the epicuticular ducts is about 12 μm long and 0.9 μm wide and is situated in an extracellular cavity of the gland cell. As in the secretory cells of g1, the extracellular cavity of the g2 cells is filled with numerous microvilli that are part of the inner gland cell membrane and that are presumably involved in the secretion of the synthesized compounds via exocytosis. The mostly electron-lucent microvilli bear an electron-dense material in their apical region, where they reach the filament layer. Electron-dense material also occurs in the cytoplasm of the whole gland cell. The gland cells are often filled with smooth endoplasmic reticulum. Sparse ovoid mitochondria with short cristae occur near the microvilli. Golgi systems, free ribosomes and rough endoplasmic reticulum are seldom. Each gland cell is coated by an 85-nm-wide basal lamina and is equipped with an ovoid nucleus of 5–6 μm in diameter. The proximal secretion-conducting part of the epicuticular duct transports the secreted compounds into the reservoir. The tubular ducts are up to 50 μm in length and 0.7 μm in diameter and exhibit numerous evaginations all over their surface. The secretion-conducting ducts of g2 are situated in at least one canal cell, which penetrates the gland cell and closely approaches the secretion-receiving part of the canal. As an extracellular structure, the conducting canal is surrounded by an extracellular cavity. In contrast to the secretion-receiving canals, the conducting canals are enclosed by a 60-nm-thick compact non-perforated epicuticle, lined with a cuticulin layer of 5–8 nm in width. The very slim canal cells are poor in internal structures and cell organelles but have an ovoid nucleus. They are surrounded by a basal lamina. Proximally, the canal cells are bundled onto a pore field where the canals open into the efferent duct of r1. The distal secretion-receiving and the proximal secretion-conducting parts of the ducts are linked by a short evagination-free transition section with non-perforated epicuticle. The gland cell and canal cell are interconnected via septate junctions. The pore field onto which the g2 canals open into the efferent duct of r1 is situated exactly opposite to the reservoir r2. In the living beetles, the r1 efferent duct collapses, and the pore plate is pressed onto the opening of r2. The r2 wall consists of epithelium cells secreting a layer of filaments covered by a puckered epicuticular intima of 8–13 nm in width. The epithelium cells do not contain any muscle fibrils as observed in the r1 wall.

In many species, r2/g2 are massively reduced and functionally altered. The best developed and most complex r2/g2 gland system can be found in *S. comma* and *S. biguttatus* (Fig. 11.15). Other selected species that have been examined show distinct reductions of either g2 or r2 or of both components. Whereas r2 of *S. comma* is up to 330 μm long, in *S. bimaculatus*, a species that is about 1 mm larger than *S. comma*, r2 only reaches a length of approximately 60 μm. Furthermore, in *S. bimaculatus*, the conical outline of r2 is lost, and it becomes a small tubular appendix of the r1 efferent duct. An even more extreme form of reduction of r2 occurs in *S. fulvicornis*, *S. juno* and *S. picipes*. Species such as *D. coerulescens*, *S. providus*, *S. solutus*, *S. pubescens* and *S. latifrons* exhibit a small reservoir r2 that keeps its conic form, but its size is dramatically reduced. *S. flavipes* is the only species that possesses a reservoir r2 with a reservoir size/body size ratio comparable with that of *S. comma* and *S. biguttatus*. In all species except *S. comma* and *S. biguttatus*, the epicuticular ducts of g2 do not gather on a pore field but open within a widespread area that does not lie opposite to r2 and from there into the efferent duct of r1. Their comparatively short conducting canals do not exhibit any evaginations, and the distinct secretion-receiving structure observed in *S. comma* and *S. biguttatus* is replaced by a filamentary receiving canal. Overall, their organization is reminiscent of that of the g1 canals. This is also the case for *S. flavipes*.

The eversible membrane parts of the anal glands do not vary in different species (Kanehisa and Tsumuki 1996). They are situated at the base of the efferent ducts of the reservoirs r1 and can be extruded laterally to the anus between the ninth tergite and sternite (Fig. 11.13). They are each made up of a tubular membrane, which is inverted at the tip and thereby forms a cylindrical double-walled structure bearing a pore at the end for secretion release. Whereas the inner membrane represents an extension of the r1 efferent duct, the outer membrane of the eversible cylinder is continuous with a membrane separating the gut and pygidial gland apparatus from the gonads and other abdominal organs. Jenkins (1957) reports more details concerning the morphology of the eversible membrane parts.

Although the secretion-receiving canals of g1 and g2 gland cells both exhibit a porous granular epicuticle for the reception of the secreted compounds, the conducting canals are lined with a continuous epicuticle and an additional cuticulin layer. This seems important in order to minimize the risk of self-intoxication with cytotoxic defence secretions (Noirot and Quennedey 1974, 1991; Quennedey 1998). As non-proteinous secretions of low molecular weight, the defence compounds of the Steninae are thought to be synthesized in the smooth endoplasmic reticulum and Golgi systems. Whereas the g2 gland cells are completely filled with smooth endoplasmic reticulum, these structures are lacking in g1 gland cells. Thus, the cellular biosynthetic origin of the alkaloids of r1/g1 remains unknown. A sponge-like or filamentous layer surrounding the receiving canal of a gland cell, as is found in both secretory tissues of the Steninae, is usually interpreted as protein or epicuticular filaments (Noirot and Quennedey 1991). In the g2 gland cells, the filament layer surrounding the receiving canal probably does not represent an accumulating secretion. Although the fine filaments do not resist maceration by 10% KOH, they are nevertheless proposed to be of epicuticular origin and are lost following the severe maceration process during preparation for SEM. In contrast to the filamentous layer of the g1 secretion-receiving apparatus, the single filaments of g2 cells are clearly distinguishable as a built-up mesh-like structure. The chemical modification might result in an increasing toxicity during passage through the canal. The conspicuous length of the g2 conducting canals supports this hypothesis. However, such processes would require enzymes or other compounds located in the canal evaginations, but no structures have been found that suggest secretory activity of the canal cells. Thus, the secretion-modifying compounds or enzymes would also have to be secreted from the g2 gland cells. All the secretion-conducting canals of g2 gather on a moderately sclerotized pore plate on the efferent duct of r1, comparable with those reported for *Eleodes longicollis* (Eisner et al. 1964), but here its position is of fundamental significance for the functionality of r2/g2. The pore plate is situated opposite to the opening of the reservoir r2, and so the secreted compounds have to pass the r1 efferent duct to enter the reservoir r2. The compounds found in r2 have not been detected in r1 (Schierling et al. 2013), and thus a mechanism must exist preventing the g2-synthesized compounds from entering r1. Because of the distinctive position of the pore plate exactly opposite the opening of r2, a transfer of the secretion from g2 to r2 might be facilitated by the collapse of the r1 efferent duct under resting conditions (no secretion release), an event that might result in the pore plate being pressed onto the opening of r2. The wall of reservoirs r1 and r2 is lined by a tight epicuticular intima that prevents the haemolymph and organs from being contaminated with toxic secretion. Furthermore, the reservoir walls are reinforced by myofilaments (r1) or epicuticular filaments (r2). If reservoir r1 is not maximally filled, the myofilaments in the epithelium cells are contracted, and the intima is puckered. A puckered intima has also been observed in the r2 wall, despite the presence of epicuticular filaments instead of muscle fibrils. However, the reservoirs are thus flexible structures that can be adjusted to the actual secretion level. The flat muscular epithelium surrounding r1 is thin and is therefore probably not responsible for the rapid gland eversion and secretion release, which is actually achieved by increasing haemolymph pressure, as proposed by Jenkins (1957). Jenkins (1957) describes the whole pygidial defence gland apparatus of the Steninae as an invagination of the pleural membrane. The duct structures of both secretory tissues g1 and g2 and the corresponding reservoirs are lined with epicuticular material supporting Jenkins' proposal of the glands' epidermal origin (Noirot and Quennedey 1974; Quennedey 1998).

The appearance and characterization of a second smaller reservoir r2 in Steninae abdominal glands were first described for *S. comma* by Schildknecht (1970). Later, it was reported for *S. biguttatus* and *S. comma* (Lusebrink 2007), but it has never been described for other Steninae. Jenkins (1957) mentioned some structural changes at the basal parts of the gland tissue r1 but was unable to identify it as being independent from g1. We have found r2 and the associated gland tissue g2 in every species examined, including in *D. coerulescens*. However, great species-dependent reductions are apparent in the morphology, ultrastructure and function of this small reservoir r2. As described above, the position of the pore plate exactly opposite to the reservoir r2 seems important for the correct transfer of secretion from g2 to r2 in *S. comma* and *S. biguttatus*. Only by this means can the pore plate be pressed onto the opening of r2 when the r1 efferent duct is collapsed. In the other species examined during this study, the secretion-conducting canals open into the efferent duct of r1 cranial to r2, and so the secretion transfer cannot work. As a consequence, reservoir r2 is reduced, as can be seen in various species. Furthermore, the epicuticular ducts of species with reduced r2/g2 are similar in structure to those of g1, and hence the presence of structurally modified g2 cells or just tightly packed and round g1 cells opposite to the small reservoir r2 is difficult to determine. The small reservoirs r2 contain terpene compounds such as α-pinene, 1,8-cineol (eucalyptol) and 6-methyl-5-hepten-2-one (Schildknecht 1970; Schildknecht et al. 1975, 1976; Lusebrink 2007). Whereas α-pinene and 1,8-cineol have been found in many *Stenus* species, 6-methyl-5-hepten-2-one has only been identified in the r2/g2 secretion of *S. comma* and *S. biguttatus* (Lusebrink 2007; Schierling et al. 2013), which show the best developed r2/g2 gland system within the whole genus. Furthermore, the amount of α-pinene and 1,8-cineol is maximized within these two species. In other Steninae, these terpenes occur, if at all, only in traces (Lusebrink 2007; Schierling et al. 2013). Because *S. comma* and *S. biguttatus* represent the only hitherto investigated species exhibiting distinct evaginations (paratergites) of the secretion-conducting canals, a correlation might be present. *S. comma* and *S. biguttatus* both reveal distinctly raised lateral margins at their abdominal tergites. This character is classified as phylogenetically plesiotypic, and so the two species are arranged at the base of the genus *Stenus* (Puthz 2006, 2010). Because of their better adaption in nearly all areas, the alkaloid compounds of r1 are effective for defence against predators or microorganisms and for locomotion via skimming (Schierling et al. 2013; Lang et al. 2012). Thus, most phylogenetically advanced species have nearly lost or at least dramatically reduced their r2 compounds and the structures responsible for their synthesis and storage. "*Dianous*" with its single Middle European species *D. coerulescens* has hitherto been classified as a phylogenetically basal taxon of the Steninae because of its lack of an adhesive prey-capture apparatus that probably represents an apomorphy in *Stenus* (Puthz 1981; Betz 1996, 1998b, 1999; Leschen and Newton 2003). The molecular and chemotaxonomic analyses of the phylogeny of the Steninae indicate that the genus "*Dianous*" should be integrated into *Stenus* as a species with a secondarily reduced prey-capture apparatus (Koerner et al. 2013; Lang et al. 2015). These analyses are in accordance with the morphology of the small gland system r2/g2 of *D. coerulescens*. It exists only in a strongly reduced form comparable with that of the phylogenetically advanced *Stenus* species.

11.6.2 Secretion Chemistry and Biosynthesis of Secretion

The reservoir r1 of investigated representatives of Steninae is filled with g1-synthesized **piperidine-** (Fig. 11.13: stenusine; norstenusine; 1′,3-dehydrostenusine; 1′,3-dehydronorstenusine),

piperideine- (cicindeloin; 2,3,4,5-tetrahydro-5-(2-methylbutylidene)-pyridine) and **pyridine**-alkaloids (Fig. 11.13: (*Z*)-3-(2′-methyl-1′-butenyl)-pyridine; (*E*)-3-(2′-methyl-1′-butenyl)-pyridine and 3-(2′-methyl-1′-propenyl)pyridine). The small gland reservoir r2 contains terpenes (e.g. Fig. 11.13; 1,8-cineol and α-pinene and, in few cases, 6-methyl-5-hepten-2-one and isopiperitenol) produced by g2 (Schildknecht 1970; Schildknecht et al. 1975; Kohler 1979; Lusebrink 2007; Lusebrink et al. 2009; Müller et al. 2012; Neumann 1993). In order to verify our trace-analytical investigations, most of the components, some together with their stereoisomers, were synthesized in the laboratory (Müller et al. 2012).

Whereas all terpenoids identified from the small glands of Steninae are widely distributed within plants and arthropods (Blum 1981), all nine alkaloids reported here are uniquely found in Steninae beetles. Nevertheless, other biologically highly active **piperidine**-alkaloids are also found in other animals and plants (Breitmaier 2008). Fire ants of the genus *Solenopsis* contain various toxic and insecticidal 2-alkyl-6-methylpiperidines and even N-alkylpiperidines (Morgan 2010). Moreover, poison dart frogs of the genera *Dendrobates*, *Epipedobates* and *Phyllobates* produce toxic bicyclic piperidine-alkaloids, such as histrionicotoxin A and its derivatives, in their skin. Since these frogs use their skin as a kind of excretion organ, Steninae alkaloids might also be found in frog skin, because these amphibians feed on Steninae beetles (see Dettner 2007). As compiled by Breitmaier (2008), a considerable number of biologically active piperidine-alkaloids are present in various plants such as piperine (*Piper nigrum*, Piperaceae), (−)-coniine (*Conium maculatum*, Apiaceae), (R)-(−)-pelletierine (*Punica granatum*, Punicaceae), (−)-sedamin (*Sedum acre*, Crassulaceae), (−)-lobeline (*Lobelia inflata*, Campanulaceae), pinidin (*Pinus* and *Picea* species) and (+) carpaine (*Carica papaya*, Caricaceae). Bicyclic piperidine-alkaloids also exist such as (+)-α-skytanthin (*Skytanthus acutus*, Apocynaceae), nitramine (*Nitraria schoberi*, Zygophyllaceae) and (−) sibirin (*Nitraria sibirica*, Zygophyllaceae). A few **piperideine**-alkaloids (see cicindeloine) such as γ-coniceine from the toxic plant *Conium maculatum* (Apiaceae) and toxic 2-methyl-6-alkyl-piperideines from the ant *Solenopsis geminata* have been reported (Breitmaier 2008; Morgan 2010). Apart from the **pyridine-alkaloids** from Steninae, various biological pyridine-alkaloids are known from plants such as ricinin (*Ricinus communis*, Euphorbiaceae), anibin (*Aniba duckei*, Lauraceae), arecoline (*Areca catechu*, Palmaceae) and nicotine and its derivatives (*Nicotiana tabacum*, Solanaceae). Animals such as various caterpillars of Saturniidae and Lymantriidae produce nicotine and other pyridine-derivatives such as nicotinic acid or nicotinamide in glandular defensive hairs (Deml and Dettner 1995, 1997). Again, a halogenated pyridine-alkaloid from the poison dart frog skin of *Epipedobates tricolor* named (−)-epibatidine acts as a powerful analgesic and is even superior to morphine.

All terpenes are biosynthesized according to the mevalonate pathway (Morgan 2010). In contrast, all *Stenus* alkaloids from the large anal glands are biogenetically derived from amino acids. In particular, piperidine-alkaloids are principally derived either from lysine, acetate or mevalonate. Lusebrink et al. (2008a) have revealed that the N-ethyl-group of stenusine and norstenusine is derived from acetate, a pathway established by using trideuterated sodium acetate (CD_3COONa), whereas L-lysine forms the piperidine ring (stenusine, norstenusine) and the sidechain originates from L-isoleucine. Furthermore, in (*Z*)-3-(2-methyl-1-butenyl) pyridine, L-lysine yields the pyridine ring, and L-isoleucine yields the 2-methyl-1-butenyl sidechain (Schierling et al. 2011). In the same way, the piperideine ring of cicindeloine is biosynthesized from L-lysine, and, again, the sidechain is biosynthesized from L-isoleucine (Schierling et al. 2012). Wittmann et al. (2015) have additionally revealed that the precursor in the biosynthesis of cicindeloine must be (*S*, *E*)-1 [= 2,3,4,5-tetrahydro-5-(2-methylbutylidene-)

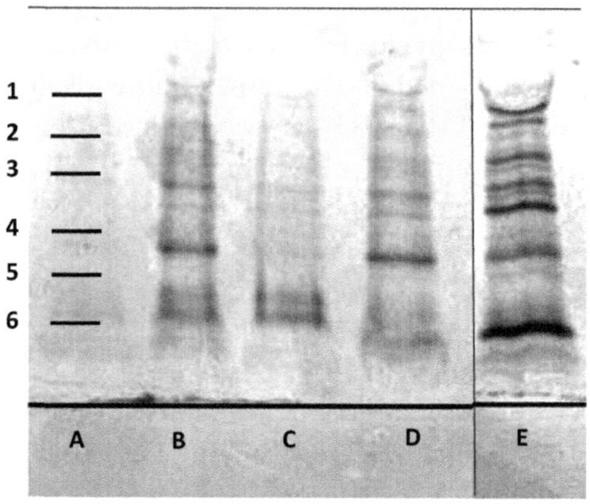

Fig. 11.16 SDS gels (Coomassie staining) of glandular compartments of the two *Stenus* species *S. comma* (**A–D**) and *S. bimaculatus* (**E**) (according to Schrüfer 2013). Vertical numbering of reference compounds (**A**): **1** phosphorylase b, **2** albumin, **3** ovalbumin, **4** carbonic anhydrase, **5** trypsin inhibitor, **6** α-lactalbumin. Horizontal lettering: (**B**) filled whole anal gland system of *Stenus comma*, (**C**) filled small glandular system of *S. comma*, (**D**) filled large glandular system of *S. comma*, (**E**) filled large glandular system of *S. bimaculatus*

pyridine] and not the enantiomer (*S*, *Z*)-1. These results amply indicate that alkaloids not only and exclusively represent so-called secondary plant compounds but are also produced and sequestered by many animals and especially arthropods.

As is well known from the anal defensive glands of carabid beetles (Bombardier beetles; Schildknecht et al. 1968) or recently from chrysomelid larvae (Burse and Boland 2015), biosynthesizing enzymes that are responsible for producing intermediate and end products of the secretions are localized either within glandular cells or within the gland reservoir or are found in both structures. Therefore, the Steninae enzymes for terpenoid biosynthesis via mevalonate should be present in the reservoir or gland cells of the small gland. In contrast, those Steninae enzymes that are responsible for alkaloid biosynthesis should be found in the large gland system. These enzymes have been postulated to process L-lysine and L-isoleucine (decarboxylation of L-lysine; oxidative deamination) and should be responsible for subsequent oxidations, reductions, N-ethylations and epoxidations. When the proteins from the various glandular components of gland 1 and gland 2 of *S. comma* were separated electrophoretically, different protein patterns were evident (Fig. 11.16). Another species, *S. bimaculatus*, with reduced small glands showed a similar electrophoretic pattern to that of *S. comma*, although typical bands of the reduced small gland were absent. These investigations need to be continued by using other species and by applying enzyme assays in order to differentiate between proteins from alkaloid biosynthesis and structural proteins.

11.6.3 Biology and Role of Behaviour-Modifying Chemicals

Nearly all Steninae secretion compounds show significant antibiotic and deterrent activities (Lusebrink et al. 2009; Schierling et al. 2013) and thus can serve as potent chemical defence compounds. When molested, the beetles bend their abdomen towards the source of irritation, evert their glands and moisten the aggressor with their secretion. In addition, the beetles show typical cleaning movements and use their pygidial gland secretion to coat their body surface, thus avoiding infection by microorganisms (Betz 1999). Moreover, some species of Steninae living on the banks of water use the pygidial defence gland secretion for an exceptional form of locomotion called skimming. This phenomenon was first described for Steninae by Piffard (1901) and for *Stenus cicindeloides* and *S. tarsalis* by Billard and Bruyant (1905). Supported on the water surface by their hydrophobic tarsi, the beetles touch the surface with the tip of their abdomen and release small amounts of secretion from the everted pygidial glands (see next section).

The secretions of most Staphylinidae, such as Oxytelinae and Aleocharinae, are mainly defensive, but the *Stenus*/"*Dianous*" pygidial gland secretions appear to be multifunctional. This becomes apparent, when the beetles are observed in nature. When molested, the beetles evert the basal parts of their glands and moisten the aggressor with secretion by bending their flexible abdomen (Jenkins 1957; Betz 1999), a reaction that clearly is defensive. In this regard, the piperidine and pyridine-alkaloids act in particular as feeding deterrents in ant bioassays (Honda 1983; Lusebrink et al. 2009; Schierling et al. 2013). Furthermore, extensive cleaning behaviour is often shown by the Steninae, whereby pygidial gland secretion is dispensed over the entire body surface. Via an analysis of the behavioural inventory of *Stenus*, Betz (1999) observed the beetles spending a large amount of their time budget rubbing their hind legs over the abdominal tip with the everted glands emitting small amounts of secretion. Subsequently, the secretion was transferred to the middle and fore legs and then dispensed over the entire body surface. In view of the wet habitat of most Steninae, stenusine and the pyridine-alkaloid have unsurprisingly been reported as antimicrobial compounds (Schildknecht 1976; Lusebrink et al. 2008b, 2009).

Complete analyses of the pygidial gland secretions have revealed great quantitative and qualitative differences between the analysed species (see Lusebrink et al. 2009; Schierling et al. 2013). Stenusine is usually abundant in most of the species investigated, most often as the main constituent. Analogous to stenusine, the corresponding norstenusine is almost always detected, albeit in lesser amounts. According to Lusebrink et al. (2009), the pyridine- and piperideine-alkaloids are restricted to the subgenera *Metatesnus* and *Hypostenus*. The piperideine derivative has only been detected in *S. binotatus* (*Metatesnus*), *S. solutus* and *S. cicindeloides* (*Hypostenus*). Although the existence of terpenes is variable in the secretion of the analysed species, some of them seem to be restricted to species of the subgenera *Hemistenus* and *Stenus* s. str. and *Dianous coerulescens*. 6-Methyl-5-hepten-2-one is found only in *S. comma* and *S. biguttatus*, whereas isopiperitenol only occurs in *S. comma*. Except for a few trace terpenes, the results are in agreement with the findings reported by Schildknecht (1976), Lusebrink (2007), Lusebrink et al. (2009) and Müller et al. (2012). The natural isomeric compositions of the alkaloids investigated also seem to be of great interest, as seen in Fig. 11.17 for stenusine.

Ants whose mouthparts come in contact with an alkaloid-containing milk drop (Fig. 11.18) immediately retreat from the drop with panic-like movements, often rubbing their mandibles on the substrate. This action is followed by extensive cleaning activities of the antenna and mouthparts. Among the terpenes, 6–8 (Fig. 11.18a) especially also show significant deterrent activities against ants but do not elicit the drastic reactions evoked by the alkaloids (Fig. 11.18). Additionally, the deterrent responses of the secretion main components have been compared among one another (Fig. 11.18b): the differences in deterrencies between stenusine and the pyridine-alkaloid and the piperideine-alkaloid as compared with the pyridine-alkaloid are extreme.

In order to examine potential synergistic effects of the secretion, we have also tested the three main alkaloids 1, 3 and 5 against naturally occurring secretions (Fig. 11.18c). The secretion of *S. comma* (CO) and that of *S. similis* (Si) is significantly less deterrent than the single main compounds. In addition, the difference in the reaction of *Lasius flavus* to the secretion of *S. solutus* (SO) as compared with the piperideine derivative is remarkable (Fig. 11.18c).

In order to complete the determination of the potential biological activities of Steninae secretions against arthropods, equimolar amounts of Steninae

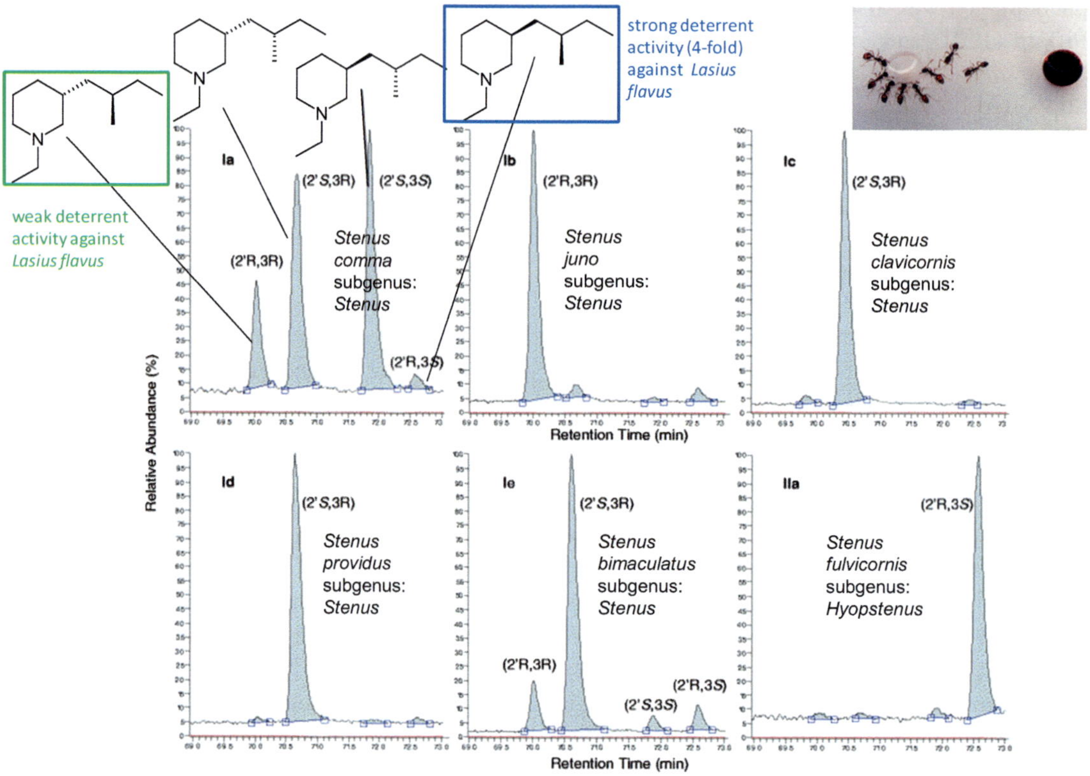

Fig. 11.17 Intrageneric differences of the four stenusine stereoisomers among *Stenus* beetles as shown by enantio-MD/GCMS of selected species of the subgenera *Stenus* and *Hypostenus* (Lusebrink et al. 2007). Concerning the enantiomers (2′R,3R) and (2′R,3*S*), the last mentioned exhibits a stronger deterrent activity against *Lasius flavus* ants that were fed with sugar solutions or sugar solutions mixed with an enantiomer (right inlet figure; Schierling et al. 2013)

components were used in a scratch assay involving *Periplaneta americana* cockroaches. An immediate scratching by the hind legs was observed when the terpenes 1,8-cineol and α-pinene were applied (Fig. 11.19). In contrast, stenusine and piperideine-alkaloid were not so effective.

Deterrent activities of Steninae compounds have also been registered against vertebrates such as the fish species *Xiphophorus hellerii* by applying test pellets (Fig. 11.19). In particular, 1,8-cineol, α-pinene, stenusine and the piperideine derivatives show remarkable effects (Rupprecht 2011; Schierling et al. 2013).

Most of the tested Steninae compounds exhibit antimicrobial activities against bacteria (gram positive; gram negative) such as *Escherichia coli, Serratia entomophila* and *Bacillus sphaericus*, as determined by the detection of inhibition zone diameters. As compared with the commercially available Tygacil stenusine, the pyridine derivative and α-pinene are significantly active (Lusebrink et al. 2008b).

Finally, a few Steninae compounds and especially α-pinene are also significantly able to inhibit the growth of *Saccharomyces cerevisiae* (Schierling et al. 2013).

Both bioassays (*S. cerevisiae*, three bacteria species) have revealed that synthetic mixtures of Steninae secretions exhibit similar effects to those of single-tested compounds, indicating that synergistic effects are not as important as compared with the defensive secretions of other staphylinids (Francke and Dettner 2005).

In order to interpret the intrageneric differences in the four stereoisomers of stenusine (Lusebrink et al. 2007), bioassays have been

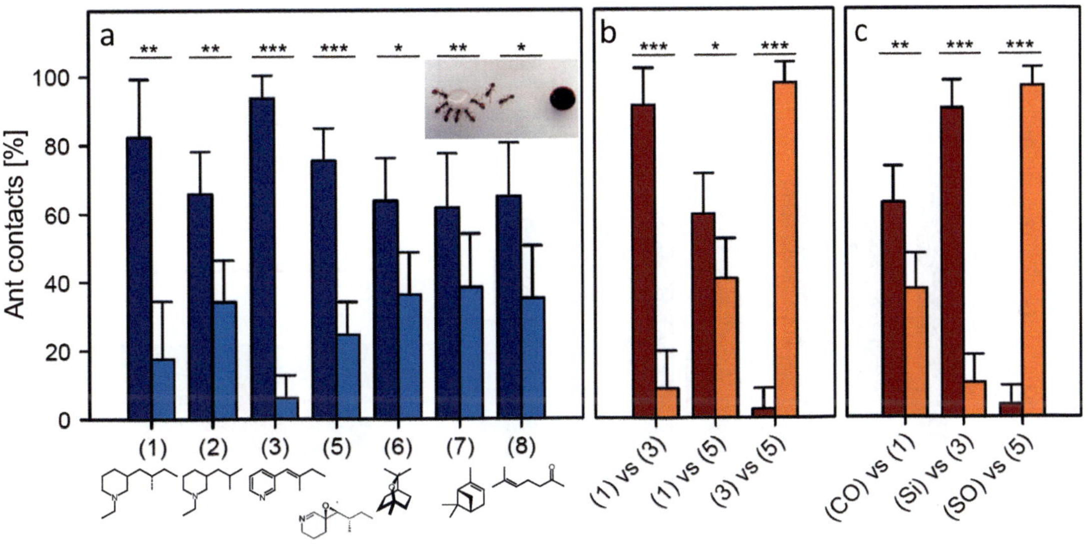

Fig. 11.18 Two-choice bioassays with *Lasius flavus* and synthetic *Stenus* pygidial gland secretion components. The diagrams show the distribution of ant contacts with the control and the test solutions including the standard deviation. (**a**) Test of the pygidial gland secretion components (light blue) versus controls (dark blue). (**b**) comparison of the deterrencies of three of the main components of the secretion of *S. comma*, *S. similis* and *S. solutus*. (**c**) Comparison of the deterrencies of one of the main components of the secretion with naturally occurring secretions. **1** Stenusine, **2** norstenusine, **3** 3-(2-methyl-1-butenyl) pyridine, **4** cicindeloine, **5** 1,8-cineole, **6** α-pinene, **7** 6-methyl-5-hepten-2-one. **CO** Secretion of *S. comma*, **Si** secretion of *S. similis*, **SO** secretion of *S. solutus*. Concentration of compounds or secretion in the drops offered: (**a**) 100 μg/ml, (**b**) 30 μg/ml, (**c**) 80 μg/ml (according to Schierling et al. 2013)

performed with two stereoisomers of stenusine. Stereoisomer 2′R,3S exhibits similar inhibition zones against *E. coli*, *S. entomophila* and *B. sphaericus* as stereoisomer 2′R,3R. In contrast, the deterrence of 2′R,3S is stronger than that of isomer 2′R,3R, when *Lasius flavus* ants are used as test organisms (see Fig. 11.17; Schierling et al. 2013).

In nature, Steninae beetles are confronted with several predacious and probably also parasitoid organisms that would not hesitate to attack the small beetles. The list of potential predators includes arthropods such as ants, spiders and carabid beetles and vertebrates such as fish, amphibians, birds and small mammals, all of which have to be deterred by the pygidial gland secretion if the beetles are to survive in their environment. Piperidine- and pyridine-derived alkaloid compounds and terpenes are known from the chemical defence systems of various arthropods (Ekpa et al. 1984; Deml and Dettner 1995; MacConnel et al. 1971; Blum 1981; Huth and Dettner 1990; Stoeffler et al. 2007). The piperidine-, pyridine- and epoxypiperidine-alkaloids, however, are restricted to the genera *Stenus* and "*Dianous*." The distinct deterrent activity of the secretion of living Steninae on attacking ants has previously been demonstrated (Neumann 1993). Furthermore, the pure selected alkaloids and terpenes have been described as ant deterrents by Honda (1983) and Lusebrink et al. (2009). In our experiments, we have been able to confirm the previous observations and to extend these results with the newly available secretion compounds. Ants touching the compound solutions in our tests most often retreat from the milk drops without feeding. Additionally, these ants often exhibit distinct reactions, such as (1) panic movements, (2) mandible rubbing and (3) intensive cleaning that can be classified as typical rejection behaviour (Eisner et al. 1961; Dettner et al. 1996). Since all tested secretion constituents show significant deterrence against *L. flavus* (Fig. 11.18) according to the definition

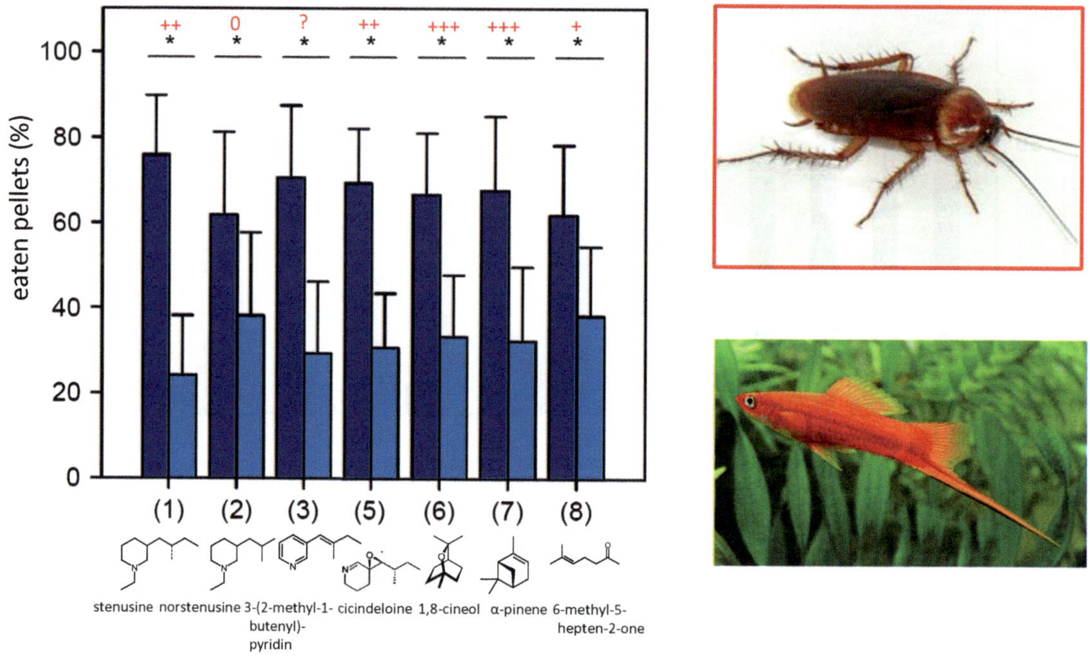

Fig. 11.19 Deterrent activities of synthetic compounds from anal glands of *Stenus* against fish *Xiphophorus hellerii* (Poeciliidae, two-choice-bioassays), as examined by using test pellets. The number of control (dark blue) and test pellets (light blue) are shown (Schierling et al. 2013; Rupprecht 2011). In addition, note the preliminary results of the scratch assay with *Periplaneta americana* cockroaches (Question mark, not tested; zero, no reaction; plus, weak reaction; double plus, medium reaction; triple plus, strong reaction)

of Schoonhoven et al. (2005), the secretion is with certainty capable of protecting the beetles from attacks by arthropod predators. In our tests, we have applied solutions of maximal 100 μg compound per millilitre milk. The content of stenusine in the glands of a *Stenus comma* (syn. *S. bipunctatus*) beetle (Fig. 11.13) is only about 8,35 μg/beetle (Lusebrink et al. 2008b), but, if a predator attacks a *Stenus*, the beetle emits pure and undiluted secretion. If this secretion is applied to sensitive sensory organs such as mouthparts, the aggressor is deterred immediately. A comparison of the deterrence of naturally occurring secretions of *S. comma* and *S. similis* with the respective main alkaloidal components has revealed that the mixture is less active in both cases. This might be because of the lower proportion of the high-active compounds in the secretion mixture, since the same total amount of compound or mixture is used in both the tests with the pure compounds and those with the secretion mixture. Hence, the highly active compounds are "diluted" by the less-active ones. The secretion of *S. solutus* turns out to be more active than its main component 4. The better activity of the secretion, however, is more probably attributed to the strong deterrence of 3, which is significantly higher than that of 4. A synergism or quasi-synergism effect of the compounds in nature, as reported for Oxytelinae (Staphylinidae) by Dettner (1987, 1993) and Francke and Dettner (2005), obviously does not exist in Steninae. In the context of the skimming locomotion of some Steninae on the water, a fish deterrence of the secretion seems to be a reasonable goal. Moving objects on the water surface are always attractive to fish, as confirmed by observations of gyrinid beetles on fish-inhabited waters (Benfield 1972; Dettner 1985, 2014; Eisner and Aneshansley 2000). Although *X. hellerii* is not present in Central Europe (Frey 1973), *Stenus* species are also expected to exist in the natural habitat of the tested fish (Hermann 2001; Puthz 1974). Thus, *X. hellerii* can be considered as an appropriate test organism for deterrent experiments with the secretion. Indeed, intact individuals of *S. providus* offered to *X. hellerii* in early tests were always caught by the fish but were spat out immediately and never swallowed. Thus, unsurprisingly, the pure gland secretion compounds possess significant deterrent activities against the fish *X. hellerii* in experiments. The high concentration of the test compounds necessary for significant avoidance of the test pellets is attributable to the fact that only the amount of the compounds contained in the surface of the pellet can trigger an effect. The main part of the test compounds is embedded in the pellet and is not accessible to the fish. In contrast to the ant bioassays, the alkaloids 1, 3 and 5 do not reveal any different levels of deterrent activity in the fish bioassays. To defeat fish predators, the amount of emitted secretion seems to be more relevant than its composition. Steninae show distinct cleaning behaviour, whereby they take up secretion from their abdominal tip and subsequently dispense it over their entire body surface (Betz 1999). This behaviour, which is called "secretion grooming," has previously been observed in several Hydradephaga and water bug families (Maschwitz 1967; Kovac and Maschwitz 1989, 1990). The secretion when spread over the body acts as an antimicrobial agent and prevents the beetles from infestation with microorganisms (Schildknecht and Weis 1962; Maschwitz 1967; Schildknecht 1970; Dettner 1985; Kovac and Maschwitz 1990). When kept at very high humidity over several weeks in the laboratory, *S. bimaculatus* beetles partly show a whitish microbial growth on the thorax and elytra, whereas the other parts of the body surface appear clean and glossy. Assuming an antimicrobial activity of the secretion, this observation can be explained on the basis that the thorax and elytra represent those body regions that are most difficult to reach by the legs of the beetle during secretion grooming and therefore are covered by an amount of secretion not sufficient for inhibiting microbial growth (Lusebrink et al. 2008b). Because the pygidial gland secretion of the Steninae has previously been reported as potentially antimicrobial (Dettner 1985, 1993; Betz 1999), Lusebrink et al. (2008b) tested the effects of selected piperidine-alkaloids on several microorganisms. In their bioassays, both tested compounds revealed pronounced antimicrobial activity comparable with that of commercially available antibiotics. Later we attempted to copy exactly the testing conditions we used previously, but we were never able to reproduce these impressive results unless we applied a 200-fold amount of substance, whereas the commercial antibiotic Tygacil (15 lg) produced comparable inhibition zone diameters in both studies. The tendencies observed in Lusebrink et al. (2008b), however, could be reproduced, whereby stenusine revealed higher antibacterial activity than norstenusine. As with the tests for deterrencies, the single secretion compounds showed better antimicrobial activity than the naturally occurring secretions. Thus, a synergistic effect can be excluded concerning the defence against microorganisms. Although the antimicrobial activity of the secretion was not as strong as that reported in the literature, it should have been capable of inhibiting infestation by bacteria and fungi, because it exists pure and undiluted on the body surface of the beetle. Furthermore, the high frequency of cleaning behaviour shown by the beetles indicates that the amount of antimicrobial compounds is sufficient to inhibit microbial growth on the body surface in nature (Betz 1999). The secretion of the Steninae contains various alkaloids that occur as different configurational isomers. The ratio of the four isomers of stenusine (Fig. 11.17) shows great interspecific differences but is always constant within individuals of one single species (Lusebrink 2007; Lusebrink et al. 2007). Accordingly, the ratio is adjusted by the beetles and is not attributable to random synthesis. Concerning the antimicrobial activity, no differences have been observed with the tested stereoisomers, but 2′*R*,3*S* was more effective than 2′*R*,3*R* in ant bioassays (Fig. 11.17). In *S. comma*, (20R,3S)-1 is only present in minor amounts, whereas it represents the main isomer of 1 in *S. similis*

(Lusebrink 2007; Lusebrink et al. 2007). However, to gain complete insight into the activity of stenusine in natural isomeric composition, further tests should be performed with the remaining isomers. Stereoselective biosynthesis of a compound requires a separate enzyme for each stereoisomer (Morgan 2010). Furthermore, the constancy in the occurrence of the single stereoisomers of stenusine in the secretion demands the exact regulation of every single enzyme involved in the synthesis. This raises the question of selective advantage, because the most active stereoisomer or a random mixture of all stereoisomers should be sufficient for both defence and skimming. Males and females of *S. bimaculatus* are able to perceive the isomeric mixture of alkaloid 1 as determined by electrophysiological tests (Schierling, unpublished results). Thus, a potential pheromone function of some secretion components should be kept in mind because, in this case, the absolute configuration of the compounds is of the highest relevance (Mori 1999; Morgan 2010). However, this possibility is for the moment, speculative, since no confirming experiments have been performed. Corresponding tests are in preparation.

11.6.4 Skimming Behaviour, Evolution of Secretion and Chemotaxonomic Significance

The most unusual function of the pygidial gland secretion in the Steninae is the locomotion on the water surface, called "skimming" (Piffard 1901; Billard and Bruyant 1905; Linsenmair and Jander 1963; Lang et al. 2012), a behaviour typical for many species of the genera *Stenus* and *"Dianous."* The beetles are equipped with hydrophobic tarsi and are thus able to stand on the water surface. For locomotion, they touch the surface of the water with the tip of their abdomen and release a small amount of secretion by everting the pygidial glands. The secretion spreads immediately on the water surface, forming a monomolecular film and thus propelling the beetles forward with considerable speed (Schildknecht 1970; Dettner 1991). The alkaloids (Fig. 11.13) presumably represent the main spreading agents for skimming, but some of the terpenes (Fig. 11.13) also show a distinct spreading activity (Schildknecht et al. 1976; Lang et al. 2012). The Steninae alkaloid components occur as various configurational isomers in the secretion. Although great interspecific differences occur in the stereoisomer ratio of stenusine, their composition is constant within different individuals of a single species (Lusebrink et al. 2007). This probably also applies to all other chiral *Stenus* alkaloids. All stereoisomers of stenusine should be equally effective as spreading agents, despite the slight differences in their biological activity (see above) such as deterrence. However, no information is available about a potential pheromone function of any *Stenus* secretion compound.

While hunting for springtails or other arthropods, the hydrophobic beetles might accidently fall into water. Behaviour of spreading has evolved in these insects as a mechanism to save themselves from drowning and to escape predators such as water striders (Linsenmair 1963; Schildknecht et al. 1975). Piffard (1901) was the first to discover that some species of *Stenus* beetles are able to move over the water surface in a rapid and extraordinary manner. The beetle glides on the water at high velocity without using its legs. A few years later, Billard and Bruyant (1905) observed this locomotion in *Stenus tarsalis* and *S. cicindeloides*. They report an emission of chemicals that interact with the water surface like a surfactant and propel the beetle rapidly forward comparable with a "soap" boat. This locomotion is driven by Marangoni propulsion (Scriven and Sternling 1960). Marangoni flows are those forced by surface tension gradients. Surfactants such as soap and the emitted substances of the beetles are molecules that find it energetically favourable to reside at the free surface and that act to decrease the local surface tension (Bush and Hu 2006). Jenkins (1960) investigated the spreading behaviour of *Stenus* and *"Dianous"* beetles. He named the spreading action of these two genera "skimming." Moreover, he observed the negative phototactic navigation of *Stenus* and "*Dianous*" on the water surface towards the dark bank of a pool. The beetles waste no time in gaining the safe waterside. Schildknecht et al. (1975) found, for the first time, that a secretion of the pygidial glands in the anal region is responsible for the typical skimming action. The gland compounds form a monomolecular film whose front pushes the beetle forward (Dettner 1991). By this kind of locomotion, *Stenus comma*, for example, can achieve a velocity of 0.75 ms^{-1}, and, if the secretion is continuous, a distance up to 15 m can be covered (Linsenmair and Jander 1963). Apart from *Stenus* and "*Dianous*" beetles, such extraordinary movement is only shown by the water cricket *Velia caprai* Tamanini, which uses its rostrum to spread active saliva for skimming (Linsenmair and Jander 1963). No other animal is known to have this unique kind of locomotion. The driving force for skimming is the spreading potential of the gland compounds. Thereby, the gland compounds must be barely soluble in water; otherwise, they could not form a monolayer. Additionally, the spreading agent has to exhibit a lower surface tension than the layer-carrying substance, which is water in case of *Stenus* and "*Dianous*" (Adamson and Gast 1997). The spreading pressures of the compounds can be measured experimentally, for instance, at an interfacial tensiometer (Schildknecht et al. 1976). After the measurement of the surface tension σ of the compounds against air and the interfacial tension γ against water, the spreading pressure P can be calculated according to the following equation defined by Wolf (1957):

$$P = \sigma_{\text{water}} - \left(\sigma_{\text{substance}} + \gamma_{\text{substance/water}}\right)$$

Spreading action can only be observed if the difference of σ_{water} and $(\sigma_{\text{substance}} + \gamma_{\text{substance/water}})$ is positive, i.e. work is obtained (Wolf 1957). Otherwise, the applied liquid remains as a lens on the water surface. Although almost all Steninae and representatives of "*Dianous*" contain the spreading-active stenusine in their pygidial glands, not every species exhibits skimming behaviour (Jenkins 1960; Linsenmair 1963; Dettner 1991). The ability to skim is assumed to be linked to the diverse kinds of habitats that are colonized by *Stenus* (Jenkins 1960).

As shown in Fig. 11.20 for ten Steninae species, the skimming behaviour is species-specific and can be characterized by variable features. Whereas some species can skim voluntarily and show a typical linear skimming behaviour with a continuous depletion of secretion (Fig. 11.20), other species skim not always voluntarily and are characterized by a non-linear, partly circular skimming, stepwise probably because of interrupted discontinuous secretion release. Some species exhibit no skimming behaviour at all, whereas other species have to be forced to skim in order to obtain usable values.

Apart from ethological investigations of skimming Steninae beetles, the physicochemical properties regarding the spreading potential of single secretion compounds and mixtures have also been addressed (Lang et al. 2012). An aqueous oversaturated solution was assembled for measurements by a drop volume tensiometer. Differences of spreading pressures between the substances could be revealed, as shown for the pygidial gland secretion of *S. comma* (Fig. 11.20). The monoterpenoid α-pinene is characterized by the highest spreading pressure followed by 3-(2-methyl-1-butenyl) pyridine (22.22 mN m^{-1}; not present in *S. comma*), norstenusine, stenusine and 6-methyl-5-hepten-2-one (15.63 mN m^{-1}; not present in *S. comma*). Cicindeloine (4.27 mN m^{-1}; not present in *S. comma*) and the monoterpenoid 1,8-cineol (Fig. 11.20) are the least surface active. With respect to the first compound, the low value may be attributable both to its extremely low water solubility and its resinous consistency. For the investigation of naturally identical secretion compositions by the tensiometer, four beetle species were chosen based on their gland content (Lang et al. 2012). The species typify representatives of the different *Stenus* groups: *S. clavicornis*, a species that exhibits no skimming behaviour, is characterized by a stenusine-based secretion composition (the so-called piperidine group; see below) and a spreading pressure of 20.67 mN m^{-1}. The second

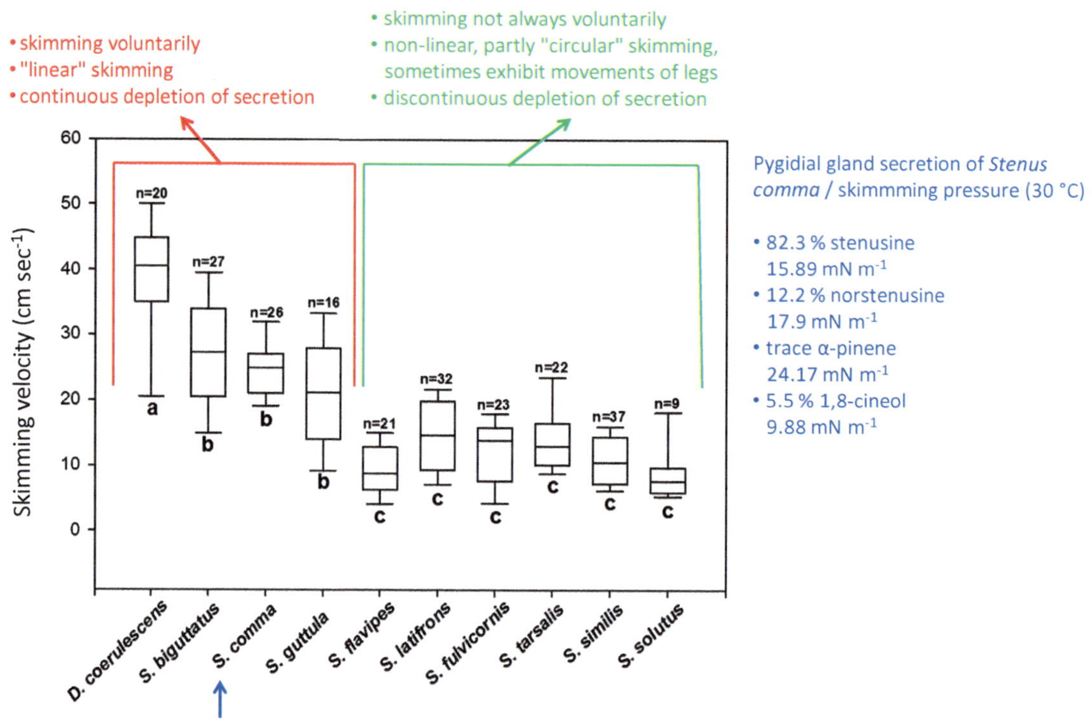

Fig. 11.20 Skimming velocities (cm sec^{-1}) of nine studied *Stenus* species and *Dianous coerulescens*. The boxplots are shown with the median and corresponding standard deviation bars. *D. coerulescens* showed the highest spreading rate, followed by *S. biguttatus*, *S. comma* and *S. guttula*. The spreading behaviour of the analysed remaining species (*S. flavipes* to *S. solutus*) cannot be significantly differentiated statistically. The lowercase letters a, b and c indicate significant differences based on calculated *p*-values. Number of specimens investigated is indicated by *n*. Orange-bracketed species skim voluntarily and show a linear skimming and a continuous depletion of secretions. Green-bracketed species do not always skim voluntarily; show a non-linear, partly circular spreading; and sometimes exhibit movements of their legs. These species are characterized by discontinuous secretion depletions. Right inset: percentual quantitative secretion constituents from the anal glands of *Stenus comma* (arrow) are indicated together with spreading pressures of the single compounds (Lang et al. 2012)

species chosen, *S. comma*, also belongs to the piperidine group and skims voluntarily with high velocities and 19.71 mN m^{-1}. *S. similis* from the so-called pyridine group is characterized by discontinuous secretion release and non-linear skimming (Fig. 11.20) but, nevertheless, has a secretion with a spreading pressure of 20.97 mN m^{-1}. These data illustrate that skimming behaviour is not positively correlated with the spreading pressure of species-specific formulations. In contrast, in *S. solutus*, skimming actions can only rarely be observed, i.e. the beetles skim always involuntarily and very slowly by using their legs (Fig. 11.20). In the same way, the secretion has a low spreading pressure of 11.19 mN m^{-1}.

As a whole, all Steninae species hitherto investigated chemically possess spreading-active substances in their pygidial glands, but not every *Stenus* species actually shows skimming behaviour. As Steninae inhabit interstices (Horion 1963) and banks of open waters (Dettner 1987), the disposition of the beetles in displaying skimming behaviour might be linked to the different habitat claims of the species. *D. coerulescens*, which is characterized by the significant highest skimming velocity in this study (Fig. 11.20) inhabits the immediate neighbourhood of waterfalls, weirs and fast-flowing mountain streams with declines (Horion 1963). Therefore, for the beetles to survive, they need to be able to skim rapidly in the fast-moving waters. Consequently, these beetles exhibit distinct skimming behaviour and velocities. Other *Stenus* species, which always display a skimming disposition and high skimming velocities, such as *S. comma* and *S. biguttatus*, can be found in habitats close to the banks of stagnant open waters (Horion 1963). These biotopes also require voluntary and persistent skimming action. Several other *Stenus* species investigated (Fig. 11.20) are characterized by a proportionally slow skimming velocity and a less disposition to skim, such as *S. flavipes* (*Metatesnus*) and representatives of the subgenus *Hypostenus* such as *S. latifrons*. Other species such as *S. fulvicornis*, *S. tarsalis*, *S. similis* and *S. solutus* can be predominantly found in swampy biotopes and wet places within woods, meadows and marshes. If these beetles live in wet places near open waters, they stay in rotten leaves and other detritus material or in places abundantly covered with vegetation (Horion 1963). Therefore, the beetles rarely accidently fall into open water, and the occasion to skim is uncommon. This might be a reason for the beetles not having developed a distinctive skimming behaviour such as the investigated species *D. coerulescens*, *S. comma* and *S. biguttatus*. Since many Steninae predominantly prefer moist habitats, the danger of infestation by microorganisms is increased. In order to avoid this infestation, *Stenus* displays secretion grooming (Kovac and Maschwitz 1990; Betz 1999) in which the beetle spreads its antimicrobial secretion of the defensive pygidial glands over the entire body surface. *Stenus* beetles living in interstices in plant debris or detritus might primarily use their pygidial gland secretion for secretion grooming, instead of using it for skimming action. Moreover, another aspect seems to be of importance. In 1960, Jenkins reported, in the context of his ethological studies on "*Dianous*" and *Stenus* species, a negative phototactic skimming behaviour to the dark bank of a pool. Once fallen on the water surface, the beetles at once orient themselves into the "correct" direction and waste no time in gaining the safety of the bank by skimming. These results are supported by the observations of Linsenmair (1963). As *Stenus* beetles are equipped with extraordinary large eyes in relation to their body size and compared with related staphylinid beetles, the visual sense is evidently essential for their orientation. Furthermore, the beetles not only might orient themselves in the correct direction but also might estimate the distance to the bank. The beetles might be able to balance the distance to the safe bank against the amount of secretion in their glands. Only in exceptional circumstances, do the beetles "decide" to use their precious secretion, e.g. complete physical exhaustion or repeated attacks by predators (Linsenmair 1963). These circumstances might be the reason that skimming velocities of 40–75 cm s^{-1} and a covered distance of up to 15 m (Linsenmair and Jander 1963 for *S. comma*) have usually not been achieved by the beetles in our studies. In summary, tested *Stenus* species from vegetation-rich habitats show less precise skimming behaviour than Steninae from open habitats possibly because of to the lower optical capacity of the former.

The chemical gland content of the different Steninae species and their relationship allows a chemosystematic arrangement to be developed (Francke and Dettner 2005). In the case of Steninae whose gland secretion consists predominantly of stenusine, norstenusine and minor terpenoids, the species can be regarded as phylogenetically basal ("piperidine group" according to Schierling et al. 2012). These species, such as *S. clavicornis* and *S. comma*, use their piperidine-alkaloids mainly for the optionally shown skimming behaviour and as predator avoidance (Connert 1974; Dettner et al. 1996; Hesse 2000). Furthermore, the terpenoids from the small glands are pharmacologically active substances and are used for antimicrobial protection against bacteria and fungi (Schildknecht et al. 1976) and as an insect repellent (Honda 1983; Blum 1981).

The phylogenetically more derived *Stenus* species in this chemotaxonomically based study, as represented by *S. similis* ("pyridine group" according to Schierling et al. 2012), still possess stenusine as the main component in their glands, but the "new" pyridine-alkaloid also comprises a large part of the secretion. Not only is the spreading pressure of the pyridine higher than the pressure of stenusine, but also the antimicrobial effect is higher against *Bacillus sphaericus*, *Escherichia coli* and *Serratia entomophila* (Lusebrink et al. 2009).

The next phylogenetic level in this chemotaxonomical study is represented by *S. solutus* ("epoxypiperideine group" according to Schierling et al. 2012), which also includes *S. binotatus* and *S. cicindeloides*. These beetles contain remarkable amounts neither of piperidine-alkaloids nor of terpenes but produce the pyridine- and the piperideine-alkaloids. Although the fluid pyridine might be used to gain at least a minor spreading potential together with deterrence against ants together with bactericidal and fungicidal effects, the role of the resinous piperideine cicindeloine cannot be assessed.

In summary, the secretion composition of *Stenus* is subject to evolutionary optimization. From phylogenetically plesiotypic to derived taxa, new alkaloids in the large gland were gradually developed, whereas the small gland containing the terpenes was gradually reduced. The alkaloids were biosynthetically produced from the amino acids L-lysine and L-isoleucine. This observed evolutionary trend in Steninae contrasts with defensive glands of adults of other Staphylinidae such as the Oxytelinae (Francke and Dettner 2005) or Aleocharinae (Steidle and Dettner 1990), where one or a few active principles such as solid quinones are dissolved within different solvents and solvent mixtures. During the evolution of the Oxytelinae or Aleocharinae, these formulations have been optimized in order to increase the insecticidal activities of the glandular mixtures. Another difference exists between the secretions of the Steninae and Oxytelinae/Aleocharinae. In the Steninae, no synergistic or quasi-synergistic effects have ever been observed between the single compounds within the secretion as seen in the Oxytelinae/Aleocharinae (Francke and Dettner 2005). Moreover, all Steninae secretion components represent biologically active natural compounds, whereas secretions from the two other subfamilies contain many inactive solvents. Thus, the *Stenus* secretion meets many different demands. The spreading potential of the gland compounds is essential not only as a locomotory or escape mechanism but also because of the antimicrobial and protective mixture that can be distributed over the body surface of the beetle while secretion grooming. Whereas the defensive secretions of other insects also serve as surfactants and are able to spread [e.g. the oral secretions regurgitated by *Spodoptera exigua* (Lepidoptera; Rostás and Blassmann 2009) and the secretion of whirligig beetles (Coleoptera, Gyrinidae; Vulinec 1987)], no other insect has developed such striking skimming behaviour. Although extraordinary skimming is a secondary effect of the multifunctional *Stenus* secretion that primarily was evolved for chemical defence, it is essential for the beetles' survival and helps both *Stenus* and "*Dianous*" beetles to be optimally adapted to their various habitats and the challenges of life among staphylinids.

Based on three genes (COI, 16S rRNA and Histone), we have obtained further and comprehensive insights into the molecular phylogeny of the subfamily of Steninae (Fig. 11.21), traditionally comprising the genera *Stenus* and "*Dianous*". These results indicate that "*Dianous*" belongs to *Stenus* (Koerner et al. 2013; Lang et al. 2015) (see Sect. 11.2). Furthermore, we have shown that our chemotaxonomic approach presented previously by Schierling et al. (2013) is consistent with the obtained molecular data. The piperidine group is the most extensive species group listed. Examples are presented in Fig. 11.21 in red. The next chemotaxonomic level is represented by the pyridine group members *S. similis* and *S. tarsalis* possessing the chemotaxonomic-derived gland compound 3-(2-methyl-1-butenyl) pyridine in their pygidial glands (Schierling et al. 2013; yellow colour). Although these two species cannot be found in the same clade, they are part of a major cluster comprising the species *S. bifoveolatus*, *S. binotatus* and *S. solutus*. Interestingly, the basal piperidine group member, *S. bifoveolatus* (*Metatesnus*), belongs to this cluster, which contains chemotaxonomic-derived species. In this case, the subgenus concept has also not been maintained: *Hypostenus* species (*S. similis*,

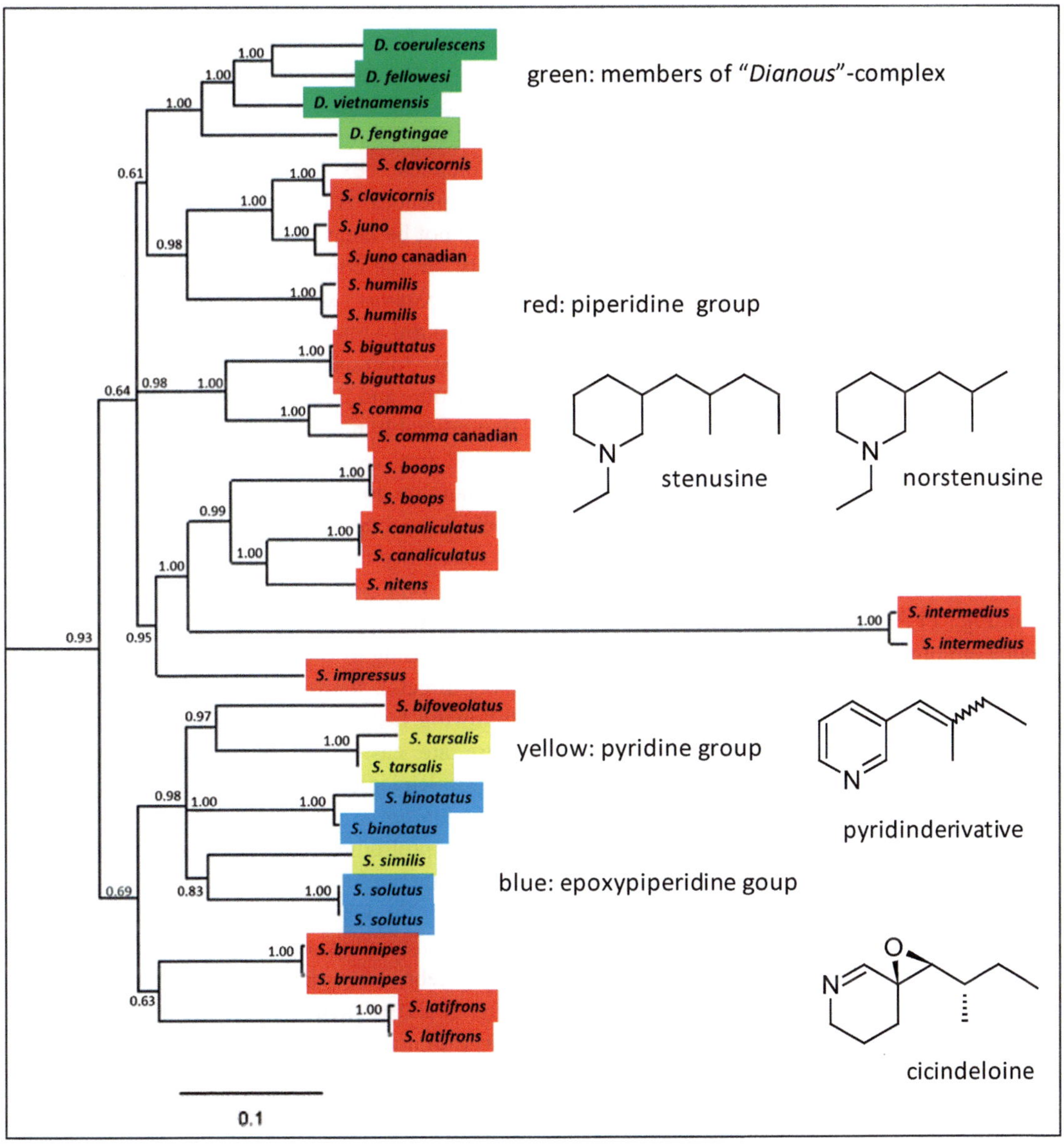

Fig. 11.21 Maximum posterior (MAP) tree for the genera *Stenus* and "*Dianous*" resulting from Bayesian analysis (Ronquist and Huelsenbeck 2003; Lang et al. 2015). The numbers above the branches indicate Bayesian posterior probabilities $\geq$0.50. The molecular data area is based on cytochrome oxidase I (COI), 16S rRNA and the histone H3-gene. Members of chemotaxonomic groups according to Schierling et al. (2013), Puthz (1981, 2000b, 2005a), Shi and Zhou (2011) and Tang et al. (2011) (*Stenus*, red, yellow, blue; "*Dianous*," green)

S. tarsalis and *S. solutus*) cluster together with *Metatesnus* species (*S. bifoveolatus* and *S. binotatus*).

The most evolved chemotaxonomic level is represented by the epoxypiperideine group comprising *S. binotatus* and *S. solutus* in our study. Species of this group are characterized by an extension of the pygidial gland repertoire to the new alkaloid cicindeloine as the main gland constituent (Schierling et al. 2013). Piperidine-alkaloids such as stenusine and norstenusine are completely replaced by pyridine- and epoxypiperideine-alkaloids. In the tree constructed (Fig. 11.21), *S. solutus* is positioned in the direct neighbourhood of the pyridine group member *S. similis*, which also possesses the chemotaxonomically derived pyridine-alkaloid but lacks the most evolved piperideine epoxide in its glands. With regard to the species groups of Puthz (2008), the *S. similis* group consisting of *S. similis* and *S. solutus* is supported (BPP = 0.83) in the present chemotaxonomic classification. The other epoxypiperideine group member *S. binotatus* occupies an isolated position regarding *S. solutus* within a major cluster of both chemotaxonomically derived and most evolved species. However, all *Stenus* species possessing chemotaxonomically evolved gland compounds show a close relationship within the tree constructed.

In Fig. 11.21, the various chemotaxonomic groups are highlighted in colour [red, piperidine group (primitive); yellow, pyridine group (derived); blue, epoxypiperideine group (most derived)]. However, future investigations of a wider range of *Stenus* and "*Dianous*" species will be necessary to reveal fully the chemotaxonomy and molecular phylogeny of Steninae.

11.7 Ecomorphological Diversity

When morphological features that form direct interfaces to the environment are considered, major interspecific differences occur with respect to the compound eyes, the labial prey-capture apparatus, the legs including the tarsi and the structure of the abdominal tergites. In order to attain a better understanding of the evolution of ecomorphs across the Steninae, we need to understand the functional and ecological consequences of these different morphologies.

11.7.1 Compound Eyes

Steninae beetles have large protruding eyes that are typical for visual hunters among diurnal predatory beetles (e.g. Bauer et al. 1998). Such eyes favour peripheral vision and may be associated with an array of ommatidia improving the resolution in the frontal field of view (e.g. Burkhardt and de la Motte 1983). Although all Steninae seem to belong to this type, species-specific differences exist that render functional consequences. Although interommatidial angles could not be measured to reconstruct their visual fields, simple comparative measures of the eyes of 18 *Stenus* species correlated well with behavioural features in the study of Betz (1998a). Interspecific differences in approach behaviour and reaction ability in response to unpredictably moving springtails in standardized behavioural experiments were, to some extent, attributable to differences in eye morphology, such as the total number of ommatidia (indicative of the ability to resolve fine details and to detect moving objects at a distance) and the extent of the lateral projection of the eyes (usually corresponding to wide frontal acute zones in predatory insects). Narrow acute zones are unsuited for following fast, unpredictably moving prey, since these are easily lost from the zone (Horridge 1978). Ripicolous species such as *S. comma*, *S. biguttatus*, *S. fossulatus* and *S. guttula* with widely protruding eyes that are composed of many ommatidia are agile and highly responsive (Fig. 11.22a), whereas the representatives of many species mounting the vegetation or inhabiting plant debris are not or behave intermediately (Fig. 11.22b). Their eyes are flatter and consist of fewer ommatidia (Betz 1998a, b).

The "*Dianous*" species also exhibit notable differences in the eye size, all of which have fed into the definition of the species groups (Puthz 2015, 2016).

11.7.2 Labial Prey-Capture Apparatus

The elongate labia of the various *Stenus* beetles differ with respect to their length and the morphology of the sticky cushions (i.e. the paraglossae) at the tip of the prementum. In 220 measured *Stenus* species, the labium length differed between ca. 0.4 mm (*S. canaliculatus*) and 3.4 mm (*S. bilunatus*). Although the absolute

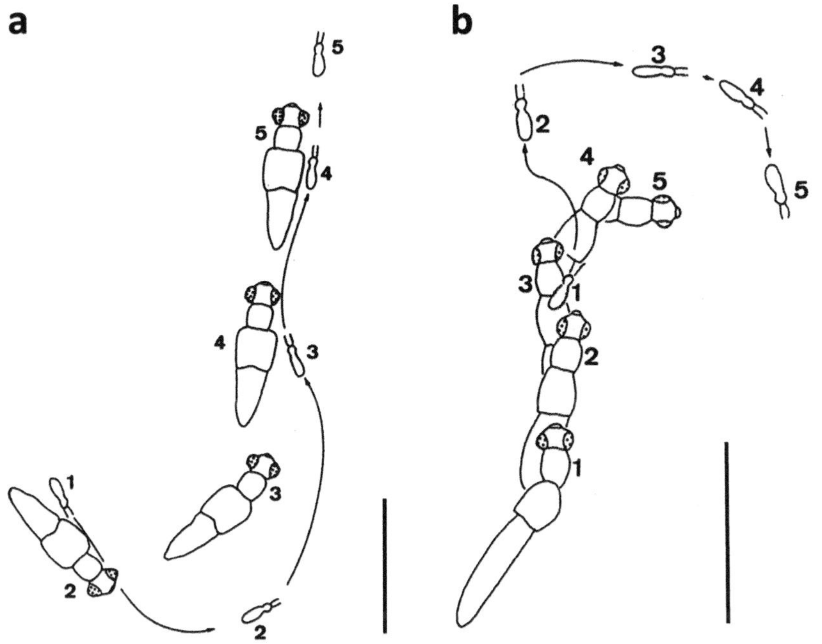

Fig. 11.22 Two opposed forms of optically controlled approaches towards a springtail moving on a plain surface of plaster of Paris. Sequence drawn from video footage. Identical numbers at prey and predator indicate simultaneous constellations. (**a**) *Stenus comma*: time course of the depicted sequence [seconds that lapsed from the start (= picture 1); the first number refers to the prey, the second one to the predator], (**1**) 0; (**2**) 1/0, (**3**) 3/0; (**4**) 5/5; (**5**) 8/5. The behaviour of the beetle is characterized by a high locomotion speed and a barely delayed orientation towards the prey. (**b**) *Stenus pubescens*: time course of the depicted sequence [seconds that lapsed from the start (= picture 1); the first number refers to the prey, the second one to the predator], (**1**) 0; (**2**) 1/0; (**3**) 3/0; (**4**) 5/8; (**5**) 9/8. The behaviour of the beetle is characterized by a low locomotion speed and a temporally delayed responsiveness towards the prey. Scale bars = 5 mm

labium length (i.e. the prementum plus the membranous connecting tube) correlates with body length, the comparison of the relative labium lengths (i.e. the labium length divided by the pronotum length) shows that, in some groups of species, the labia have become especially elongate independently of body size, whereas in others, they have become largely shortened (Fig. 11.23). In the members of the *S. canaliculatus* group, the shortened labium length is the result of a secondary reduction of the labium (Betz 1996, 1998b, 2006; Koerner et al. 2013). Large species generally have longer labia and attain higher attack distances than small ones. Since 10–30% of the attack distance is overcome by a forward strike of the entire beetle, the relationship between labium length and attack distance becomes even stronger, when one considers the difference between the attack distance and the length of the forward strike instead of the pure attack distance (Betz 1996). Variations of the attack distance set by the beetles prior to the predatory strike are probably responsible for interspecific differences in the compressive forces exerted by the labium (Betz 1996, 1998a; Koerner et al. 2012a, b, 2017). Differences in the haemolymph pressure generated to eject the labium for prey-capture might also be responsible for variations in the compressive forces.

An interspecific morphological comparison of the sticky cushions (paraglossae) of the labium of >200 *Stenus* species has established that, in most of the investigated species, they represent a general type (i.e. the ventral surface of the sticky cushions has the shape of an ellipsoid; Fig. 11.24a) that has apparently been modified in different ways during the course of evolution

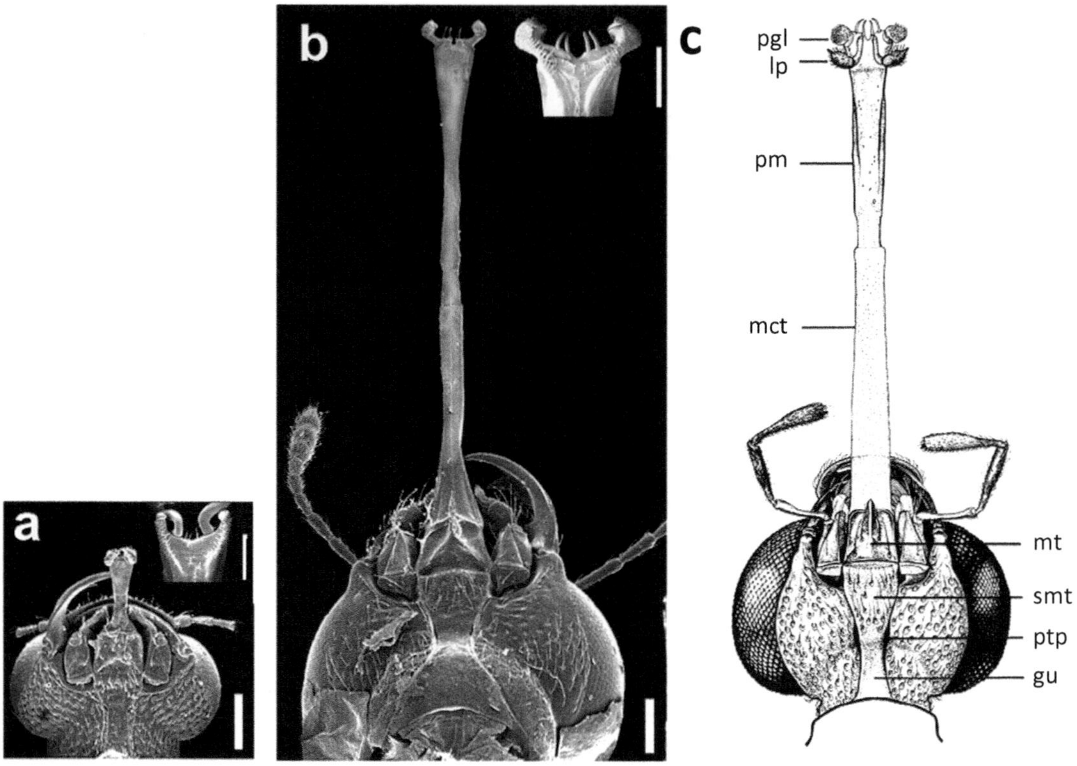

Fig. 11.23 Ventral aspects of the protruded labium in (**a**) *S. canaliculatus* and (**b**) *S. comma*. *S. canaliculatus* beetles show a largely shortened labium with vestigialized sticky cushions. Scale bars = 200 μm. Dorsal aspects of the tip of the prementum are shown as insets (scale bar (**a**), 50 μm; scale bar (**b**), 100 μm). In (**b**), the labial palps were removed to expose the paraglossae (**c**) ventral aspect of the head with protruded labium of *S. comma*. From Weinreich (1968). Abbreviations: *gu* gula, *lp* labial palpus, *mct* membranous connecting tube, *mt* mentum, *pgl* paraglossa, *pm* prementum, *ptp* posterior tentorial pit, *smt* submentum

(Betz 1996). These changes primarily concern the number of adhesive trichomes on a cushion, the degree of branching of a single adhesive seta and the manner in which both modes of alteration are combined. Out of the 220 *Stenus* species measured, the area of the sticky pads ranged from 181 μm^2 in *S. canaliculatus* to 10.760 μm^2 in *S. bimaculatus* (L.K. & O.B., unpublished data). As shown experimentally, larger sticky cushions and higher numbers of adhesive trichomes or adhesive contacts improve adhesion and thus lead to enhanced capture success. Hence, the morphology of the sticky cushions is thought to have been exposed to a high selective force that led to an improvement of their selective advantage (Betz 1996). In this regard, the modifications of the sticky cushions in some *Stenus* species (Fig. 11.24) are probably the result of a progressive selection that led to an extension of the feeding niche by an increase of the maximum prey size or that is connected to a particular suite of potential prey species. For example, the paraglossae of some species within the subgenera *Hypostenus* (*S. latifrons*, *S. fulvicornis*) and *Hemistenus* (*S. persicus*, *S. turk*, *S. parcior*, *S. glacialis*) are considerably elongated (longiform paraglossae, Figs. 11.24b, c), and, hence, a remarkable enlargement of their surface area is attained. Extreme modifications of the paraglossae are also shown by some Asian species with coniform paraglossae [e.g. *S. ananda*, *S. stigmatias* (Fig. 11.24d), *Stenus flammeus* group; Betz 1996; Puthz 1998]; the sticky cushions of these species largely extend in a dorsoventral direction, with continuous narrowing. Again, the surface of the sticky cushions is considerably enlarged. It is also entirely covered by adhesive trichomes (ca 600 per sticky cushion),

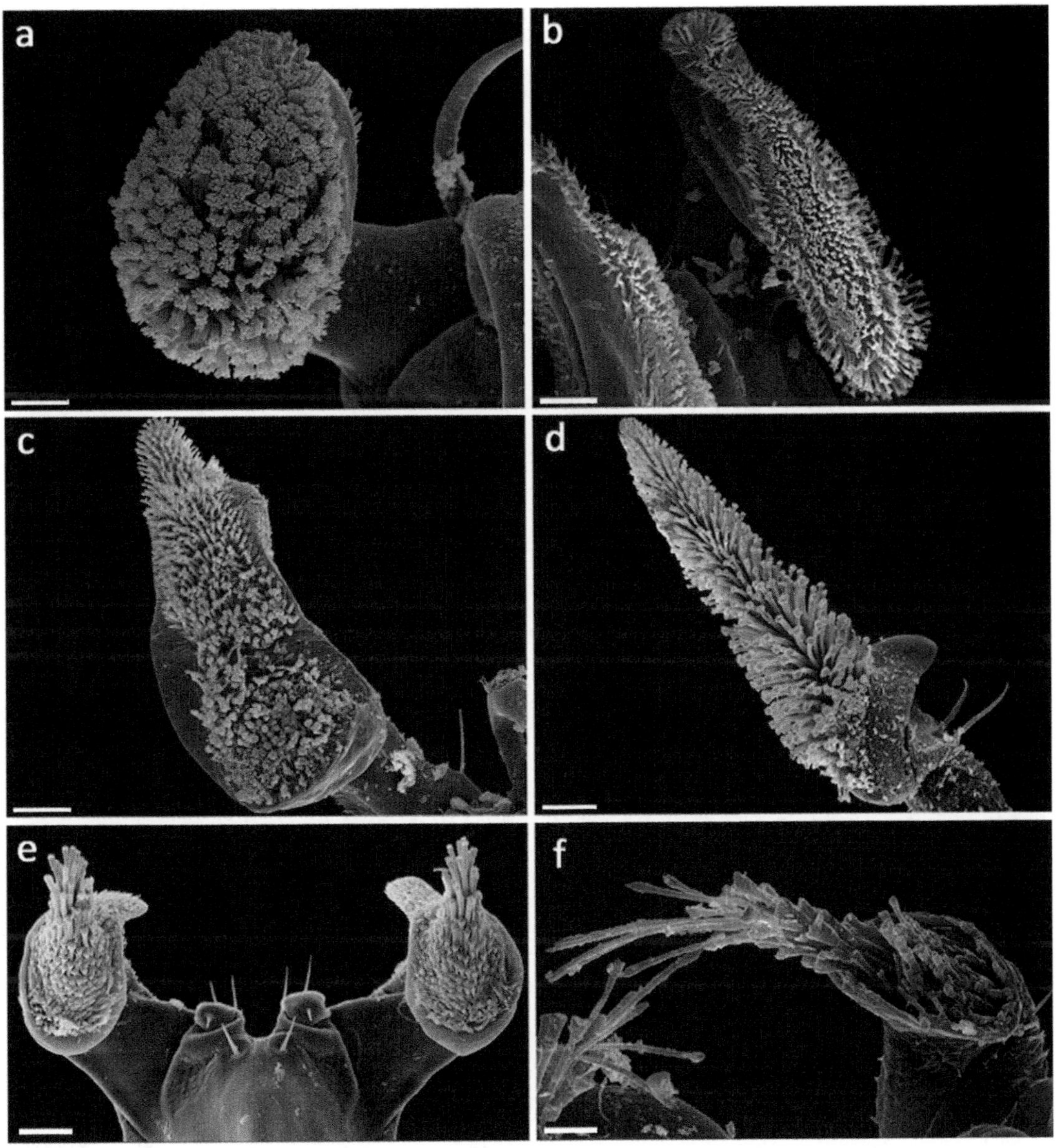

Fig. 11.24 The various paraglossa types as hitherto established in *Stenus* spp. (**a**) Ellipsoid (*S. biguttatus*). Scale bar = 10 μm. (**b**, **c**) Longiform (*S. fulvicornis*, *S. persicus*). Scale bar = 20 μm. (**d**) Coniform (*S. stigmaticus*). Scale bar = 20 μm. (**e**) Actiniform (*S. alpaca*). Scale bar = 20 μm. F: Sileniform (*S. nepalensis*), Scale bar = 10 μm

some being elongated. Some Middle and South American species [e.g. *S. emily*, *S. electriger*, *S. alpaca* (Fig. 11.24e); Puthz 2005b] have actiniform, and some Asian species have sileniform [e.g. *S. pilicornis*, *S. nepalensis* (Fig. 11.24f); Puthz 2013] paraglossae; both these types are characterized by very long adhesive trichomes that become longer towards the middle of the sticky pad.

In contrast, some closely related species within the *S. canaliculatus* species group (e.g. *S. canaliculatus*, *S. nitens*, *S. labilis*, *S. caseyi*, *S. dolosus*, *S. vinnulus*) are provided with comparatively simple short labia [i.e. the labium length ranging from 0.4 mm in *S. canaliculatus* (Fig. 11.23) to 0.9 mm in *S. vinnulus*] that have only very small paraglossae without modified trichomes (i.e. the surface area ranging from 181 μm^2 in *S. canaliculatus* to 640 μm^2 in *S. labilis*). Ryvkin (2012) suggested that the reduction of the labium is a characteristic feature of all the members of the *Stenus canaliculatus* group, but this must be verified by further investigations. Molecular, anatomical and ethological findings indicate that these simple labia are vestigial (Betz 1996, 1998b, 2006; Koerner et al. 2013). The development of the presumably vestigial labia of these species might result from a specialization of the beetles involving prey-capture with their mandibles (i.e. these species have comparatively long sabre-like mandibles, which are appropriate for catching large prey).

Species-specific differences have been found in the adhesion of the labia and are thought to result from divergencies in the sticky cushions, especially the surface area of their adhesive surface and the number of adhesive trichomes or adhesive contacts. The influence of the morphology of the sticky cushions on adhesive performance during the predatory strike has been measured in 14 Central European *Stenus* by using micro-force sensors (Koerner et al. 2017). The measurements have revealed strong interspecific differences in the adhesive forces generated during the predatory attack; these differences vary from 0.3 mN in *S. morio* to 1.1 mN in *S. bimaculatus*.

The variations in the adhesive performance are functionally correlated with the morphology of the sticky cushions, i.e. their surface area (Fig. 11.25b) and their number of adhesive

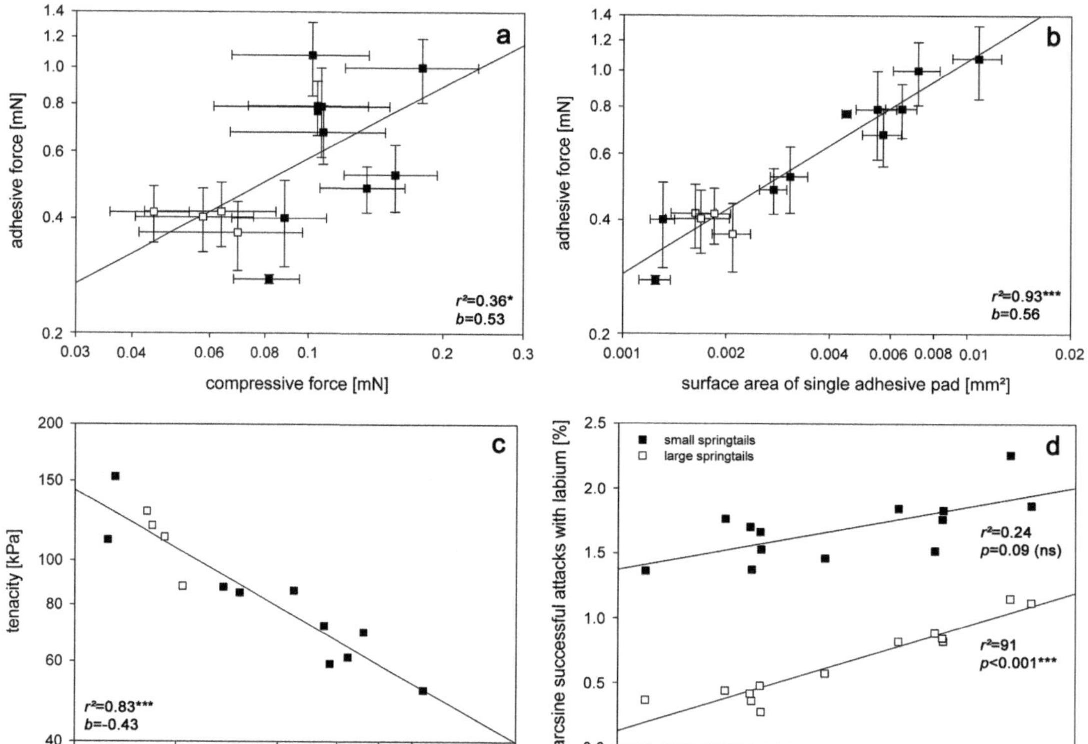

Fig. 11.25 Relationships between morphological and performance parameters of the prey-capture apparatus of 14 *Stenus* species (log-log plots). Trend lines were computed by using phylogenetic regression (PGLS) for gradual model of character evolution (see Koerner et al. 2017). Filled squares in figures (**a**)–(**c**) represent inhabitants of moist humus or plant debris near the ground, whereas open squares represent "surface runners" on bare ground [Adapted from Koerner et al. (2017). With kind permission from Oxford University Press]

outgrowths, and divergences in the compressive (impact) forces generated during the predatory strike, ranging from 0.05 mN in *S. biguttatus* to 0.2 mN in *S. juno*. These experiments strongly suggest that the involved adhesive mechanism is pressure-sensitive, i.e. that higher compressive forces result in better adhesive performances (Fig. 11.25a).

The differences in the adhesive forces significantly influence the prey-capture success towards large springtails of *Heteromurus nitidus* (Fig. 11.25d), whereas this relationship is significant only at a 0.1 significance level for small springtails of this species (Fig. 11.25d). Thus, higher adhesive forces lead to higher prey-capture success rates in *Stenus* beetles; this effect is more pronounced towards large-sized springtails (Koerner et al. 2017).

The mean tenacity, as calculated by dividing the mean adhesive force by the mean surface areas of both adhesive pads, amounted to between 51.9 kPa in *S. bimaculatus* and 153.2 kPa in *S. humilis*. Allometric analyses revealed that the tenacity shows negative correlations with the surface area of the adhesive pad and the body size (Fig. 11.25c). Thus, although species with smaller adhesive pads (which usually correspond to smaller body sizes) generated lower absolute values of the adhesive forces, they attained higher tenacities and therefore had more adhesively efficient pads. These negative relationships might mean that the action of the adhesive secretion alone plays a more important role than the size and the morphological complexity of the adhesive pads, i.e. the adhesive strength of the secretion probably overrides the adhesive impact of the pad size and the intercorrelated number of adhesive contacts. This would represent an important functional feature, especially for species with smaller adhesive pads, since it enables them to achieve a relatively high adhesive performance.

11.7.3 Legs Including the Tarsi

Apart from a slight positive allometry of the leg length with respect to the pronotum length, some *Stenus* species (mostly riparian surface runners on bare ground) show especially elongate slender legs (Betz 1994). Elongate legs usually increase the possibility to effectively and rapidly accelerate/decelerate and improve maneuverability.

The tarsi of all three leg pairs are composed of five tarsomeres. Across the Steninae, a striking diversity of tarsal morphologies exists that manifests itself in (1) the general shape of the single tarsomeres (these are slender non-bilobed, sub-bilobed or wide bilobed[3]) (Fig. 11.26) and (2) the diversity of morphological types of tenent setae that cover their ventral surface (e.g. Puthz 1971; Betz 1998b, 2000, 2002, 2003, 2006; Puthz 2016).

Wide bilobed tarsi within several species groups are associated with a considerable augmentation of tarsal ventral setae (Betz 2003). About 70% of *Stenus* species and 36% of "*Dianous*" species show bilobed tarsi, whereby these numbers also include species with only weakly bilobed tarsi (cf. Puthz 2016). Contact angle measurements at the underside of the tarsi have revealed that all the investigated species are well supported by the surface of water while walking on it. In this context, the non-wettability of the underside of the tarsi and their total circumference play a major role and not whether they are bilobed or not. At least in *Stenus*, wide bilobed tarsi might have evolved in the context of expanding the adaptive zone towards living plants as hunting sites. Indeed, comparative measurements of the pulling forces attained on vertical substrates have demonstrated that widened bilobed tarsi exhibit a significantly higher adhesive performance on smooth (plant) surfaces than do slender tarsi. This is attributable to the number of tarsal tenent setae and their specific morphology (Betz 2002).

In 18 investigated Central European *Stenus* and one "*Dianous*" species, nine different tarsal seta types could be distinguished; indeed, even within a single species, up to six of these different types might be present. This diversity is partly attributable to some setae representing (mechano) sensilla; others must be considered

[3] The lobed state is especially established on the antepenultimate and the penultimate tarsomeres.

Fig. 11.26 Ventral aspects of the hind tarsi of various morphological types, i.e. (**a**) slender (e.g. *Stenus comma*), (**b**) sub-bilobed (*S. brunnipes*) and (**c**) wide bilobed (*S. pubescens*). Scale bars = 20 μm. From Betz (2003)

as tenent setae and release an adhesive secretion that is produced by underlying unicellular glands (Betz 2003). In particular, the spatulate type of tenent setae seems to be important for generating a considerable adhesive performance; it has only been found in species with wide bilobed tarsi (Betz 2003).

Most "*Dianous*" species show simple tarsi (all "*Dianous*" group I and most "*Dianous*" group II species), whereby in the "*Dianous*" group II, 16 species with deeply bilobed penultimate tarsi also occur (Puthz 2000b, 2016). One peculiar tarsal character occurring in many "*Dianous*" group II beetles is the "tarsal shoe" that distally projects from the tarsomeres and is formed by a dense array of elongated setae, which may be modified into tape-like or willow-leaf-like structures (Rougemont 1985; Puthz 2000b; Betz 2003) (Fig. 11.27). These structures might trap air and thus support the beetles on the surface of water (Rougemont 1985). Such a structure might be of special importance, as "*Dianous*" beetles typically live in the immediate vicinity of powerful streams and cascades, where they permanently run the risk of being swept away by the current.

11.7.4 Abdominal Tergites

Stenus species may bear one pair of distinctly raised margins (i.e. the latero- or paratergites) at

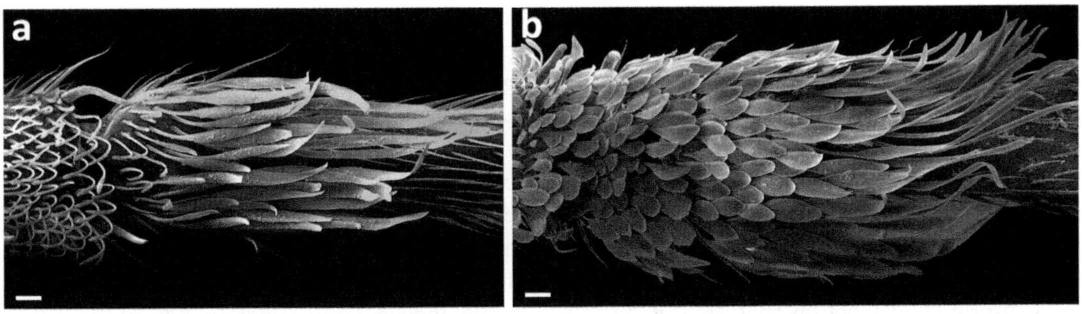

Fig. 11.27 Ventral aspects of the fourth tarsomeres of the hind legs showing distinct "tarsal shoes." (**a**) *Dianous obliquenotatus*, (**b**) *D. fornicifrons*. Scale bars = 20 µm. For further explanations, see text

their tergites IV–VI. However, such margins are absent in about half the *Stenus* species (Puthz 2017), which then have ring-like cylindrical segments. Many other *Stenus* species represent an intermediate situation, since they show only a delicate tergosternal suture. Such reductions of the laterotergites might have entailed improved abdominal mobility (Puthz 1971) providing an advantage in many biological contexts in which the abdomen is involved (Betz 1999): terrestrial locomotion, swimming and skimming on the water surface, prey-capture (for building up the haemolymph pressure that is necessary for the ejection of the labium), mating, oviposition, self-grooming, chemical defence and the folding of the hind wings. To assist wing folding, the tergites may bear medial and/or lateral keels (at the first tergites), membranous apical rims (seventh and, occasionally, also eighth tergite), combs (ninth tergite) and spicule patches (tenth tergite and, occasionally, ninth tergite) (Puthz 1971). The folding pattern of the alae is described in Blum (1979).

In "*Dianous*," the paratergites are usually well established, and only a few species show slight reductions in their widths.

11.7.5 Ecomorphology

For an understanding of the ecomorphological radiation that has occurred in the Steninae, we need to consider the way that morphological, behavioural and ecological traits interact with each other to form certain ecomorphs and the relationship of this to the phylogeny of the group. Habitat choice experiments on 16 Central European *Stenus* species revealed that these beetles forage in three different zones, i.e. they are (1) inhabitants of moist humus and plant debris near the ground, (2) plant climbers or (3) surface runners on bare ground. A comparison of 18 Central European *Stenus* species has revealed different complexes of adaptations concerning predatory behaviour and linked morphological characters. Surface runners (3) such as *S. comma* have laterally protruding eyes with a large number of ommatidia, long legs and slender tarsi. Standardized prey-capture experiments with springtails have revealed that these predators are highly agile and are capable of pursuing prey that moves quickly and unpredictably. Their searching behaviour is of the ambush-searching type. Moreover, they are less reliant on their specialized labium for prey-capture but have refined the mandible-attack mechanism. The results of Koerner et al. (2017) are indicative of a process of the secondary reduction of their labial prey-capture apparatus, since these species possess, in relation to their body size, only small and simple-structured adhesive pads and generate exceptionally low compressive (impact) forces during their predatory strike (Fig. 11.25a). The improvement of the mandible-attack mechanism in these specialists seems to compensate for the limitation of the adhesive strength of the labium towards larger prey.

However, such ecomorphs are an exception and may have evolved from (1) detriticolous or (2) planticolous predator types that make up the majority of recent *Stenus* species. These beetles are not particularly agile and stalk in plant debris or in the vegetation in order to hunt stationary or slow moving prey. Consequently, they have flat eyes and wide bilobed tarsi and depend to a higher degree on the labium for prey-capture, since it permits these predators, despite their limited agility, to catch prey in a sudden and surprising manner.

A discriminant function analysis (DCA) considering 91 *Stenus* species has been used here to examine those morphological variables that determine the three general hunting sites of the species under study (Fig. 11.28). Plant climbers (green triangles) are distinct from soil dwellers (red squares and orange circles) mainly by their widened tarsi, in agreement with experimental results (Betz 2002). This feature might thus represent a key innovation that has made possible the expansion of the adaptive zone to live plants, contributing to the tremendous radiation of this genus. Indeed, approximately 70% of the >3000 *Stenus* species described belong to groups whose representatives have bilobed tarsi. Our DCA

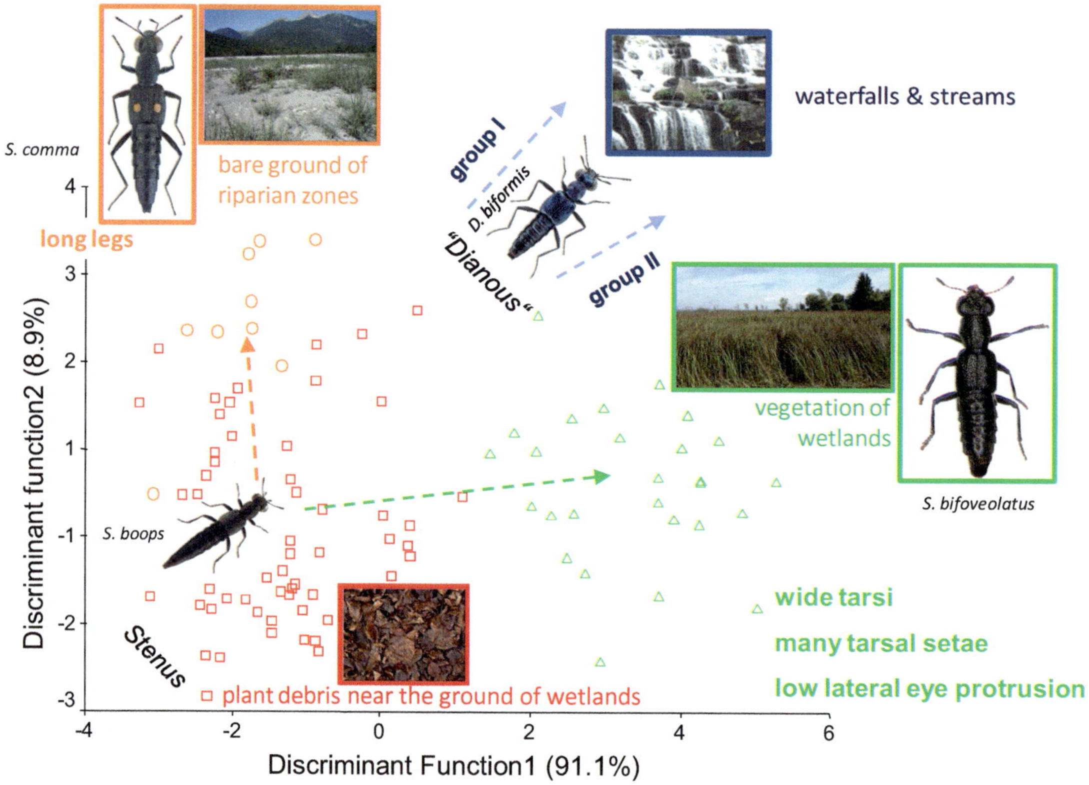

Fig. 11.28 Discriminant analytical separation of the three major hunting sites established in *Stenus* species. Each dot represents a different *Stenus* species. Values were log-transformed and corrected for body size before the analysis. The tarsus width is the most important predictor for the separation of the ground-dwelling versus the vegetation-dwelling species. Proceeding from the debris near the ground (red squares) as the presumed ancestral habitat, the arrows are indicative of the evolutionary shifts of the hunting sites in some phyletic lines towards the vegetation (green triangles) and the bare ground of open riparian habitats (orange circles). Such shifts may have occurred several times independently. The arrows starting from "*Dianous*" are indicative of its hypothesized (iterative) ecological radiation repeating the ecomorphological diversification found across the *Stenus* subgenera, although this seems to be widely restricted to waterfall environments. For further explanation, see text. Discriminant function 1: relative tarsus width, 0.957*; relative number of tarsal setae, 0.718*; lateral eye protrusion, −0.296*; relative leg length, −0.116. Discriminant function 2: relative tarsus width, −0.109; relative number of tarsal setae, −0.158; lateral eye protrusion, 0.147; relative leg length, 0.952*. Beetle images courtesy of Udo Schmidt (www.kaefer-der-welt.de)

indicates that, among soil dwellers, surface runners on bare ground (orange circles) differ from debris-dwelling species (red squares) by having longer legs (Fig. 11.28) and by adjusting their prey-capture technique to elusive prey, such as springtails, i.e. although they predominantly use their labium for small springtails, they shift to the mandible prey-capture technique for large ones (see above).

If we consider the moist debris on the ground as the ancestral *Stenus* habitat, our analysis suggests that, in some phylogenetic lineages, an ecological radiation has occurred into two different habitat types (arrows in Fig. 11.28). On the one hand, an extension of the hunting habitat has occurred into open, sparsely vegetated sites such as river banks. On the other hand, in connection with widened tarsi, these beetles have effectively explored the vegetation (e.g. reeds) (in the tropics, also the canopy) as a permanent novel adaptive zone. In order to substantiate such an evolutionary scenario further, additional ecomorphological investigations, field studies and molecular phylogenetic analyses need to be conducted taking into consideration the tropical fauna. This also is true for the members of the genus "*Dianous*," which most probably represents a branch within the genus *Stenus* (Koerner et al. 2013; Lang et al. 2015) and whose biology and evolution have been largely neglected. Although lacking a protrudable elongated labium, "*Dianous*" comprises a diversity of ecomorphs that are similar to those found in *Stenus*, i.e. plant climbers characterized by widened bilobed tarsi, dwellers of bare ground with slender tarsi and largely protruding eyes and inhabitants of moist detritus on the ground (Puthz 2000b). These differences are partly mirrored by the division of this genus into two main species groups (Puthz 1981). Based on the well-supported phylogenetic hypothesis suggesting probable monophyletic "*Dianous*" branches within *Stenus*, this diversity of ecomorphs within "*Dianous*" is indicative of an iterative convergent ecological radiation. Species flocks produced by iterative convergent radiations are produced not by single but by multiple adaptive radiations during their evolutionary history, where upon subclades radiate across similar ecomorphs related to similar adaptive zones. As a result, within a lower taxonomic unit such as a tribe or a genus, we expect to find ecomorphological diversifications, within certain subclades, similar to those found across the subclades of the superordinate taxon. As possible reasons for such a repeated radiation scenario, Frédérich et al. (2013) mention constraints caused by developmental processes, pleiotropic effects, morphological integration and competition.

Acknowledgements Our own work that was included in this chapter was supported by the Deutsche Forschungsgemeinschaft to O.B. and K.D. (PAK 478: BE 2233/10-1, BE 2233/11-1, DE 258/12-1, SE 595/14-1) and the Bundesministerium für Bildung und Forschung (Bionics Competition, BNK2-052) to O.B. We thank K. H. Seifert (Bayreuth) for support in identifying and synthesizing Steninae alkaloids. We thank Volker Puthz for critical reading of the manuscript and continuous support of our research. The English was corrected by Theresa Jones. Monika Meinert helped with the SEM work.

References

Adamson AW, Gast AP (1997) Physical chemistry of surfaces. Wiley-VCH, New York

Anderson R (1984) Staphylinidae (Coleoptera) in Ireland-3: Steninae. Irish Nat J 21(6):242–251

Bauer T, Pfeiffer M (1991) 'Shooting' springtails with a sticky rod: the flexible hunting behaviour of *Stenus comma* (Coleoptera; Staphylinidae) and the counter-strategies of its prey. Anim Behav 41:819–828

Bauer T, Desender K, Morwinsky T, Betz O (1998) Eye morphology reflects habitat demands in three closely related ground beetle species (Coleoptera: Carabidae). J Zool 245:467–472

Benfield EF (1972) Defensive secretion of *Dineutes discolor* (Coleoptera: Gyrinidae). Ann Entomol Soc Am 65:1324–1327

Benick L (1922) Paarungsvorgang bei *Stenus*-Arten. Entomol Bl 18:94

Benick L (1929) Steninae (Staphyl.). In: Bestimmungstabellen europäischer Coleoptera, Bd. 96. Karl Skrobanek & Söhne, Troppau, pp 5–10

Betz O (1994) Der Fangapparat bei *Stenus* spp. (Coleoptera, Staphylinidae): Bau, Funktion, Evolution. Dissertation, Universität Bayreuth

Betz O (1996) Function and evolution of the adhesion-capture apparatus of *Stenus* species (Coleoptera, Staphylinidae). Zoomorphology 116:15–34

Betz O (1998a) Comparative studies on the predatory behaviour of *Stenus* spp. (Coleoptera: Staphylinidae):

the significance of its specialized labial apparatus. J Zool 244:527–544

Betz O (1998b) Life forms and hunting behaviour of some central European *Stenus* species (Coleoptera, Staphylinidae). Appl Soil Ecol 9:69–74

Betz O (1999) A behavioural inventory of adult *Stenus* species (Coleoptera: Staphylinidae). J Nat Hist 33:1691–1712

Betz O (2000) Zum Anpassungswert unterschiedlicher Tarsusformen bei *Stenus* spp. (Coleoptera, Staphylinidae). Verh Westdeut Entomol 1998:127–133

Betz O (2002) Performance and adaptive value of tarsal morphology in rove beetles of the genus *Stenus* (Coleoptera, Staphylinidae). J Exp Biol 205:1097–1113

Betz O (2003) Structure of the tarsi in some *Stenus* species (Coleoptera, Staphylinidae): external morphology, ultrastructure, and tarsal secretion. J Morphol 255:24–43

Betz O (2006) Der Anpassungswert morphologischer Strukturen: Integration von Form, Funktion und Ökologie am Beispiel der Kurzflügelkäfer-Gattung *Stenus* (Coleoptera, Staphylinidae). Entomol Heute 18:3–26

Betz O (2010) Adhesive exocrine glands in insects: morphology, ultrastructure, and adhesive secretion. In: von Byern J, Grunwald I (eds) Biological adhesive systems. From nature to technical and medical application. Springer, Berlin, pp 111–152

Betz O, Fuhrmann S (2001) Life history traits in different life forms of predaceous *Stenus* beetles (Coleoptera, Staphylinidae), living in waterside environments. Neth J Zool 51(4):371–393

Betz O, Kölsch G (2004) The role of adhesion in prey capture and predator defence in arthropods. Arthropod Struct Dev 33:3–30

Betz O, Koerner L, Gorb S (2009) An insect's tongue as the model for two-phase viscous adhesives? adhes Adhes Sealants 3:32–35

Beutel RG, Molenda R (1997) Comparative morphology of selected larvae of Staphylinoidea (Coleoptera, Polyphaga) with phylogenetic implications. Zool Anz 236:37–67

Billard G, Bruyant G (1905) Sur un mode particulier de locomotion de certains *Stenus*. C R Soc Biol 59:102–103

Blair KG (1917) A note on the biology of *Stenus similis* Herbst. Ent Mon Mag 53:175

Blum P (1979) Zur Phylogenie und ökologischen Bedeutung der Elytrenreduktion und Abdomenbeweglichkeit der Staphylinidae (Coleoptera). Vergleichend- und funktionsmorphologische Untersuchungen. Zool Jahrb Abt Anat Ontog Tiere 102(8):533–582

Blum MS (1981) Chemical defenses of arthropods. Academic Press, New York

Breitmaier E (2008) Alkaloide, 3. Aufl. Vieweg & Teubner, Wiesbaden

Burkhardt D, de la Motte I (1983) How stalk-eyed flies eye stalk-eyed flies: observations and measurements of the eyes of *Cyrtodiopsis whitei* (Dopsidae, Diptera). J Comp Physiol 151:407–421

Burse A, Boland W (2015) RNAi based functional analysis of biosynthetic enzymes and transport proteins involved in the chemical defense of juvenile leaf beetles. In: Hoffmann KH (ed) Insect molecular biology and ecology. CRC Press, Boca Raton, pp 351–375

Bush JWM, Hu DL (2006) Walking on water: biolocomotion at the interface. Annu Rev Fluid Mech 38:339–369

Cai CY, Clarke DJ, Huang DY, Nel A (2014) A new genus and species of Steninae from the late Eocene of France (Coleoptera, Staphylinidae). Alcheringa 38:557–562

Cameron M (1930) The Fauna of British India including Ceylon and Burma. Coleoptera: Staphylinidae, Vol. 1 (Micropeplinae, Oxytelinae, Oxyporinae, Steninae and Euaesthetinae). Taylor & Francis, London

Caron E, Ribeiro-Costa CS, Newton AF (2008) New position of an abdominal defensive gland complex in Staphylinidae (Coleoptera) with redescription of *Piestus heterocephalus* Fauvel, 1902 (Piestinae). Zootaxa 1895:1–9

Champion GC (1919) The genus *Dianous* Samouelle, as represented in India and China (Coleoptera). Ent Mon Mag 55:41–55

Chatzimanolis S (2018) A review of the fossil history of Staphylinoidea. In: Betz O, Irmler U, Klimaszewski J (eds) Biology of rove beetles (Staphylinidae) – life history, evolution, ecology and distribution. Springer International Publishing, Cham, pp 27–45

Clarke D, Chatzimanolis S (2009) Antiquity and long-term morphological stasis in a group of rove beetles (Coleoptera: Staphylinidae): description of the oldest *Octavius* species, from cretaceous Burmese amber and review of the 'Euaesthetine subgroup' fossil record. Cretac Res 30(6):1426–1434

Clarke DJ, Grebennikov VV (2009) Monophyly of Euaesthetinae (Coleoptera: Staphylinidae): phylogenetic evidence from adults and larvae, review of austral genera, and new larval descriptions. Syst Entomol 34:346–397

Connert J (1974) Zur Strukturaufklärung des Stenusins. Dissertation, University of Heidelberg

Cuccodoro G (2017) Review of the observations of aggregates of Steninae reported since 1856 (Coleoptera Staphylinidae). Biodiv J 8(1):123–144

Delahon P (1927) Kleine Mitteilung. Ent Bl 23(2):90

Deml R, Dettner K (1995) "Ballon hairs" of gipsy moth larvae (Lep., Lymantriidae): morphology and comparative chemistry. Comp Biochem Physiol 112B: 673–681

Deml R, Dettner K (1997) Chemical defence of emperor moths and tussock moths (Lepidoptera: Saturniidae, Lymantriidae). Entomol Gen 21:225–251

Dettner K (1985) Ecological and phylogenetic significance of defensive compounds from pygidial glands of Hydradephaga (Coleoptera). Proc Acad Natl Sci Phila 137:156–171

Dettner K (1987) Chemosystematics and evolution of beetle chemical defense. Annu Rev Entomol 32:17–48
Dettner K (1991) Chemische Abwehrmechanismen bei Kurzflüglern (Coleoptera: Staphylinidae). Jber natwiss Ver Wuppertal 44:50–58
Dettner K (1993) Defensive secretions and exocrine glands in free-living staphylinid beetles and their bearing on phylogeny (Coleoptera: Staphylinidae). Biochem Syst Ecol 21:143–162
Dettner K (2007) Gifte und Pharmaka aus Insekten-Ihre Herkunft, Wirkung und ökologische Bedeutung. Entomol Heute 19:3–28
Dettner K (2014) Chemical ecology and biochemistry of Dytiscidae. In: Yee DA (ed) Ecology, systematics, and the natural history of predaceous diving beetles (Coleoptera: Dytiscidae). Springer, New York, pp 235–306
Dettner K, Scheuerlein A, Fabian P, Schulz S, Francke W (1996) Chemical defense of giant springtail *Tetrodontophora bielanensis* (Waga) (Insecta: Collembola). J Chem Ecol 22:1051–1074
Dietz AA, Hofmann MJ, Motschmann H (2016) The role of surface viscosity in the escape mechanism of the *Stenus* beetle. J Phys Chem B 120(29):7143–7147
Eisner T, Aneshansley DJ (2000) Chemical defense: aquatic beetle (*Dineutes hornii*) vs fish (*Micropterus salmonides*). Proc Natl Acad Sci U S A 97:11313–11318
Eisner T, Meinwald J, Monro A, Ghent R (1961) Defense mechanisms of arthropods – I. The composition and function of the spray of the whipscorpion, *Mastigoproctus giganteus* (Lucas) (Arachnida, Pedipalpida). J Insect Physiol 6:272–298
Eisner T, McHenry F, Salpeter MM (1964) Defense mechanisms of arthropods XV. Morphology of the quinine-producing glands of a tenebrionid beetle (*Eleodes longicollis* Lec.) J Morphol 115:355–400
Ekpa O, Wheeler JH, Cokendolpher JC, Duffield RM (1984) *N,N*-dimethyl-β-phenylethylamine and bornyl esters from the harvestman *Sclerobunus robustus* (Arachnida: Opiliones). Tetrahedron Lett 25:1315–1318
Francke W, Dettner K (2005) Chemical signalling in beetles. In: Schulz S (ed) Topics in current chemistry, vol 240. Springer, Heidelberg, pp 85–166
Frank JH (1991) Staphylinidae (Staphylinoidea). In: Stehr FW (ed) Immature insects, vol 2. Kendall/Hunt, Dubuque, IA, pp 341–352
Frank JH (2018) A worldwide checklist of parasites of Staphylinidae. In: Betz O, Irmler U, Klimaszewski J (eds) Biology of rove beetles (Staphylinidae) – life history, evolution, ecology and distribution. Springer International Publishing, Cham, pp 183–225
Frank JH, Thomas MC (1984) Cocoon-spinning and the defensive function of the median gland in larvae of Aleocharinae (Coleoptera, Staphylinidae): a review. Quaest Entomol 20(1):7–24
Frédérich B, Sorenson L, Santini F, Slater GJ, Alfaro ME (2013) Iterative ecological radiation and convergence during the evolutionary history of damselfishes (Pomacentridae). Am Nat 181:94–113
Frey H (1973) Das Aquarium von A bis Z. Verlag J. Neumann-Neudamm, Melsungen
Grebennikov VV, Newton AF (2009) Good-bye Scydmaenidae, or why the antlike stone beetles should become megadiverse Staphylinidae sensu latissimo (Coleoptera). Eur J Entomol 106:275–301
Hammond PM (1975) The Steninae (Coleoptera, Staphylinidae) of south-western Africa with special reference to the arid and semi-arid zones. Cimbebasia (A) 4(1):1–33
Hansen M (1997) Phylogeny and classification of the staphyliniform beetle families (Coleoptera). Biol Skr 48:1–339
Heethoff M, Koerner L, Norton RA, Raspotnig G (2011) Tasty but protected – first evidence of chemical defense in oribatid mites. J Chem Ecol 37:1037–1043
Hermann LH (2001) Catalogue of the Staphylinidae (Insecta: Coleoptera): 1758 to the end of the second millennium. I. Introduction, history, biographical sketches and omaliine group and IV. Staphylinidae group (part 1). Bull Am Mus Nat Hist 265:1–650, 1807, 2440
Hesse M (2000) Alkaloide, Fluch oder Segen der Natur. Wiley-VCH, Weinheim
Honda K (1983) Defensive potential of components of the larval osmeterial secretion of papilionid butterflies against ants. Physiol Entomol 8:173–179
Horion A (1963) Faunistik der mitteleuropäischen Käfer. Band IX. Staphylinidae. 1. Teil Micropeplinae bis Euaesthetinae. Kommissionsverlag Buchdruckerei August Feyel, Überlingen-Bodensee
Horridge GA (1978) The separation of visual axes in apposition compound eyes. Philos Trans R Soc London Ser B 285:1–59
Huth A, Dettner K (1990) Defense chemical from abdominal glands of 13 rove beetle species of subtribe Staphylinina (Coleoptera: Staphylinidae, Staphylininae). J Chem Ecol 16:2691–2711
Jałoszyński P, Peris D (2016) Cretaceous amber inclusions of Spain and Myanmar demonstrate early diversification and wide dispersal of Cephenniitae (Coleoptera: Staphylinidae: Scydmaeninae). Cretac Res 57:190–198
Jałoszyński P, Yamamoto S, Takahashi Y (2016) *Scydmobisetia* gen. nov., the first definite Glandulariini from upper cretaceous Burmese amber (Coleoptera: Staphylinidae: Scydmaeninae). Cretac Res 65:59–67
Jenkins MF (1957) The morphology and anatomy of the pygidial glands of *Dianous coerulescens* Gyllenhal (Coleoptera: Staphylinidae). Proc R Entomol Soc Lond 32:159–169
Jenkins MF (1958) Cocoon building and the production of silk by the mature larva of *Dianous coerulescens* Gyllenhal (Coleoptera: Staphylinidae). Trans R Entomol Soc Lond 110:287–301

Jenkins MF (1960) On the method by which *Stenus* and *Dianous* (Coleoptera: Staphylinidae) return to the banks of a pool. Trans R Entomol Soc Lond 112:1–14

Kanehisa K, Tsumuki H (1996) Pygidial secretion of *Stenus* rove beetles (Coleoptera: Staphylinidae). Bull Res Inst Bioresour Okayama Univ 4:25–31

Kasule FK (1966) The subfamilies of the larvae of Staphylinidae (Coleoptera) with keys to the larvae of the British genera of Steninae and Proteininae. Trans R Entomol Soc Lond 118:261–283

Koch K (1989) Die Käfer Mitteleuropas. Ökologie Band 1. Goecke & Evers, Krefeld

Kölsch G (2000) The ultrastructure of glands and the production and function of the secretion in the adhesive capture apparatus of *Stenus* species (Coleoptera: Staphylinidae). Can J Zool 78:465–475

Kölsch G, Betz O (1998) Ultrastructure and function of the adhesion-capture apparatus of *Stenus* species (Coleoptera, Staphylinidae). Zoomorphology 118:263–272

Koerner L, Gorb SN, Betz O (2012a) Adhesive performance of the stick-capture apparatus of rove beetles of the genus *Stenus* (Coleoptera, Staphylinidae) toward various surfaces. J Insect Physiol 58:155–163

Koerner L, Gorb SN, Betz O (2012b) Adhesive performance and functional morphology of the stick-capture apparatus of the rove beetles *Stenus* spp. (Coleoptera, Staphylinidae). Zoology 115:117–127

Koerner L, Laumann M, Betz O, Heethoff M (2013) Loss of the sticky harpoon – COI sequences indicate paraphyly of *Stenus* with respect to *Dianous* (Staphylinidae, Steninae). Zool Anz 252:337–347

Koerner L, Braun V, Betz O (2016) The labial adhesive pads of rove beetles of the genus *Stenus* (Coleoptera: Staphylinidae) as carriers of bacteria. Entomol Gen 36:33–41

Koerner L, Garamszegi L, Heethoff M, Betz O (2017) Divergent morphologies of adhesive predatory mouthparts of *Stenus* species (Coleoptera, Staphylinidae) explain differences in adhesive performance and resulting prey-capture success. Zool J Linenean Soc 181(3):500–518

Kohler P (1979) Die absolute Konfiguration des Stenusins und die Aufklärung weiterer Inhaltsstoffe des Spreitungsschwimmers *S. comma*. Dissertation, University of Heidelberg

Kovac D, Maschwitz U (1989) Secretion-grooming in the water bug *Plea minutissima*: chemical defense against microorganisms interfering with the hydrofuge properties in the respiratory region. Ecol Entomol 14:403–411

Kovac D, Maschwitz U (1990) Secretion-grooming in aquatic beetles (Hydradephaga): a chemical protection against contamination of the hydrofuge respiratory region. Chemoecology 1:131–138

Lang C (2014) Die chemische Ökologie der Steninae (Coleoptera: Staphylinidae) mit einem Beitrag zur molekularen Phylogenie. Inauguraldissertation, Lehrstuhl Tierökologie II, Universität Bayreuth, 98 pp. urn:nbn:de:bvb:703-epub-2022-5

Lang C, Seifert K, Dettner K (2012) Skimming behaviour and spreading potential of *Stenus* species and *Dianous coerulescens* (Coleoptera: Staphylinidae). Naturwissenschaften 99(11):937–947

Lang C, Koerner L, Betz O, Puthz V, Dettner K (2015) Phylogenetic relationships and chemical evolution of the genera *Stenus* and *Dianous* (Coleoptera: Staphylinidae). Chemoecology 25:11–24

Larsen EB (1959) Træk af sten
inernes biologi. XI. Not Entomol 39:87–88

Larsen EB (1963) Bidrag til Steninernes biologi (Col.) Entomol Medd 32:37–39

Latreille PA (1797) Précis des caractères généraux des insectes disposés dans un ordre naturel. Brive 1797:77–78

Lawrence J, Newton AF (1982) Evolution and classification of beetles. Annu Rev Ecol Syst 13:261–290

Lecoq J-C (1991) Observation d'un comportement rare chez un staphylin: le grégarisme (Col. Staphylinidae). L'Entomologiste 47(3):173–174

Leschen RAB, Newton AF (2003) Larval description, adult feeding behavior, and phylogenetic placement of *Megalopinus* (Coleoptera: Staphylinidae). Coleopt Bull 57:469–493

Li Y, Zheng F (2005) Studies on the predatory functional response and searching efficiency of *Stenus* (*Stenus*, s. str.) sp.1 on *Lipaphis erysimi* Kaltenbach. J Shenyang Norm Univ (Nat Sci) 23(2):200–203

Linsenmair KE (1963) Eine bislang unbekannte Fortbewegungsart bei Insekten: Das Entspannungsschwimmen. Kosmos, pp 331–334

Linsenmair KE, Jander R (1963) Das Entspannungsschwimmen von *Velia* und *Stenus*. Naturwissenschaften 50:231

Lohse GA (1964) Staphylinidae I (Micropeplinae bis Tachyporinae). In: Freude H, Harde KW, Lohse GA (eds) Die Käfer Mitteleuropas, vol 4. Goecke & Evers, Krefeld

Lott DA, Anderson R (2011) Handbooks for the identification of British insects, Vol. 12, part 7. The Staphylinidae (rove beetles) of Britain and Ireland; Parts 7 and 8: Oxyporinae, Steninae, Euaesthetinae, Pseudopsinae, Paederinae, Staphylininae. Royal Entomological Society, St Albans

Lusebrink I (2007). Stereoisomerie, Biosynthese und biologische Wirkung des Stenusins sowie weitere Inhaltsstoffe der Pygidialdrüsen der Kurzflüglergattung *Stenus* (Staphylinidae, Coleoptera). Dissertation, University of Bayreuth

Lusebrink I, Burkhardt D, Gedig T, Dettner K, Seifert K, Mosandl A (2007) Intrageneric differences in the four stereoisomers of stenusine in the rove beetle genus, *Stenus* (Coleoptera: Staphylinidae). Naturwissenschaften 94:143–147

Lusebrink I, Dettner K, Seifert K (2008a) Biosynthesis of Stenusine. J Nat Prod 71:743–745

Lusebrink I, Dettner K, Seifert K (2008b) Stenusine, an antimicrobial agent in the rove beetle genus *Stenus* (Coleoptera, Staphylinidae). Naturwissenschaften 95:751–755

Lusebrink I, Dettner K, Schierling A, Müller T, Daolio C, Schneider B, Schmidt J, Seifert K (2009) New pyridine alkaloids from rove beetles of the genus *Stenus* (Coleoptera: Staphylinidae). Z NatForsch 64c:271–278

MacConnel JG, Blum MS, Fales HM (1971) The chemistry of fire ant venom. Tetrahedron 26: 1129–1139

Maschwitz U (1967) Eine neuartige Form der Abwehr von Mikroorganismen bei Insekten. Naturwissenschaften 54:649

McKenna DD, Farrell BD, Caterino MS, Farnum CW, Hawks DC, Maddison DR, Seago AE, Short AEZ, Newton AF, Thayer MK (2015) Phylogeny and evolution of Staphyliniformia and Scarabaeiformia: forest litter as a stepping stone for diversification of nonphytophagous beetles. Syst Entomol 40:35–60

Meinert F (1884) Tungens Udskydelighed hos Steninerne, en Slægt of Staphylinernes Familie. Vid Medd Dansk naturhist Foren, pp 180–207

Meinert F (1887) Die Unterlippe der Käfergattung *Stenus*. Zool Anz 10:136–139

Morgan ED (2010) Biosynthesis in insects, adv edn. Cambridge, RSC

Mori K (1999) Miscellaneous natural products including marine natural products, pheromones, plant hormones and aspects of ecology. In: Barton D, Nakanishi K, Meth-Cohn O (eds) Comprehensive natural product chemistry, vol 8. Elsevier, Oxford

Motschulsky V (1857) Enumération des nouvelles espèces de Coléoptères, IV Staphylinides. Bull Soc Imp Nat Mosc 30:490–517

Motschulsky V (1860) Énumération des nouvelles espèces de Coléoptères rappartiées de ses voyages. Bull Soc Nat Mosc 33:539–588

Müller T, Göhl M, Lusebrink I, Dettner K, Seifert K (2012) Cicindeloine from *Stenus cicindeloides* – isolation, structure elucidation and total synthesis. Eur J Org Chem 12:2323–2330

Naomi S-I (2018) Structures and functions of the endophallic copulatory tube in the family Staphylinidae (Insecta: Coleoptera). In: Betz O, Irmler U, Klimaszewski J (eds) Biology of rove beetles (Staphylinidae) – life history, evolution, ecology and distribution. Springer International Publishing, Cham, pp 299–319

Naomi S-I, Nomura S, Puthz V (2017) The subfamily Steninae MacLeay, 1825 (Coleoptera: Staphylinidae) of Japan. Part 1. *Dianous* and *Stenus* (*S. comma* group to *S. guttalis* group). Natl Mus Nat Sci Monogr 46:1–339

Neumann S (1993) Die Analdrüsen der Spreitungsschwimmer (Coleoptera: Staphylinidae). Diploma thesis, University of Bayreuth

Newton AF Jr (1990) Insecta: Coleoptera: Staphylinidae adults and larvae. In: Dindal DL (ed) Soil biology guide. Wiley, New York, pp 1137–1174

Newton AF, Thayer MK, Ashe JS, Chandler DS (2001) Staphylinidae Latreille, 1802. In: Arnett RH Jr, Thomas MC (eds) American beetles. CRC Press, Boca Raton, pp 272–418

Noirot C, Quennedey A (1974) Fine structure of insect epidermal glands. Annu Rev Entomol 19:61–80

Noirot C, Quennedey A (1991) Glands, gland cells, glandular units: some comments on terminology and classification. Ann Soc Entomol Fr 27:123–128

O'Brien WJ, Evans BI, Browman HI (1989) Flexible search tactics and efficient foraging in saltatory searching animals. Oecologia 80:100–110

Piffard A (1901) Steni gliding on the surface of water. Ent Mon Mag 12:99

Puthz V (1971) Revision der afrikanischen Steninenfauna und Allgemeines über die Gattung *Stenus* Latreille (Coleoptera Staphylinidae) (56. Beitrag zur Kenntnis der Steninen). Ann R Mus Afr Centr Ser 8(187):1–376

Puthz V (1974) Neue mexikanische *Stenus*-Arten: 135. Beitrag zur Kenntnis der Steninen (Coleoptera, Staphylinidae). Dtsch Entomol Z 21:203–216

Puthz V (1980) Die *Stenus*- Arten (*Stenus* + *Nestus* REY) der Orientalis: Bestimmungstabelle und Neubeschreibungen (Coleoptera, Staphylinidae) 178. Beitrag zur Kenntnis der Steninen. Reichenbachia 18:23–41

Puthz V (1981) Was ist *Dianous* Leach, 1819, was ist *Stenus* Latreille, 1796? Oder: Die Aporie des Stenologen und ihre taxonomischen Konsequenzen (Coleoptera, Staphylinidae). Ent Abh St Mus Tierk, Dresden 44:87–132

Puthz V (1998) Die Gattung *Stenus* Latreille in Vietnam (Coleoptera, Staphylinidae). Rev Suisse Zool 105:383–394

Puthz V (2000a) Beiträge zur Kenntnis der Steninen CCLXII Neue und alte neotropische Arten der Gattung *Stenus* Latreille (Staphylinidae, Coleoptera). Philippia 9:165–213

Puthz V (2000b) The genus *Dianous* Leach in China (Coleoptera: Staphylinidae) 261. Contribution to the knowledge of Steninae. Rev Suisse Zool 107:419–559

Puthz V (2001) Beiträge zur Kenntnis der Steninen CCLXIX Zur Ordnung in der Gattung *Stenus* LATREILLE (Staphylinidae, Coleoptera). Philippia 10:53–64

Puthz V (2005a) Notes on Chinese *Dianous* Leach (Coleoptera, Staphylinidae). Entomol Rev Jpn 60:137–152

Puthz V (2005b) Neue und alte neotropische *Stenus* (*Hemistenus*-) Arten (Coleoptera: Staphylinidae). Mitt Int Entomol Ver Suppl XI:1–60

Puthz V (2006) Revision der neotropischen *Stenus* (*Tesnus*) Arten (Coleoptera: Staphylinidae) 288. Beitrag zur Kenntnis der Steninen. Rev Suisse Zool 113:617–674

Puthz V (2008) *Stenus* Latreille und die segenreiche Himmelstochter (Coleoptera, Staphylinidae). Linz Biol Beitr 40(1):137–230

Puthz V (2010) *Stenus* Latreille, 1797 aus dem Baltischen Bernstein nebst Bemerkungen über andere fossile *Stenus*-Arten (Coleoptera, Staphylinidae). Ent Bl 106:265–287

Puthz V (2012) Steninae. In: Freude H, Harde KW, Lohse A (eds) Die Käfer Mitteleuropas. Band 4, Zweite Auflage, pp 286–317

Puthz V (2013) Übersicht über die orientalischen Arten der Gattung *Stenus* Latreille 1797 (Coleoptera, Staphylinidae) 330. Beitrag zur Kenntnis der Steninen. Linz Biol Beitr 45(2):1279–1470

Puthz V (2015) Übersicht über die Arten der Gattung *Dianous* LEACH group I (Coleoptera, Staphylinidae) 345. Beitrag zur Kenntnis der Steninen. Linz Biol Beitr 47(2):1747–1783

Puthz V (2016) Übersicht über die Arten der Gattung *Dianous* Leach group II (Coleoptera, Staphylinidae) 347. Beitrag zur Kenntnis der Steninen. Linz Biol Beitr 48(1):705–778

Puthz V (2017) Übersicht über die neotropischen Arten der Gattung *Stenus* Latreille mit seitlich ungerandetem Abdomen und gelappten Tarsen (Coleoptera, Staphylinidae) 351. Beitrag zur Kenntnis der Steninen. Linz Biol Beitr 49(1):749–883

Quennedey A (1998) Insect epidermal gland cells: ultrastructure and morphogenesis. In: Harrison FW, Locke M (eds) Microscopic anatomy of invertebrates, vol 11A. Wiley-Liss, London, pp 177–207

Quennedey A, Drugmand D, Deligne J (2002) Morphology and ultrastructure of paired prototergal glands in the adult rove beetle *Philonthus varians* (Coleoptera, Staphylinidae). Arthropod Struct Dev 31:173–183

Qui G-H, Zheng F-K (2006) Effects of three insecticides on predation function of (*Stenus* sp.) J Shenyang Norm Univ (Nat Sci Edn) 24(1):84–87

Renkonen O (1934) Über das Vorkommen der *Stenus*-Arten (Col. Staph.) an verschiedenen Wohnorten in Finnland. Ann Zool Soc Zool Bot Fenn Vanamo 1 (4):1–33

Renkonen O (1950) Zur Autökologie einiger *Stenus*-Arten im Lichte synökologischer Betrachtungen. In: 8th International congress of entomology, pp 1–6

Rey C (1884) Tribu des Brévipennes. Deuxième groupe: Micropéplides. Troisième groupe: Sténides. Annls Soc Linn Lyon 30:153–415

Ronquist F, Huelsenbeck JP (2003) MrBayes 3: Bayesian phylogenetic inference under mixed models. Bioinformatics 19:1572–1574

Rostás M, Blassmann K (2009) Insects had it first: surfactants as a defence against predators. Proc R Soc B 276:633–638

de Rougemont GM (1983) More stenine beetles from Thailand (Coleoptera, Staphylinidae). Nat Hist Bull Siam Soc 31(1):9–54

de Rougemont GM (1985) In the footsteps of H. G. Champion: new *Dianous* species from the Himalaya (Coleoptera, Staphylinidae). Entomol Basiliensia 10:123–144

Rupprecht J (2011) Biologische Aktivität ausgewählter Abwehrstoffe von *Stenus*-Kurzflügelkäfern (Coleoptera, Staphylinidae) gegen Schwertträger der Art *Xiphophorus helleri* (Poeciliidae) und den Pilz *Verticillium lecanii*. Bachelor thesis, University of Bayreuth

Ryvkin AB (1988) New cretaceous Staphylinidae (Insecta) from the Far East. Paleontol J 22(4):100–104

Ryvkin AB (2012) New species and records of *Stenus* (*Nestus*) of the *canaliculatus* group, with the erection of a new species group (Insecta: Coleoptera: Staphylinidae: Steninae). Eur J Taxon 13:1–62

Schatz I, Steinberger K-H, Kopf T (2003) Auswirkungen des Schwellbetriebes auf uferbewohnende Arthropoden (Aranei; Insecta: Coleoptera: Carabidae, Staphylinidae) am Inn im Vergleich zum Lech (Tirol, Österreich). Natur in Tirol. Naturkundliche Beiträge der Abteilung Umweltschutz. Ökologie und Wasserkraftnutzung. Raggl, Innsbruck, pp 202–231

Schierling A (2013) Die chemische Ökologie von Kurzflügelkäfern der Gattungen *Dianous* und *Stenus* (Coleoptera, Staphylinidae). Dissertation, University of Bayreuth

Schierling A, Dettner K (2013) The pygidial defense gland system of the Steninae (Coleoptera, Staphylinidae): morphology, ultrastructure and evolution. Arthropod Struct Dev 42:197–208

Schierling A, Schott M, Dettner K, Seifert K (2011) Biosynthesis of the defensive alkaloid (Z)-3-(2-Methyl-1-butenyl)-pyridine in *Stenus similis* beetles. J Nat Prod 74:2231–2234

Schierling A, Dettner K, Schmidt J, Seifert K (2012) Biosynthesis of the defensive alkaloid cicindeloine in *Stenus solutus* beetles. Naturwissenschaften 99:665–669

Schierling A, Seifert K, Sinterhauf SR, Rieß JB, Rupprecht JC, Dettner K (2013) The multifunctional pygidial gland secretion of the Steninae (Coleoptera: Staphylinidae): ecological significance and evolution. Chemoecology 23:45–57

Schildknecht H (1970) The defensive chemistry of land and water beetles. Angew Chem 9:1–9

Schildknecht H (1976) Chemical ecology – a chapter of modern natural products chemistry. Angew Chem 15:214–222

Schildknecht H, Weis KH (1962) Zur Kenntnis der Pygidialblasensubstanzen vom Gelbrandkäfer (*Dytiscus marginalis* L.) XIII. Mitteilung über Insektenabwehrstoffe. Z NatForsch 17B:448–452

Schildknecht H, Maschwitz E, Maschwitz U (1968) Die Explosionschemie der Bombardierkäfer: Struktur und Eigenschaften der Brennkammerenzyme. Z NatForsch 23B:1213

Schildknecht H, Krauss D, Connert J, Essenbreis H, Orfanides N (1975) The spreading alkaloid stenusine from the staphylinid *S. comma* (Coleoptera: Staphylinidae). Angew Chem 14:427

Schildknecht H, Berger D, Krauss D, Connert J, Gehlhaus J, Essenbreis H (1976) Defense chemistry of *Stenus comma* (Coleoptera: Staphylinidae). J Chem Ecol 2:1–11

Schlüter T (1978) Zur Systematik und Palökologie harzkonservierter Arthropoda einer Taphozönose aus dem Cenomanium von NW-Frankreich. Berl Geowiss Abh Reihe A 9:1–150

Schmitz G (1943) Le labium et les structures bucco-pharyngiennes du genre *Stenus* LATREILLE. Cellule 49:291–334

Schomann AM, Solodovnikov A (2016) Phylogenetic placement of the austral rove beetle genus *Hyperomma* triggers changes in classification of Paederinae (Coleoptera: Staphylinidae). Zool Scr 46 (3):336–347

Schoonhoven LM, van Loon JJA, Dicke M (2005) Insect plant biology, 2nd edn. Oxford University Press, Oxford

Schrüfer T (2013) Elektrophoretische Charakterisierung von Proteinen und Enzymen aus den Pygidial-wehrdrüsen ausgewählter Steninae (Coleoptera, Staphylindae). Bachelor Thesis, University of Bayreuth

Schülke M, Smetana A (2015) Staphylinidae [Omaliinae – Scydmaeninae]. In: Löbl I, Löbl D (eds) Catalogue of Palearctic Coleoptera, vol 2/1. Brill, Leiden, pp 304–900

Scriven L, Sternling C (1960) The Marangoni effect. Nature 187:186–188

Serri S, Frisch J, von Rintelen T (2016) Genetic variability of two ecomorphological forms of *Stenus* Latreille, 1797 in Iran, with notes on the infrageneric classification of the genus (Coleoptera, Staphylinidae, Steninae). Zookeys 626:67–86

Shi K, Zhou H-Z (2011) Taxonomy of the genus *Dianous* (Coleoptera: Staphylinidae: Steninae) in China and zoogeographic patterns of its distribution. Insect Sci 18:363–378

Solodovnikov AY, Schomann A (2009) Revised systematics and biogeography of 'Quediina' of Subsaharan Africa: new phylogenetic insights into the rove beetle tribe Staphylinini (Coleoptera: Staphylinidae). Syst Entomol 34:443–446

Solodovnikov A, Yue Y, Tarasov S, Ren D (2013) Extinct and extant rove beetles meet in the matrix: early cretaceous fossils shed light on the evolution of a hyperdiverse insect lineage (Coleoptera: Staphylinidae: Staphylininae). Cladistics 29:360–403

Steidle J, Dettner K (1990) Die Tergaldrüse der Aleocharinae (Staphylinidae, Coleoptera): Chemie, Morphologie und phylogenetische Bedeutung. Mitt Dtsch Ges Allg Angew Entomol 7:541–545

Stoeffler M, Maier TS, Tolasch T, Steidle JML (2007) Foreign language skills in rove-beetles? Evidence for chemical mimicry of ant alarm pheromones in myrmecophilous *Pella* beetles (Coleoptera: Staphylinidae). J Chem Ecol 33:1382–1392

Tang L, Li L-Z, Cao G-H (2011) On Chinese species of *Dianous* group I (Coleoptera, Staphylinidae Steninae). ZooKeys 111:67–85

Thayer MK (2005) Staphylinidae. In: Beutel RG, Leschen RAB (eds) Handbook of zoology, Coleoptera, vol 1. De Gruyter, Berlin, pp 296–344

Thayer MK (2016) 14. Staphylinoidea. 14.7. Staphylinidae Latreille, 1802. In: Beutel RG, Leschen RAB (eds) Handbook of zoology, Arthropoda: Insecta; Coleoptera, beetles: morphology and systematics; Archostemata, Adephaga, Myxophaga, Polyphaga partim, vol 1, 2nd edn. De Gruyter, Berlin, pp 394–442

Thayer MK, Newton AF, Chatzimanolis S (2012) *Prosolierius*, a new mid-cretaceous genus of Solieriinae (Coleoptera: Staphylinidae) with three new species from Burmese amber. Cretac Res 34:124–134

Urban C (1928) *Stenus* auf dem Wasser laufend. Ent Bl 24:45

Voris R (1934) Biologic investigations on the Staphylinidae (Coleoptera). Trans Ac Sc St. Louis 28 (8):232–261

Vulinec K (1987) Swimming in whirligig beetles (Coleoptera: Staphylinidae): a possible role of the pygidial gland secretion. Coleopt Bull 41:151–153

Weber H (1933) Lehrbuch der Entomologie. Gutsav Fischer, Jena

Weinreich E (1968) Über den Klebfangapparat der Imagines von *Stenus* LATR. (Coleopt., Staphylinidae) mit einem Beitrag zur Kenntnis der Jugendstadien dieser Gattung. Z Morphol Ökol Tiere 62:162–210

Welch RC (1966) A description of the pupa and third instar larva of *Stenus canaliculatus* Gyll. (Col., Staphylinidae). Ent Mon Mag 101:246–250

Whitman DW, Blum MS, Alsop DW (1990) Allomones: chemicals for defense. In: Evans DL, Schmidt JO (eds) Insect Defenses. State University of New York Press, Albany, pp 289–351

Wittmann I, Schierling A, Dettner K, Göhl M, Schmidt J, Seifert K (2015) Detection of a new piperideine alkaloid in the pygidial glands of some *Stenus* beetles. Chem Biodivers 12:1422–1434

Wolf KL (1957) Physik und Chemie der Grenzflächen. Springer, Berlin

Yamamoto S, Solodovnikov A (2016) The first fossil Megalopsidiinae (Coleoptera: Staphylinidae) from upper cretaceous Burmese amber and its potential for understanding basal relationships of rove beetles. Cretac Res 59:140–146

Yang L-H (2003) Preliminary study on life habit of *Stenus* sp. (*Stenus* s. str.) J Sichuan Teach Coll (Nat Sci) 24 (2):217–221

Zhang X, Zhou H-Z (2013) How old are the rove beetles (Insecta: Coleoptera: Staphylinidae) and their lineages? Seeking an answer with DNA. Zool Sci 30 (6):490–501

Zhao CY, Zhou HZ (2004) Five new species of the subgenus *Hemistenus* (Coleoptera: Staphylinidae, Steninae) from China. Pan-Pac Entomol 80:93–108

Żyła D, Yamamoto S, Wolf-Schwenninger K, Solodovnikov A (2017) Cretaceous origin of the unique prey-capture apparatus in megadiverse genus: stem lineage of Steninae rove beetles discovered in Burmese amber. Sci Rep 7:45904. https://doi.org/10.1038/srep45904

Biology of Acarophagous Scydmaeninae 12

Paweł Jałoszyński

Abstract

Some Scydmaeninae are strict specialists that feed exclusively on heavily sclerotized oribatid or uropodine mites. The chapter reviews the available literature on the feeding habits of *Euconnus*, *Stenichnus*, *Scydmaenus*, and *Cephennium* beetles and presents previously unpublished observations on *Neuraphes* and *Microscydmus* species. Species with unspecialized mouthparts attack the mite's gnathosoma, removing movable parts to gain access to soft tissues. They also often remove genital or anal plates to feed through the resulting openings. In *Euconnus* that are specialized to feed on ptyctimous (i.e., capable of encapsulating) oribatids, a sticky droplet of digestive juice exuded onto the predator's mouthparts is used to capture mites. The prey is then lifted and covered with noxious digestive juice, which weakens or kills the encapsulated mite. Once the muscles responsible for maintaining the encapsulation are relaxed, the prey's prodorsum opens, and *Euconnus* beetles use their mandibles to crush the mite's ventral plates and gain access to the flesh. In *Scydmaenus* that are specialized to feed on non-ptyctimous Oribatida and Uropodina, the mandibles play a major role both in capturing prey and in breaching the mite's defenses. The prey's legs are often cut off if they are long or spiny, which facilitates the subsequent attack on the gnathosoma. Cephenniini are the "hole scrapers": they have paired labial suckers on the prementum, which are used to immobilize their prey. Once the mite adheres to the suckers, the predator's mandibles slowly grind a small hole in the prey's cuticle. Digestive juices are then injected; through the same puncture, liquefied tissues are ingested. The entire feeding process can take many hours. Some species show preferences toward particular mite taxa and may play a significant role in the oribatid or uropodine mite population dynamics.

12.1 Introduction

Over a century ago, Reitter (1909) noticed that Scydmaeninae (Scydmaenidae at that time) seemed to feed on mites. This view was supported by Schuster (1966a, b), who observed the feeding of adult *Cephennium majus* Reitter and larvae of *C. majus* and *C. thoracicum* Müller and Kunze; Schuster carried out the first published prey preference experiments with these minute (1–1.5 mm) beetles. Scydmaeninaes were given a broad spectrum of potential prey,

P. Jałoszyński (✉)
Museum of Natural History, University of Wrocław, Wrocław, Poland
e-mail: scydmaenus@yahoo.com

O. Betz et al. (eds.), *Biology of Rove Beetles (Staphylinidae)*,
https://doi.org/10.1007/978-3-319-70257-5_12

including Oribatida, Uropodina, and Gamasida mites, as well as Collembola, Protura, and oligochaete Enchytraeidae. They were found to feed mostly on the armored Oribatida, and only rarely on Uropodina and Gamasida.

A classic work was published by Schmid (1988), who made systematic observations of the feeding preferences and techniques of adults of many species belonging to the genera *Cephennium* Müller and Kunze (Cephenniini), *Neuraphes* Thomson, *Scydmoraphes* Reitter, *Stenichnus* Thomson, *Microscydmus* Saulcy and Croissandeau, *Euconnus* Thomson (Glandulariini), and *Scydmaenus* Latreille (Scydmaenini), as well as larvae of *Cephennium*, *Stenichnus*, and *Scydmoraphes*. A broad spectrum of Oribatida and Uropodina species was tested (approximately 200 species). Various structures of the mouthparts and legs of Scydmaeninae were interpreted as adaptations for feeding on these heavily armored Acari.

Later, Molleman and Walter (2001) demonstrated that some Australian Glandulariini and Scydmaenini showed strong preferences toward armored mites. However, they also scavenged on dead ants, beetles, springtails, and unarmored nymphs of galumnid mites (Oribatida). Jałoszyński and Olszanowski (2013, 2015, 2016) carried out prey choice experiments with a broad spectrum of possible prey mites offered to four species of Scydmaeninae under laboratory conditions. They obtained results concerning their prey preferences and interesting details of their feeding techniques.

The data obtained so far demonstrate that Scydmaeninae not only use different methods to breach their prey's defenses but also show relatively narrow preferences toward certain taxa or particular body forms of oribatids or (less frequently) uropodines. Although our knowledge is still fragmentary, two distinct feeding techniques can be defined, depending on the morphological specialization of the predator's mouthparts. Some behavioral variants were also discovered, which evolved to cope with the different and often sophisticated defense systems of armored mites.

It should be noted that not all Scydmaeninae are specialist predators feeding on heavily sclerotized mites. Leleup (1968) noticed that South African Mastigini carry small larvae and springtails in their mandibles. Furthermore, O'Keefe and Monteith (2000) mentioned observations of the only Australian Clidicini species carrying large neanurine springtails in their mandibles. Jałoszyński (2012a, b) demonstrated that two European species of *Scydmaenus* preferred either springtails or soft-bodied Acari and, under laboratory conditions, showed no interest in Oribatida; scavenging dead arthropods and cannibalism were also reported by the same author.

Because many other arthropods feed on soft-bodied prey and very few are strictly specialized to utilize heavily protected armored mites as the only source of food, the prey preferences and feeding techniques of acarophagous Scydmaeninae have attracted much attention. Oribatida were once believed to have evolved their defense systems in response to predation by prostigmatan and mesostigmatan mites; because this pressure is now low, they currently live in an "enemy-free space" (e.g., Jeffries and Lawton 1984; Peschel et al. 2006). Oribatids are indeed well protected against most invertebrate predators. Their defense systems include thick cuticles (which are often reinforced by mineralization, carinae, or reticulation). Furthermore, depending on taxon, they also have long setae on the idiosoma or spiny legs that make it difficult for a predator to attack the mite's body. Some produce repellents or toxins in the so-called oil glands, whereas others accumulate soil particles on their body surface that form an additional protective crust. Many oribatids have pteromorphs, which are lateral laminar projections over their coxae that protect the legs from being cut off by predators. The so-called ptyctimous mites can "encapsulate"—that is, they adopt a compact defensive posture with all appendages and vulnerable ventral membranous structures hidden under their closed prodorsum, which is shield-like and can move to open/close the encapsulation (Pachl et al. 2012; Schmelzle et al. 2008, 2009, 2010). A combination of several defensive mechanisms or structures in one species is not uncommon. Predators that have adopted to feed on this kind of prey are expected to use unusual techniques or to have unique tools to breach defenses of their prey. Such adaptations are summarized in this chapter on the basis of the available literature—mainly studies published by the author (Jałoszyński 2016; Jałoszyński and Olszanowski 2013, 2015, 2016) but also previously unpublished observations concerning the genera *Microscydmus* and *Neuraphes*.

12.2 Cephenniini, the "Hole Scrapers"

All known species of Cephenniini have highly modified mouthparts (Fig. 12.1a, b), with the labium transformed into a prey immobilizing device. The head is strongly declined, such that the mouthparts are directed downward. The labrum is typically semicircular with a membranous marginal velum and a membranous epipharynx; the mandibles are variable in shape but often short and relatively blunt; and the maxillae are generalized, as those in all Scydmaeninae. The most unusual is the labium (Fig. 12.1b), which has a highly movable prementum, with its anterior surface capable of tilting dorsally, ventrally, or/and laterally. The labial palps are exceptionally small and broadly separated, and the area between them is occupied by four or six symmetrically distributed suckers. Often, the anterior surface of the prementum is additionally divided by a median longitudinal groove, so that the lateral halves can move independently and better fit to the convex surface of oribatid mites.

The ultrastructure of the suckers was studied by Jałoszyński and Beutel (2012), who found that each sucker is composed of an outer oval plate connected by a circumferential ring with the inner plate bearing a median perforation; thus, the lumen between the plates is continuous with the inner space of the labium. The suckers, the labial cuticle, and the internal sclerotized scaffold of the labium, including the hypopharyngeal suspensorium, form a continuous functional unit operated by labial muscles, which can modulate the degree of concavity of the outer plates of the suckers. This sophisticated system of structures requires muscle contractions only during the attack; however, when the mite adheres to the suckers by suction forces, the muscles relax. It was suggested that the complex structural components of the suction discs have evolved by invagination of exocuticular structures and that the development of suckers was preceded by local sclerotizations of the prementum, induced by contact with soft-bodied or partly armored prey (Jałoszyński and Beutel 2012).

The labial modifications found in all extant Cephenniini suggest that they all feed on armored mites and presumably cannot feed on other types of prey. It is possible that this is an ancient adaptation: the oldest unambiguously identified fossil of Cephenniini, from Upper Cretaceous (Cenomanian) Burmese amber, is morphologically nearly identical with extant representatives of this tribe (Jałoszyński and Peris 2016). However, mouthparts of this fossil are not exposed. Thus, further study is needed to shed more light on the evolution of this intriguing, narrow feeding adaptation that most likely enabled Cephenniini to avoid competition with other small invertebrate predators of the forest floor.

Prey preferences, feeding techniques, and the functional morphology of mouthparts were studied using several species of European *Cephennium* as model organisms (Jałoszyński and Beutel 2012; Jałoszyński and Olszanowski 2016). Jałoszyński and Olszanowski (2016) described three phases of the feeding process on oribatid prey: (1) attack and stabilization of the attachment site (about 4–5 min), (2) penetration of the mite's cuticle (about 40 min), and (3) feeding (7–8 h). The mite is attacked from behind or from above (Fig. 12.3a) and lifted using the beetle's prementum (Fig. 12.3b, c). The adhesion between the beetle's mouthparts and the mite is so strong that it is possible to kill and preserve them, then subsequently take scanning electron microscopy images, without disrupting the connection (Fig. 12.1c, d). Within the first 1–5 min, the captured mite is rotated using the protarsi, apparently to find the best attachment site. The manipulations take place without detaching the prey; the mouthparts of the beetles appear to slide over the surface of the mite's cuticle.

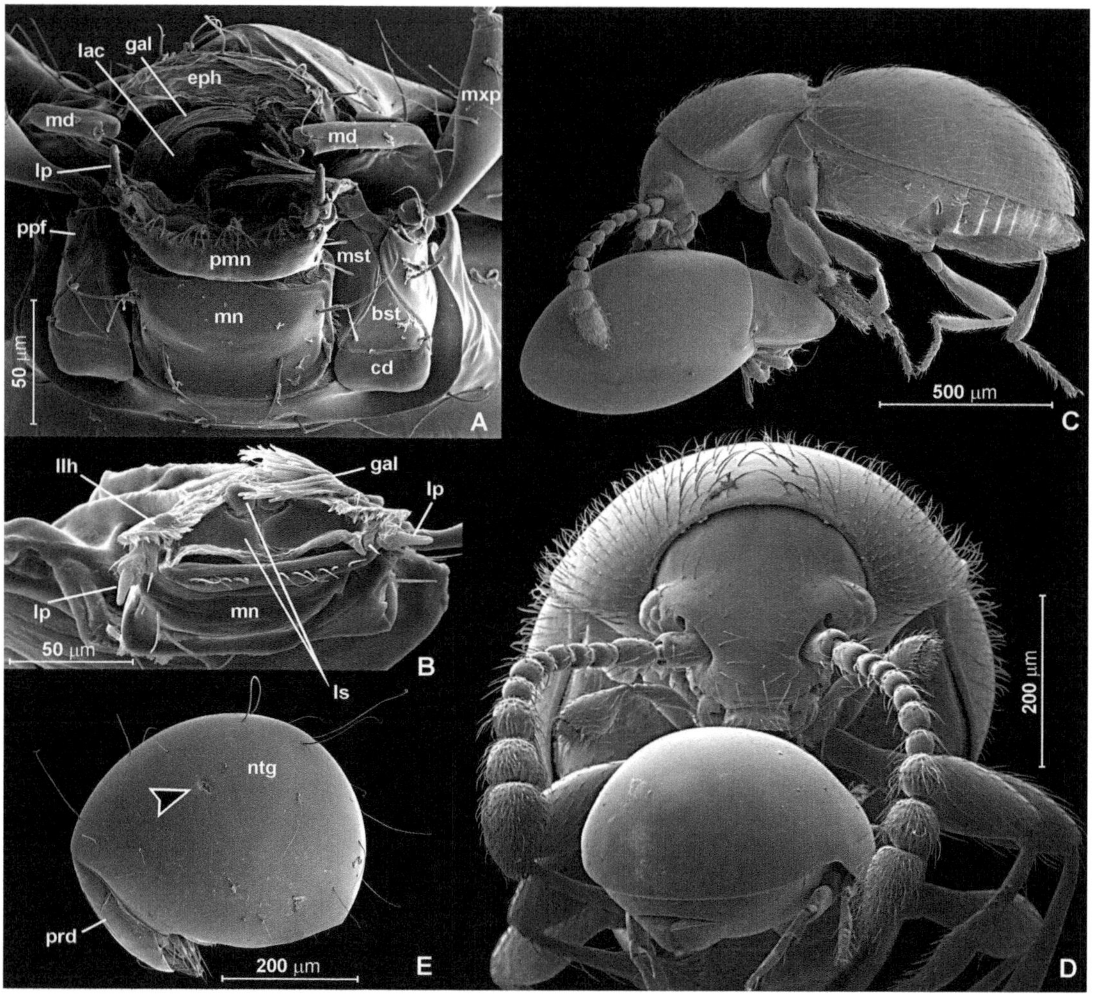

Fig. 12.1 Examples of Cephenniini (**a–d**) and their prey (**e**). (**a**) *Cephennium majus*, mouthparts in anteroventral view. (**b**) Separated labium of *Cephennium majus* in anterior view. (**c**) *Cephennium majus* preserved during feeding on ptyctimous oribatid mite. (**d**) *Cephennium ruthenum* preserved during feeding on non-ptyctimous oribatid mite. (**e**) *Phthiracarus* sp. (Oribatida, Phthiracaridae), prey of *Cephennium majus*, showing feeding damage (arrow). Abbreviations: *bst* basistipes, *cd* cardo, *eph* epipharynx, *gal* galea, *lac lacinia*, *llh* lateral lobe of hypopharynx, *lp* labial palp, *ls* labial sucker, *md* mandible, *mn* mentum, *mst* mediostipes, *mxp* maxillary palp, *ntg* notogaster, *pmn* prementum, *ppf* palpifer, *prd* prodorsum

When an apparently suitable site has been chosen, both mandibles start to rhythmically spread and close, and the second phase begins. The beetles broadly open one mandible, while the other mandible makes short scraping movements within the area delimited by the labrum and the prementum. For most of the time, the prey remains lifted; however, when the grip is in the posterodorsal region of the idiosoma, the mite often manages to reach for the ground with some legs, pulls itself closer to the arena, and starts crawling forward. Beetles counteract by lifting the mite higher. The scraping movements of one mandible can be observed for 10–15 min. Then, the position of the mite is changed, with the working mandible being replaced by the previously resting one. After about 20 min, the mite's leg movements weaken to suddenly become very rapid. This increased activity of the mite marks the moment of completing the perforation of the cuticle; however, the prey remains alive and erratically moves its legs for about half an hour. *Cephennium* keeps on working with one mandible, presumably broadening the hole or trying to insert a tip of the mandible deeper.

When the movements of the prey's legs definitively stop, it is usually possible to see that the tip of one mandible of the beetle is inserted into the hole and the other mandible is still broadly open. This is when the third phase starts. The beetle can now feed through the tiny hole, which is often only about 20 μm wide (Fig. 12.1e). When attached to the mite's cuticle, the surface of the prementum that bears the suckers is parallel to the body surface of the mite, and the labrum is strongly flipped dorsally. To feed, *Cephennium* beetles close their mandibles, which are bent and short enough to be contained between the labrum and labium. The attachment site is completely surrounded by the beetle's mouthparts, which tightly seal the hole margins to enable injection of digestive juices and subsequent ingestion of liquefied tissues. The connection is sealed by the membranous marginal velum of the labrum, which laterally fits tightly into the dorsomesal notch on each closed mandible. Ventrally and laterally, the contact zone is sealed by the flexible lateral lobes of the hypopharynx, the prementum, and the maxillae; the latter fits into the concave ventral surface of the closed mandibles. The feeding can take several hours.

The only damage caused by the beetles is a tiny hole scraped through the mite's cuticle (Fig. 12.1e). Because the attack is directed onto the dorsolateral or posterodorsal surface of the idiosoma, *Cephennium* can successfully feed on both ptyctimous and non-ptyctimous mites, as the encapsulation of the former does not protect against this type of attack. Furthermore, pteromorphs are an inefficient protection against this feeding technique. However, the unique morphological and behavioral adaptation of *Cephennium* requires a subglobose prey with smooth and sparsely setose body to ensure the adhesion by suction forces; deeply sculptured, reticulate, or spiny mites are not attacked. In addition, those that accumulate soil particles on the body surface avoid being captured by *Cephennium*.

Even among "morphologically acceptable" oribatids, *Cephennium* species select only particular taxa, showing strong preferences toward ptyctimous Phthiracaridae and non-ptyctimous Ceratozetidae, Achipteriidae, and Liacaridae. Jałoszyński and Olszanowski (2016) observed also significant differences in prey preferences between two morphologically similar *Cephennium* species that differ slightly in their body size. Their choice of prey was clearly affected by this difference, as the larger *C. majus* was able to feed on larger prey than the smaller *C. ruthenum* Machulka. The feeding process clearly depends on the prey's body size (or volume). Presumably, the structure of the mite's cuticle also plays an important role, as the time from attack to the killing of the mite in some instances depends on the mite taxon rather than its body length.

12.3 Glandulariini and Scydmaenini: Brutal Force and Leg-Cutting

Species of Glandulariini and Scydmaenini known to feed on Oribatida and/or Uropodina have unspecialized mouthparts, except for typically sharp and slender mandibular apices (Fig. 12.2a, b) that are well adapted to insert into the mite's natural body openings. It was demonstrated that adults of different species within the same genus, not differing in the structure of mouthparts, can feed on soft-bodied arthropods or on armored mites only. Such a pair of morphologically very similar species is *Scydmaenus tarsatus* Müller and Kunze and *Scydmaenus rufus* Müller and Kunze; the former feeds on weakly sclerotized Acaridae and springtails, whereas the latter feeds on Oribatida and Mesostigmata (Uropodina) (Jałoszyński 2012a; Jałoszyński and Olszanowski 2015). Moreover, although the adults and larvae of *Sc. tarsatus* have strikingly different mandibles (asymmetrical and with mesal teeth in adults vs. symmetrical, falciform, and lacking teeth in larvae), their prey choices are similar (Jałoszyński 2012a; Jałoszyński and Kilian

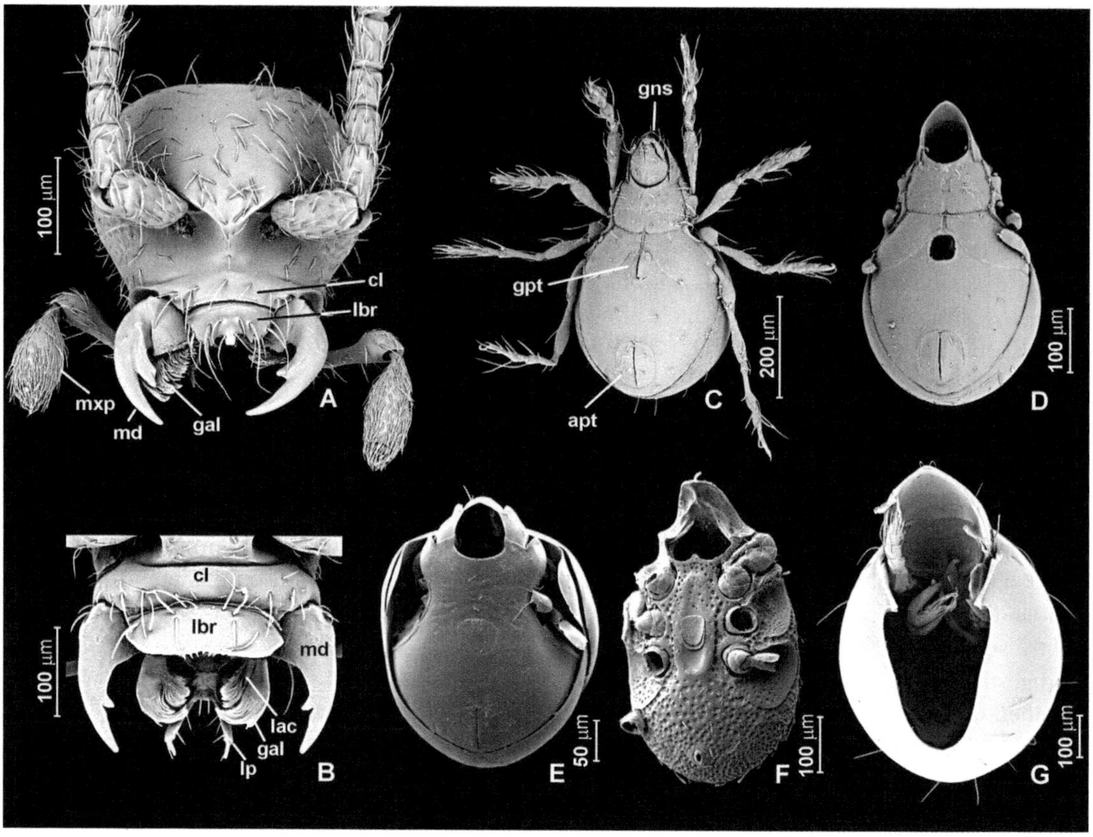

Fig. 12.2 Examples of Scydmaenini (**a**) and their prey (**c–f**) and Glandulariini (**b**) and their prey (**g**). (**a**) Head of *Scydmaenus rufus* in anterior view (after Jałoszyński and Olszanowski 2015, modified). (**b**) Mouthparts of *Euconnus pubicollis* in anterior view (after Jałoszyński and Olszanowski 2013, modified). (**c–d**) *Oppia nitens* (Oribatida, Oppiidae) before (**c**) and after (**d**) feeding of *Scydmaenus rufus*. (**e–g**) Prey of *Scydmaenus rufus* (**e**, **f**) and *Euconnus pubicollis* (**g**) showing feeding damage: *Punctoribates punctum* (Oribatida, Mycobatidae) (**e**), *Uroobovella pyriformis* (Uropodina, Urodinychidae) (**f**), and *Phthiracarus* sp. (Oribatida, Phthiracaridae) (**g**). Abbreviations: *apt* anal plate, *cl* clypeus, *gal* galea, *gns* gnathosoma, *gpt* genital plate, *lac* lacinia, *lbr* labrum, *lp* labial palp, *md* mandible, *mxp* maxillary palp

2012). It seems that behavioral rather than morphological adaptations play the key role in feeding for Glandulariini and Scydmaenini. Consequently, it is not possible to infer their preferred prey by studying the structure of the mouthparts; only direct behavioral observations can address the question concerning the prey choice and feeding technique.

The compost-inhabiting European *Scydmaenus rufus* feeds predominantly on oribatids belonging to Scheloribatidae and Oppiidae and Urodinychidae uropodines (Jałoszyński and Olszanowski 2015). Scheloribatids and urodinychids are short-legged mites with either smooth or distinctly reticulate cuticle. Cuticular structures do not protect them against *Sc. rufus* because the primary target of its attack is the gnathosoma, with the secondary target being the genital or anal opening. Therefore, typical feeding damage is restricted to these body regions (Fig. 12.2e, f). The beetles attack the anterior body region of their prey, inserting one mandible into the mite's mouth opening and breaking off all of its mouthparts by rotating the mite. The feeding takes place through the resulting opening by external digestion. The beetles inject digestive juices into the gnathosomal opening and ingest the liquefied tissues. In abandoned empty mite shells, usually some or all the legs are also removed. However, this is a secondary process that results from rotating and manipulating the dead mite during feeding; the legs, with their internal soft tissues already dissolved, are brittle and easily break off.

A modification of this simple mechanism is required when *Sc. rufus* attacks Oppiidae mites, which have long and spiny legs (Figs. 12.2c and 12.3g). The legs are a part of the mites' defense system and mechanically interfere with a predator's attempts to get close to the vital regions of the gnathosoma or ventral structures. Consequently, the attack typically begins with the cutting off of some legs. An experiment with 60 beetles, each attacking a single *Oppia* mite, took observations 10–100 min after the attack (Jałoszyński and Olszanowski 2015). After just 10 min, one or two of the mite's legs or at least some podites were removed in 40% of mites. After 30 min, several legs were removed in 70% cases and the first successful attempts to attack the gnathosoma were noticed. After 100 min, all of the mite prey had some or all of their legs cut off; their mouthparts were also completely removed. In some cases, the secondary targets—the genital plates—were also removed at this phase (Fig. 12.2d).

Beetles that manage to successfully attack the gnathosoma can be observed exuding a droplet of digestive juices onto their prey and then sucking it back; one mandible remains inserted into the gnathosomal (or genital) opening during the entire feeding process. Beetles select their prey based on unknown factors. However, body size is certainly one of them because attacks on too-large prey (e.g., some Liacaridae oribatids) are usually unsuccessful. It is worth noting, however, that beetles do make such attempts despite poor chances to succeed. Tactile or chemical (and less so, visual) stimuli seem to play an important role in initiating attacks.

A different technique of attack, but a similar method of feeding, is used by a common European inhabitant of the forest leaf litter, *Euconnus pubicollis* (Müller & Kunze), a glandulariine ant-like stone beetle (Jałoszyński and Olszanowski 2013). When presented with a broad spectrum of oribatid and uropodine mite taxa, this species strongly prefers the ptyctimous Phthiracaridae (*Phthiracarus* spp.). Several other oribatid families (both ptyctimous and non-ptyctimous) represented less than 8% of its diet in laboratory prey choice experiments (based on 30 beetles observed for a month that chose their prey from more than 1400 living mites belonging to 24 families and 50 species; Jałoszyński and Olszanowski 2013). The mouthparts of *E. pubicollis* are similar to those of most Glandulariini (Fig. 12.2b), with unmodified labrum, maxillae, and labium, as well as elongated, curved mandibles that each have a small preapical mesal tooth and a slender apical portion. The preferred prey of this species is ptyctimous and therefore is capable of encapsulation; consequently, when attacked, the mites retract and protect all vulnerable body parts under their shield-like prodorsum. This defensive posture does not leave any externally accessible grip sites or intersegmental membranes that could be pierced. *Euconnus pubicollis* uses a different technique to capture its prey and breach its defenses than does *Cephennium* or *Sc. rufus*.

Euconnus pubicollis, when confronted with its preferred prey, rapidly moves its head and mouthparts toward the cuticle of the mite, exudes a droplet of sticky liquid from its mouth, and lifts the mite; furthermore, it often additionally uses the protibial apices with spatulate, adhesive setae to manipulate its prey. However, the beetles are able to lift their prey without using their fore legs, only by means of the sticky properties of the liquid on their mouthparts or/and capillary forces. The anterior portion of the beetle's labrum or its dorsal surface adheres to the mite cuticle by means of the liquid exuded from the mouth; the mandibles remain widely spread and only their apices touch the mite. The maxillae are protruded anteriorly, with galea and lacinia covered with the liquid and adhering to the prey cuticle. The maxillary palps are spread laterally and occasionally touch the prey's cuticle. The labium is protruded anteriorly; its anterior part is also covered with the liquid and adheres to the mite. This arrangement of mouthparts allows the formation of a large contact area from the labrum

Fig. 12.3 Scydmaeninae feeding on armored mites. (**a**–**c**) *Cephennium majus* feeding on ptyctimous (**a**, **c**) and non-ptyctimous (**b**) Oribatida. (**d**, **e**) *Euconnus pubicollis* feeding on *Phthiracarus* sp. (**f**) *Microscydmus* sp. feeding on Oribatida. (**g**) *Scydmaenus rufus* feeding on *Oppia nitens*. (**h**) *Stenichnus godarti* feeding on Uropodina. (**i**) *Neuraphes elongatulus* feeding on juvenile Damaeidae

to the galea and lacinia, bearing dense trichia covered with the liquid.

The beetles typically adopt a posture of a raised head and prothorax while standing on their middle and hind legs; alternatively, they use substrate particles to attack from above, standing head down, with hind and middle legs on the side of a soil or wood particle, and the anterior part of the body with the captured mite hanging down above the ground (Fig. 12.3d, e). The attack phase, if undisturbed by other beetles, is immediately followed by manipulating the prey and searching for access to fragile or movable parts of the integument. If there are other beetles frequently disturbing the successful predator, the latter walks for minutes or even hours with the prey held in its mouthparts, searching for shelter under soil particles or in narrow spaces between them, where it could continue to manipulate the mite to overcome its encapsulation.

Adult *Euconnus* beetles that have already captured a phthiracarid mite frequently rotate their prey using their protibiae, and often detach and attach their mouthparts to the prey. A voluminous droplet of liquid is regularly produced and sucked back in cycles of a few seconds, with the droplet first increasing in size then rapidly decreasing in volume. This is repeated from about 90 min to more than 20 h, during which time the manipulated mite remains motionless and encapsulated. During this prolonged activity, *Euconnus* covers the entire body surface of its prey with the liquid exuded from the mouthparts. The moving and rotating of the mite is occasionally accompanied by movements of the mandibles; apparently, the beetle searches for a grip on the margins of the closed prodorsum or the genital and anal plates. It seems that toxic or otherwise noxious properties of the exuded liquid (presumably digestive juice) are responsible for slowly weakening the mite; eventually, the muscles that maintain the encapsulation relax.

The moment when the prodorsum is at least partly lifted marks the end of the long struggle and the beginning of the last phase. The time from the attack to the lifting of the prodorsum can range from approximately 1 to more than 20 h and depends on the body length of the prey. To gain access to the flesh after opening the prodorsum of the prey, the *Euconnus* beetle presses the mite's ventral (genital and anal) plates. To this aim, the mite is usually pressed against the ground; the beetle adopts a posture with its head and pronotum lowered toward the prey while standing on all six legs. The mandibles are used to grip the margin of genital or anal valves. Gradually, the entire complex of ventral plates is pressed into the notogaster and often crushed, exposing the flesh. The mouthparts and nearly the entire head of *Euconnus* are gradually inserted deeper and deeper into the notogaster, if the mite is large enough. During feeding, the mandibles rapidly chew away the soft internal tissues. At this stage, the beetle exudes a small amount of digestive juice from its mouth. When the prey is too small for the beetle's head to be inserted into the opening, only the mandibles (or even only one mandible) are inserted into the mite's body. The mite is rotated around the inserted mandible while copious amounts of digestive juice are exuded and then ingested.

Empty phthiracarid mite shells that were abandoned after feeding by *E. pubicollis* have characteristic feeding damage patterns (Fig. 12.2g). The prodorsum is either broadly open or (frequently) completely removed. The ventral plates are typically deeply pressed inside the idiosoma, often crushed and fragmented. The entire feeding process shows a strong linear correlation with the prey length; it can take from several hours to more than 30 h.

Euconnus pubicollis can also feed on some non-ptyctimous oribatid mites (but not on uropodines). Species that were successfully attacked by beetles in prey choice experiments belong to Achipteriidae, Chamobatidae, Oribatellidae, Ceratozetidae, and Galumnidae (Jałoszyński and Olszanowski 2013). Their defense systems do not rely on encapsulation, but solely on the strength of their smooth and sparsely setose cuticle. Additionally, some of them (e.g., galumnids) have pteromorphs—lateral cuticular lobes that protect their legs from being cut off by predators. These structures do not protect mites from being killed by *E. pubicollis*, whose technique does not involve cutting the legs prior to attacking other body parts. Non-ptyctimous oribatids are captured and lifted in a similar way as the beetles handle phthiracarids (i.e., by a droplet of sticky liquid produced from the predator's mouth). However, further manipulations are clearly different. The mite, which adheres to the mouthparts (and often to the protibiae) of *E. pubicollis*, is moved, rotated, and frequently pressed against the ground or soil particles when the predator detaches its tibiae to change its grip. *Euconnus* tries to insert the slender and pointed tip of one mandible into the gnathosoma, the genital or anal valves. When successful, the leg movements of the prey stop, apparently marking the moment of death; the beetle rotates the mite around the inserted mandible to tear off movable structures around the opening. The time from the attack to inserting one mandible into the prey is 25–190 min. During feeding, beetles usually remove all or most of the legs of the mite. However, this is secondary damage: the legs are broken off during the last 5–20 min by the mandible, which remains outside the mite during rotations. Purposeful severing of the legs with both mandibles, as a prerequisite for attacking the body openings, was not observed.

Euconnus beetles often take short breaks during feeding on both ptyctimous and non-ptyctimous oribatids. The dead mite is placed on the ground and the predator spends a short time (up to slightly over 2 min) with self-grooming. The cleaning is restricted to the antennae and distal portions of the fore legs, which are passed through the mouthparts. Additionally, the middle legs are used to clean the elytra. Then, the beetles resume their feeding.

Thus far, all observed Scydmaeninae that feed on armored mites and have an unmodified labium feed in a similar way as *Scydmaenus rufus* or *Euconnus pubicollis* (although lifting of the prey using a droplet of sticky liquid was observed only for the latter species). Many observations were made by the author of this chapter; however, most of them remain unpublished because collecting a sufficient number of specimens for conclusive prey choice experiments is a difficult task. Acarophagous Glandulariini include one of the smallest predaceous beetles; among them are the genus *Microscydmus* Saulcy and Croissandeau (Fig. 12.3f), which comprise species with adults as small as 0.6–0.8 mm in body length. Such small beetles attack only oribatids with the smallest adults; however, they feed in the same way as the much larger *Scydmaenus* or *Euconnus* (i.e., through the damaged gnathosoma of non-ptyctimous mites).

A notable example of acarophagous Glandulariini showing a mixed mechanism of capturing mites is the genus *Stenichnus* Thomson. These middle-sized (typically 1.5–2.5 mm) Holarctic beetles have long and very slender falciform mandibles, usually with finely serrated mesal margins, and one pair of membranous suckers or adhesive discs on their prementum (Jałoszyński 2013). The ultrastructure of these organs remains unknown, but they seem to be simpler than those in specialized Cephenniini. Moreover, the labial palps in *Stenichnus* are large and not reduced, as those in Cephenniini. Little is known about feeding habits of this genus; most observations so far have remained unpublished. It seems that *Stenichnus* shows preferences toward feeding on armored and relatively large Uropodina. Jałoszyński (2016) reared an adult of the European *St. godarti* (Latreille) ex larva and fed it with uropodines; this single beetle ate 112 individuals of *Trichouropoda* sp. within 92 days of its life. Adults of *Stenichnus* seem to use their labial suckers only during the initial phase of capturing the prey to adhere to the mite's cuticle and lift the uropodine mite. Then, the long mandibles take over and further manipulations lead to their insertion into the mite's gnathosoma. This mixed mechanism can be expected to allow for feeding on various mites. Indeed, some observations of several species of *Stenichnus* showed that they can feed on smooth and finely reticulated Oribatida and finely or coarsely reticulated Uropodina (Jałoszyński, unpublished data).

Schmid (1988) suggested that the mouthparts of *Neuraphes* Thomson are used to grasp the legs of Damaeoidea (as Belboidea) oribatids, but no further details were given. Damaeoidea include mites that were not attacked by any Scydmaeninae species tested by Jałoszyński and Olszanowski (2013, 2015, 2016); they seem to be especially well protected against predators by their morphological structures. Some species are exceptionally large and have particularly thick and hard cuticles. Some Damaeidae accumulate soil particles on their idiosoma to form an additional protecting crust, and many have very long and spiny legs. Previously unpublished observations made by the author of this chapter show that several Central European species of *Neuraphes* indeed feed exclusively on Damaeidae—but on juveniles, not on heavily sclerotized adults (Fig. 12.3i). Juveniles are spiny but soft-bodied; their main protection are long and spiny legs, which prevent predators from getting close to the vulnerable body. *Neuraphes* beetles grasp the mite's legs to turn their prey upside down, then attack soft ventral structures. Neither living juveniles nor their remains abandoned after feeding can be identified to the genus or species level. Thus, it is especially difficult to study prey preferences of *Neuraphes*. Besides the general technique they use, nothing else is known about their prey choice.

12.4 Problems and Perspectives

Defensive adaptations of Oribatida—and to a lesser extent, those of similarly armored mesostigmatan Uropodina—are relatively well studied. They seem so efficient that acarologists proposed the hypothesis of an "enemy-free space" where extant mite taxa live after having developed impenetrable protection during co-evolution with predatory prostigmatan and mesostigmatan mites (e.g., Jeffries and Lawton 1984; Peschel et al. 2006). Indeed, oribatids are particularly difficult prey because of their thick and mineralized cuticle, which is often reinforced by a system of grooves, carinae, or reticulation; they are also protected by long spines or accumulated soil particles, as well as the presence of defensive glands in many taxa. However, it is well-known that various oribatids can be successfully attacked and eaten by some rove beetles, as Pselaphinae and Scydmaeninae, and by some ants.

Park (1947) mentioned that *Batrisodes* Reitter (Pselaphinae, Batrisini) feeds on oribatids, but no further details concerning the feeding technique or mite taxa were given. Two species of Japanese ants in the genus *Myrmecina* Curtis (Myrmicinae, Crematogastrini) showed some behavioral and morphological adaptations to use oribatids as a major or sole source of food. The worker ants crush and tear off a large portion of the mite's cuticle to feed larvae; the latter have elongate and narrow heads that can be easily inserted into the partly damaged mite shell to feed on the flesh (Masuko 1994). Early reports concerning the featherwing beetles (Ptiliidae) being capable of feeding on Oribatida (Riha 1951) have never been confirmed and seem dubious, as ptiliids are currently recognized as a group of fungivorous or spore-feeding beetles (e.g., Betz et al. 2003; Jałoszyński 2015). Therefore, the Scydmaeninae are currently the best studied examples of arthropod predators specialized to feed on armored mites, which are one of the best protected prey among thousands of soft-bodied invertebrates that inhabit the soil, leaf litter, rotten wood, or decomposing plant remains.

There are currently more than 5300 nominal species of Scydmaeninae known. Prey preferences and feeding-related behaviors have been studied under laboratory conditions in a few of them, including only four that feed on oribatid or uropodine mites (Jałoszyński and Olszanowski 2013, 2015, 2016). However, already in such a tiny fraction of known scydmaeninae diversity, the observed spectrum of behavioral and morphological adaptations and differences in prey preferences are astounding. *Cephennium* species are "hole scrapers" and use sophisticated structures of their modified, specialized mouthparts to capture subglobose and smooth oribatids. When given a choice between more than 40 species representing more than 20 families of Oribatida and Uropodina, they predominantly fed on Phthiracaridae, Ceratozetidae, and Achipteriidae; the larger of two tested species also fed on Liacaridae. The choice of prey was apparently not affected by the ptyctimous versus non-ptyctimous body form of the prey, and the entire feeding process took place through a tiny hole ground in the mite's cuticle (Jałoszyński and Olszanowski 2016).

Euconnus pubicollis, when given a choice between mites belonging to 50 species and representing 25 families of Oribatida and Uropodina, predominantly fed on one family only—the Phthiracaridae—showing strong preferences toward the ptyctimous body form of its prey. This species captures mites using a droplet of sticky liquid exuded from its mouth, to which the prey adheres and can be further manipulated and "opened" by a slow process in which copious amounts of digestive juices weaken the mite; mandibles are only used in the final *coup de grâce* (Jałoszyński and Olszanowski 2013). *Scydmaenus rufus*, when offered more than 20 species representing 15 families of Oribatida and Uropodina, predominantly fed on the oribatid Scheloribatidae and Oppiidae, and only marginally on the uropodine Urodinychidae and other taxa. This species also has unspecialized mouthparts, which are used to attack the mite's gnathosoma to feed through a large opening left after tearing off the prey's mouthparts. If the prey mites have long and spiny legs, they are partly removed before the predator can gain access to the gnathosoma (Jałoszyński and Olszanowski 2015). Unpublished observations of the author of this chapter on several other Scydmaeninae species show an even broader spectrum of adaptations and narrower prey preferences, as those of *Neuraphes*, which seems to feed exclusively on juvenile Damaeidae.

It seems that gaining access to armored mites as a source of food might have been an important event in the evolution of Scydmaeninae. An unnamed species that is morphologically very similar to the extant acarophagous Cephenniini is known from the Cenomanian (Jałoszyński and Peris 2016), and a *Stenichnus*-like glandulariine species with a specialized prementum bearing a pair of suckers was recently discovered in Turonian amber (Jałoszyński et al. 2017). Oribatids are beyond doubt a much more ancient group than scydmaeninaes; the oldest fossils of Oribatida date to the Middle and Upper Devonian (e.g., Norton et al. 1988; Subías and Arillo 2002; reviewed by Arillo et al. 2012), whereas ant-like stone beetles are known from the Upper Cretaceous (reviewed by Jałoszyński and Peris 2016). It remains unknown how scydmaeninaes adapted to feed on armored mites or what was the food of their ancestors. Oribatids—and to a lesser extent, uropodines—are very rich food sources in terrestrial ecosystems; however, they are so well protected against predators that only few can feed on these mites. Species that are able to breach the defenses of this prey, can escape the competition that shapes relationships between numerous small soil predators, such as ants, ground beetles, spiders, pseudoscorpions, mesostigmatan mites, and others. Furthermore, various species of Scydmaeninae that co-occur in the forest floor can avoid competition by specializing to feed on particular mite taxa or mite body forms. This seems to be a major achievement for a large group of predators that live in highly competitive environments.

Have oribatid mites evolved defense mechanisms during at least 100 Ma of co-evolution with specialized predators? To date, it has not been possible to answer this question. Apparently, some of the most efficient defense mechanisms, such as the encapsulation of ptyctimous mites, are easily overcome by scydmaeninaes, and oribatids seem helpless during attacks. Even toxic secretions of their defensive glands do not protect them against scydmaeninaes; for example, *Scheloribates laevigatus* (Koch), readily eaten by *Scydmaenus rufus*, is a well-known producer of highly toxic alkaloids—among others the infamous pumiliotoxins, which are components of skin secretions of dendrobatid poisonous frogs (Saporito et al. 2007, 2011). On the other hand, Jałoszyński and Olszanowski (2016) analyzed the morphological characters of mites not eaten by any species of Scydmaeninae tested so far. They concluded that adults of oribatid taxa with particularly thick and typically densely sculptured cuticles, such as Carabodidae, Nothridae, Damaeidae, and Hermanniellidae, avoid predation by ant-like stone beetles.

It seems that Scydmaeninae may exert some pressure on the local population dynamics of their prey. Although under laboratory conditions *Euconnus pubicollis* consumed on average only 1 mite per 3.7 days, *Scydmaenus rufus* was able to consume approximately 1.4 mites per day (Jałoszyński and Olszanowski 2013, 2015). Assuming that *Sc. rufus* is active only during the warm season in Central Europe and feeding rates remain constant over time, then 100 beetles might consume nearly 26,000 mites from April to September (Jałoszyński and Olszanowski 2015). *Sc. rufus* commonly inhabits compost, from which more than 50 beetles were collected from 10 L of the substrate taken only from the upper compost layer (Jałoszyński and Olszanowski 2015). Thus, it seems possible that a population of this species contained within a typical garden compost heap may significantly affect the population dynamics of their most preferred prey—that is, scheloribatids and oppiids. Because oribatids are known to alter the chemistry and nutrient cycling in decomposing plant matter (e.g., Wickings and Grandy 2011), these processes may also be affected by their dedicated predators.

A major open research question in studies of the specialized feeding of scydmaeninaes on armored mites is the astonishingly long feeding process. It may take over 10 h to complete feeding by *Cephennium* beetles and more than 30 h for *Euconnus* (Jałoszyński and Olszanowski 2013, 2016). During this process, the mandibles of the beetle may be buried deeply in the idiosoma (or one mandible in the gnathosoma) of the mite (*Euconnus*), or the tip of one mandible may be inserted into the tiny hole drilled by *Cephennium*. This is not a good position to escape from larger generalist predators of soil and leaf litter, such as ants or ground beetles, which are common in this habitat. The effort and energy investment made into the slow process of penetrating the mite's cuticle or breaking off its mouthparts must be awarded by feeding long enough to gain energy, not to lose it. A disturbance from numerous soil invertebrates, and especially predators that could attack scydmaeninaes, is likely to disrupt the feeding before the energy balance reaches a positive value. How the beetles protect themselves while being attached to prey that is often nearly as large as themselves, and how they manage to complete their feeding undisturbed, remain major questions in the study of Scydmaeninae biology.

References

Arillo A, Subías LS, Shtanchaeva U (2012) A new species of fossil oribatid mite (Acariformers, Oribatida, Trhypochthoniidae) from the Lower Cretaceous amber of San Just (Terruel Province, Spain). Syst Appl Acarol 17(1):106–112

Betz O, Thayer M, Newton AF (2003) Comparative morphology and evolutionary pathways of the mouthparts in spore-feeding Staphylinoidea (Coleoptera). Acta Zool 84(3):179–238

Jałoszyński P (2012a) Adults of European ant-like stone beetles (Coleoptera: Staphylinidae: Scydmaeninae) *Scydmaenus tarsatus* Müller & Kunze and *S. hellwigii* (Herbst) prey on soft-bodied arthropods. Entomol Sci 15:35–41

Jałoszyński P (2012b) Observations on cannibalism and feeding on dead arthropods in *Scydmaenus tarsatus* Müller & Kunze. Genus 23(1):25–31

Jałoszyński P (2013) Revision of subgenera of *Stenichnus* Thomson, with review of Australo-Pacific species (Coleoptera, Staphylinidae, Scydmaeninae). Zootaxa 3630(1):39–79

Jałoszyński P (2015) *Ptenidium pusillum* (Gyllenhal, 1808) from egg to pupa (Coleoptera: Ptiliidae). Zootaxa 3948(3):361–421

Jałoszyński P (2016) Mature larva of *Stenichnus godarti* (Latreille) (Coleoptera: Staphylinidae, Scydmaeninae): redescription, hypothesis of displaced epicranial sutures and alternative interpretation of homology between chaetotaxic structures. Zootaxa 4196(1):77–94

Jałoszyński P, Beutel R (2012) Functional morphology and evolution of specialized mouthparts of Cephenniini (Scydmaeninae, Staphylinidae). Arthr Str Dev 41:593–607

Jałoszyński P, Kilian A (2012) Larval morphology of *Scydmaenus tarsatus* and *S. hellwigii*, with notes on feeding behavior and a review of bibliography on preimaginal stages of ant-like stone beetles (Coleoptera: Staphylinidae, Scydmaeninae). Eur J Entomol 109:587–601

Jałoszyński P, Olszanowski Z (2013) Specialized feeding of *Euconnus pubicollis* (Coleoptera: Staphylinidae, Scydmaeninae) on oribatid mites: prey preferences and hunting behaviour. Eur J Entomol 110:339–353

Jałoszyński P, Olszanowski Z (2015) Feeding of *Scydmaenus rufus* (Coleoptera: Staphylinidae, Scydmaeninae) on oribatid and uropodine mites: prey preferences and hunting behaviour. Eur J Entomol 112:151–164

Jałoszyński P, Olszanowski Z (2016) Feeding of two species of Scydmaeninae 'hole scrapers', *Cephennium majus* and *C. ruthenum* (Coleoptera: Staphylinidae), on oribatid mites. Eur J Entomol 113:372–386

Jałoszyński P, Peris D (2016) Cretaceous amber inclusions of Spain and Myanmar demonstrate early diversification and wide dispersal of Cephenniitae (Coleoptera: Staphylinidae: Scydmaeninae). Cret Res 57:190–198

Jałoszyński P, Perrichot V, Peris D (2017) Ninety million years of chasing mites by ant-like stone beetles. Gondwana Res 48:1–6

Jeffries MJ, Lawton JH (1984) Enemy free space and the structure of ecological communities. Biol J Linn Soc 23:269–286

Leleup N (1968) Révision des Mastigini de l'Afrique du Sud. Ann Mus Roy Afr Centr, Tervuren 166:1–107

Masuko K (1994) Specialized predation on oribatid mites by two species of the ant genus *Myrmecina* (Hymenoptera: Formicidae). Psyche 101:159–173

Molleman F, Walter DE (2001) Niche segregation and can-openers: Scydmaenid beetles as predators of armoured mites in Australia. In: Halliday RB, Walter DE, Proctor HC, Norton RA, Colloff MJ (eds) Acarology: Proceedings of the 10th international congress. CSIRO Publishing, Melbourne, pp 283–288

Norton RA, Bonamo PM, Grierson JD, Shear WA (1988) Oribatid mite fossils from a terrestrial deposit near Gilboa, New York. J Paleontol 62(2):259–269

O'Keefe ST, Monteith GB (2000) *Clidicus abbotensis* O'Keefe, a new species of Scydmaenidae (Coleoptera: Staphylinoidea) from Australia with description of the larva. Mem Queensland Mus 46:211–223

Pachl P, Domes K, Schulz G, Norton RA, Scheu S, Schaefer I, Maraun M (2012) Convergent evolution of defense mechanisms in oribatid mites (Acari, Oribatida) shows no "ghosts of predation past". Mol Phylogenet Evol 65:412–420

Park O (1947) Observations on *Batrisodes* (Coleoptera: Pselaphidae), with particular reference to the American species east of the Rocky Mountains. Bull Chicago Acad Sci 8:43–132

Peschel K, Norton RA, Scheu S, Maraun M (2006) Do oribatid mites live in enemy-free space? Evidence from feeding experiments with the predatory mite *Pergamasus septentrionalis*. Soil Biol Biochem 38:2985–2989

Reitter E (1909) Fauna Germanica. Die Käfer des Deutschen Reiches, vol 2. Lutz KG, Stuttgart

Riha G (1951) Zur Ökologie der Oribatiden in Kalksteinböden. Zool Jb Syst 80:408–450

Saporito RA, Donnelly MA, Norton RA, Garraffo HM, Spande TF, Daly JW (2007) Oribatid mites as a major dietary source for alkaloids in poison frogs. Proc Natl Acad Sci U S A 104:8885–8890

Saporito RA, Norton RA, Andriamaharavo NR, Garraffo HM, Spande TF (2011) Alkaloids in the mite *Scheloribates laevigatus*: further alkaloids common to oribatid mites and poison frogs. J Chem Ecol 37:213–218

Schmelzle S, Helfen L, Norton RA, Heethoff M (2008) The ptychoid defensive mechanism in Euphthiracaroidea (Acari, Oribatida): a comparison of exoskeletal elements. Soil Org 80:233–247

Schmelzle S, Helfen L, Norton RA, Heethoff M (2009) The ptychoid defensive mechanism in Euphthiracaroidea (Acari: Oribatida): a comparison of muscular elements with functional considerations. Arthr Struct Dev 38:461–472

Schmelzle S, Helfen L, Norton RA, Heethoff M (2010) The ptychoid defensive mechanism in *Phthiracarus longulus* (Acari, Oribatida, Phthiracaroidea): exoskeletal and muscular elements. Soil Org 82:253–273

Schmid R (1988) Morphologische Anpassungen in einem Räuber-Beute-System: Ameisenkäfer (Scydmaenidae, Staphylinoidea) und gepanzerte Milben (Acari). Zool Jahrb, Abt Syst, Ökol Geogr Tiere 115:207–228

Schuster R (1966a) Über den Beutefang des Ameisenkäfers *Cephennium austriacum* Reitter. Naturwiss 53:113

Schuster R (1966b) Scydmaeniden-Larven als Milbenräuber. Naturwiss 53:439–440

Subías LS, Arillo A (2002) Oribatid fossil mites from the Upper Devonian of South Mountain, New York and the Lower Carboniferous of County Antrim, North Ireland (Acariformes, Oribatida). Estud Mus Cienc Nat Álava 17:93–106

Wickings K, Grandy AS (2011) The oribatid mite *Scheloribates moestus* (Acari: Oribatida) alters litter chemistry and nutrient cycling during decomposition. Soil Biol Biochem 43:351–358

Structures and Functions of the Endophallic Copulatory Tube in the Family Staphylinidae (Insecta: Coleoptera) 13

Shun-Ichiro Naomi

Abstract

This paper deals with the structures and functions of the endophallic copulatory tube in Staphylinidae, which has been previously called "copulatory piece," "endophallus," "flagellum," etc. in descriptive studies. First, the general morphology of the copulatory tube and the pseudocopulatory tube is discussed, together with the characterizations and terminology. Second, the distribution of the copulatory tubes in Staphylinidae is described. Third, the structures of the seven basic and some other unique forms of the copulatory tubes in Staphylinidae are described in detail, paying special attention to those of Steninae. Fourth, the functions of the copulatory tubes in Staphylinidae are described or hypothesized, paying special attention to those of Steninae. Namely, the function of the copulatory tube is simply a "spermatophore (or sperm) depositor" in some cases (e.g., some *Stenus*), whereas it has double function: a "spermatophore (or sperm) depositor" and an "extension tube (or sperm) guiding rod in some other cases (e.g., some *Aleochara*).

13.1 Introduction

In Coleoptera, the aedeagus consists externally of the tegmen and the median lobe; and the tegmen comprises of the phallobase (basal piece) and the parameres (lateral lobes) (Sharp and Muir 1912; Lawrence and Britton 1994). Since the phallobase is almost or completely missing in most species of Staphylinidae, the aedeagus consists of the median lobe and paired parameres (Crowson 1981); and in the median lobe, there is internal endophallus. The external and internal structures of aedeagi are highly diverse morphologically, and they are considered to be very useful both for classifying the species and also for classifying the species groups of genera in Staphylinidae; and thus, taxonomists have paid special attention to them when attempting to do the revisional studies of a group in question.

In Staphylinidae, the endophallus comprises of a copulatory tube, sclerites, sclerotized bands, etc.; and the copulatory tube is located inside a membranous reversible internal sac, and the latter connects the rim of the apical foramen (i.e., the

S.-I. Naomi (✉)
Natural History Museum and Institute, Chiba, Japan
e-mail: naostenus@hb.tp1.jp

O. Betz et al. (eds.), *Biology of Rove Beetles (Staphylinidae)*,
https://doi.org/10.1007/978-3-319-70257-5_13

ostium) with the base of the copulatory tube. The simple form of the endophallus is made up of a simple reversible internal sac. However, the endophallus very often includes the following three components: (a) paired expulsion hooks that have a function of triggering the copulatory tube, (b) paired longitudinal bands that have a function of reinforcing the membrane of the internal sac, and (c) a copulatory tube. These relatively complex endophalli are found in such genera as *Stenus* (Naomi 2006a, b; Naomi et al. 2017) and *Sepedophilus* (Naomi and Maruyama 1997). In some genera, e.g., *Scopaeus* (Frisch 1997, 1998, 1999), *Fenderia* (Puthz 2003a), and *Aploderus* (Shimada 2002), the endophallus is highly complex, because in addition to all or two of those three components, it is also armed with a variety of sclerites, tufts, and other structures. Crowson (1981:70) suggested that in some groups of Coleoptera, the sclerotizations of the endophallus may play a role in shaping a spermatophore, but the function of the endophallic sclerites in Staphylinidae still remains unclear in most cases.

Out of the various structures of the endophallus in Staphylinidae, the copulatory tube can be regarded as the structure that plays an important role in copulation, because it serves as a tube for transmitting the sperm or spermatophore to the vagina or spermatheca. Thus, it is worthy of studying from the morphological point of view. After Sharp and Muir (1912) first studied in detail the basic structures of the aedeagus in Coleoptera (including the Staphylinidae), Blackwelder (1936) and Naomi (1990) each studied the aedeagus of Staphylinidae in a comparative perspective. However, the endophallic copulatory tubes in Staphylinidae have not been so far studied in detail probably due to the three reasons described below.

First, not all staphylinids possess the copulatory tube as a component of the aedeagus. Second, the morphological diversity of the copulatory tubes is very high. For example, in some cases (e.g., some *Dianous*), it is composed of a very thick stick or rod; and thus, it looks as a different structure. In other cases, it is composed of a very long flexible, whiplike tube called "flagellum." To make the matter more complex and inextricable, such a whiplike flagellum has evolved parallely many times in different species in Steninae, Euaesthetinae, Aleocharinae, etc. Consequently, the copulatory tubes in Staphylinidae are difficult to homologize. And, third, it is also difficult to understand the homologous relations of the parts of a copulatory tube among the various groups of Staphylinidae.

Here, I present a detailed study of the structures and functions of the endophallic copulatory tube in the family Staphylinidae. First, I discuss the general structures of the copulatory tube and the pseudocopulatory tube, together with the characterizations and terminology. Second, the distribution of the copulatory tubes in Staphylinidae is described and briefly discussed. Third, I describe the structures of the seven basic and some other unique forms of the copulatory tubes in detail in Staphylinidae, paying special attention to those of Steninae. Fourth, I describe or hypothesize the functions of the copulatory tubes in Staphylinidae, paying special attention to those of Steninae.

13.2 Some Morphological Terms

Some morphological terms are not consistently used; and then the term (e.g., "endophallus") has several different meanings because each has been previously used differently by different authors. Thus, they are ambiguous and difficult to use in a precise way. Some other terms also are so because some morphological parts (that are designated by those terms) are complex in structure and position. In order for readers to understand the precise meanings of some important morphological terms used in this paper, I discuss here some important details.

Endophallus Following Nichols (1989: 239), the "endophallus" is described as follows: "In the male insects, the internal sac or tube of the phallus invaginated at the end of the aedeagus,... ." (Note here that the terminal membranous part of the aedeagus in insects has been in general called "vesica," but Sharp and Muir (1912: 585) called it "internal sac.") It seems that this characterization is incomplete. This is because, according to this characterization, the endophallus comprises of "internal sac" or "internal tube," but in many cases, it is composed of "the internal sac and tube." Namely, there exist within an aedeagus, both the internal tube (which is connected anteriorly with the reversible internal sac) and the internal sac (which is connected anteriorly with the rim of the apical foramen). Thus, I consider here that the endophallus consists of the internal structures of aedeagus containing the copulatory tube, internal sac, and its related sclerites.

Basal chamber of copulatory tube Klimaszewski (1984) and Naomi (2006a, b) called the basal swollen portion of the endophallic flagellum "sperm(a) sac," while Gack and Peschke (1994) called the inner sac of the spermatophore "sperm sac." Thus, presently, the term "sperm sac" is a homonym. The latter usage of the term "sperm sac" seems to be adequate, given that the sperm is deposited within the inner sac of a spermatophore. Thus, to avoid the terminological confusion caused by the term "sperm sac," I call the basal swollen portion of the copulatory tube "basal chamber."

Dorsal side of aedeagus In general, the dorsal side of the aedeagus in Staphylinidae is the side opposite to the parameral attachment; and thus it is sometimes called "non-parameral" or "abparameral" (Gusarov 2003: 9) side. However, in Steninae the dorsal side of the aedeagus is the upper side of the aedeagus on which the parameres are attached, because the aedeagus of Steninae is positioned in situ within the abdomen, in such a way that the parameres face dorsally. This is the reverse of the situation in many other Staphylinidae.

13.3 Endophallic Copulatory Tube and Pseudocopulatory Tube

In the taxonomic studies of staphylinids, the intromittent sclerite of the aedeagal median lobe has been called as "basal tube" (Naomi 2006a, b; Naomi et al. 2017), "copulatory piece" (Maruyama and Klimaszewski 2004; Maruyama 2008), "dorsal copulatory piece" (Brunke et al. 2016), "endophallus" (Nomura 2001a; Irmler 2005), "flagellum" (Ashe 1984, 1992), "guide sclerite" (Löbl and Calame 1996), "internal structure of male genitalia" (Hammond 1973), etc. These endophallic structures are certainly highly diverse morphologically within Staphylinidae, but due to the existence of various intermediate conditions, we can trace the morphological transformation series from the original forms (simple, short setal structures) to one extremity (very thick sticks) or to the other extremity (whiplike, very long flagella). Furthermore, as mentioned later, we can also understand that these structures basically consist of same morphological components. Thus, they are considered homologous structures from the genuine morphological point of view. Note however that all endophallic sclerites that have been previously called variously by the aforementioned terms in descriptive studies are not necessarily copulatory tubes.

First, all these intromittent sclerites are used and function during a copulation; and second, although they are highly morphologically diverse in Staphylinidae, ranging from the thick and stiff rods to the whiplike, very long, flexible flagella, they all are tubular structures. Note here that the tube is open only at the dorsal side in *Aleochara* (Gack and Peschke 2005: 309, Fig. 1C) or only at the ventral side in *Stenus* (Naomi 2006a, b), so that it is a U-shaped tube in cross section. Given the function and structure of the endophallic intromittent sclerites, they are here collectively called "copulatory tubes."

Precisely, what is the endophallic copulatory tube in Coleoptera? In order to identify the copulatory tube, we need to settle the landmark for its circumscription. The ejaculatory duct is formed by an ectodermal invagination; and the position of its opening is supposed to be settled in the early stage of genital morphogenesis (Hemming 2003). Thus, the position of the gonopore (i.e., the opening of the ejaculatory duct: Fig. 13.1a, b) can be considered an important landmark for identifying the endophallic copulatory tube in Coleoptera (see Matsumura and Yoshizawa 2012 in the case of the flagellum). By using the position of gonopore as its landmark, we will be able to characterize the coleopterous copulatory tubes. When characterizing it in an appropriate way, we must consider the following two points: first, the copulatory tube can be regarded as an endophallic tube that is distal from the opening of the ejaculatory duct; and second, the characterization must be applied to the morphologically diverse forms of the tube located at the apex of the endophallus. The copulatory tube of Coleoptera can be thus characterized as "a sclerotized, rod-, tube-, or whiplike terminal extension of the ejaculatory duct which is distally located from the opening of the ejaculatory duct (i.e., primary gonopore"; Fig. 13.1a, b) in such a way that we can regard a thick, stiff stick as well as a long, whiplike tube of the endophallus, as the copulatory tube.

A major problem here is that the endophallic copulatory tubes are highly diverse morpholog-

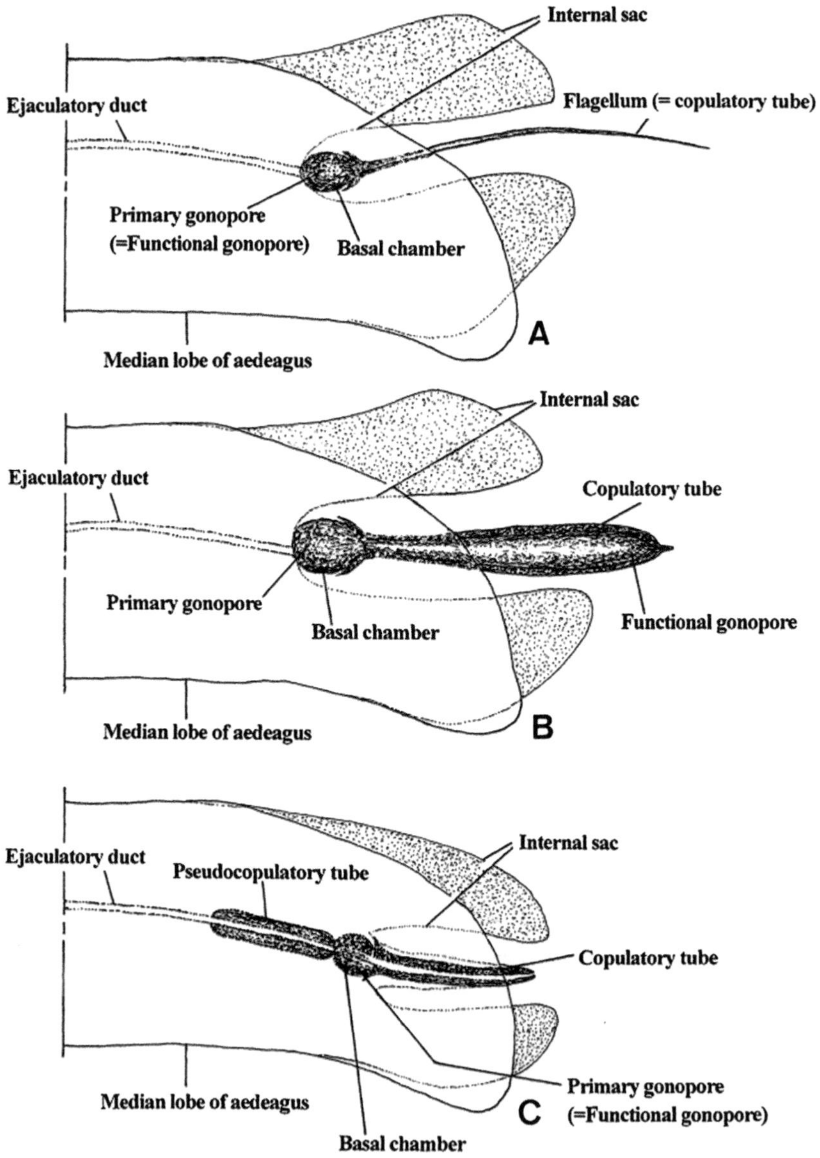

Fig. 13.1 Diagrams of the endophalli in Staphylinidae (lateral views). (**a**) Copulatory tube (= flagellum) in which the primary gonopore corresponds in place to the functional gonopore; (**b**) copulatory tube in which the primary gonopore does not correspond in place to the functional gonopore; (**c**) pseudocopulatory tube and copulatory tube

ically; and thus, coleopterists have used various descriptive terms that are suitable for the conditions of copulatory tubes that they studied. This is really the case for the "flagellum" (Fig. 13.1a), which is in some cases (e.g., Peschke 1978) called "virga" (Snodgrass 1935: 622). The endophallic flagellum of insect is characterized by Nichols (1989) as "sclerotized terminal prolongation of the ductus ejaculatorius, usually concealed within the internal sac when in repose, but sometimes very long and constantly protruding through the ostium of the penis." The flagellum of Coleoptera is similarly described as "internal sac of a very long, slender eversible type" (Crowson 1981) or by other slightly different wordings in textbooks. I think that these traditional characterizations of the flagellum represent reasonable attempts made during the expansion of the knowledge of insect morphology; and in those senses, they can apply to the terminal, tubular, or whiplike modifications of the ejaculatory duct and thus are useful for descriptive purposes.

However, those characterizations are not precise, because we do not know, by using such wordings, how to homologize the flagellum or how to distinguish the flagellum from the ejaculatory duct. Note here that the so-called flagellum is a form of the copulatory tube; and consequently, it is the long or very long, whiplike form of the copulatory tube. The proper characterization of a flagellum may be obtained by slightly modifying the characterization of the copulatory tube as shown above. The endophallic flagellum is therefore characterized as "a whiplike, flexible terminal extension of the ejaculatory duct which is distally located from the opening of ejaculatory duct (i.e., the primary gonopore"; Fig. 13.1a). Note also that the term flagellum, by its characterization, designates the whiplike tube that includes, if present, the basal (swollen or ovoidal) chamber.

The above characterizations seem to be sufficient for identifying the endophallic copulatory tubes of Staphylinidae, but an important point is that, in the endophallus of a few Staphylinidae (e.g., some *Stenaesthetus* species), there exist another terminal sclerotizations of the ejaculatory ducts which are nonhomologous to the copulatory tubes. This endophallic sclerotization (which I call "pseudocopulatory tube" in this paper; Fig. 13.1c) is similar in structure to the true copulatory tube because it is tubular. In insects, the epithelial wall of the ejaculatory duct is surrounded by a strong muscular sheath (Snodgrass 1935: 572). Given that the position and structure of the pseudocopulatory tube is the same as in the muscular sheath, the pseudocopulatory tube may be identical with a special form of the muscular sheath, which may aid in ejaculating the sperm (enveloped in a spermatophore). It just forms the apical portion of the ejaculatory duct, but it is distinctly different from the true copulatory tube because of the following two reasons: first, the opening of the ejaculatory duct (i.e., primary gonopore) is located at its tip, but not at its base (Fig. 13.1c), and, second, it is located just proximal to the true copulatory tube (e.g., *Stenaesthetus apterus* Puthz 1988a, Fig. 6); in other words, the pseudocopulatory tube is directly connected at its most distal portion with the base of the true copulatory tube.

Since this paper strictly deals with the true copulatory tubes (but not with the various tubular structures of the endophallus including pseudocopulatory tubes), the pseudocopulatory tube should be clearly distinguished from the true copulatory tube, to correctly identify the latter in Staphylinidae.

13.4 Distribution of Endophallic Copulatory Tubes in Staphylinidae

Since, in Staphylinidae, the distribution of the species that have the aedeagus with an endophallic copulatory tube is very characteristic and interesting, I describe it here before describing and dealing with the matter of structures and functions of the copulatory tubes in Staphylinidae.

In Staphylinidae, the simple form of the endophallus is composed of a simple reversible internal sac (Fig. 13.3a); and it is often furnished (or covered) internally with spines and/or spinules (Fig. 13.3e). These primitive conditions are widely found in many groups of staphylinids including *Cerapeplus* (Löbl and Burckhardt 1988), *Dasycerus* (Löbl and Calame 1996), *Glypholoma* (Thayer and Newton 1978; Thayer 1997), *Neophonus* (Thayer 1987), *Habrocerus* and *Nomimocerus* (Assing and Wunderle 1995), *Pseudopsis* (Herman 1975), *Trichophya* (Ashe and Newton 1993), *Xantholinus* (Bordoni 2002, 2011), etc. The endophallic copulatory tube is not found in the aedeagus of these staphylinids.

However, the species that have the aedeagus with a copulatory tube are widely scattered in the various subfamilies of Staphylinidae (Table 13.1): omaliine group (Omaliinae, Proteininae, Dasycerinae, Pselaphinae); oxyteline group (Trigonurinae, Oxytelinae, Osoriinae, Scaphidiinae); tachyporine group (Tachyporinae, Phloeocharinae, Aleocharinae), and staphylinine group (Oxyporinae, Paederinae, Staphylininae, Scydmaeninae, Megalopsidiinae, Euaesthetinae, Steninae, Leptotyphlinae). Major characteristics regarding the distribution of the copulatory tubes in Staphylinidae are as follows:

1. The copulatory tubes are found in both, the basal subfamilies (e.g., Trigonurinae, Omaliinae, Tachyporinae, Oxyporinae) and more advanced ones (e.g., Pselaphinae, Aleocharinae, Euaesthetinae, Steninae, Leptotyphlinae). However, the subfamilies with many component species having the aedeagus with a copulatory tube belong to the evolutionally advanced subfamilies (i.e., Pselaphinae, Aleocharinae, Euaesthetinae, Steninae, Leptotyphlinae). As far as I know, there is no staphylinid subfamily, with all species having the aedeagus with a copulatory tube.
2. The distribution of the species with a copulatory tube in a given subfamily (in which species with a copulatory tube are found) is rather biased; that is, the copulatory tube is in general restrictedly found in the species of some particular groups of the subfamily. For example, in Paederinae, the copulatory tube is found in many species of *Scopaeus* (Frisch 1997, 1998, 1999, 2003). In Aleocharinae, the copulatory tube/flagellum is found in many species of Gyrophaenina (Ashe 1984), Bolitocharina (Ashe 1992), Aleocharini (Klimaszewski 1984; Yamamoto and Maruyama 2012), etc. In Staphylininae, the copulatory tube is found in many species of Othiini (Assing 1999) and *Cyrtoquediina* (Brunke et al. 2016).
3. In small-sized genera, only one or several species have the aedeagus with a copulatory tube. For example, out of 17 worldwide species of *Dasycerus*, only one species *D. angulicollis* has the copulatory tube (Löbl and Calame 1996) among the large genera. The genus *Ocypus* seems to be also an example of this. Out of 56 species of the genus *Ocypus* distributed in the west part of Palaearctic region, only one species *O. similis* possesses the aedeagus with a copulatory tube (Coiffait 1974).
4. When many species of a genus possess the copulatory tube, then various forms of the copulatory tube are found within the genus under consideration. This is, for example, true for *Stenus* and *Dianous* (Table 13.1).

13.5 Structures of Endophallic Copulatory Tubes in Staphylinidae

13.5.1 General Structure of the Copulatory Tube

In Staphylinidae, the endophallic copulatory tube is in general composed of a basal chamber and a main tube (Fig. 13.2). The basal chamber is very small to moderate or large in size; and it is usually ovoidal or fusiform when it is swollen. It is, however, often missing (e.g., *Kathetopodion*, Leschen and Löbl 2005). If present, the basal chamber is usually demarcated by a basal constriction from the main tube (Fig. 13.2). The main tube is very often divided into two components: the basal tube and the apical tube (Fig. 13.2); in other words, the copulatory tube is tripartite (basal chamber, basal tube, and apical tube) in such cases. The copulatory tube is easily distinguishable from the ejaculatory duct in many cases,

Table 13.1 Representatives of the higher taxa of Staphylinidae, in which the species with an endophallic copulatory tube are included

Aleocharinae[a]: *Adelarthra* [4]: Ashe (2003); *Aleochara* [3v, 4]: Gack and Peschke (1994, 2005), Yamamoto and Maruyama (2012); *Aphaenochara* [3v]: Maruyama and Hlavác (2003); *Aspidobactrus* [2ra]: Maruyama (2000); *Autalia* [4]: Hoebeke and Ashe (1994); *Baeosthethus* [2v]: Steel (1964); Bolitocharina [2ra, 2rb, 3v]: Ashe (1992); *Creochara* [4]: Maruyama (2004a); *Diaulota* [1r, 2ra]: Ahn (1996); *Dinusa* [2v]: Assing (2001); *Giraffaenictus* [2rb]: Maruyama (2008); *Goniusa* [2ra]: Maruyama and Klimaszewski (2004); Gyrophaenina [2ra, 3v, 4, 5]: Ashe (1984); *Halorhadinus* [2ra, 2rb]: Ahn (2001); *Hygropetrophila* [4]: Wunderle and Assing (2000); *Kistnerella* [4]: Kanao et al. (2011); *Leptusa* [2ra, 2rb, 3, 4]: Pace (1999), Smetana (1973); *Liparocephalus* [2rb]: Ahn (1997); *Myllaena* [4]: Pace (2009); *Myrmecopella* [2ra, 3]: Maruyama (2004b); *Myrmecosticta* [4]: Maruyama et al. (2011); *Oligota* [2ra, 2rb, 3v, 4]: Williams (1976, 1978); *Oreokklina* [2ra]: Assing (2002); *Orphnebius* [2rb]: Assing (2006); *Oxypoda* [2ra, 2rb, 3, 4]: Pace (2010), Assing (2012); *Pella* [1r, 2rb]: Maruyama (2006); *Phanerota* [3, 4]: Ashe (1986); *Rothium* [2rb, 3v]: Ahn and Ashe (1996); *Tetrasticta* [4]: Maruyama and Sugaya (2002); *Zoosetha* [2rb]: Assing (2003)
Dasycerinae: *Dasycerus* [2v]: Löbl and Calame (1996)
Euaesthetinae[b]: *Edaphosoma* [2ra, 2rb, 3v]: Puthz (1986a, 2010); *Edaphus* [2ra, 2rb, 3, 4]: Puthz (1985a, 1986b, 1992); *Euaesthetus* [2ra, 2rb, 3, 3v, 4, 5, 6]: Puthz (1998, 2014); *Kiwiaesthetus* [2ra]: Puthz (2008b); *Nothoesthetus* [2rb]: Puthz (2012b); *Octavius* [2ra, 3, 3v, 4, 6]: Puthz (1977, 1985b, 1986c, 1989c); *Orosthetus* [4]: Puthz (1979); *Schatzmayrina* [3, 3v, 4]: Puthz (1978, 1989a); *Stenaesthetus* [2rb, 2v, 4]: Puthz (1988a, 2011a); *Stictocranius* [3v, 4]: Puthz (1989b, 2011b); *Tamotus* [2rb, 3]: Puthz (1973, 2002); *Turellus* [3, 4]: Puthz (1974, 1976)
Leptotyphlinae: *Cyrtotyphlus* [3v]: Coiffait (1972); *Entomoculia* [3, 3v, 6]: Coiffait (1972); *Mesotyphlus* [2ra, 3, 6]: Coiffait (1972); *Paratyphlus* [2ra, 2v]: Coiffait (1972); *Hesperotyphlus* [5, 6]: Coiffait (1972)
Megalopsidiinae: *Megalopinus* [2v, 3v, 5, 6]: Puthz (2012a, c)
Omaliinae: *Eusphalerum* [1r, 2ra, 2rb]: Zanetti (2014); *Geodromicus* [3v, 4]: Shavrin (2012)
Osoriinae: *Allotrochus* [3]: Naomi and Irmler (2012); *Apotocnemius* [3v]: Naomi (1986); *Holotrochus* [3v, 4]: Irmler (1981, 2005); *Lispinus* [4]: Naomi (1996); *Nacaeus* [4]: Naomi (1997a); *Osorius* [3v, 4]: Naomi (1986); *Thoracophorus* [4]: Irmler (1985)
Oxytelinae: *Thinobius* [1r, 2rb]: Makranczy and Schülke (2001)
Oxyporinae: *Oxyporus* [2ra]: Hwang and Ahn (2000)
Paederinae: *Acaratopus* [4]: Herman (1981); *Micrillus* [3v, 4]: Assing (2013); *Pinobius* [3, 3v, 4]: Assing (2014); *Pinophilus* [4]: Abarbanell and Ashe (1989); *Scopaeus* [2ra, 2rb, 3, 4]: Frisch (1997, 1998, 1999, 2003)
Phloeocharinae: *Charhyphus* [4]: Herman (1972)
Proteininae: *Nesoneus* [4]: Steel (1966); *Paranesoneus* [4]: Steel (1966)
Pselaphinae: *Articerodes* [2ra]: Nomura (2001b); Batrisina [3, 3v, 5, 6]: Nomura (1991); *Megatyrus* [3v, 5]: Nomura et al. (2011); *Octomicrus* [6]: Nomura (2010); *Odontalgus* [6]: Arai and Nomura (2003); *Paralasinus* [3v, 5]: Hlaváč and Nomura (2001); *Pselaphogenius* [5, 6]: Nomura (2001a)
Scaphidiinae: *Baeocera* [1r, 2ra, 3, 4]: Löbl (2012, 2015); *Bertiscapha* [2v, 3v]: Leschen and Löbl (2005); *Kathetopodion* [1sa]: Leschen and Löbl (2005); *Scaphisoma* [1r, 2ra, 3v, 4]: Löbl (2002, 2015); *Xotidium* [3v, 4]: Ogawa and Löbl (2016)
Scydmaeninae: *Cephennodes* [2ra, 3, 3v]: Jałoszyński and Nomura (2009); *Cephennomicrus* [1r]: Jałoszyński (2010); *Hlavaciellus* [2ra]: Jałoszyński (2010)
Staphylininae: *Ocypus* [4]: Coiffait (1974); *Othius* [4]: Assing (1999); *Xantholinus* [1sa, 1sb, 2s]: Coiffait (1972)
Steninae[c]: *Stenus*: [2ra, 2rb, 3, 3v, 4, 5, 6, 7]: Puthz (2000b, c, 2003b, 2008a, 2013), Naomi (2006a, b); *Dianous* [2ra, 2rb, 3, 3v, 4, 5, 6, 7]: Puthz (1988b, 2000a)
Tachyporinae: *Ischnosoma* [1r, 2ra, 2rb]: Kocian (1996); *Sepedophilus* [2rb, 3, 4]: Hammond (1973), Naomi and Maruyama (1998); *Tachinus* [1r, 2ra]: Kim and Ahn (2000)
Trigonurinae: *Trigonurus* [3v]: Kishimoto (2000)

[1r], [1s], [2ra], [2rb], [2s], [2v], [3], [3v], [4], [5], [6], and [7] mean the form 1r, form 1s, form 2ra, form 2rb, form 2s, form 2 (variation), form 3, form 3 (variation), form 4, form 5, form 6, and form 7 of copulatory tube, respectively. The numbers refer to those shown in Figs. 13.3 and 13.4

[a]A characteristic of the Aleocharinae copulatory tube is that the apical part of the main tube sometimes strongly curves (e.g., *Bolitochara*: Ashe 1992; *Myrmecopella*; Maruyama 2004b). Another characteristic is that the flagellum goes out from the median lobe at its dorso-basal part (e.g., *Adelarthra*: Ashe 2003; *Sternotropa*, *Pseudoligota*: Ashe 1984) or at its ventro-basal part (e.g., *Pseudoligota*: Ashe 1984) when it is longer than the whole length of the aedeagus

[b]The structure of the copulatory tube is highly diverse in Euaesthetinae; and there are found various forms including the very thin, extremely long flagella (form 4) and thick, sticklike tubes (forms 5 and 6)

[c]The structure of the copulatory tube is highly diverse in Steninae (forms 2–7), as in Euaesthetinae. The Euaesthetinae and Steninae are presently considered to have the sister-group relationship (Clarke and Grebennikov 2009; Mckenna et al. 2015), and thus the highly morphological diversity of the endophallus seems to be a characteristic of the clade comprising the Euaesthetinae and Steninae

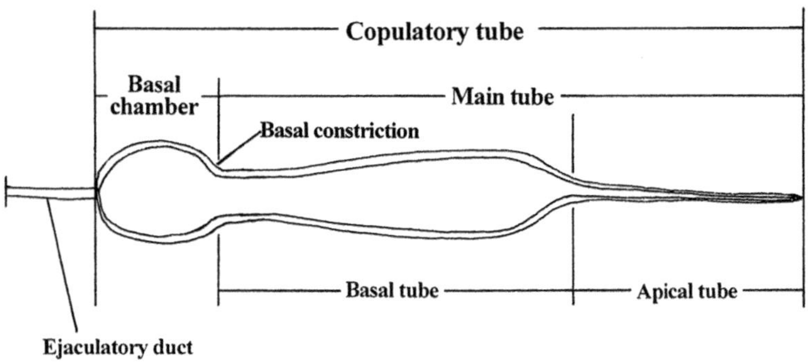

Fig. 13.2 Diagram of the general structure of the copulatory tube in Staphylinidae

because the base of copulatory tube (i.e., the basal chamber) is more or less swollen (Fig. 13.2). Even in cases of the copulatory tube without a basal chamber (e.g., *Pinophilus*; Abarbanell and Ashe 1989), it may be easily distinguishable from the ejaculatory duct in most cases, because the proximal portion of the copulatory tube is more or less thick than the ejaculatory duct.

In the copulatory tube of Staphylinidae (Fig. 13.1a, b), the opening of the ejaculatory duct (first gonopore) is not necessarily identical with the opening, through which the spermatophore (or sperm) is ejaculated during copulation (second gonopore). In some species of staphylinids, the first gonopore seems to substantially correspond in its position to the second (Fig. 13.1a). Namely, the first gonopore is located at the bottom of the basal chamber, and it is through this opening that the spermatophore (or sperm) is transferred to the vagina during copulation (e.g., *Aleochara*; Gack and Pescke 1994). However, in some other species of Staphylinidae, the first gonopore does not correspond in position to the second (Fig. 13.1b). Namely, the first gonopore is located at the bottom of the basal chamber, whereas the second is at the middle of the copulatory tube or at or near its tip (e.g., some *Stenus*: Fig. 13.6g; Naomi 2006b). Thus, these two openings should be precisely distinguished in the morphological studies of Staphylinidae. In this paper, the first gonopore is termed "primary gonopore," whereas the second one is termed "functional gonopore."

13.5.2 Seven Basic and Some Other Unique Forms of Copulatory Tubes

The endophallic copulatory tubes are highly morphologically diverse in Staphylinidae, as mentioned above, but they seem to be basically classified into seven forms. In this subsection, the structures of these seven basic forms are first described in detail; and then some other unique forms of the copulatory tubes are described.

13.5.2.1 Form 1 (or Original Form) of Copulatory Tube

Concerning the original forms of the copulatory tube in Staphylinidae, empirical data suggest that there seems to be two candidates. First, a simple bulbous chamber (Fig. 13.3b) should be considered an original form of a copulatory tube, because it is an extension of the ejaculatory duct which is distal to the primary gonopore. It consists only of the basal chamber; and it is here regarded as "form 1r." It is found in some *Thinobius* (Makranczy and Schülke 2001), etc. Second, a simple, seta-like sclerite (Fig. 13.3f, g) is also regarded as the other original form of a copulatory tube ("form 1s"). It consists of the simple main tube (i.e., a seta-like sclerite), with (Fig. 13.3f) or without (Fig. 13.3g) setulae around it. These two are here regarded as "form 1sa" and "form 1sb," respectively. The form 1s is found in the endophallus which is covered densely with spines and spinules. It is rare in Staphylinidae; and it occurs in some *Xantholinus* (Coiffait 1972:

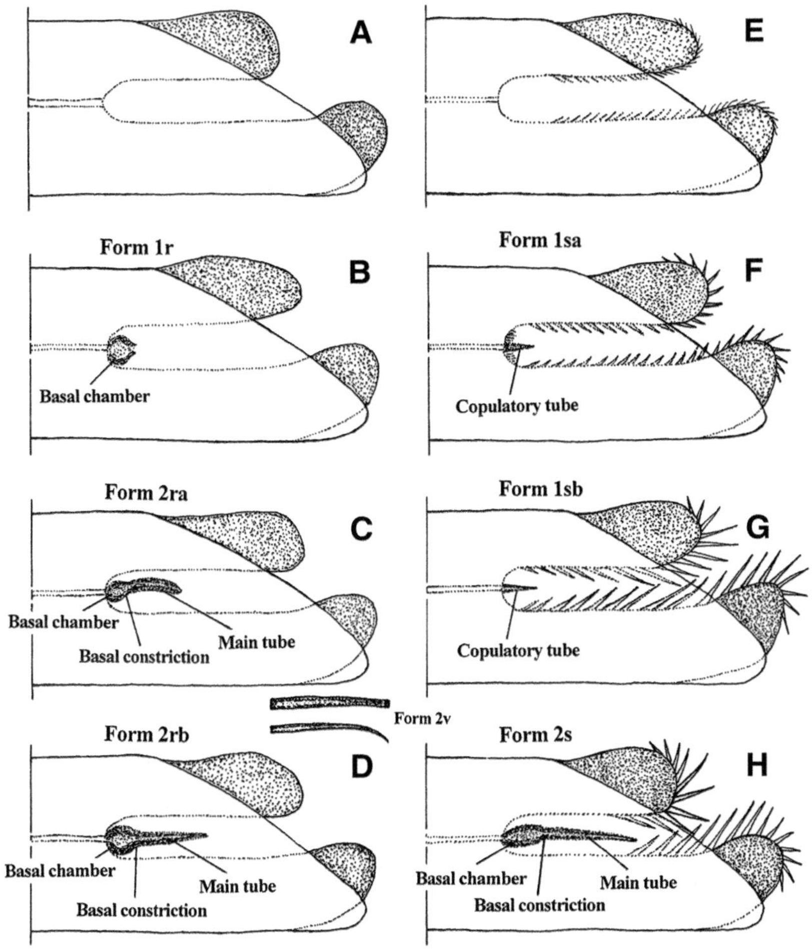

Fig. 13.3 Diagrams of the endophalli in Staphylinidae (lateral views). (**a**) Internal sac without modification; (**b**) internal sac with the copulatory tube (form 1r); (**c**) internal sac with the copulatory tube (form 2ra); (**d**) internal sac with the copulatory tube (form 2rb); (**e**) internal sac with setulae; (**f**) internal sac with the setulae and copulatory tube (form 1sa); (**g**) internal sac with the setae and copulatory tube (form 1sb); (**h**) internal sac with the setae and copulatory tube (form 2s)

241) and some Scaphidiines (Leschen and Löbl 2005: 34).

13.5.2.2 Form 2 of Copulatory Tube ("Copulatory Piece Auctorum")

The form 2 of a copulatory tube is composed of the basal chamber and a simple main tube (Fig. 13.3c, d, h). The basal chamber is small (Fig. 13.3c), medium (Fig. 13.3d, h), or large in size (e.g., some *Diaulota*; Ahn 1996). The basal constriction is indistinct (Fig. 13.3c, h) or distinct (Fig. 13.3d). The main tube is basically short and thin or moderately thick; it simply tapers toward the pointed apex (Fig. 13.3d, h) or is more or less curved (Fig. 13.3c). There often exist variations of form 2. They vary in structure, namely, they are thin to moderately thick, short to moderately long, and straight or curved, but they do not have a basal chamber (e.g., *Dasycerus*: Löbl and Calame 1996). They are classified into the "form 2v" in Fig. 13.3 and Table 13.1. The form 2 (and also 3) are rather different in structure from the typical whiplike flagellum and also from the thick, sticklike tube;

and they have been called by various terms (e.g., "copulatory piece," "guide sclerite," and "endophallus") in descriptive studies.

The form 2 of the copulatory tube is common in Staphylinidae (Table 13.1), but it is considered that the origins of these copulatory tubes must be different, because there are two different original forms described above. The "form 2ra" (Fig. 13.3c) and "form 2rb" (Fig. 13.3d) are here considered as derived from the form 1r, by extending posteriorly the distal portion of basal chamber. The "form 2s" (Fig. 13.3h) and its variations are here considered as derived from the form 1sa or 1sb, by enlarging its base and extending the main tube posteriorly (e.g., *Xantholinus linearis*; Coiffait 1972: 236). The form 2s and its variations are found in the endophallus which is covered densely with spines and spinules (e.g., in Xantholinine genera; Bordoni 2002, 2011; Coiffait 1972). (Note here that once a copulatory tube has evolved into the form 3 or other advanced forms (i.e., whiplike form 4 and thick forms 5–7), we cannot see in general whether the copulatory tube is derived from a simple seta-like sclerite or from a simple basal chamber.)

13.5.2.3 Form 3 of Copulatory Tube ("Copulatory Piece Auctorum")

The form 3 of the copulatory tube is in general tripartite; and it is composed of the basal chamber and the main tube (basal tube + apical tube) (Figs. 13.4 and 13.5c–e). It is basically long; and it is sometimes a little shorter than the whole length of the aedeagus. The proximal part of the copulatory tube is swollen to form a basal chamber; and when seen from the ventral side, it consists of "two thin rods" in some *Stenus* (Naomi 2006a, b). The main tube is U-shaped in cross section; and the basal tube is thin (Fig. 13.5c, d) or moderately thick (Fig. 13.5e), while the apical tube is whiplike (Fig. 13.5c) or moderately thick (Fig. 13.5e). The main tube is almost straight (Fig. 13.5c) or weakly (Fig. 13.5d) or strongly curved, but it basically tapers apically, and the demarcation between the basal tube and apical tube is indistinct.

There are variations of form 3. The "form 3v" is slightly different from the typical form 3 (Figs. 13.4 and 13.5e). For example, in some cases, the main tube is hardly divided into the basal and apical tubes (Fig. 13.5c, d); or it is

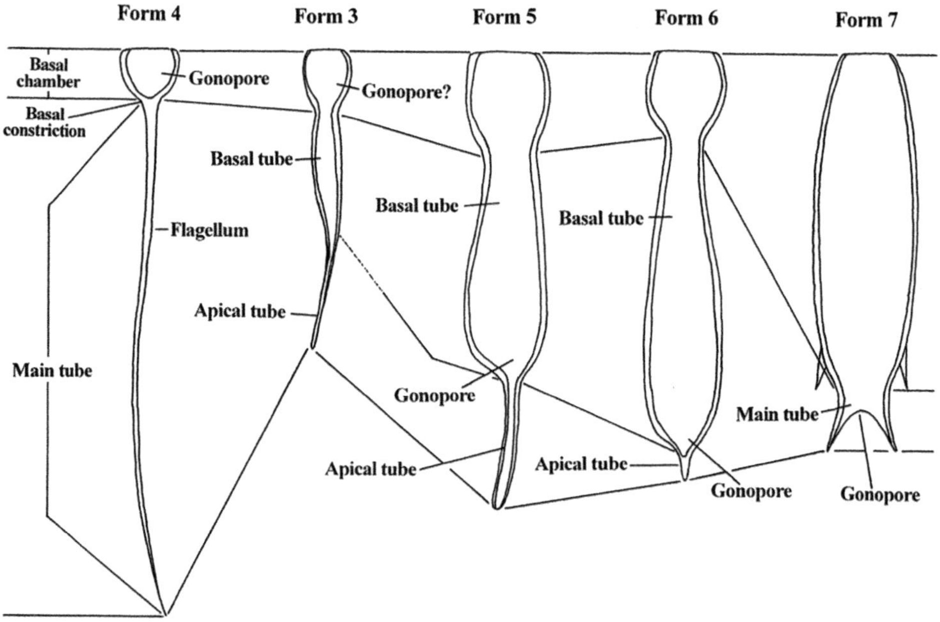

Fig. 13.4 Diagrams of the endophallic copulatory tubes (forms 3–7) in Staphylinidae. The lines are drawn, which show homologous parts of the different forms of the copulatory tubes

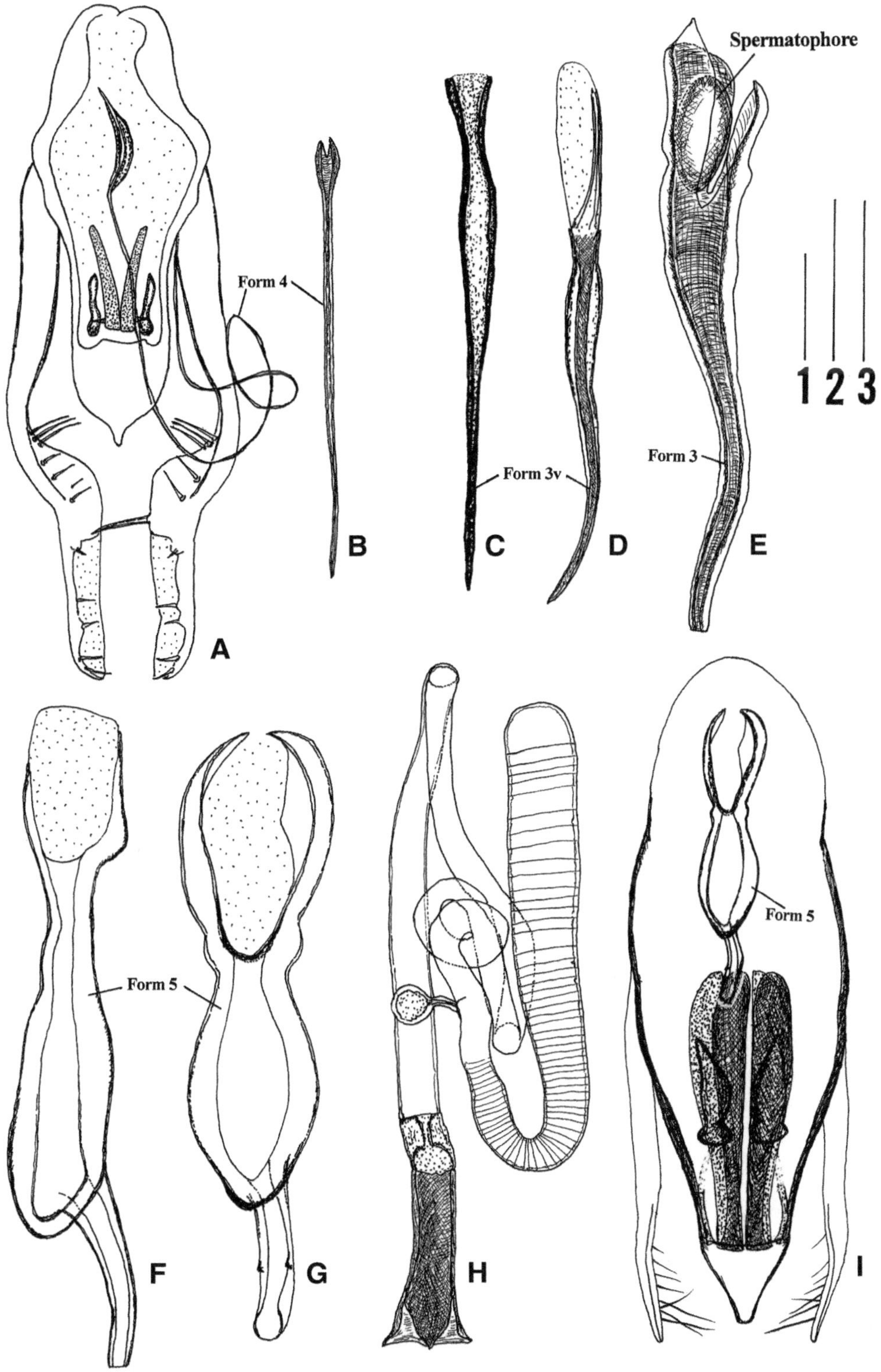

Fig. 13.5 (**a**, **i**) Aedeagi (ventral view); (**b**–**g**) copulatory tubes (ventral view); (**h**) spermatheca. (**a**) *Stenus riukiuensis* Puthz; (**b**) *S. miroku* Naomi; (**c**) *S. unagi* Hromádka; (**d**) *S. gagyumontis* Naomi; (**e**) *S. ohtoensis* Naomi; (**f**) *S. olliformis* Naomi; (**g**)–(**i**) *S. ebisu* Naomi (**a**, **c**: original illustration; **b**, **d**–**i**: Naomi 2006b) Scale 1: 0.1 mm for (**a**); scale 2: 0.1 mm for (**b**)–(**h**); scale 3: 0.2 mm for (**i**)

entirely sticklike so that it has the same or similar thickness from the base to the apex. The basal chamber is missing in some other cases. In cases where a copulatory tube is thin, moderately long, and weakly tapers apically (e.g., *Aphaenochara*; Maruyama and Hlavác 2003), the copulatory tube may be still regarded as a variation of form 3, but due to the various intermediate conditions, it is sometimes difficult to separate the form 3 from the form 2 or 4. The form 3 and its variations are, as in the form 2, common in Staphylinidae (Table 13.1).

13.5.2.4 Form 4 of Copulatory Tube ("Flagellum Auctorum")

The form 4 of the copulatory tube has been called "flagellum" in descriptive studies because it is whiplike. It consists of the basal chamber and the main tube (Fig. 13.4). The proximal part of the flagellum is more or less swollen (Klimaszewski 1984; Gack and Peschke 2005: 309, Fig. 1A: bf; Naomi 2006a, b) to form a basal chamber. The main tube is thin (Fig. 13.5b) to very thin (Fig. 13.5a), moderately long (Fig. 13.5b) to very long (Fig. 13.5a), straight (Fig. 13.5b), weakly or strongly curved, or irregularly, loosely coiled several times (Fig. 13.5a). In rare cases (e.g., *Othius bhutanensis*: Assing 1999; *Aleochara tristis*: Gack and Peschke 2005; *Stenaesthetus afer*: Puthz 2011a), it is very thin, extremely long, and regularly coiled many times. The tube is almost even in thickness from the base to the apex (Fig. 13.5a), or it weakly becomes thinner toward the apex (Fig. 13.5b). A unique flagellum is found in *Stenus paludivagus* Puthz 2000b. It is very long, wide, flat, and loosely coiled many times; and furthermore, it gradually widens toward the apex (Puthz 2000b). The form 4 is sporadically found in some species and genera of Aleocharinae, Euaesthetinae, Steninae, Othiini, etc. (Table 13.1).

13.5.2.5 Form 5 of Copulatory Tube

The form 5 of a copulatory tube is tripartite; and it is composed of the basal chamber and the main tube (basal tube + apical tube) (Fig. 13.4). Forms 3 and 5 are relatively similar in structure and thus sometimes difficult to distinguish from each other, but in the form 5 (Fig. 13.4), the basal tube is much thicker, and the basal and the apical tubes are in general easily distinguishable by a more or less distinct constriction between them. The basal chamber is small to large (Fig. 13.5f) or very large (Fig. 13.5g). The basal tube is thick (Fig. 13.5f) to very thick (Fig. 13.5g) and short to moderately long (Fig. 13.5f, g); and the apical tube is thin to feebly thick (Fig. 13.5f, g), feebly tapers toward its tip (Fig. 13.5f), or has almost the same thickness (Fig. 13.5g). The form 5 and its variations are most frequently found in Steninae, often in Pselaphinae, and in some genera of Euaesthetinae and Leptotyphlinae, etc. (Table 13.1).

13.5.2.6 Form 6 of Copulatory Tube

The form 6 of a copulatory tube consists of the basal chamber and the main tube (Fig. 13.4). Forms 5 and 6 are difficult to classify in some cases, but in the form 6, the apical tube is reduced into a small apicomedian protuberance. The basal chamber is medium to large (Fig. 13.6g) or very large sized (Fig. 13.6a, e). The basal tube is thick (Fig. 13.6a, 8) to very thick (Fig. 13.6d, e) and short to moderately long (Fig. 13.6f) or long (Fig. 13.6c). The apicomedian protuberance is small, thin, and simply pointed (Fig. 13.6a, e, g) or minutely bifurcate (Fig. 13.6d); and it is single (Fig. 13.6a, e, g) or rarely double (Fig. 13.6d). The form 6 is found in Pselaphinae, Steninae, Euaesthetinae, etc. (Table 13.1).

13.5.2.7 Form 7 of Copulatory Tube

The form 7 of a copulatory tube consists of the basal chamber and the main tube (Fig. 13.4). The basal chamber is in general stout and large (Fig. 13.6h) to very large (Fig. 13.6i). In some cases, it is modified with very thin, apicolateral projections (Fig. 13.6i), and/or others. The main tube is moderately thick (Fig. 13.6i) to thick or very thick (Fig. 13.6h) and large (Fig. 13.6h) or very large (some *Stenus*; Naomi 2010); and it usually lacks the constriction between the basal and the apical tube. It is sometimes reduced into a short and small tube or rod (Fig. 13.6i). The main tube has accessory lobes, protuberances, etc. in some cases. For example,

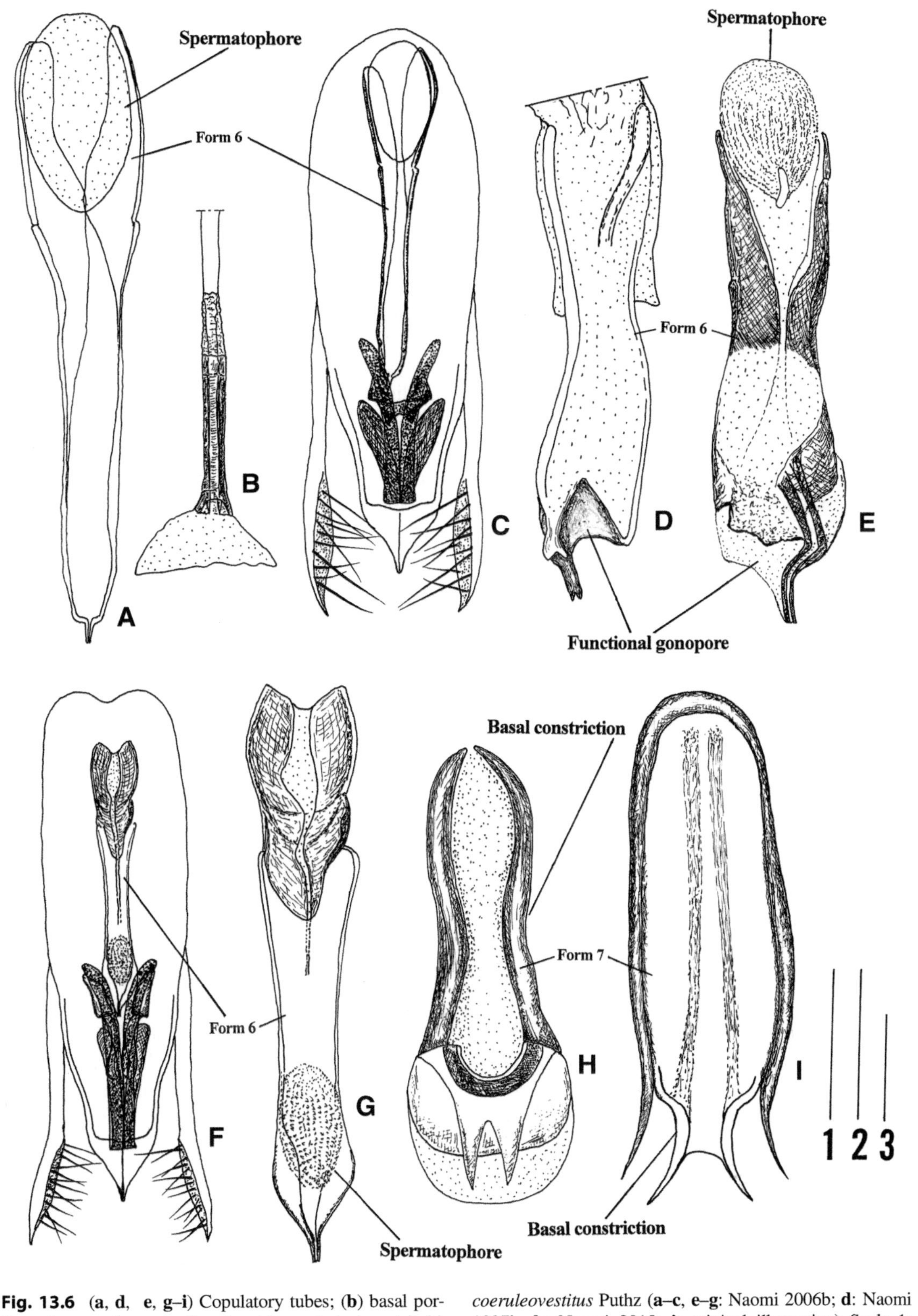

Fig. 13.6 (**a**, **d**, **e**, **g**–**i**) Copulatory tubes; (**b**) basal portion of spermatheca; (**c**, **f**) aedeagi in ventral view. (**a**)–(**c**) *Stenus ichihashii* Naomi; (**d**, **e**) *S. nakanei* Hromádka; (**f**, **g**) *S. gyrosus* Naomi; (**h**) *S. yasuhikoiellus* Naomi; (**i**) *Dianous coeruleovestitus* Puthz (**a**–**c**, **e**–**g**: Naomi 2006b; **d**: Naomi 1997b; **h**: Naomi 2010; **i**: original illustration) Scale 1: 0.1 mm for (**a**, **b**, **d**, **e**, **g**, **h**); scale 2: 0.2 mm for (**c**, **f**); scale 3: 0.1 mm for (**i**)

in *Stenus yasuhikoiellus*, the main tube has ventrally a deeply bifurcate lobe (Fig. 13.6h); and in *Dianous coeruleovestitus*, it has a pair of pointed lateral projections (Fig. 13.6i). The form 7 is, as far as I know, found only in Steninae.

13.5.2.8 Other Forms of Copulatory Tube

There certainly exist some other unique forms of copulatory tubes in Staphylinidae. Some representatives of the unique forms are as follows:

Corkscrew Form The copulatory tube is strongly, tightly coiled in some *Holotrochus* (Irmler 1981, 2005); and in some *Scaphobaeocera* (Hoshina and Sugaya 2003; Löbl 2015), it is thin and very loosely coined like a corkscrew. In some *Octavius* (Puthz 1989c), the copulatory tube is long and moderately thick; and just like a corkscrew, the tube is strongly, tightly coiled at least in its basal half, although the entire tube is almost straight. In some *Brachida* (Ashe 1984), the basal part of the main tube is thick and strongly, tightly coiled.

Bifurcate Form In some *Octavius* (Puthz 1977), the copulatory tube is long and moderately thick, with the apical half of main tube split longitudinally; the right lobe is thicker than the left one, and the mesial margin of the right lobe is furnished with an irregular line of 4, 5, 6, or 7 denticles. In *Cyrtoquedius* (Brunke et al. 2016) and *Euplectus lapponicus* (Löbl and Mattila 2010), the apical part of the main tube is split longitudinally.

Deformed Form In Pselaphinae (e.g., Jeannel 1959; Nomura 1991), the copulatory tube is often morphologically highly deformed, for example, in *Pselaphogenius* (Nomura 2001b) and *Octomicrus* (Nomura 2010), the main tube branches with spine-like lobes of various forms; and it is spatulate, strongly curved, asymmetric and/or sinuous. However, even in these cases, the copulatory tube yet retains some basic characteristics. Namely, it is rodlike at least at its base; and the cross section of the tube is U-shaped at least at its base.

13.6 Functions of Endophallic Copulatory Tubes in Staphylinidae

The status quo is that the functions of the endophallic copulatory tube are not studied in most groups of Staphylinidae. However, the function of the copulatory tube in *Aleochara* (as "extension tube guiding rod") was unraveled by Gack and Peschke (1994, 2005), whereas in some *Stenus*, the other function (as "spermatophore or sperm depositor") was inferred from the information obtained during my morphological studies by discovering that the spermatophores were retained in the male copulatory tube (Figs. 13.5e and 13.6e, g) and also a spermatophore was found in the female genital chamber, which was probably pasted by a male during copulation (Fig. 13.7; Naomi 2006b). Therefore, based on the present observations and descriptions of the copulatory tubes, together with available information from the published papers, it is certainly possible to reasonably hypothesize the functions of the copulatory tubes in many cases of Staphylinidae. Thus, in this section, the functions of the seven forms of the copulatory tubes are separately described and discussed. I hope that the descriptions, observations, and possible hypotheses about them in this section are not only useful for descriptive studies of Staphylinidae but also form a discussion basis for further morphological studies on them.

In this section, the term "gonopore" means the "functional gonopore" (but not the "primary gonopore""), because the position of the functional gonopore (Fig. 13.1a, b) is more useful for efficiently classifying the various forms of staphylinid copulatory tubes.

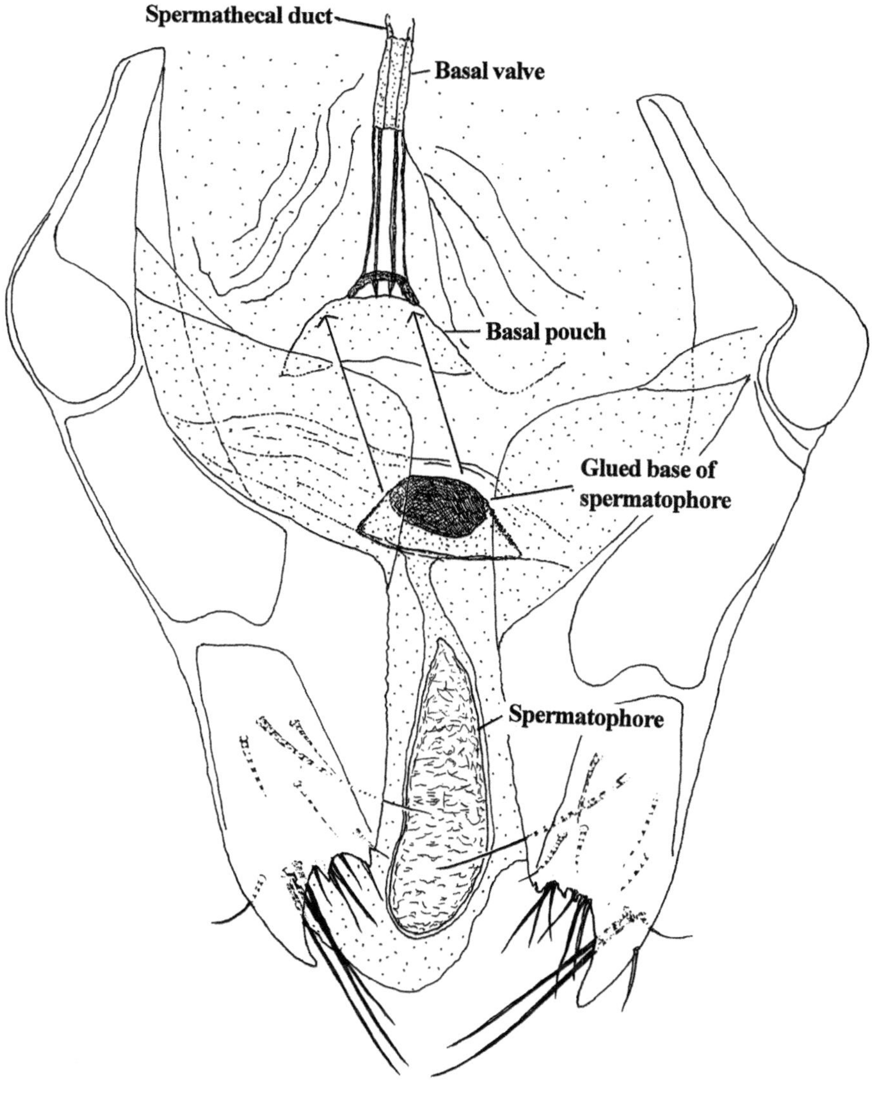

Fig. 13.7 Gonocoxites and the basal part of the spermatheca, with a spermatophore deposited in the vagina by a male (ventral view). Note that the spermatophore (which was once pasted to the basal pouch of the spermatheca by a male) comes out from the basal pouch. *Stenus ichihashii* Naomi (2006a). Scale: 0.1 mm

13.6.1 Functions of the Seven Forms of Copulatory Tubes

13.6.1.1 Form 1 (or Original Form) of Copulatory Tube

Regarding the form 1r (Fig. 13.3b), the gonopore is located at the base of the basal chamber, which functions as a chamber for storing a spermatophore (or sperms). Given its bulbous structure, the copulatory tube is supposed to play a role as a "spermatophore (or sperm) depositor" (see "form 6" of this subsection with respect to its characterization). Regarding the form 1sa and 1sb (Fig. 13.3f, g), the gonopore is supposed to be located at or near the base of the seta-like sclerite. Given its thin structure, it must function as an intromittent organ, which is inserted into the basal portion of the spermathecal duct during copulation.

13.6.1.2 Form 2 of Copulatory Tube ("Copulatory Piece Auctorum")

In the form 2 (2ra, 2rb, 2s) (Fig. 13.3c, d, h), the gonopore is located at the base of the basal chamber. In cases where the copulatory tube has a thin, attenuate main tube (2rb, 2s), it is supposed to be inserted into the spermathecal duct during copulation. However, the function of the form 2ra with the relatively thick main tube is unclear.

13.6.1.3 Form 3 of Copulatory Tube ("Copulatory Piece Auctorum")

The apical tube is whiplike so that it is too thin to transmit a large spermatophore to its tip (Fig. 13.5c, d); and thus, the gonopore is supposed to be located at or near the base of the copulatory tube. Since the whiplike apical tube is obviously thinner than the basal part of the spermathecal duct of female in some *Stenus* species (Naomi 2006a, b), the apical tube seems to have a role as an intromittent organ, which is inserted into the spermathecal duct during copulation.

13.6.1.4 Form 4 of Copulatory Tube ("Flagellum Auctorum")

The function of the form 4 (i.e., the flagellum) of *Aleochara* was studied in detail by Gack and Peschke (1994, 2005). The gonopore is located at the base of the flagellum. The flagellum is an intromittent tube that is inserted into the spermathecal duct of the female during copulation. It plays the role as the *guiding rod of an extension tube (or sperm)*. Note here that the extension tube is a very thin tube growing out from the spermatophore, to go through the spermathecal duct (Gack and Peschke 1994: Fig. 3).

In *Phanerota* (Ashe 1986), the copulatory tube is basically thin, and about as long as or distinctly longer than the whole length of the aedeagus, but the main tube seems to weakly differentiate into the stiff basal tube and the flexible, whiplike apical tube (e.g., Ashe 1986, Figs. 1, 8A and 9A). It belongs to form 3 or 4 (Fig. 13.4). At the dorsal side of the demarcation part of the copulatory tube between the basal and the apical tube, there exists a small pointed hook. On the other hand, the spermatheca of the female is moderately long and loosely coiled; and there exists a small triangular chamber at one side of the opening of the spermathecal duct (Ashe 1986, Fig. 7). Since the hook of the male copulatory tube nearly fits the triangular chamber of the spermatheca in size, the triangular chamber seems to function as the container for receiving the hook during copulation. It is thus hypothesized that the flexible, whiplike apical tube only is inserted into the spermathecal duct; and the pointed hook of the copulatory tube is pocketed into the triangular chamber of the spermatheca so that the copulatory tube tightly fixes in position during copulation.

13.6.1.5 Form 5 of Copulatory Tube

Given the thickness of the basal tube of the main tube in the form 5 (Fig. 13.4), the basal tube is considered to function as the chamber for storing a spermatophore (or sperm), as in the form 6 (e.g., Fig. 13.6g). It means that a spermatophore goes through the basal tube until near its tip before it is transmitted to the female genital chamber. Thus, the gonopore is supposed to be located at or near the tip of the thick basal tube in *Stenus* (Fig. 13.4). In *Stenus ebisu* (Fig. 13.5g, i) and its allied species (e.g., *S. olliformis*; Fig. 13.5f), the apical tube seems to play a role as an intromittent organ into the spermathecal duct; and it may also serve as a rod guiding an extension tube (growing out from a spermatophore), because of the following two reasons: first, the length and width of the apical tube (Fig. 13.5g) just fit the length and width of the female spermathecal duct from the opening to the base of the basal valve (Fig. 13.5h); and second, the cross section of the apical tube is - U-shaped (Fig. 13.5g) as in the flagellum of *Aleochara*, which functions as the guiding rod of an extension tube.

13.6.1.6 Form 6 of Copulatory Tube

It was observed during my morphological studies that a spermatophore is deposited near the apex of the main tube in a male of *Stenus gyrosus* Naomi 2006b (Fig. 13.6f, g). This certainly implies that the gonopore of this *Stenus* species is located at or near the tip of the main tube (Fig. 13.4). The copulatory tube of *S. gyrosus* is so strongly sclerotized and rigid that a spermatophore may safely pass through the main tube to the gonopore, without its deformation.

Given the thickness of the main tube, it is apparent that the form 6 (Fig. 13.4) is not an intromittent organ into the spermathecal duct. In a female of *S. ichihashii* (whose male has a form 6 of a copulatory tube), a large spermatophore was observed that is probably pasted by a male at the opening of the spermathecal duct during copulation (Fig. 13.7). This observation suggests that the copulatory tube is supposed to function as the *spermatophore (or sperm) depositor*; here the spermatophore (or sperm) depositor means a rod or stick that directly deposits a spermatophore (or sperm) in the vagina or in the female basal pouch or infundibulum (i.e., a bowl-like pouch located at the opening of the spermathecal duct; Fig. 13.6b). Thus, in *Stenus* the sperm is transferred from the vagina to the spermatheca after the copulation, as in *Aleochara* (Gack and Peschke 1994). The apicomedian protuberance (Fig. 13.6a, d, e, g) probably has a function of fixing the main tube in position during copulation, by putting it into the opening of the spermathecal duct, because the size of the apicomedian protuberance just matches the size of the opening of the spermathecal duct. One might compare, for example, the size of the apicomedian protuberance of the copulatory tube in *S. ichihashii*, (Fig. 13.6a) with the size of the opening of spermathecal duct of the same species (Fig. 13.6b).

13.6.1.7 Form 7 of Copulatory Tube

In *Dianous* (Fig. 13.6i; Puthz 2000a) and *Stenus* (Fig. 13.6h; Naomi 2010), there exists a large opening at the apicomedian part of the copulatory tube (Fig. 13.4), which is considered the gonopore. A spermatophore (or sperm) is stored in the large basal chamber, whereas the main tube is supposed to play the role of fixing a copulatory tube in position during copulation, given the occurrence of various modifications (e.g., accessory lobes and projections) at the lateral and/or apical parts of the main tube. When considering the aforementioned function, together with the thickness of the main tube, the form 7 is considered to function as a spermatophore (or sperm) depositor.

In some *Dianous* species with an atrophied main tube (Fig. 13.6i), the basal chamber seems to become very large as if it were the main tube; and thus a possible interpretation of it is that the atrophy of the main tube is compensated by the enlargement of the basal chamber. If my interpretation is correct, then it seems in such cases that the basal chamber of form 7 functionally plays a similar role in the basal tube of form 6 (e.g., Fig. 13.6g), while the reduced main tube of form 7 plays a similar role in the apicomedian protuberance (as a reduced apical tube) of form 6 (e.g., Fig. 13.6g) in a sense that it has a role of fixing the copulatory tube in position during copulation.

13.6.2 Summary on the Functions of Endophallic Copulatory Tubes

What we understood and hypothesized about the functions of copulatory tubes in Staphylinidae are here summarized.

The functional gonopore is located at the base of the copulatory tube in the form 1, 2, and 4, at or near the base of the copulatory tube in the form 3, at or near the apex of the basal tube in the form 5, near the apex of the main tube in the form 6, and at the apicomedian part of the main tube in the form 7.

We can certainly demonstrate that the forms 1–7 of the copulatory tube all have the function of transferring a spermatophore (or sperm) to the vagina and then to the spermatheca of a female.

However, strictly, the presented analyses suggest the following functions of copulatory tubes: (1) the form 1r (e.g., some *Thinobius*) and 6 and 7 (e.g., some *Stenus*, *Dianous*), which each comprises of a thick or very thick copulatory tube, are not intromittent organs into the spermathecal duct. Each serves simply as the spermatophore (or sperm) depositor, that is, a stick or a swell that directly deposits the spermatophore (or sperm) into the vagina or into the basal pouch located at the opening of the spermathecal duct; (2) the forms 1sa, 1sb, 2rb, 2s, and 3–5 each play a role as an intromittent organ into the spermathecal duct, because the apical portion or the apical tube of the copulatory tube is made up of a thin tube. The spermatophore (or sperm) is deposited in the vagina by the copulatory tubes of these forms. At least the forms 3–5 each are, given their structures (Fig. 13.4), supposed to serve also as an extension tube (or sperm) guiding rod, that is, a rod that guides the spermatophore extension tube (or sperm) into the spermathecal duct, as shown in *Aleochara* by Gack and Peschke (1994). Thus, the functions of the copulatory tubes are double in such cases (e.g., some *Aleochara*, *Stenus*): a "spermatophore (or sperm) depositor" and an "extension tube (or sperm) guiding rod"; (3) the function of the form 2ra is unclear.

Acknowledgments I would like to express my sincere gratitude to Dr. Alessandro Minelli (Padova University) and to Dr. Margaret Thayer (Field Museum) for their constructive and also critical comments on my earlier versions of the draft of this paper and to Dr. Yoko Matsumura (Hokkaido University) for her useful discussion with me about the morphology of the flagellum in Coleoptera. I sincerely thank Dr. Volker Puthz (Schlitz) for his kindness in sending me some valuable *Stenus* specimens for this study and also for informing me of some interesting forms of *Stenus* flagella, Dr. Alexey Shavrin (Daugavpils University) for letting me know valuable information on the copulatory tubes in Omaliinae, and Dr. Volker Assing (Hannover) for informing me of the matter with the paederine flagellum. My hearty thanks are due to Dr. Munetoshi Maruyama (Kyushu University Museum) for providing important information on the morphology of the coleopterous flagellum and to Dr. Hiroyuki Yoshitomi (Ehime University) for his advices for this study in various ways. Finally I sincerely thank the editors of this book for improving the manuscript of my paper in various ways.

References

Abarbanell NR, Ashe JS (1989) Revision of the species of *Pinophilus* Gravenhorst (Coleoptera: Staphylinidae) of America north of Mexico. Field Zool 54:1–32

Ahn KJ (1996) A review of *Diaulos* Casey (Coleoptera: Staphylinidae: Aleocharinae), with description of a new species and known larvae. Col Bull 50:270–290

Ahn KJ (1997) A review of *Liparocephalus* Mäklin (Coleoptera: Staphylinidae: Aleocharinae), with descriptions of larvae. Pan-Pacif Ent 73:79–92

Ahn KJ (2001) Phylogenetic relationships of the intertidal genus *Halorhadinus* Sawada and key to the genera of the Liparocephalini (Coleoptera: Staphylinidae: Aleocharinae). Insect Syst Evol 32:123–132

Ahn KJ, Ashe JS (1996) A revision of *Rothium* Moore and Legner (Coleoptera: Staphylinidae: Aleocharinae) with a discussion of its phylogenetic relationships. J Kansas Ent Soc 69:234–256

Arai S, Nomura S (2003) Discovery of the tribe Odontalgini (Coleoptera: Staphylinidae; Pselaphinae) in Japan. Ent Sci 6:193–197

Ashe JS (1984) Generic revision of the subtribe Gyrophaenina (Coleoptera: Staphylinidae: Aleocharinae) with review of the described subgenera and major features of evolution. Quaest Ent 20:129–349

Ashe JS (1986) *Phanerota cubensis* and *Phanerota brunnessa* n. sp., with a key to the species of *Phanerota* occurring in Florida (Coleoptera: Staphylinidae). Florida Ent 69:236–245

Ashe JS (1992) Phylogeny and revision of genera of the subtribe Bolitocharina (Coleoptera: Staphylinidae: Aleocharinae). Univ Kansas Sci Bull 54:335–406

Ashe JS (2003) First record and four new species of *Adelarthra* Cameron 1920 from Australia, and a related new genus and species *Notiomerinx zealandica*, from New Zealand (Coleoptera: Staphylinidae: Aleocharinae). In: Cuccodoro G, Leschen RA (eds) Systematics of Coleoptera: Papers celebrating the retirement of Ivan Löbl. Memoirs on Entomology International 17:617–636

Ashe JS, Newton AF Jr (1993) Larvae of *Trichophya* and phylogeny of the tachyporine group of subfamilies (Coleoptera, Staphylinidae) with a review, new species and characterization of the Tachyporinae. Syst Ent 18:267–286

Assing V (1999) A revision of *Othius* Stephens (Coleoptera, Staphylinidae). VIII. Further records, new species, and a new synonym. Linzer biol Beitr 31:661–691

Assing V (2001) The first record of *Dinusa* Saulcy 1864 from Turkey (Coleoptera: Staphylinidae, Aleocharinae). Linzer biol Beitr 33:187–190

Assing V (2002) On some micropterous species of Athetini from Nepal and China (Coleoptera: Staphylinidae, Aleocharinae). Linzer biol Beitr 34:953–969

Assing V (2003) A revision of *Zoosetha* Mulsant & Rey and *Poromniusa* Ganglbauer. II. A new species from Spain and additional records (Coleoptera: Staphylinidae: Aleocharinae). Zootaxa 310:1–8

Assing V (2006) A revision of the Palaearctic species of *Orphnebius* Motschulsky (Coleoptera: Staphylinidae: Paederinae). Ent Prob 36:1–26

Assing V (2012) On the taxonomy and zoogeography of some *Oxypoda* species of the West Palaearctic region (Coleoptera: Staphylinidae: Aleocharinae). Linzer biol Beitr 44:365–399

Assing V (2013) On the Palaearctic and Oriental species of *Scymbalium* and *Micrillus* (Coleoptera: Staphylinidae: Paederinae). Linzer biol Beitr 45:1479–1520

Assing V (2014) A revision of the species of *Pinobius* MacLeay, 1871 of the Oriental, Palaearctic, and Australian Regions (Coleoptera: Staphylinidae: Paederinae). Kol Rund 84:115–191

Assing V, Wunderle P (1995) A revision of the species of the subfamily Habrocerinae (Coleoptera: Staphylinidae) of the world. Rev Suisse Zool 102:307–359

Blackwelder R (1936) Morphology of the coleopterous family Staphylinidae. Smith misc coll 94:1–102

Bordoni A (2002) Xantholinini della Regione Orientale (Coleoptera: Staphylinidae). Classificazione, filogenesi e revisione tassonomica. Museo Regionale di Scienze Naturali Monografie Turin 33:1–998

Bordoni A (2011) Notes on Palaearctic Xantholinini. VI. New species and new records from Caucasus (Coleoptera: Staphylinidae). Fragm ent 43:41–56

Brunke AJ, Chatzimanolis S, Schillhammer H, Solodovnikov A (2016) Early evolution of the hyperdiverse rove beetle tribe Staphylinini (Coleoptera: Staphylinidae: Staphylininae) and a revision of its higher classification. Cladistics 32:427–451

Clarke DJ, Grebennikov VV (2009) Monophyly of Euaesthetinae (Coleoptera: Staphylinidae): phylogenetic evidence from adults and larvae, review of austral genera, and new larval description. Syst Ent 34:346–397

Coiffait H (1972) Coléoptères Staphylinidae de la région Paléarctique occidentale. I. Suppl Nouv Rev Ent 2:1–651

Coiffait H (1974) Coléoptères Staphylinidae de la région Paléarctique occidentale. III. Suppl Nouv Rev Ent 4:1–593

Crowson RA (1981) The biology of the Coleoptera. Academic Press, London

Frisch J (1997) A revision of some West Palaearctic species of *Scopaeus* Erichson (Coleoptera, Staphylinidae, Paederinae). Rev Suisse Zool 104:523–557

Frisch J (1998) A revision of some West Palaearctic species of *Scopaeus* Erichson (Coleoptera, Staphylinidae, Paederinae). Rev Suisse Zool 105:89–124

Frisch J (1999) A revision of the *Scopaeus debilis* species group, with description of a new species from Madagascar (Coleoptera, Staphylinidae, Paederinae). Rev Suisse Zool 106:361–383

Frisch J (2003) A revision of the *Scopaeus laevigatus* species group, with descriptions of ten new species from the East Palaearctic, the Oriental and the Australian regions (Coleoptera, Staphylinidae, Paederinae). In: Cuccodoro G, Leschen RA (eds) Systematics of Coleoptera: Papers celebrating the retirement of Ivan Löbl. Memoirs on Entomology International 17:649–726

Gack C, Peschke K (1994) Spermathecal morphology, sperm transfer and a novel mechanism of sperm displacement in the rove beetle, *Aleochara curtula* (Coleoptera, Staphylinidae). Zoomorphology 114:227–237

Gack C, Peschke K (2005) 'Shouldering' exaggerated genitalia: a unique behavioural adaptation for the retraction of the elongate intromittent organ by the male rove beetle (*Aleochara tristis* Gravenhorst). Biol J Linn Soc 84:307–312

Gusarov V (2003) Revision of some types of North American aleocharines (Coleoptera: Staphylinidae: Aleocharinae), with synonymic notes. Zootaxa 353:1–134

Hammond PM (1973) Notes on British Staphylinidae 3. The British species of *Sepedophilus* Gistel (*Conosomus* auctt.) Ent Mon Mag 108:130–165

Hemming BS (2003) Insect development and evolution. Cornell University Press, New York

Herman LH (1972) A revision of the rove-beetle genus *Charhyphus* (Coleoptera, Staphylinidae, Phloeocharinae). Am Mus Nov 2496:1–16

Herman LH (1975) Revision and phylogeny of the monogenetic subfamily Pseudopsinae for the world (Staphylinidae, Coleoptera). Bull Am Nat Hist 155:245–317

Herman LH (1981) Revision of the subtribe Dolicaonina of the New World, with discussions of phylogeny and the old world genera (Staphylinidae, Paederinae). Bull Am Mus Nat Hist 167:327–520

Hlaváč P, Nomura S (2001) A taxonomic revision of Tyrini of the Oriental region 1. *Paralasinus* (Coleoptera, Staphylinidae, Pselaphinae), a new genus of Tyrina from Indochina. Elytra 29:163–174

Hoebeke ER, Ashe JS (1994) New species of *Autalia* Leach 1819 from the Neotropics (Coleoptera Staphylinidae Aleocharinae), with new distributional data for *A. phricotrichosa* Hoebeke 1988 and key to the Neotropical species of *Autalia*. Tropical Zool 7:191–208

Hoshina H, Sugaya H (2003) New records of the genera *Scaphobaeocera* and *Scaphoxium* (Coleoptera: Staphylinidae) from the Ryukyus, Japan, with descriptions of two new species. Ent Rev Jpn 58:35–41

Hwang WS, Ahn KJ (2000) Taxonomy of the Korean Oxyporinae (Insecta, Coleoptera, Staphylinidae). Korean J Syst Zool 16:191–202

Irmler U (1981) Descriptions of new neotropical *Holotrochus* and a key to the species of the genus (Coleoptera: Staphylinidae). Col Bull 35:379–397

Irmler U (1985) Neue Arten der Gattungen *Aueucamptus* und *Thoracophorus* (Col., Staphylinidae) aus der Neotropis. Ent Blät 81:41–58

Irmler U (2005) New neotropical species of the genus *Holotrochus* (Coleoptera: Staphylinidae: Osoriinae). Sci Pap Nat Hist Mus Univ Kan 36:1–18

Jałoszyński P (2010) Cephenniini of the Philippines. Part 3. New species of *Cephennomicrus* Reitter and *Hlavaciellus* Jałoszyński from Palawan (Coleoptera: Staphylinidae: Scydmaeninae). Genus 21:13–20

Jałoszyński P, Nomura S (2009) The Cephenniini (Coleoptera, Scydmaenidae) of Vietnam. Bull Natl Mus Sci Ser A 35:167–225

Jeannel R (1959) Révision des Psélaphides de l'Afrique intertropicale. Ann Mus r Congo Belge Tervuren Ser 8 Sci Zool 75:1–742

Kanao T, Maruyama M, Hashim R (2011) A new genus of Compactopediina (Coleoptera: Staphylinidae) associated with *Longipeditermes* (Isoptera: Termitidae) in Peninsular Malaysia with a key to and phylogenetic analysis of all genera of the subtribe. Insect Syst Evol 42:349–364

Kim HJ, Ahn KJ (2000) A taxonomic review of the genus *Tachinus* Gravenhorst in Korea (Insecta: Coleoptera: Staphylinidae). Korean J Env Biol 18:125–131

Kishimoto T (2000) A new *Trigonurus* (Coleoptera: Staphylinidae, Trigonurinae) discovered in Sichuan, China. Elytra 28:305–309

Klimaszewski J (1984) A revision of the genus *Aleochara* Gravenhorst of America north of Mexico (Coleoptera: Staphylinidae, Aleocharinae). Mem Ent Soc Can 129:1–211

Kocian M (1996) A revision of Western Palearctic species of the genus *Ischnosoma* Stephens (Coleoptera: Staphylinidae, Tachyporinae). Acta Univ Carol Biol 40:241–299

Lawrence JF, Britton EB (1994) Australian beetles. Melbourne University Press, Carlton

Leschen RAB, Löbl I (2005) Phylogeny and classification of Scaphisomatini (Staphylinidae: Scaphidiinae) with notes on mycophagy, termitophyly, and functional morphology. Col Soc Monogr 3:1–63

Löbl I (2002) New species of *Scaphisoma* Leach (Coleoptera: Staphylinidae: Scaphidiinae) from Mt. Wilhelm, Papua New Guinea. Acta Zool Hung 48:181–189

Löbl I (2012) On Taiwanese species of *Baeocera* Erichson (Coleoptera: Staphylinidae: Scaphidiinae). Zool Stud 51:118–130

Löbl I (2015) On the Scaphidiinae (Coleoptera: Staphylinidae) of the Lesser Sunda Islands. Rev Suisse Zool 122:75–112

Löbl I, Burckhardt D (1988) *Cerapeplus* gen. n. and the classification of micropeplids (Coleoptera: Micropeplidae). Syst Ent 13:57–66

Löbl I, Calame FG (1996) Taxonomy and phylogeny of the Dasycerinae (Coleoptera: Staphylinidae). J Nat Hist 30:247–291

Löbl I, Mattila J (2010) *Euplectus lapponicus* (Coleoptera: Staphylinidae: Pselaphinae), a new species from boreal Finland. Ent Fenn 21:181–186

Makranczy G, Schülke M (2001) Typenstudien an den mitteleuropäischen Vertreten der Artengruppe des *Thinobius linearis* Kraatz, 1857 (Coleoptera, Staphylinidae, Oxytelinae). Ent Blät 97:185–193

Maruyama M (2000) A revision of the myrmecophilous genus *Aspidobactrus* (Coleoptera: Staphylinidae: Aleocharinae). Sociobiology 35:149–173

Maruyama M (2004a) Redescription of the genus *Creochara* (Coleoptera: Staphylinidae: Aleocharinae: Aleocharini) and its systematic position. Can Ent 136:621–637

Maruyama M (2004b) Taxonomy of the myrmecophilous genus *Myrmecopella* Kistner with descriptions of seven new species (Coleoptera: Staphylinidae: Aleocharinae). Kol Rund 74:201–219

Maruyama M (2006) Revision of the Palearctic species of the myrmecophilous genus *Pella* (Coleoptera: Staphylinidae: Aleocharinae). Nat Sci Mus Monogr Tokyo 32:1–207

Maruyama M (2008) *Giraffaenictus eguchii* (Coleoptera, Staphylinidae, Aleocharinae), a new genus and species of fully myrmecophile from a colony of *Aenictus binghami* (Hymenoptera, Formicidae, Aenictinae) in Vietnam. Esakia 48:51–56

Maruyama M, Hlavác P (2003) A new genus and species of myrmecophilous Aleocharini from Morocco (Coleoptera, Staphylinidae, Aleocharinae), associated with *Aphaenogaster* (Hymenoptera: Formicidae: Myrmicinae). Ann Soc ent Fr (new ser) 39:225–228

Maruyama M, Klimaszewski J (2004) A new species of the myrmecophilous genus *Goniusa* (Coleoptera, Staphylinidae, Aleocharinae) from Canada. Elytra 32:315–320

Maruyama M, Matsumoto T, Itioka T (2011) Rove beetles (Coleoptera: Staphylinidae) associated with Aenictus laeviceps (Hymenoptera: Formicidae) in Sarawak: strict host specificity, and first myrmecoid Aleocharini. Zootaxa 3102:1–26

Maruyama M, Sugaya H (2002) A new species of *Tetrasticta* (Coleoptera, Staphylinidae, Aleocharinae) from Japan and Taiwan. Jpn J Syst Ent 8:17–21

Matsumura Y, Yoshizawa K (2012) Homology of the internal sac components in the leaf beetle subfamily Criocerinae and evolutionary novelties related to the extremely elongated flagellum. J Morph 273:507–518

Mckenna DD, Farrell BD, Caterino MS, Farnum CW, Hawks DC, Maddison DR, Seago AE, Short AEZ, Newton AF, Thayer MK (2015) Phylogeny and evolution of Staphyliniformia and Scarabaeiformia: forest litter as a stepping stone for diversification of nonphytophagous beetles. Syst Ent 40:35–60

Naomi S (1986) Taxonomic study on the subfamily Osoriinae (Coleoptera, Oxytelidae) from Japan, I. Elytra 14:33–42

Naomi S (1990) Comparative morphology of the Staphylinidae and the allied groups (Coleoptera, Staphylinoidea) XI. Abdominal glands, male genitalia and female spermatheca. Jpn J Ent 58:16–23

Naomi S (1996) Revision of the genus *Lispinus* Erichson and *Neolosus* Blackwelder (Coleoptera: Staphylinidae) from Japan. Jpn J Ent 64:762–771

Naomi S (1997a) A revision of the genus *Nacaeus* Blackwelder (Coleoptera: Staphylinidae) from Japan. Jpn J Ent 65:127–142

Naomi S (1997b) Taxonomic study on *Stenus wasabi* Hromádka and its allied species (Coleoptera; Staphylinidae; Steninae), with discussion on the internal structure of the aedeagus. Jpn J Ent 65:600–611

Naomi S (2006a) Taxonomic revision of the genus *Stenus* Latreille, 1797 (Coleoptera, Staphylinidae, Steninae) of Japan: species group of *S. indubius* Sharp. Jpn J Syst Ent 12:39–120

Naomi S (2006b) Taxonomic revision of the genus *Stenus* Latreille, 1797 (Coleoptera, Staphylinidae, Steninae) of Japan: species group of S. (*Hypostenus*) *rufescens* Sharp. Nat Hist Res Chiba Spec Issue 9:1–81

Naomi S (2010) Descriptions of two new species of the genus *Stenus* Latreille (Coleoptera: Staphylinidae: Steninae) from Chubu District, Central Japan. Nat Hist Res Chiba 11:35–40

Naomi S, Irmler U (2012) First record of the genus *Allotrochus* Fagel, 1955 (Coleoptera: Staphylinidae: Osoriinae) from the Oriental region, with description of the new species A. takahashii from Ishigaki Is., Nansei Islands, Japan. Biogeography 14:105–110

Naomi S, Maruyama M (1997) Revision of the genus *Sepedophilus* Gistel (Coleoptera, Staphylinidae, Tachyporinae) from Japan: species group of *S. glabratus* and *S. exiguus*. Ent Rev Jpn 52:61–71

Naomi S, Maruyama M (1998) A revision of the genus *Sepedophilus* Gistel (Coleoptera, Staphylinidae, Tachyporinae) from Japan: species group of *S. pedicularius*. Jpn J Syst Ent 4:51–75

Naomi SI, Nomura S, Puthz V (2017) The subfamily Steninae Macleay, 1825 (Coleoptera: Staphylinidae) of Japan. Part 1. *Dianous* and *Stenus* (*S. comma* group to *S. guttalis* group). Nat Sci Mus Monogr Tokyo 46:1–339

Nichols SW (1989) The Torre-Bueno glossary of entomology. New York Entomological Society, New York

Nomura S (1991) Systematic study on the genus *Batrisoplisus* and its allied genera from Japan (Coleoptera, Pselaphidae). Esakia 30:1–462

Nomura S (2001a) A taxonomic revision of the Japanese species of the genus *Pselaphogenius* (Coleoptera, Staphylinidae, Pselaphinae). Part 3, Species from central Honshu. Mem Natn Sci Mus Tokyo 37:279–291

Nomura S (2001b) Descriptions of two new species of the Clavigerine genus *Articerodes* (Coleoptera, Staphylinidae, Pselaphinae) from the Ogasawara Islans, Japan. Elytra 29:343–351

Nomura S (2010) A taxonomic revision of Asian genera and species of the tribe Dimerini (Coleoptera, Staphylinidae, Pselaphinae). Jpn J syst Ent 16:221–247

Nomura S, Sakcoowong W, Chanpaisaeng J (2011) Three new species of the genus *Megatyrus* (Coleoptera, Staphylinidae, Pselaphinae) from Thailand. Spec Publ Jpn Soc Scarab 1:121–129

Ogawa R, Löbl I (2016) A review of the genus *Xotidium* Löbl, 1992 (Coleoptera, Staphylinidae, Scaphidiinae), with descriptions of five new species. Dt Ent Zeit 63:155–169

Pace R (1999) Nuove specie del genere *Leptusa* Kraatz raccolte da Manfred Kahlen. Veröff Tiroler Land Ferd 79:207–214

Pace R (2009) Aleocharinae del cile raccolte dal Dr. P. M. Giachino (Coleoptera, Staphylinidae). Lavori Soc Ven Sc Nat 34:25–32

Pace R (2010) Thamiaraeini, Lomechusini, Oxypodini, Hoplandriini e Aleocharini di Taiwan (Coleoptera, Staphylinidae). Boll Mus Civ Stor Nat Verona 34:19–54

Peschke K (1978) Funktionsmorphologische Untersuchungen zur Kopulation von *Aleochara curtula* Goeze (Coleoptera, Staphylinidae). Zoomorphologie 89:157–184

Puthz V (1973) On some Neotropical Euaesthetinae (Coleoptera, Staphylinidae). Stud Neotr Fauna 8:51–73

Puthz V (1974) Bemerkungen über die Gattung *Edaphus* Motschulsky und über Kistnerische *Edaphus*-Arten nebst neuen und alten afrikanischen Euaesthetinen. Rev Zool Afr 88:749–770

Puthz V (1976) Zwei neue Gattungssynonyme und ein Nomen novum bei den Euaesthetinen (Coleoptera: Staphylinidae). Ent Scand 7:239–240

Puthz V (1977) Die Gattung *Octavius* Fauvel (Coleoptera: Staphylinidae) weltweit verbreitet! Opusc Zool Budap 14:105–126

Puthz V (1978) Bemerkungen über *Schatzmayrina* Koch (Coleoptera, Staphylinidae). Ent Blät 74:47–51

Puthz V (1979) Eine neue Euaesthetinengattung aus dem Himalaya (Coleoptera, Staphylinidae). Reichenbachia 17:9–14

Puthz V (1985a) Neue und alte neotropische *Edaphus*-Arten aus dem Genfer Museum (Coleoptera, Staphylinidae). Rev Suisse Zool 92:351–357

Puthz V (1985b) Neue *Octavius*-Arten aus Nordindien und aus dem Himalaya-Gebiet (Coleoptera: Staphylinidae). Revue suisse Zool 92:697–705

Puthz V (1986a) Eine neue Art der Gattung *Edaphosoma* Scheerpeltz. Reichenbachia 23:135–138

Puthz V (1986b) Fünf neue *Edaphus*-Arten aus Afrika (Staphylinidae, Coleoptera). Philippia 5:311–317

Puthz V (1986c) Neue Arten der Gattung *Octavius* Fauvel, 1872, aus Südafrika nebst einer Bestimmungstabelle der in der Äthiopis lebenden Gattungsvertreter (Coleoptera: Staphylinidae). Ann Trans Mus 34:185–201

Puthz V (1988a) Three new *Stenaesthetus* species from Ethiopia (Insecta, Coleoptera: Staphylinidae). Reichenbachia 26:9–13

Puthz V (1988b) Die Steninen von Borneo II (Coleoptera, Staphylinidae). Neue und alte Arten, vorwiegend aus den Ausbeuten des Genfer Naturhistorischen Museums. Rev Suisse Zool 95:631–679

Puthz V (1989a) Zwei neue Euaesthetinen von der Elfenbeinküste (Coleoptera, Staphylinidae). Ent Blät 85:75–77

Puthz V (1989b) The male of *Stictocranius mariae* (Hatch) (Coleoptera, Staphylinidae). Ent Blät 85:165–166

Puthz V (1989c) Eine neue *Octavius* -Art aus Sumatra (Coleoptera, Staphylinidae). Rev Suisse Zool 96:115–116

Puthz V (1992) Neue und alte afrikanische *Edaphus*-Arten (Insecta, Coleoptera, Staphylinidae: Euaesthetinae). Ent Abhand 55:13–33

Puthz V (1998) Neuweltliche *Euaesthetus*-Arten (Staphylinidae, Coleoptera). Phillipia 8:223–244

Puthz V (2000a) The genus *Dianous* Leach in China (Coleoptera, Staphylinidae). Rev Suisse Zool 107:419–559

Puthz V (2000b) Neue und alte neotropische Arten der Gattung *Stenus* Latreille (Staphylinidae, Coleoptera). Philippia 9:165–213

Puthz V (2000c) Aptere *Stenus*-Arten aus den mexikanischen Hochgebirgen (Coleoptera: Staphylinidae). Mitt internat-entomol Ver 25:103–127

Puthz V (2002) Revision der Gattung *Tamotus* Schaufuss, 1872 (Coleoptera, Staphylinidae). Z Arb Gem Öst Ent 54:115–125

Puthz V (2003a) Revision of the genus *Fenderia* Hatch, 1957 (Coleoptera, Staphylinidae). In: Cuccodoro G, Leschen RA (eds) Systematics of Coleoptera: Papers celebrating the retirement of Ivan Löbl. Memoirs on Entomology International 17:637–648

Puthz V (2003b) Neue und alte Arten der Gattung *Stenus* Latreille aus China (Insecta: Coleoptera: Staphylinidae: Steninae). Ent Abhand 60:139–159

Puthz V (2008a) *Stenus* Latreille und die segenreiche Himmelstochter (Coleoptera, Staphylinidae). Linzer biol Beitr 40:137–230

Puthz V (2008b) *Kiwiaesthetus*, a new genus of Euaesthetinae from New Zealand (Coleoptera: Staphylinidae). Z Arb Gem Öst Ent 60:59–69

Puthz V (2010) Die Gattung *Edaphosoma* Scheerpeltz, 1976 in China (Coleoptera, Staphylinidae). Ent Blät 106:289–306

Puthz V (2011a) Afrikanische *Stenaesthetus*-Arten (Coleoptera: Euaesthetinae). Linzer biol Beitr 43:793–812

Puthz V (2011b) Neue und alte Euaesthetinen (Coleoptera: Staphylinidae). Z Arb Gem Öst Ent 63:13–31

Puthz V (2012a) Über die neuweltlichen *Megalopinus*-Arten (Coleoptera: Staphylinidae). Linzer biol Beitr 44:613–834

Puthz V (2012b) Zwei neue Arten der Gattung *Nothoesthetus* Saíz aus Chile (Coleoptera: Staphylinidae). Mitt int ent Ver Frankfurt a M 37:159–162

Puthz V (2012c) Über die *Megalopinus*-Arten der Orientalis (Coleoptera: Staphylinidae). Linzer biol Beitr 44:1373–1430

Puthz V (2013) Übersicht über die orientalischen Arten der Gattung *Stenus* Latreille 1797 (Coleoptera, Staphylinidae). Linzer biol Beitr 45:1279–1470

Puthz V (2014) Nordamerikanische Arten der Gattung *Euaesthetus* Gravenhorst (Coleoptera, Staphylinidae). Linzer biol Beitr 46:845–876

Sharp D, Muir F (1912) The comparative anatomy of the male genital tube in Coleoptera. Trans Ent Soc Lond 1912:477–642

Shavrin A (2012) On the taxonomic placement of *Geodromicus pusillus* Coiffait, 1983 (Coleoptera: Staphylinidae: Omaliinae). Zootaxa 3552:66–68

Shimada T (2002) A new *Aploderus* (Coleoptera, Staphylinidae, Oxytelinae) discovered in central Japan. Spec Bull Jpn Soc Coleopt 5:233–239

Smetana A (1973) Die *Leptusa*-Arten der Tschechoslovakischen Republik einschließlich Karpatorußlands (Col., Staphylinidae). Stutt Beitr Naturk Ser A Biol 255:1–46

Snodgrass RE (1935) Principles of insect morphology. McGraw-Hill, New York

Steel WO (1964) Insects of Campbell Island. Coleoptera: Staphylinidae. Pac Ins Monogr 7:340–375

Steel WO (1966) A revision of the staphylinid subfamily Proteininae (Coleoptera) I. Trans R Ent Soc Lond 118:285–311

Thayer MK (1987) Biology and phylogenetic relationships of *Neophonus bruchi*, an anomalous south Andean staphylinid (Coleoptera). Syst Ent 12:389–404

Thayer MK (1997) *Proglypholoma aenigma* gen. et sp. nov., *Glypholoma* spp. nov. and new records, and a phylogenetic analysis of Glypholomatinae (Coleoptera: Staphylinidae). Ann Zool Warszawa 47:157–174

Thayer MK, Newton AF Jr (1978) Revision of the south temperate genus *Glypholoma* Jeannel, with four new species (Coleoptera: Staphylinidae: Omaliinae). Psyche 85:25–63

Williams SA (1976) The genus *Oligota* (Coleoptera, Staphylinidae) in New Zealand. New Zeal J Zool 3:247–255

Williams SA (1978) The genus *Oligota* Mannerheim (Col., Staphylinidae) in the Ethiopian region. Ent Mon Mag 114:177–190

Wunderle P, Assing V (2000) A revision of the European species of *Hygropetrophila* Bernhauer (Coleoptera, Staphylinidae, Aleocharinae). Ent Blät 96:31–40

Yamamoto S, Maruyama M (2012) Revision of the seashore-dwelling subgenera *Emplenota* Casey and *Triochara* Bernhauer (Coleoptera: Staphylinidae: genus Aleochara) from Japan. Zootaxa 3517:1–52

Zanetti A (2014) Taxonomic revision of North American *Eusphalerum* Kraatz, 1857 (Coleoptera, Staphylinidae, Omaliinae). Ins Mundi 379:1–80

Morphological Diversity of Immature Scydmaeninae

14

Paweł Jałoszyński

Abstract

Larvae of ant-like stone beetles are exceptionally poorly known. Scydmaeninae comprises over 5000 species, but the immature stages have been described for less than 0.5% of them; the pupa has been illustrated for four species, and the first larval instar for only one species. In several tribes, larvae still remain unknown, and many descriptions are inaccurate or poorly illustrated. Chaetotaxic structures have been coded in only six species. Our knowledge of larval Scydmaeninae is so fragmentary that even for the largest, most common, and abundant genus, *Euconnus* Thomson (nearly 2500 nominal species!), the immature stages have never been adequately described. Known larvae of Scydmaeninae show a great diversity of body shapes and structures, more than expected within one subfamily of Staphylinidae. Known larvae of Eutheiini, Scydmaenini, Glandulariini, Mastigini, Clidicini, and Leptomastacini have a ten-segmented abdomen, whereas those of Cephenniini have only nine abdominal segments. Larvae of Eutheiini and Mastigitae are campodeiform, subcylindrical, or flattened, resembling those of other subfamilies of Staphylinidae. However, larvae of *Scydmaenus* s. str. are nearly onisciform, with demarcated laterotergites of thoracic segments, densely and asymmetrically covered with setae and microtrichia. Larvae of Glandulariini are also onisciform, but without demarcated laterotergites; they are sparsely setose and have nearly smooth cuticles. Larvae of some tribes have short, unsegmented urogomphs, whereas in others the urogomphs are absent. This chapter summarizes known data on the immature stages of Scydmaeninae, including their biology, indicating major problems and future directions.

14.1 Introduction

Ant-like stone beetles were treated as a separate family, Scydmaenidae, for nearly 200 years until Grebennikov and Newton (2009) placed this group within Staphylinidae. This is a large subfamily, currently comprising over 5300 species classified into 1 extinct and 3 extant supertribes, including 10 tribes and over 100 genera. Within rove beetles, scydmaenines were hypothesized to represent a sister group of a clade Steninae + Euaesthetinae (Grebennikov and Newton 2009) or Solieriinae (Thayer et al. 2012; McKenna et al. 2015), although their true relationships remain unclear. Within Scydmaeninae, the

P. Jałoszyński (✉)
Museum of Natural History, University of Wrocław, Wrocław, Poland
e-mail: scydmaenus@yahoo.com

O. Betz et al. (eds.), *Biology of Rove Beetles (Staphylinidae)*,
https://doi.org/10.1007/978-3-319-70257-5_14

monophyly of Cephenniitae was well supported, but the largest supertribe, Scydmaenitae, is almost certainly not monophyletic (Jałoszyński 2012a, 2014). Larval morphological characteristics are expected to help solve still problematic relationships within Scydmaeninae, but our knowledge of immature ant-like stone beetles is surprisingly scarce, considering there has been two centuries of research.

The published record of immature Scydmaeninae began with the most spectacular misidentification ever made regarding beetles. Tömösvary (1882) described a family Anisosphaeridae in Zygentoma (then Thysanura) to accommodate a new genus, *Anisosphaera*, an odd-looking arthropod discovered in Central Europe. Silvestri (1899) did not agree that *Anisosphaera* belonged to the same group as the silverfish and established a new insect order, Anisosphaeridia, discussing its possible affinities to springtails (Collembola). Tömösvary's description and illustrations were fairly accurate, and Bergroth (1899) recognized in *Anisosphaera* a larva belonging to a common western Palaearctic scydmaenine genus, *Cephennium* Müller & Kunze (Cephenniini).

Larvae of merely 13 genera (out of over 100) have been described or illustrated so far, accounting for less than 0.5% of known species. Some figures, especially those in older publications, are highly simplified and inaccurate, and some identifications require verification. Therefore, even within these 13 genera whose larvae are known, only a few have been characterized with satisfactory accuracy. There is only one description based on immature stages obtained ex ovo by rearing adults; consequently, details of the life cycle are known for one species. Moreover, all immature stages, from egg to pupa, have only been described for one species (De Marzo 1983, 1984). Larvae of four species have been unambiguously identified by rearing, including two described on the basis of the shed last larval skin, obtaining the pupa and, subsequently, the adult (Jałoszyński 2013, 2016). All remaining descriptions were based on identifications made by associating larvae with adults collected in nature. Detailed descriptions of pupae have been published for only two species (De Marzo 1984; Jałoszyński 2012b); pupae of two more species are illustrated by photographs, showing only some details (Jałoszyński 2013, 2016). Details of the chaetotaxy were coded using modern methods for only six species (Wheeler and Pakaluk 1983; Jałoszyński 2013, 2015a, b, 2016; Jałoszyński and Kilian 2016). Not surprisingly, available data on larval biology are exceptionally scarce and mostly restricted to occasional observations of feeding on soil arthropods (Jałoszyński and Kilian 2012; Jałoszyński 2016).

14.2 Eggs

Only one species oviposited under laboratory conditions; its eggs have also been found in nature (De Marzo 1983). Eggs of *Palaeostigus pilifer* (Mastigini) are orange, elongate, 1.5 × 1.0 mm, laid in clusters of 2–9, and covered with a white, granulose secretion from the female's abdominal gland. The eggs are relatively small in relation to the female's body, which is about 6 mm long. According to De Marzo (1983), oviposition takes place in autumn; the eggs are deposited under stones, and larvae hatch a few days later. The number of eggs laid in one egg cluster is often larger than the number of ovarioles, which was found to be only two in each ovary in South African (Jałoszyński et al. 2015) and European (Jałoszyński, unpublished observations) species of Mastigini.

14.3 Larval Morphology

Larvae of Scydmaeninae (Figs. 14.1 and 14.2) show such a diversity of structures that to date it has not been possible to provide a clear-cut diagnosis of the subfamily based on larval morphology. The only constant structures shared by all known larvae are the labrum indistinguishably fused with frontoclypeus, broadly separated insertions of labial palps, and annuliform spiracles. The known range of structural

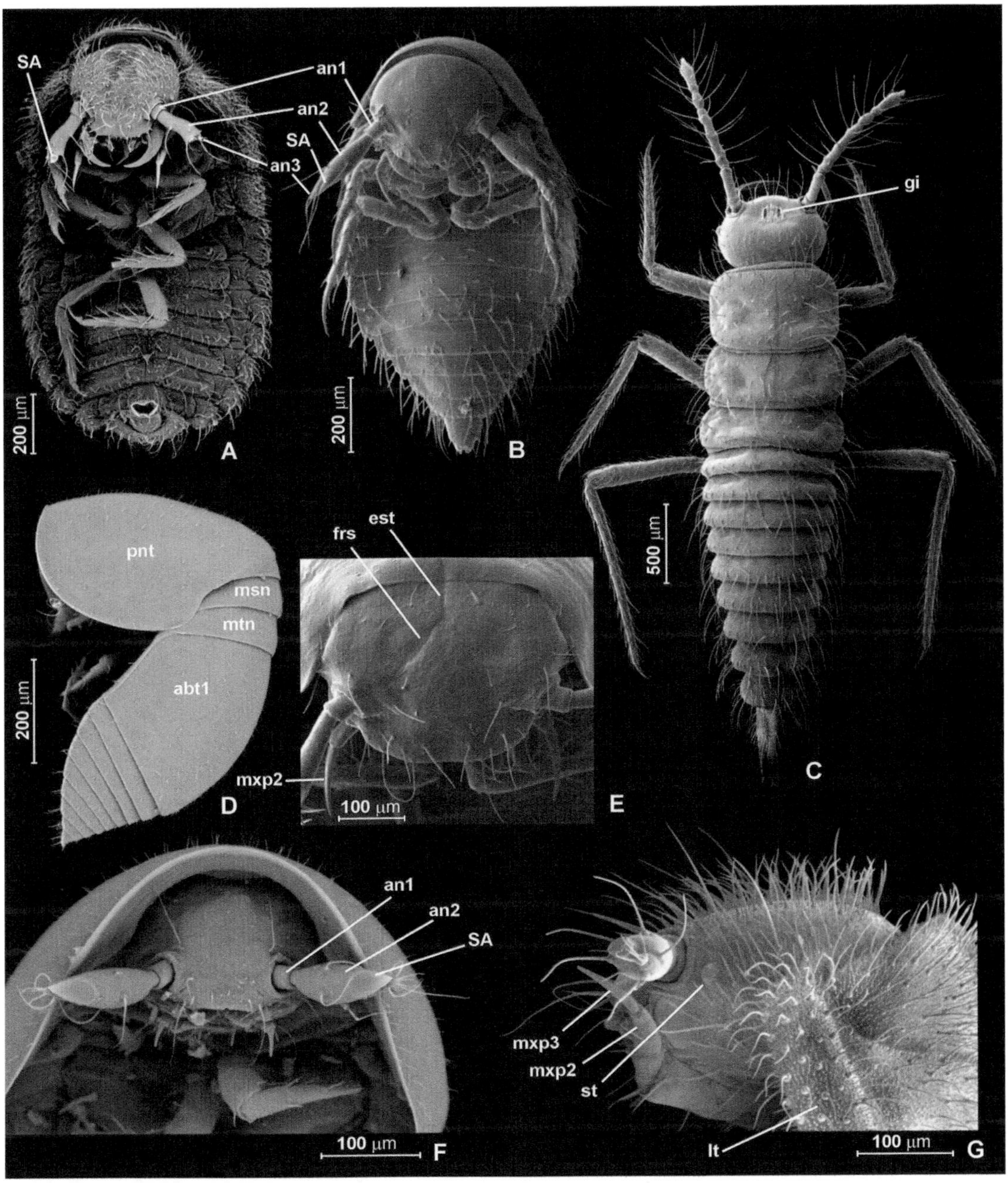

Fig. 14.1 Scanning electron micrographs of Scydmaeninae larvae. (**a**) Mature larva of *Scydmaenus tarsatus* (Scydmaenini), ventral view. (**b**) Mature larva of *Stenichnus* sp. (Glandulariini), ventral view. (**c**) Mature larva of *Palaeostigus* sp. (Mastigini), dorsal view. (**d**) Mature larva of *Cephennium* sp. (Cephenniini), lateral view. (**e**) Head of *Stenichnus* sp. larva, anterodorsal view. (**f**) Head and prothorax of *Cephennium* sp. larva, anteroventral view. (**g**) Head of *Scydmaenus tarsatus* larva, lateral view. Abbreviations: *abt1* abdominal tergite I, *an1–3* antennomere I–III, *est* epicranial stem, *frs* frontal suture, *gi* glandular impression, *lt* laterotergite, *msn* mesonotum, *mtn* metanotum, *mxp2–3* maxillary palpomere II–III, *pnt* pronotum, *SA* sensory appendage, *st* stemma

Fig. 14.2 Scydmaeninae larvae and pupae, living individuals. (**a**–**c**) Larvae of *Stenichnus* spp. (Glandulariini) feeding on Oribatida. (**d**) Larva of *Scydmaenus tarsatus* (Scydmaenini) feeding on *Ceratophysella* springtail. (**e**) Larva of *Scydmaenus rufus* (Scydmaenini). (**f**) Larva of *Palaeostigus palpalis* (Mastigini) feeding on beetle pupa. (**g**) Larva of *Stenomastigus longicornis* (Mastigini). (**h**) Prepupa of *Stenichnus godarti* (Glandulariini). (**i**) Freshly emerged pupa of *Stenichnus godarti*. (**j**–**k**) Pupa of *Stenichnus godarti* shortly before emergence of the adult

diversity comprises onisciform (Fig. 14.1a) or campodeiform (Fig. 14.1c) body shape; head with or without demarcated "neck"; epicranial stem and frontal sutures present (Fig. 14.1e) or absent (Fig. 14.1f); antennae short and clubbed (Fig. 14.1a, b, f) or long and not thickened (Fig. 14.1c); dome-shaped, subconical, or even spatulate sensory appendage on antennomere II; nasale with or without teeth; maxillae with undivided mala or with distinct galea and lacinia; abdomen with ten (Fig. 14.1a–c) or nine segments (Fig. 14.1d); and urogomphs present or absent.

Even the number of functional spiracles differs among taxa. Larval characteristics are therefore given for each tribe separately.

14.3.1 Eutheiini

This tribe currently comprises six extant genera. Incomplete descriptions of larvae representing only two of these are available: *Eutheia* Stephens is illustrated for the European *E. schaumi* Kiesenwetter, whereas the larva of the sympatric *E. linearis* Mulsant & Rey was only mentioned as similar (Brown and Crowson 1980). An undetermined Nearctic species of *Veraphis* Casey has also been described (Newton 1991).

Larvae are strongly elongate and slender, nearly parallel-sided, and sparsely setose; pigmentation was not described. The head capsule is subrectangular or suboval. The epicranial suture in *Eutheia* was described by Brown and Crowson (1980) as indistinct, but the head was illustrated as entirely lacking stem and frontal arms of epicranial suture. In *Veraphis*, the epicranial stem is present, shorter than half the length of the head; frontal sutures are distinct but incomplete and obliterated anteriorly (Newton 1991). The anterior margin of the frontolabral region (nasale) has fine paired teeth; stemmata are absent. The antenna has short antennomere I, elongate antennomere II that broadens distally, and tiny antennomere III about as long as dome-shaped but broader than long accessory appendage of antennomere II. Mandibles are falciform, curved and pointed, with two tiny submedian mesal teeth (*Eutheia*) or several submedian mesal and dorsomesal denticles (*Veraphis*). Stipital projection of maxilla has two apical lobes, one short and the other several times longer, possibly representing galea and lacinia, both with sparse bristles. Maxillary palp for *Eutheia* was illustrated as two-segmented (Brown and Crowson 1980), which might have been a result of overlooking very short, annuliform palpomere I, illustrated for *Veraphis* (Newton 1991). In both genera, palpomere III is strongly elongate with rounded apex. Labial palp is short and composed of two palpomeres, of which II is distinctly longer than I, and has a rounded apex. Thoracic tergites were illustrated only for *Eutheia* (Brown and Crowson 1980) as subtrapezoidal (prothorax) or subrectangular (meso- and metathorax), transverse, and undivided. Abdomen is composed of ten segments; all abdominal segments in *Eutheia* are transverse; in *Veraphis* only tergites VIII–X were illustrated (Newton 1991), of which VIII and IX are strongly transverse and X weakly transverse. Urogomphs are present, composed of one segment fused with lateroposterior margin of abdominal segment IX, slightly upturned and pointed. Abdominal segment X is strongly elongate. Legs are short. Spiracles (presumably nine pairs) are present lateroventrally on mesothorax and abdominal segments I–VIII.

14.3.2 Cephenniini

Larval characters are known for 2 genera out of 19 currently recognized within this tribe. Immature stages of the European *Cephennium carnicum* Reitter, *C. gallicum* Ganglbauer, *C. laticolle* (Aubé), *C. majus* Reitter, and *C. thoracicum* (Müller & Kunze) have been adequately described and illustrated (Tömösvary 1882; Peyerimhoff 1899; Jeannel 1909; Scholz 1926; Ionesco 1937; Paulian 1941; Schuster 1966; Brown and Crowson 1980; Schmid 1988a, b; Jałoszyński and Beutel 2012). Additionally, illustrations of the general habitus and a few morphological details of an undetermined Japanese and Australian species of *Cephennodes* Reitter were published, respectively, as an unidentified genus by Morimoto and Hayashi (1986) and as *Coatesia* Lea, a junior synonym of *Cephennodes*, by Newton (1991).

Larvae (Fig. 14.1d, f) are onisciform; creamy white, yellowish, or brownish, suboval; and typically capable of curling into a compact ball (illustrated by Ionesco 1937). Setae are sparse, unmodified, and often very short. The head (Fig. 14.1f) is prognathous, strongly declined, and lacking stemmata. The epicranial stem, if present, is very short with frontal sutures variously distinct; nasale lacks teeth. The antenna has short antennomere I, long antennomere II that is broadest near the distal third or middle; and tiny antennomere III that is typically shorter (sometimes only slightly) than the asymmetrical, subconical, and pointed sensory appendage of antennomere II. Mandibles are short and subtriangular, with strongly curved distal portion and numerous denticles along outer margin. Maxillary mala forms a single protruding apical lobe with sparse bristles, maxillary palp with short palpomeres I and II, and elongate palpomere III. Prementum has one large median adhesive pad and a pair of small lateral suckers; labial palps are short, with elongate palpomere II. Ecdysial line on thoracic tergites is variously distinct (in *Cephennium* typically faint, barely discernible). Pronotum is extremely enlarged in *Cephennium* (Fig. 14.1d), usually much longer than meso- and metanotum combined, but less enlarged in *Cephennodes*; meso- and metanotum are very short and strongly transverse.

Nine abdominal segments are visible. The abdomen has greatly enlarged tergite I in *Cephennium* (usually longer than all remaining tergites combined); it is unmodified in *Cephennodes*; urogomphs absent; abdominal sternites are undivided. Legs are short and robust. Spiracles are annular, one pair ventrally on mesothorax and an unclear number on proximal abdominal segments. Ionesco (1937) listed four abdominal pairs of spiracles in *Cephennium*, but there are spiracles in small impressions located lateroventrally on segment I and laterodorsally on segments II–IV. At least on segments V–VII, in the same places, there are smaller, presumably rudimentary, spiracles lacking the surrounding impression (Jałoszyński, unpublished observations).

14.3.3 Glandulariini

Glandulariini is the largest tribe of Scydmaeninae, currently comprising 71 extant genera. Larval characters have been illustrated for only four genera: *Stenichnus* Thomson, *Scydmoraphes* Reitter, *Neuraphes* Thomson, and *Euconnus* Thomson. However, some species were identified only putatively.

The first truly modern description of scydmaenine larvae was given for a species of this tribe, with detailed mapping and coding of chaetotaxic characteristics (Wheeler and Pakaluk 1983). The study was focused on the larva of Nearctic *Stenichnus turbatus* (Casey) and became a standard description for comparative studies published later, with detailed redescriptions of the immature western Palaearctic *St. collaris* (Müller & Kunze) and *St. godarti* (Latreille) (Jałoszyński 2013, 2016). In all these papers, modern methods of coding chaetotaxic structures were applied and, for the first time, provided data on the serial and interspecific homology within Scydmaeninae.

Paulian (1941) was the first to present data on selected structures of an immature glandulariine species, *St. collaris*. Larvae of the same *Stenichnus* were also studied to a limited extent by Kasule (1966), Brown and Crowson (1980), and Schmid (1988a, b). Franz (1965) gave simplified illustrations of the immature *St. gomerae* Franz from the Canary Islands. Wheeler and Pakaluk (1983) described some characters of the Nearctic *St. conjux* (Casey), and Schmid (1988a) illustrated various structures of another European species, *St. godarti* (redescribed in detail by Jałoszyński 2016) and a putative larva of *St. scutellaris* (Müller & Kunze). Moreover, Brown and Crowson (1980) mentioned, but did not illustrate, larval mandibular structures similar to those of *St. collaris* but collected in Great Britain in association with adults of *St. bicolor* (Denny), and others collected with adults of *St. pusillus* (Müller & Kunze).

Other glandulariine species with larvae adequately described or illustrated include the western Palaearctic *Scydmoraphes sparshalli* Denny (Brown and Crowson 1980; Jałoszyński 2015b) and *Sc. helvolus* (Schaum) (Schmid 1988a). Brown and Crowson (1980) described a putative larva of European *Neuraphes elongatulus* (Müller & Kunze) and mentioned that it was similar to that of an undescribed but studied larva of *N. angulatus* (Müller & Kunze) (misspelled as *angularis*). Schmid (1988a) illustrated two different larvae putatively attributed to *Neuraphes* (s. str.) and *Neuraphes* (*Pararaphes*); Jałoszyński (2015b) demonstrated that the latter ("Typ 2-Larve" of Schmid) was in fact a misidentified *Scydmoraphes*. Finally, a putative larva of an Iberian *Euconnus*, presumably *E. eppelsheimi* Croissandeau, was partly described and illustrated by Brown and Crowson (1980). Structures similar to those found in the latter larva were also illustrated by Newton (1991) and putatively attributed to a Nearctic *Euconnus* sp. It should be noted, however, that "*Euconnus*" is a conglomerate of various genera; some subgenera have been recently removed from *Euconnus* and, consequently, older identifications without subgeneric placement should be treated as uncertain.

In the larvae of *Stenichnus*, it was recently found that a similar pattern of dorsal and lateral cephalic setae was present in taxa with very large or very small frontolabral region. Consequently, some setae located anteriorly to frontal sutures (i.e., on the frontolabral area) in one species may be homologous with setae located behind the sutures (i.e., on the epicranial halves) in another species. A hypothesis of "shifted" epicranial suture was proposed in order to draw homologies between groups of cephalic setae, an alternative approach to the traditional comparisons made strictly within the limits of head regions defined by the epicranial stem and frontal arms of the epicranial suture (Jałoszyński 2016).

Larvae (Figs. 14.1b and 14.2a–c) are typically onisciform or elongate suboval, yellowish to dark brown and glossy, sparsely to densely setose, and with setae unmodified. Head (Fig. 14.1e) is prognathous, weakly declined in living larvae, with one or three stemmata in a close group; epicranial stem and frontal sutures are distinct; nasale lacks teeth. The antenna has short antennomere I, long and clubbed antennomere II, which is usually broadest in the distal half, and small but not rudimentary antennomere III that is comparable in length to the long, subconical, typically slightly curved and pointed sensory appendage of antennomere II. Mandibles are falciform, slender, and pointed, with mesal margins often serrated. Maxillary mala is undivided and elongate, apically with bristles or spatulate setae; maxillary palp is long and usually with all palpomeres elongate, palpomere III being the longest. The prementum in some species has a pair of membranous adhesive pads or suckers that are difficult to notice, labial palps with short palpomere I, and long, slender palpomere II. The ecdysial line is typically distinct on all thoracic tergites; the pronotum is usually the longest. The abdomen is composed of ten segments; tergites I–IX are transverse and X usually elongate; abdominal sternites are undivided; urogomphs are absent. Legs are variously long; the pretarsus sometimes has ventral barbs. Spiracles are annular, ventral, or ventrolateral. There are nine pairs: one on the mesothorax and eight pairs on abdominal segments I–VIII, typically located on small protuberances. The spiracles on abdominal segment VIII are much smaller than the others, barely discernible, and nonfunctional.

14.3.4 Scydmaenini

Seven genera are known, but the larvae have been described only for *Scydmaenus* Latreille. Meinert (1888) was the first to describe immature *Scydmaenus* (s. str.) *tarsatus* Müller & Kunze (Europe). Additional (but partly misinterpreted) details were given by Brown and Crowson (1980), and larvae of the same species were redescribed in detail by Jałoszyński and Kilian (2012). Jałoszyński (2012b) later described and illustrated the pupa. Jeannel and Paulian (1945) gave some morphological details of the Afrotropical *Scydmaenus* (*Pseudeumicrus*) *tachyoryctidis* (Jeannel & Paulian), and Böving and Craighead (1931) illustrated larva of the Nearctic *Scydmaenus* (presumably s. str.) *longicollis* (Casey). Some structures of an undetermined *Scydmaenus* (presumably s. str.) from Panama were illustrated by Newton (1991). Beutel and Molenda (1997) illustrated the head of *Scydmaenus* sp. of unknown origin. Larvae of the European *Scydmaenus* (*Cholerus*) *hellwigii* (Herbst) and *Scydmaenus* (*Parallomicrus*) *rufus*

(Müller & Kunze) were described and illustrated in detail by Jałoszyński and Kilian (2012) and Jałoszyński (2015a), respectively.

Larvae are nearly onisciform, broad and suboval (*Scydmaenus* s. str.; Figs. 14.1a and 14.2d) or campodeiform, strongly elongate and nearly parallel-sided (subgenera *Cholerus* and *Parallomicrus*; Fig. 14.2e), creamy white to yellowish brown, typically with dense and often asymmetrically distributed setae inserted on small papillae and with dense fine microtrichia on the head, thorax, and abdomen (especially in *Scydmaenus* s. str.). The head (Fig. 14.1g) is prognathous, with one pair of stemmata; epicranial stem and frontal sutures are distinct. The nasale has a row of several small teeth, often of unequal length and slightly asymmetrically distributed; sometimes teeth are located under the anterior margin of the labrum and not projecting in dorsal view. The antenna has short antennomere I; long and only slightly broadened apically antennomere II that is typically broadest near the apex (sometimes thickened indistinctly); and very small but distinct antennomere III that is shorter than the slightly asymmetrical sensory appendage of antennomere II, which can be subconical with blunt apex or barrel-shaped in the proximal half with a subconical apex. Mandibles are falciform and slender, pointed or blunt, with smooth mesal margin. Stipital projection of maxilla with two small apical lobes can be identified as lacinia and galea, both densely setose. The maxillary palp is long, with all palpomeres elongate; palpomere III is typically longest; prementum is distinctly emarginate anteriorly; labial palps have elongate palpomere I and II of various lengths. Thoracic tergites in onisciform larvae have fleshy laterotergites, each demarcated by a longitudinal groove; tergites are undivided in campodeiform larvae; ecdysial lines are not visible. Ten abdominal segments are present, with all except IX and X being transverse; segment X is elongate; abdominal sternites are undivided; urogomphi are absent. Legs are relatively short. Spiracles are annular and ventrolateral. One of the nine pairs is on the mesothorax and eight pairs are on abdominal segments I–VIII.

14.3.5 Clidicini

Three extant genera are known; the larva is described only for the Australian *Clidicus abbotensis* O'Keefe (O'Keefe and Monteith 2000).

The larva is campodeiform, strongly elongate, subparallel-sided and slightly flattened, whitish but with reddish brown head and tergal plates (however, O'Keefe and Monteith (2000) stated that "in life the larvae were the same bright color as the adults," which are orange-reddish brown), moderately densely setose, and with unmodified setae. Head is prognathous, weakly declined, with demarcated short, annuliform "neck" and one pair of stemmata. Epicranial stem and frontal sutures are distinct; nasale lacks teeth. The antenna is longer than the head and very slender, not clubbed; long antennomeres I and II are of similar diameter; antennomere III is small but not rudimentary, strongly elongate, longer than strongly elongate, spatulate accessory appendage of antennomere II. Mandibles are falciform, moderately slender and pointed, each with one submedian mesal tooth. The maxillary mala is undivided and densely setose at apex; the maxillary palp is long, with all palpomeres elongate; palpomeres II and III are comparable in length. For labial palps, palpomere I is longer than II, inserted on a long palpifer resembling an additional palpomere. Thoracic tergites have distinct ecdysial lines. Ten abdominal segments are present; all except X are transverse; segment X is elongate; urogomphi are present, composed of one elongate segment fused with posterolateral margin of abdominal segment IX; abdominal sternites are not described. Legs are very long and slender. Spiracles are annular and ventrolateral; one of the nine pairs is on the mesothorax and eight pairs are on abdominal segments I–VIII.

14.3.6 Leptomastacini

Three extant genera are known; only larva of the European (Mediterranean) *Leptomastax hypogaea* Pirazzoli has been described (Vít and De Marzo 1989).

The larva is campodeiform, with subparallel and flattened body, whitish with testaceus head, and moderately densely setose; setae were illustrated and described as "rigid, spine-shaped." The head is prognathous, unusually shaped, rhomboidal, strongly transverse, strongly narrowing anteriorly and posteriorly from the broadest place located near the middle, with demarcated annuliform "neck" and one pair of stemmata; epicranial stem and frontal sutures are distinct; nasale has two long robust teeth separated by deep U-shaped emargination. The antenna is only slightly longer than the head; antennomeres I and II are long and of similar length, but antennomere II is slightly broadened distally, being broadest near the apex; antennomere III is small but not rudimentary, strongly elongate, shorter than the very long, slightly asymmetrical, subconical accessory appendage, which is slightly bent near the middle. Mandibles are falciform and extremely long and slender, longer than the head or antennae, pointed, with smooth mesal margins. Stipital projection of maxilla has two lobes: subapical lobe is elongate and densely setose; apical lobe is broad and with two modified, thick setae. The maxillary palp is long, with all palpomeres elongate; palpomeres II and III are comparable in length; in labial palps palpomere II is longer than I. Thoracic tergites have distinct ecdysial lines, at least on pronotum and mesonotum. Ten abdominal segments are present, all except X are transverse; segment X is elongate. Urogomphi are present, composed of one elongate segment fused with the posterolateral margin of abdominal segment IX. Legs are long and slender, densely covered with relatively short, spiny setae. Spiracles are annular and lateral; one of the nine pairs is on the mesothorax and eight pairs are on abdominal segments I–VIII.

14.3.7 Mastigini

This tribe includes three extant genera; larvae are known for *Palaeostigus* Newton and *Stenomastigus* Leleup. De Marzo (1983, 1984) described eggs, all larval instars, and the pupa of the European *Palaeostigus pilifer* (Kraatz). Newton (1991) illustrated some structures of the Turkish *Palaeostigus ruficornis schimitscheki* (Machulka). Grebennikov and Newton (2009) illustrated some larval structures of the South African *Palaeostigus bifoveolatus* (Boheman). Most recently, Jałoszyński and Kilian (2016) gave a detailed description of the second- and third-instar larvae of the South African *Stenomastigus longicornis* (Boheman).

Mature larvae (Figs. 14.1c and 14.2f, g) are campodeiform, subparallel or with a strongly narrowing abdomen, flattened or subcylindrical. Membranous areas are whitish or yellowish, heavily sclerotized regions (head, tergal, and sternal plates) are variously dark brown to nearly black (Fig. 14.2f). Larvae of *Stenomastigus* are orange (Fig. 14.2g). The body is covered with sparse, long setae and dense asperities, typically granulose or pointed and forming patterns among smooth areas of tergal plates. Most setae are unmodified; additionally, short, flat, leaf-like setae with fine elongate ribs are distributed along anterior and posterior margins of tergal plates, often partly covered by folds of intersegmental membranes. The head is prognathous and slightly declined, lacking a demarcated "neck," with one pair of stemmata. Epicranial stem and frontal sutures are distinct but short, together with antennal insertions shifted to the posterior half of the head capsule; nasale has a row of several short setae inserted on papillate protuberances. The head has a large glandular impression at the junction of the epicranial stem and frontal sutures, filled with or surrounded by modified setae to form a presumed evaporation apparatus. The antenna is much longer (often several times longer) than the head and very slender, not broadened; long antennomeres I and II are of similar diameter; antennomere II is subdivided into three sections; antennomere III is rudimentary, developed as a tiny, often barely discernible, papilla adjacent to the base of strongly elongate, slightly asymmetrical, subconical, and pointed accessory appendage. Mandibles are falciform, moderately slender, pointed, each with one submedian mesal tooth.

The stipital projection of the maxilla is divided into two very short and broad, densely setose lobes, but it is unclear whether they represent lacinia and galea or one slightly emarginate mala. The maxillary palp is long, with palpomere I being short and II and III strongly elongate; labial palps have palpomere I longer than II. Thoracic tergites are in transparent mounts with a distinct ecdysial line, which in SEM images is visible as a smooth longitudinal median line among lateral fields of dense asperities. Ten abdominal segments are present, all except X (or IX and X) being transverse; segment X is elongate; urogomphi are absent. Sternal plates on the thorax and abdomen are reduced to small, paired (2–4), and setose sclerites. Legs are very long and slender, with particularly densely setose tibiotarsi. Spiracles are annular and lateral; one of the nine pairs is on the mesothorax and eight pairs are on abdominal segments I–VIII.

First instar larvae are known for *Palaeostigus*; they differ from older instars in having a short, stout, and orange body with strikingly thick legs and antennae; setae on all body parts are very short and the frons lacks glandular impression.

Some data from the cited literature have been used by various authors in identification keys, reviews, and summaries, and some illustrations were redrawn and reproduced several times, often losing (or even gaining) details (e.g., Arnolde et al. 1964; Kasule 1966; Klausnitzer 1978, 1997; Newton and Franz 1998; O'Keefe 2005; Jałoszyński and Kilian 2016).

14.4 Biology of Larvae

Very little is known about the behavioral ecology or synecology of Scydmaeninae larvae. They can be found in moist leaf litter, soil, rotten wood, decomposing plant debris (compost, flood remains, rice straw, etc.), and sometimes also in ant and termite nests, typically together with adults and during all seasons, even in temperate climates. However, larvae of *Scydmaenus* species that inhabit compost can be most frequently collected during April to June in Central Europe, and mature larvae of southern Iberian *Palaeostigus* can be found only in winter and early spring (Jałoszyński, unpublished observations). Notable exceptions to finding larvae and adults together by sifting leaf litter are some species of South African Mastigini, especially *Stenomastigus*, whose adults climb tall grasses, bushes, and even trees and can be efficiently collected using a sweeping net or entomological umbrella; their larvae are associated with humid leaf litter, often along water bodies (Jałoszyński and Kilian 2016). However, a single observation (made by Rafał Ruta, University of Wrocław) known to the author was documented with photos showing a larva of *Stenomastigus* found in a South African forest climbing the moist trunk of a living tree during the day, half a meter above the ground. Larvae of *Clidicus abbotensis* have also been observed to show diurnal and exposed behavior similar to that of adults, "walking in close proximity to foraging adults and presumably seeking the same prey" (O'Keefe and Monteith 2000).

Larvae of Scydmaeninae kept under laboratory conditions typically hide among substrate particles, but rapid exposure to direct light (even a camera flash) does not disturb feeding in larvae of Glandulariini, Scydmaenini, and Mastigini (Jałoszyński, unpublished observation). All scydmaenine larvae observed alive by the author use their pygopod (i.e., abdominal segment X) extensively during walking to adhere to particles of substrate with the anal membrane. The pygopod is often used to stabilize the position of larva during feeding, by adhering to a substrate, leaving front and middle leg pairs free to manipulate prey (Fig. 14.2b).

Females of *Palaeostigus pilifer* cover their eggs with a glandular secretion during oviposition; larvae stay at the secretion deposit, and the second instars have been observed feeding on it (De Marzo 1983). Mature larvae of this species under laboratory conditions feed on pieces of dead caterpillars (De Marzo 1983); mature larvae of other European species of *Palaeostigus* kept in captivity readily accept beetle pupae (Fig. 14.2f), dead soft-bodied arthropods, and pieces of raw or boiled meat but were not observed hunting any living prey (Jałoszyński, unpublished observations).

Mature larvae of *Stenichnus* feed on armored mites (Oribatida) (Fig. 14.2a–c). They attack the gnathosoma, remove mite's mouthparts, and feed through the resulting opening (Jałoszyński 2016). Mature larvae of *Scydmaenus* (Fig. 14.2d) show strong feeding preference for springtails, especially slow-moving Hypogastruridae, but were also observed feeding on Tomoceridae (Jałoszyński and Kilian 2012). Larvae of *Cephennium* were observed feeding on Oribatida or similarly heavily sclerotized Uropodina (Mesostigmata) by using their labial suckers to immobilize prey and short mandibles to slowly scrape through the mite's cuticle (Schuster 1966; Schmid 1988b). The entire feeding process takes place through the resulting tiny hole, by injecting digestive juices and then ingesting liquefied tissues (Jałoszyński and Beutel 2012). In the larva of *Cephennium galllicum*, Jałoszyński and Beutel (2012) found a system of pharyngeal valves, presumably facilitating the flow of digestive juices and liquefied tissue between the predator's mouth opening and the interior of prey.

When feeding, onisciform larvae of Cephenniini and Glandulariini (Fig. 14.2b) often curl around their prey; the unusual body shape may have evolved as an adaptation for such a feeding posture. Alternatively, the onisciform body of *Cephennium* larvae, their enlarged pronotum, and the first abdominal tergite enable them to curl into a tight ball, hiding the head, appendages, and ventrites inside (Ionesco 1937), a clearly defensive behavior that helps protect them against small predators.

14.5 Pupation and Morphological Structures of Pupae

Under laboratory conditions, pupation takes place either between substrate particles or on any available surface (e.g., on a plaster of Paris surface). In *Stenichnus*, the prepupa (Fig. 14.2h) shows a darker pigmentation than that in active larva. Pupae are initially lightly pigmented (Fig. 14.2i). They have very long, paired spines on the head, pronotum, and sides of abdomen; mature pupae are typically brownish (Fig. 14.2j, k). In *Stenichnus godarti*, the prepupal stage lasts 2 days and the pupal stage 7 days (Jałoszyński 2016).

14.6 Discussion

The larval stages of Scydmaeninae are exceptionally poorly studied, and most older descriptions and illustrations are inaccurate, lacking important details, or based on putatively identified material. Nevertheless, larval characters are now known for all large tribes. Only immature stages of small and monogeneric Chevrolatiini, Marcepaniini, and Leptoscydmini have yet to be found or obtained by rearing. Rearing, however, proved successful for only one species. A more promising strategy seems to be obtaining identifiable adults by rearing mature larvae collected in nature and describing morphological structures on the basis of shed larval skins (Jałoszyński 2013, 2016). Alternatively, larvae can be associated with adults using molecular markers.

Recent studies have provided novel data and allowed correction of some previous misidentifications or misinterpretations of morphological structures. Among the most interesting new results are those concerning the South African *Stenomastigus*, a genus closely related to the western Palaearctic *Palaeostigus*. In the previously known larvae of the European *Palaeostigus pilifer*, the first and second instars were orange (a pigmentation interpreted by De Marzo (1983) as coming from subcuticular deposits of an egg yolk) and spent all their time close to egg shells covered with a nutritious secretion, their only source of food. Only the third-instar larvae were capable of dispersal and feeding on other sources of food. Field observations in South Africa and morphological analysis of collected larvae of *Stenomastigus longicornis* demonstrated that two larval instars, presumably the second and third, were orange and capable of dispersal. Both were commonly collected by sifting leaf litter and also by finding them walking on the surface of soil particles (Jałoszyński and Kilian 2016; Jałoszyński, unpublished observations).

The known larvae, although representing merely a tiny fraction of nominal genera, show a greater diversity of body forms and structures than any other subfamily of Staphylinidae. The diversity is so great that it is not possible to give larval diagnostic characters to define Scydmaeninae, neither apomorphic nor a unique combination of synapomorphies. The most puzzling characteristic is the presence or absence of urogomphs in larvae of closely related tribes of Mastigitae, which otherwise show many similarities. The most enigmatic feature is the reduction in number of abdominal segments in Cephenniini from ten to nine. Moreover, three features of Mastigini are also unusual or unique among Staphylinidae: the antennomere II subdivided into three sections, antennae with very long spines, and the presumably glandular frontal impression (suggested by De Marzo 1983 to play a role in grooming). Future studies should be focused on the functional morphology and evolution of specialized modifications in Scydmaeninae larvae.

References

Arnolde LV, Bizova UB, Chilarov MS, Dolin VG, Gantiev RD, Zinovieva LA, Keleinikova SI, Kurcheva DF, Krivocheina NP, Mamaev BM, Medvedev LM, Tarsis VA, Chadaeva ZV, Sharova IC (1964) Opredeletelitel' obitayushchikh v pochve lichinok nasekomykh (Key to soil-dwelling insect larvae). Izdatel'stvo 'Nauka', Moscow

Bergroth E (1899) *Anisosphaera problematica* Töm. = *Cephennium*, larva [Col.]. Bull Soc Ent France 295

Beutel RG, Molenda R (1997) Comparative morphology of selected larvae of Staphylinoidea (Coleoptera, Ployphaga) with phylogenetic implications. Zool Anz 41:392–394

Böving AG, Craighead FC (1931) An illustrated synopsis of the principal larval forms of the order Coleoptera. Ent Am 11:1–351

Brown C, Crowson RA (1980) Observations on Scydmaenid (Col.) larvae with a tentative key to the main British genera. Ent Mon Mag 115:49–59

De Marzo L (1983) Osservazioni sulla ovideposizione e sul ciclo larvale in *Mastigus pilifer* Kraatz (Coleoptera, Scydmaenidae). Entomologica (Bari) 18:125–136

De Marzo L (1984) Morfologia delle larve e pupa in *Mastigus pilifer* Kraatz (Coleoptera, Scydmaenidae). Entomologica (Bari) 19:61–74

Franz H (1965) Beitrag zur Bodenfauna der Kanarischen Inseln. Zur Kenntnis der Coleopterenfauna von Tenerife und La Gomera. Eos Rev Esp Ent 41:59–66

Grebennikov VV, Newton AF (2009) Good-bye Scydmaenidae, or why the ant-like stone beetles should become megadiverse Staphylinidae sensu latissimo (Coleoptera). Eur J Ent 106:275–301

Ionesco MA (1937) La larve de *Cephennium carnicum* Rtt, Etude morphologique. In: Hommage au Prof. E.C. Teodoresco, Bukarest, pp 101–109

Jałoszyński P (2012a) Beetles with 'trochantelli': phylogeny of Cephenniini (Coleoptera: Staphylinidae: Scydmaeninae) with focus on Neotropical genera. Syst Ent 37:448–477

Jałoszyński P (2012b) Morphology of pupa of *Scydmaenus tarsatus* Müller & Kunze (Coleoptera: Staphylinidae: Scydmaeninae). Genus 23(3):363–366

Jałoszyński P (2013) Mature larva of *Stenichnus collaris* (Müller & Kunze) (Coleoptera: Staphylinidae, Scydmaeninae). Zootaxa 3745(1):19–34

Jałoszyński P (2014) Phylogeny of a new supertribe Cephenniitae with generic review of Eutheiini and description of a new tribe Marcepaniini (Coleoptera: Staphylinidae: Scydmaeninae). Syst Ent 39:159–189

Jałoszyński P (2015a) Late-instar larva of *Scydmaenus (Parallomicrus) rufus* Müller & Kunze (Coleoptera: Staphylinidae, Scydmaeninae). Zootaxa 3973 (2):579–590

Jałoszyński P (2015b) Redescription of late-instar larva of *Scydmoraphes sparshalli* (Denny) (Coleoptera: Staphylinidae, Scydmaeninae). Zootaxa 4032 (5):582–594

Jałoszyński P (2016) Mature larva of *Stenichnus godarti* (Latreille) (Coleoptera: Staphylinidae, Scydmaeninae): redescription, hypothesis of displaced epicranial sutures and alternative interpretation of homology between chaetotaxic structures. Zootaxa 4196(1):77–94

Jałoszyński P, Beutel R (2012) Functional morphology and evolution of specialized mouthparts of Cephenniini (Scydmaeninae, Staphylinidae). Arthr Str Dev 41:593–607

Jałoszyński P, Kilian A (2012) Larval morphology of *Scydmaenus tarsatus* and *S. hellwigii*, with notes on feeding behavior and a review of bibliography on preimaginal stages of ant-like stone beetles (Coleoptera: Staphylinidae, Scydmaeninae). Eur J Ent 109:587–601

Jałoszyński P, Kilian O (2016) Description of the second- and third-instar larva of South African *Stenomastigus longicornis* (Boheman) (Coleoptera: Staphylinidae, Scydmaeninae). Zootaxa 4158(2):151–182

Jałoszyński P, Matsumura Y, Beutel RG (2015) Evolution of a giant intromittent organ in Scydmaeninae

(Coleoptera: Staphylinidae): functional morphology of the male postabdomen in Mastigini. Arthr Str Dev 44:77–98
Jeannel R (1909) Biospeologica. X. Coléoptères (seconde série). Arch Zool Exp Gén 1:447–532
Jeannel R, Paulian R (1945) Mission scientifique de l'Omo, VI (57). Faune des terriers des Rats-Taupes. IV. Coléoptères. Mém Mus nat d'Hist nat 19:1–214
Kasule FK (1966) The subfamilies of the larvae of Staphylinidae (Coleoptera) with keys to the larvae of the British genera of Steninae and Proteininae. Trans Ent Soc London 118:261–283
Klausnitzer B (1978) Bestimmungstabellen für die Gattungen ausgewählter Käferfamilien. 5.8 Scydmaenidae. In: Klausnitzer B (ed) Ordung Coleoptera (larven). W. Junk, Hague
Klausnitzer B (1997) 20. Familie: Scydmaenidae. In: Klausnitzer B (ed) Die Larven der Käfer Mitteleuropas. Vol 4. Polyphaga, part 3. Goecke & Evers, Krefeld; Gustav Fischer Verlag, Jena
McKenna DD, Farrell BD, Caterino MS, Farnum CW, Hawks DC, Maddison DR, Seago AE, Short AEZ, Newton AF, Thayer MK (2015) Phylogeny and evolution of Staphyliniformia and Scarabaeiformia: forest litter as a stepping stone for diversification of nonphytophagous beetles. Syst Ent 40:35–60
Meinert F (1888) Scydmaenus-Larven. Ent Medd 1:144–150
Morimoto K, Hayashi N (1986) The Coleoptera of Japan in color, vol 1. Hoikusha, Osaka
Newton AF (1991) Scydmaenidae (Staphylinoidea). In: Stehr FW (ed) Immature insects, vol 2. Kendall/Hunt, Dubuque, IO, pp 330–334
Newton AF, Franz H (1998) World catalog of the genera of Scydmaenidae (Coleoptera). Kol Rundsch 68:137–165
O'Keefe ST (2005) Scydmaenidae Leach, 1815. In: Beutel RG, RAB L (eds) Handbook of zoology. Coleoptera, Beetles, vol 1: Morphology and systematics (Archostemata, Adephaga, Myxophaga, Polyphaga partim). De Gruyter, Berlin
O'Keefe ST, Monteith GB (2000) *Clidicus abbotensis* O'Keefe, a new species of Scydmaenidae (Coleoptera: Staphylinoidea) from Australia with description of the larva. Mem Queensland Mus 46:211–223
Paulian R (1941) Les premiers états des Staphylinoidea (Coleoptera), Étude de morphologie comparée. Mém Mus nat d'Hist nat 15:1–361
Peyerimhoff P (1899) Description de la larve de *Cephennium laticolle* Aub. (Col.). Bull Soc Ent France 170–174
Schmid R (1988a) Die Larven der Ameisenkäfer (Scydmaenidae, Staphylinoidea) – Neu- und Nachbeschreibung mit einem vorläufigen Bestimmungsschlüssel bis zur Gattung. Mitt Badischen Landesv Naturk Naturs 14:643–660
Schmid R (1988b) Morphologische Anpassungen in einem Räuber-Beute-System: Ameisenkäfer (Scydmaenidae, Staphylinoidea) und gepanzerte Milben (Acari). Zool Jahrb Abt Syst Ökol Geogr Tiere 115:207–228
Scholz R (1926) Die Larve von *Cephennium reitteri* Bris. (Col.) Ent Blätter 22:103–104
Schuster R (1966) Scydmaeniden-Larven als Milbenräuber. Naturwiss 53:439–440
Silvestri F (1899) Prima nota intorno all Anisosphaera Töm. Ann Mus Civ St Nat Genova 39:613–619
Thayer MK, Newton AF, Chatzimanolis S (2012) *Prosolierius*, a new mid-Cretaceous genus of Solieriinae (Coleoptera: Staphylinidae) with three new species from Burmese amber. Cret Res 34:124–134
Tömösvary Ö (1882) Adatok hazánk Thysanura-Faunájához. Matem Termész Közl, Magyar Tudom Ak 18:119–131
Vít S, De Marzo L (1989) Description of the larva of *Leptomastax hypogeus* Pirazzoli (Coleoptera Scydmaenidae). Arch Sci 42:569–578
Wheeler QD, Pakaluk J (1983) Descriptions of larval *Stenichnus* (*Cyrtoscydmus*): *S. turbatus* and *S. conjux*, with notes on their natural history (Coleoptera: Scydmaenidae). Proc Ent Soc Washington 85:86–97

Subject Index

O. Betz et al. (eds.), *Biology of Rove Beetles (Staphylinidae)*,
https://doi.org/10.1007/978-3-319-70257-5

Taxonomic Index (Including Common Names)

O. Betz et al. (eds.), *Biology of Rove Beetles (Staphylinidae)*,
https://doi.org/10.1007/978-3-319-70257-5

B

C

D

Q

R

S

GPSR Compliance

The European Union's (EU) General Product Safety Regulation (GPSR) is a set of rules that requires consumer products to be safe and our obligations to ensure this.

If you have any concerns about our products, you can contact us on ProductSafety@springernature.com

In case Publisher is established outside the EU, the EU authorized representative is:

Springer Nature Customer Service Center GmbH
Europaplatz 3
69115 Heidelberg, Germany

Batch number: 10732823

Printed by Printforce, the Netherlands